X-Ray and Neutron Dynamical Diffraction
Theory and Applications

NATO ASI Series

Advanced Science Institutes Series

A series presenting the results of activities sponsored by the NATO Science Committee, which aims at the dissemination of advanced scientific and technological knowledge, with a view to strengthening links between scientific communities.

The series is published by an international board of publishers in conjunction with the NATO Scientific Affairs Division

A Life Sciences **B Physics**	Plenum Publishing Corporation New York and London
C Mathematical and Physical Sciences **D Behavioral and Social Sciences** **E Applied Sciences**	Kluwer Academic Publishers Dordrecht, Boston, and London
F Computer and Systems Sciences **G Ecological Sciences** **H Cell Biology** **I Global Environmental Change**	Springer-Verlag Berlin, Heidelberg, New York, London, Paris, Tokyo, Hong Kong, and Barcelona

PARTNERSHIP SUB-SERIES

1. Disarmament Technologies	Kluwer Academic Publishers
2. Environment	Springer-Verlag
3. High Technology	Kluwer Academic Publishers
4. Science and Technology Policy	Kluwer Academic Publishers
5. Computer Networking	Kluwer Academic Publishers

The Partnership Sub-Series incorporates activities undertaken in collaboration with NATO's Cooperation Partners, the countries of the CIS and Central and Eastern Europe, in Priority Areas of concern to those countries.

Recent Volumes in this Series:

Volume 354—Physics and Chemistry of Low-Dimensional Inorganic Conductors
edited by Claire Schlenker, Jean Dumas, Martha Greenblatt, and Sander van Smaalen

Volume 355—Stability of Materials
edited by A. Gonis, P. E. A. Turchi, and Josef Kudrnovský

Volume 356—Spectroscopy and Dynamics of Collective Excitations in Solids
edited by Baldassare Di Bartolo

Volume 357—X-Ray and Neutron Dynamical Diffraction: Theory and Applications
edited by André Authier, Stefano Lagomarsino, and Brian K. Tanner

Series B: Physics

X-Ray and Neutron Dynamical Diffraction

Theory and Applications

Edited by

André Authier

Université Pierre et Marie Curie
Paris, France

Stefano Lagomarsino

Istituto Elettronica Stato Solido, CNR
Rome, Italy

and

Brian K. Tanner

University of Durham
Durham, England

Plenum Press
New York and London
Published in cooperation with NATO Scientific Affairs Division

Proceedings of a NATO Advanced Study Institute on
X-Ray and Neutron Dynamical Diffraction: Theory and Applications,
held April 9–21, 1996,
in Erice, Italy

NATO-PCO-DATA BASE

The electronic index to the NATO ASI Series provides full bibliographical references (with keywords and/or abstracts) to about 50,000 contributions from international scientists published in all sections of the NATO ASI Series. Access to the NATO-PCO-DATA BASE is possible in two ways:

—via online FILE 128 (NATO-PCO-DATA BASE) hosted by ESRIN, Via Galileo Galilei, I-00044 Frascati, Italy

—via CD-ROM "NATO Science and Technology Disk" with user-friendly retrieval software in English, French, and German (©WTV GmbH and DATAWARE Technologies, Inc. 1989). The CD-ROM contains the AGARD Aerospace Database.

The CD-ROM can be ordered through any member of the Board of Publishers or through NATO-PCO, Overijse, Belgium.

Library of Congress Cataloging-in-Publication Data

X-ray and neutron dynamical diffraction : theory and applications / edited by André Authier, Stefano Lagomarsino, and Brian K. Tanner.
p. cm. -- (NATO ASI series. Series B, Physics ; v. 357)
"Published in cooperation with NATO Scientific Affairs Division."
"Proceedings of a NATO Advanced Study Institute on X-ray and Neutron Dynamical Diffraction, theory and applications, held April 9-21, 1996, in Erice, Italy"--T.p. verso.
Includes bibliographical references and index.
ISBN 0-306-45501-3
1. X-rays--Diffraction--Congresses. 2. Neutrons--Diffraction--Congresses. 3. X-ray crystallography--Congresses. 4. Interferometry, Neutron--Congresses. I. Authier, André. II. Lagomarsino, Stefano. III. Tanner, B. K. (Brian Keith) IV. North Atlantic Treaty Organization. Scientific Affairs Division. V. NATO Advanced Study Institute on X-ray and Neutron Dynamical Diffraction (1996 : Erice, Italy) VI. Series.
QC482.D5X68 1997
620.'1'1272--dc21 96-37283
CIP

ISBN 0-306-45501-3

A Division of Plenum Publishing Corporation
233 Spring Street, New York, N. Y. 10013

http://www.plenum.com

10 9 8 7 6 5 4 3 2 1

Printed in the United States of America

PREFACE

This volume collects the proceedings of the 23rd International Course of Crystallography, entitled *"X-ray and Neutron Dynamical Diffraction, Theory and Applications,"* which took place in the fascinating setting of Erice in Sicily, Italy. It was run as a NATO Advanced Studies Institute with A. Authier (France) and S. Lagomarsino (Italy) as codirectors, and L. Riva di Sanseverino and P. Spadon (Italy) as local organizers, R. Colella (USA) and B.K. Tanner (UK) being the two other members of the organizing committee. It was attended by about one hundred participants from twenty four different countries.

Two basic theories may be used to describe the diffraction of radiation by crystalline matter. The first one, the so-called geometrical, or kinematical theory, is approximate and is applicable to small, highly imperfect crystals. It is used for the determination of crystal structures and describes the diffraction of powders and polycrystalline materials. The other one, the so-called dynamical theory, is applicable to perfect or nearly perfect crystals. For that reason, dynamical diffraction of X-rays and neutrons constitutes the theoretical basis of a great variety of applications such as:

- the techniques used for the characterization of nearly perfect high technology materials, semiconductors, piezoelectric, electrooptic, ferroelectric, magnetic crystals,
- the X-ray optical devices used in all modern applications of Synchrotron Radiation (EXAFS, High Resolution X-ray Diffractometry, magnetic and nuclear resonant scattering, topography, etc.), and
- X-ray and neutron interferometry.

A new interest has developed in the applications of the dynamical theory because of the possibilities offered by synchrotron radiation for *in situ* and real time studies, the production of polarized X-rays, by the recent developments in X-ray magnetic and nuclear scattering, and by the use of X-ray and neutron interferometry in modern metrology. The arrival of the third generation synchrotron radiation sources with their enhanced properties has made the course and the publication of its proceedings very timely.

The first part of these proceedings reviews the basic principles of the dynamical diffraction of X-rays and neutrons by perfect and nearly perfect crystals, with special attention to highly asymmetrical cases, the optical devices used in the synchrotron beamlines, polarization of X-rays and various types of crystal polarizers for X-rays, and statistical theory for highly imperfect crystals.

The second, third and fourth parts of these proceedings describe the principles of the techniques used to characterize the structural imperfections of materials. Several chapters are devoted to X-ray and neutron diffraction topography which enables the direct imaging of crystal defects (dislocations, stacking faults, low angle grain boundaries, microprecipitates or inclusions, magnetic or ferroelectric domains, etc.). The various settings are presented and the new possibilities opened up by the third generation synchrotron radiation sources are discussed, for instance, a fraction of a millisecond is enough to record a white beam topograph or a diffraction spectrum. A chapter is devoted to the theoretical interpretation of the contrast for the various types of defects. The main applications presented include the characterization of high technology materials, the *in situ* study of crystal growth and the relation of crystal defects to the growth conditions, *in situ* study of plastic deformation and phase transitions, and the analysis of the distribution and shape of domains in magnetic materials. It is, for instance, possible to follow the Molecular Beam Epitaxy growth of layers at the atomic scale, the movement of dislocations under an applied stress or the recrystallization of an alloy during a strain-anneal cycle. Another technique for the characterization of defects in imperfect materials and the analysis of strains is the high resolution diffraction of X-rays, which is applied in particular to layered or hetero-structures such as those used in electronic materials. The theory of reciprocal space mapping is presented in detail and examples of strain-analysis are given in the case of III-V multilayer compounds and ion-implanted silicon. New developments in the theory of the diffraction of X-rays by imperfect crystals and the improvement in computer power have now made possible the determination of the strain distribution in a component for microelectronics. The third type of application concerns the location of impurity atoms at crystal surfaces or interfaces by means of the fluorescence emitted at X-ray standing waves antinodes. It is possible by means of this technique to analyze the structure of the surface and the thermal vibrations of the atoms. The same technique is also applied to the study of thin films or long period structures.

The fifth part of the proceedings describes a very promising new application of dynamical theory, the use of multiple Bragg scattering for the determination of crystal structures. The phenomenon of multiple beam diffraction is well known; it occurs very often in electron diffraction and is usually avoided in X-ray diffraction. Due to the fact that electromagnetic waves are vector waves while the wave associated with an electron or a neutron beam is scalar, the theory is complicated for X-rays. New developments were presented at the course relative to the analysis of the *n*-beam diffraction of X-rays and it was shown that it is possible to determine the absolute phase of structure factors from diffraction profiles in the neighborhood of three-beam diffraction. This is a very important development and will help in solving crystal structures even for macromolecules such as small proteins.

In the last part, the principles of X-ray and neutron interferometers are described. They produce interference patterns with a fringe spacing, equal to the lattice spacing of the crystal used and therefore, much smaller than the fringe spacing available with optical interferometry. They require extremely perfect crystals and the use of dynamical theory of diffraction for the interpretation of data. X-ray and neutron interferometery permit calibration of extremely small linear or angular displacements to be made at the subnanometer scale. They have been used for the absolute evaluation of the Avogadro number and for the measurement of X-ray dispersion corrections. They have very promising applications in the fields of phase contrast microscopy and metrology in the nanometer regime.

André Authier
Stefano Lagomarsino
Brian K. Tanner

CONTENTS

BASIS OF THE DYNAMICAL THEORY OF DIFFRACTION

DIFFRACTION TOPOGRAPHY

X-RAY STANDING WAVES

THEORY AND APPLICATIONS OF HIGH RESOLUTION DIFFRACTOMETRY

MULTIPLE-BEAM DIFFRACTION

X-RAY AND NEUTRON INTERFEROMETRY

DYNAMICAL THEORY OF X-RAY DIFFRACTION - I. PERFECT CRYSTALS

A. AUTHIER

Université P. et M. Curie
4, Place Jussieu
75252 Paris Cedex 05, France

1. GEOMETRICAL VERSUS DYNAMICAL THEORY

The geometrical, or kinematical theory, of diffraction considers that each photon is scattered only once, and that the interaction of X-rays with matter is so small that it can be neglected. It can therefore be assumed that the amplitude incident on every diffracting centre inside the crystal is the same. The total diffracted amplitude is then simply obtained by adding the individual amplitudes diffracted by each diffracting centre, only taking into account the geometrical phase differences between them. The result is that the distribution of diffracted amplitudes in reciprocal space is the Fourier transform of the distribution of diffracting centres in physical space. The integrated reflected intensities calculated this way are proportional to the square of the structure factor and to the volume of crystal bathed in the incident beam. The diffracted intensity according to the geometrical theory would therefore increase to infinity if the volume of the crystal was increased to infinity, which is of course absurd. The theory only works because the strength of the interaction is very weak and if it is applied to very small crystals. The geometrical theory presents another drawback: it gives no indication as to the phase of the reflected wave. This is due to the fact that it is based on the Fourier transform of the electronic density limited by the external shape of the crystal. This is not important when one is only interested in measuring the reflected intensities. For perfect or nearly perfect crystals, and for any problem where the phase is important, such as the case of multiple reflections, interference between coherent blocks, standing waves, etc., even for thin or imperfect crystals, a more rigorous theory which takes into account the interactions of the scattered radiation with matter should be used; this is the *dynamical theory*.

X-ray and Neutron Dynamical Diffraction: Theory and Applications
Edited by Authier *et al.*, Plenum Press, New York, 1996

The form of the dynamical theory which is most widely used today and which will be presented in this Chapter is that developed by von Laue (1931) on the basis of Ewald's original theory (Ewald, 1917). Von Laue showed that the interaction could be described by solving Maxwell's equations in a medium with a continuous, triply periodic, distribution of electric susceptibility. The dynamical theory provides correct expressions for the diffracted intensities and introduces the fundamental notion of *wavefield*, due to Ewald, which is necessary to understand the specific properties of dynamical diffraction such as anomalous transmission and the propagation of X-rays in perfect or deformed crystals.

For reviews of the dynamical theory, see Zachariasen (1945), von Laue (1960), James (1963), Batterman and Cole (1964), Authier (1970, 1992), Kato (1974) and Pinsker (1978).

2. FUNDAMENTALS OF PLANE-WAVE DYNAMICAL THEORY

2.1. Propagation Equation of the Electromagnetic Wave

2.1.1. Von Laue's Basic Assumption. Von Laue's basic assumption (1931, 1960) is to consider that the electric negative and positive charges are distributed in a continuous way throughout the volume of the crystal. Since the crystal must be neutral, they cancel out and the local electric charge and density of current are equal to zero. The interaction of an electromagnetic wave with charged particles includes an electric interaction and a magnetic interaction. The magnetic interaction is weak and is neglected in classical treatments of the dynamical theory. It can nevertheless be observed and is discussed in Blume (1985, 1994). The strength of the electric interaction is classically proportional to $R = e^2/(4\pi\epsilon_o mc^2)$ where e is the electric charge of the particle, m its mass and ϵ_o is the dielectric constant of vacuum. Since the mass of the proton is 1830 times larger than that of the electron, electric interaction with the positive charges is neglected as a first approximation. Resonant nuclear scattering of X-rays is discussed in Hannon *et al.* (1988) and Hannon (1994). The coherent part of the electric interaction of the electromagnetic wave with the electronic distribution $\rho(\mathbf{r})$ is classically described by considering that the medium is polarized under the influence of the electromagnetic field and behaves like a perfect dielectric. Far from an absorption edge, the electric susceptibility χ is given classically by:

$$\chi(\mathbf{r}) = -\frac{e^2\rho(\mathbf{r})}{4\pi^2\epsilon_o\nu^2 m} = -R\lambda^2\rho(\mathbf{r})/\pi, \tag{1}$$

where ν is the frequency of the electromagnetic wave and $R = e^2/(4\pi\epsilon_o mc^2) = 2.817938\times10^{-15}\ m$ is the classical radius of the electron. The electric susceptibility is therefore negative and very small, of the order of 10^{-5} or 10^{-6}. It is a function of the space coordinates.

2.1.2. Maxwell's Equations. The electric field, **E**, the electric displacement, **D**, the magnetic field, **H**, the magnetic induction, **B**, are related locally by the material relations which describe the reaction of the medium to the electromagnetic field:

$$\left.\begin{aligned} \mathbf{D} &= \epsilon\mathbf{E} \\ \mathbf{B} &= \mu\mathbf{H} \end{aligned}\right\} \tag{2}$$

where ϵ and μ are the dielectric constant and the magnetic permeability, respectively. Here the magnetic interaction is neglected and μ is taken equal to μ_0, the magnetic permeability of vacuum.

In the absence of local electric charges and of a local current density, and if the magnetic interaction is neglected, Maxwell's equations reduce to:

$$\left.\begin{aligned} \text{curl } \mathbf{E} &= -\mu_0 \partial\mathbf{H}/\partial t \\ \text{curl } \mathbf{H} &= \partial\mathbf{D}/\partial t \\ \mathit{div}\ \mathbf{D} &= 0 \\ \mathit{div}\ \mathbf{B} &= 0 \end{aligned}\right\} \tag{3}$$

These equations are complemented by the following boundary conditions at the surface between two neighbouring media, 1 and 2:

$$\left.\begin{aligned} \mathbf{E}_{T1} - \mathbf{E}_{T2} = 0 \qquad \mathbf{D}_{N1} - \mathbf{D}_{N2} = 0 \\ \mathbf{H}_{T1} - \mathbf{H}_{T2} = 0 \qquad \mathbf{B}_{N1} - \mathbf{B}_{N2} = 0 \end{aligned}\right\} \tag{4}$$

where the indices $_T$ and $_N$ indicate the tangential and normal components, respectively.

2.1.3. Propagation Equation. In vacuum, $\mathbf{E} = \mathbf{D}/\epsilon_0$, and $\mathit{div}\ \mathbf{E}$ is equal to zero. Using the relation $\epsilon_0\mu_0 = 1/c^2$ where c is the velocity of light, one obtains the equation of propagation of an electromagnetic wave in vacuum by eliminating $\mathbf{H}$ between the first two relations of Maxwell's equations:

$$\Delta\mathbf{E} = \frac{1}{c^2}\frac{\partial^2\mathbf{E}}{\partial t^2} \tag{5}$$

Its simplest solution is a plane wave, $\mathbf{E} = \mathbf{E_0} \exp 2\pi i(\nu t - \mathbf{k}\cdot\mathbf{r})$, of which the wave number $k = 1/\lambda$ and the frequency ν are related by:

$$k = \nu/c$$

In a material, $\mathbf{E} = \mathbf{D}/\epsilon$, where $\epsilon = \epsilon_0(1 + \chi)$ varies with the space coordinates. Therefore, while $\mathit{div}\ \mathbf{D}$ is equal to zero, this is not true for $\mathit{div}\ \mathbf{E}$ and the electric displacement is a more suitable quantity to describe the state of the field inside the crystal than the electric field. If one takes into account coherent scattering only, that is scattering without frequency change, the solution is a wave of the form $\mathbf{D}(\mathbf{r}) \exp 2\pi i\nu t$. Since χ is very much smaller than 1, the electric field is, to a first approximation:

$$\mathbf{E} = \frac{\mathbf{D}}{\epsilon_0(1 + \chi)} = \frac{\mathbf{D}}{\epsilon_0}(1 - \chi)$$

Using this expression and eliminating $\mathbf{H}$ between the first two relations of Maxwell's equations, one obtains the equation of propagation of an electromagnetic wave in the crystalline medium:

$$\Delta \mathbf{D} + \text{curlcurl}\, \chi \mathbf{D} + 4\,\pi^2 k^2 \mathbf{D} = 0 \tag{6}$$

2.1.4. Fourier Expansion of the Dielectric Susceptibility. In a crystalline medium, χ is a triply periodic function of the space coordinates and can be expanded in a Fourier series:

$$\chi = \sum_{\mathbf{h}} \chi_h \exp\,(2\pi\, i\, \mathbf{h} \cdot \mathbf{r}) \tag{7}$$

where $\mathbf{h}$ is a reciprocal-lattice vector and the summation is extended over all reciprocal-lattice vectors. In the case of X-rays, χ is proportional to the electronic density (1) and the coefficients, χ_h, of the Fourier expansion of the electric susceptibility are related to the usual structure factor F_h by

$$\chi_h = -\, R\, \lambda^2 F_h / (\pi V) \tag{8}$$

where V is the volume of the unit cell and the structure factor is given by

$$F_h = \sum_j (f_j + f'_j + i f''_j) \exp\,[-M_j - 2\pi i\, \mathbf{h} \cdot \mathbf{r_j}] = |F_h| \exp\, i\varphi_h \tag{9}$$

where f_j is the form factor of atom j, f'_j and f''_j are the dispersion corrections and $\exp(-M_j)$ is the Debye-Waller factor. The summation is over all the atoms in the unit cell. The order of magnitude of the Fourier coefficients χ_h varies from 10^{-5} to 10^{-7} depending on the wavelength and the structure factor.

In an absorbing crystal, absorption is taken into account phenomenologically through the imaginary part of the index of refraction and of the wavevectors. The electric susceptibility is written $\chi = \chi_r + i\chi_i$. The real and imaginary parts of χ, χ_r and χ_i, respectively, are also triply periodic and can be expanded in Fourier series

$$\left.\begin{aligned} \chi_r &= \sum_h \chi_{rh} \exp(2\pi\, i\, \mathbf{h} \cdot \mathbf{r})\ ; \quad \text{with } \chi_{rh} = -\, R\lambda^2 F_{rh}/(\pi V) \\ \chi_i &= \sum_h \chi_{ih} \exp(2\pi\, i\, \mathbf{h} \cdot \mathbf{r})\ ; \quad \text{with } \chi_{ih} = -\, R\lambda^2 F_{ih}/(\pi V) \end{aligned}\right\}$$

where

$$\left.\begin{aligned} F_{rh} &= \sum_j (f_j + f'_j) \exp\,[-M_j - 2\pi i\, \mathbf{h} \cdot \mathbf{r_j}] = |F_{rh}| \exp\, i\varphi_{rh} \\ F_{ih} &= \sum_j (f''_j) \exp\,[-M_j - 2\pi i\, \mathbf{h} \cdot \mathbf{r_j}] = |F_{ih}| \exp\, i\varphi_{ih} \end{aligned}\right\}$$

The linear absorption coefficient is

$$\mu_o = -\, 2\pi\, k\, \chi_{io} = 2\, R\, \lambda\, F_{io}/V \tag{10}$$

2.2. Wavefields

The notion of wavefield, introduced by Ewald (1917), is one of the most fundamental concepts in dynamical theory. It results from the fact that since the propagation equation (6) is a second order partial differential equation with a periodic interaction coefficient, its solution $\mathbf{D(r)}$ has the same periodicity

$$\mathbf{D(r)} = \exp(-2\pi i\, \mathbf{K_o}\cdot\mathbf{r}) \sum_h \mathbf{D_h} \exp(2\pi i\, \mathbf{h}\cdot\mathbf{r}) \tag{11}$$

This expression can also be written

$$\mathbf{D(r)} = \sum_h \mathbf{D_h} \exp(-\, 2\pi i\, \mathbf{K_h}\cdot\mathbf{r}) \tag{12}$$

where

$$\mathbf{K_h} = \mathbf{K_o} - \mathbf{h} \tag{13}$$

Expression (12) shows that the solution of the propagation equation can be interpreted as an infinite sum of plane waves with amplitudes $\mathbf{D_h}$ and wavevectors $\mathbf{K_h}$. This sum is a *wavefield*, or *Ewald wave*. The wavevectors in a wavefield are deduced from one another by translations of the reciprocal lattice (13). They can be represented geometrically as shown on Figure 1. The wavevectors $\mathbf{K_o} = \mathbf{OP}$ and $\mathbf{K_h} = \mathbf{HP}$ are drawn *away* from reciprocal-lattice points. Their common extremity, P, called a *tiepoint* by Ewald, characterizes the wavefield. In an absorbing crystal, wavevectors have an imaginary part:

$$\mathbf{K_o} = \mathbf{K_{or}} + i\mathbf{K_{oi}}\ , \ \mathbf{K_h} = \mathbf{K_{hr}} + i\mathbf{K_{hi}}$$

and (13) shows that all wavevectors have the same imaginary part $\mathbf{K_{oi}} = \mathbf{K_{hi}}$ and therefore undergo the same absorption. This is one of the most important properties of wavefields.

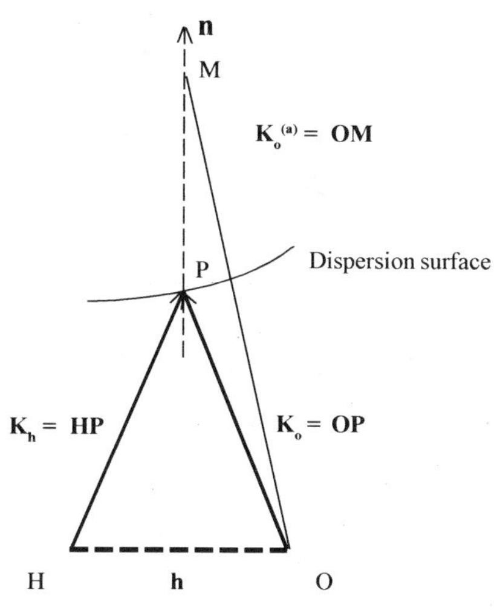

Figure 1. Boundary condition for wave-vectors at the entrance surface. **n**: normal to the entrance surface. P: tiepoint of a wavefield. $\mathbf{h} = \mathbf{OH}$: reciprocal-lattice vector.

2.3. Boundary Conditions at the Entrance Surface

The choice of the "o" component of expansion (12) is arbitrary in an infinite medium. In a semi-infinite medium where the waves are created at the interface with vacuum or air by an incident plane wave with wavevector $\mathbf{K}_\mathbf{o}^{(\mathbf{a})}$, the choice of $\mathbf{K}_\mathbf{o}$ is determined by the boundary conditions. This condition for wavevectors at an interface demands that their tangential components should be continuous across the boundary, in agreement with *Descartes-Snell*'s law. This condition is satisfied when the difference between the wavevectors on each side of the interface is parallel to the normal to that interface. This is shown geometrically on Figure 1, and formally in (14):

$$\mathbf{K}_\mathbf{o} - \mathbf{K}_\mathbf{o}^{(\mathbf{a})} = \mathbf{OP} - \mathbf{OM} = \overline{MP}.\mathbf{n} \tag{14}$$

where $\mathbf{n}$ is a unit vector normal to the crystal surface.

There is no absorption in vacuum and the incident wavevector, $\mathbf{K}^(\mathbf{a})_\mathbf{o}$, is real. Equation (14) shows that it is the component normal to the interface of wavevector $\mathbf{K}_\mathbf{o}$ which has an imaginary part

$$\mathbf{K}_\mathbf{oi} = \mathcal{I}\left(\overline{MP}\right).\mathbf{n} = -\mu\mathbf{n}/(4\pi\,\gamma_o) \tag{15}$$

where $\gamma_o = \cos(\mathbf{n}\cdot\mathbf{s}_\mathbf{o})$ and $\mathbf{s}_\mathbf{o}$ is a unit vector in the incident direction. When there is more than one wave in the wavefield, the effective absorption coefficient, μ, can differ significantly from the normal value, μ_o, given by (10) - see § 5.

2.4. Fundamental Equations of Dynamical Theory

In order to obtain the solution of dynamical theory, one puts expansions (12) and (7) in the propagation equation (6). This leads to an equation with an infinite sum of terms. It is shown to be equivalent to an infinite system of linear equations which are the *fundamental equations* of dynamical theory. These equations relate the amplitude of one of the waves in the wavefield defined by (12), that associated with wavevector $\mathbf{h}$, to those of all of them:

$$\mathbf{D}_\mathbf{hj} = \frac{K_{hj}^2}{K_{hj}^2 - k^2}\sum_{\mathbf{h}'}\chi_{h-h'}\mathbf{D}_{\mathbf{h}'[\mathbf{h}]\mathrm{j}} \tag{16}$$

where the summation is over all reciprocal-lattice vectors $\mathbf{h'}$, $\chi_{h-h'}$ is the Fourier coefficient of the dielectric susceptibility associated with reciprocal-lattice vector $\mathbf{h} - \mathbf{h'}$, and $\mathbf{D}_{\mathbf{h}'[\mathbf{h}]\mathrm{j}}$ is the projection of $\mathbf{D}_{\mathbf{h}'\mathbf{j}}$ on a plane normal to $\mathbf{K}_\mathbf{hj}$. This last term is due to the fact that electromagnetic waves are vector waves and expresses the effect of polarization. The subscript j characterizes a given particular solution of the propagation equation (6). Only those terms in (16) whose wavevector magnitudes K_h are very close to the vacuum value, k, and for which the factors, $K_{hj}^2/(K_{hj}^2 - k^2)$, called *resonance factors* by *Ewald*, are very large, have a non-negligible amplitude. These wavevectors are associated with reciprocal-lattice points which lie very close to the *Ewald sphere*. Far from any *Bragg* reflection, their number is equal to *1* and a single plane wave is propagating in the medium.

In general, for X-rays, there are only two reciprocal-lattice points on the *Ewald* sphere. This is the so-called *two-beam* case to which this treatment will be limited. There are, however, many instances where several reciprocal-lattice points lie simultaneously on the *Ewald* sphere. This corresponds to the *many-beam* case which has interesting applications for the determination of phases of reflections (see the Chapters by Colella and by Hümmer and Weckert, this Volume).

In the two-beam case, and after linearization of the resonance factors, the fundamental equations (16) of dynamical theory reduce to:

$$\left.\begin{array}{r} 2X_oD_o \; - \; kC\chi_{\bar{h}}D_h = 0 \\ -kC\chi_hD_o \; + \; 2\,X_hD_h = 0 \end{array}\right\} \qquad (17)$$

with

$$\left.\begin{array}{l} X_o = K_o \; - \; nk \\ X_h = K_h \; - \; nk \end{array}\right\} \qquad (18)$$

In the case of an absorbing crystal, K_o and K_h, X_o and X_h are complex. C is equal to *1* if $\mathbf{D_h}$ is normal to the $\mathbf{K_o}$, $\mathbf{K_h}$ plane and to cos $2\,\theta$ if $\mathbf{D_h}$ lies in that plane. This is due to the fact that the amplitude with which electromagnetic radiation is scattered is proportional to the sine of the angle between the direction of the electric vector of the incident radiation and the direction of scattering. The polarization of an electromagnetic wave is classically related to the orientation of the electric vector; in dynamical theory it is that of the electric displacement which is considered. The system (17) is therefore a system of 4 equations which admits four solutions, two for each direction of polarization.

Equations (17) and (18) are only valid with the approximation that the Ewald sphere can be replaced by its tangential plane. They are no longer valid for highly asymmetric reflections or when the Bragg angle is close to $\pi/2$. These cases are discussed in the Chapter by Holý (this Volume).

2.5. Dispersion Surface

The fundamental equations (16) of dynamical theory are a set of linear homogeneous equations whose unknowns are the amplitudes of the various waves which make up a wavefield. For the solution to be non-trivial, the determinant of the set must be put equal to zero. This provides a secular equation relating the magnitudes of the wavevectors of a given wavefield. The equation is that of the locus of the tiepoints of all the wavefields which may propagate in the crystal with a given frequency. This locus is called the *dispersion surface*. It is a constant energy surface which is the equivalent of the index surface in optics and is the X-ray analogue of constant energy surfaces in electron band theory of solids, called *Fermi* surfaces. In the two-beam case, the dispersion surface is a surface of revolution around the diffraction vector **OH**. It is made from two spheres and a connecting surface between them. The two spheres are centred at O and H and have the same radius, *nk*. Figure 2 shows the intersection of the dispersion surface with a plane passing through **OH**.

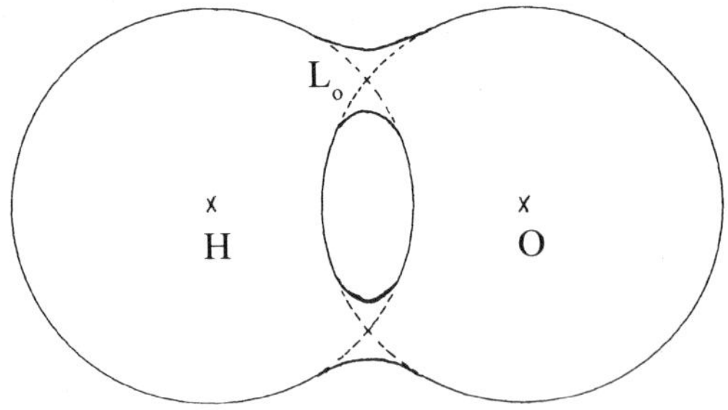

Figure 2. Intersection of the dispersion surface with the plane of incidence. L_o: Lorentz point.

When the tiepoint lies on one of the two spheres, far from their intersection, only one wavefield propagates inside the crystal. When it lies on the connecting surface, two waves are excited simultaneously. The equation of this surface is obtained by equating to zero the determinant of system (17):

$$X_o X_h = k^2 C^2 \chi_h \chi_{\bar{h}}/4 \tag{19}$$

Equations (18) show that, in the zero-absorption case, X_o and X_h are to be interpreted as the distances of the tiepoint P from the spheres centred at O and H respectively. From (19) it can be seen that they are of the order of the vacuum wavenumber times the Fourier coefficient of the electric susceptibility, that is five or six orders of magnitude smaller than k. The two spheres can therefore be replaced by their tangential planes. Equation (19) shows that the product of the distances of the tie point from these planes is constant. The intersection of the dispersion surface with the plane passing through **OH** is therefore a hyperbola (Figure 3) whose diameter is, using (8) and (19)

$$\overline{A_{02}A_{01}} = |C|\sqrt{\chi_h\chi_{\bar{h}}}/\cos\theta = |C|\; R\lambda\sqrt{F_h F_{\bar{h}}}/(\pi V\; \cos\theta) \tag{20}$$

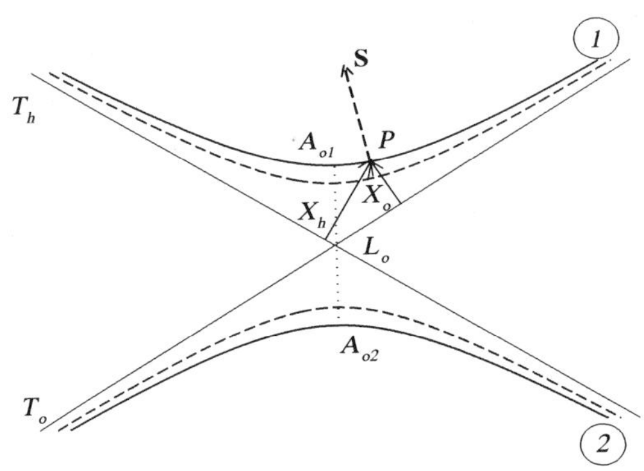

Figure 3. Intersection of the dispersion surface with the plane of incidence: close up view. The reciprocal lattice points O and H are out of the field of view. *Full curve*: polarization normal to the plane of incidence ($C = 1$). *Dashed curve*: polarization parallel to the plane of incidence ($C = \cos 2\,\theta$). **S**: Poynting vector. P: tiepoint of a wavefield.

It can be noted that the larger is the structure factor, that is the stronger the interaction of waves with matter, the larger is the diameter of the dispersion surface. The asymptotes T_o and T_h to the hyperbola are tangents to the circles centred at O and H, respectively. Their intersection, L_o, is called the *Lorentz* point.

A wavefield propagating in the crystal is characterized by a tiepoint P on the dispersion surface and two waves of wavevectors $\mathbf{K_o} = \mathbf{OP}$ and $\mathbf{K_h} = \mathbf{HP}$, respectively. The ratio, ξ, of their amplitudes D_h and D_o is given by means of (17):

$$\xi = \frac{D_h}{D_o} = \frac{2\,X_o}{k\,C\,\chi_{\bar{h}}} = \frac{-2\pi V X_o}{R\,\lambda C F_{\bar{h}}} \tag{21}$$

The hyperbola has two branches, *1* and *2*, for each direction of polarization, that is, for $C = 1$ or $\cos 2\theta$ (Figure 3). Branch *2* is the one situated on the same side of the asymptotes as the reciprocal-lattice points O and H. Given the orientation of the wavevectors which has been chosen away from the reciprocal-lattice points (Figure 1), the coordinates of the tiepoint, X_o and X_h , are positive for branch *1* and negative for branch *2*. The phase of ξ is therefore equal to $\pi + \varphi_h$ and to φ_h for the two branches, respectively, where φ_h is the phase of the structure factor (9). This π difference between the two branches has important consequences on the properties of the wavefields. As mentioned above, due to absorption, wavevectors are actually complex and so is the dispersion surface.

2.6. Polarization of the Waves

The fact that the dispersion surface has one sheet for each direction of polarization is due to the vectorial nature of electromagnetic waves. The fundamental equations of dynamical theory are also vector equations and, when there are p nodes of the reciprocal lattice close to the Ewald sphere, they correspond to $3p$ scalar equations. However, from Maxwell's equations, *div* $\mathbf{D} = 0$, and it can be shown that the waves of expansion (12) are transverse; namely, $\mathbf{D_h}$ is normal to $\mathbf{K_h}$. The system of $3p$ equations can therefore be reduced to a system of $2p$ equations and the dispersion surface has two sheets. The polarization phenomena are discussed in the Chapter by Malgrange (this Volume).

2.7. Propagation Direction of the Wavefields

The energy of all the waves in a given wavefield propagates in a common direction, which is obtained by calculating either the group velocity or the Poynting vector. It can be shown (von Laue, 1960) that, averaged over time and the unit cell, the Poynting vector of a wavefield is

$$\mathbf{S} = (c/\epsilon_o)\exp(4\pi\,\mathbf{K_{oi}}\cdot\mathbf{r})\left[|D_o|^2\,\mathbf{s_o} + |D_h|^2\,\mathbf{s_h}\right] \tag{22}$$

where $\mathbf{s_o}$ and $\mathbf{s_h}$ are unit vectors in the $\mathbf{K_o}$ and $\mathbf{K_h}$ directions, respectively, c is the velocity of light and ϵ_o the dielectric permittivity of vacuum. From (22) and equation (19) of the dispersion surface, it can be shown that the propagation direction of the wavefield lies along the normal to the dispersion surface at the tiepoint (Figure 3). The angle, α, between the propagation direction and the lattice planes is given by:

$$\tan\alpha = \frac{1 - |\xi|^2}{1 + |\xi|^2}\tan\theta \tag{23}$$

3. SOLUTIONS OF PLANE-WAVE DYNAMICAL THEORY

3.1. Departure From Bragg's Law of the Incident Wave

The wavefields excited in the crystal by the incident wave are determined by applying the boundary condition already mentioned for the continuity of the tangential component of the wavevectors. The wavenumber of waves propagating in vacuum in the incident or reflected directions is $k = 1/\lambda$ and the common extremity, M, of their wavevectors $\mathbf{OM} = \mathbf{K_o^{(a)}}$ and $\mathbf{HM} = \mathbf{K_h^{(a)}}$ lies on spheres of radius k and centred at O and H, respectively. The intersections of these spheres with the plane of incidence are two circles which can be approximated by their tangents T_o' and T_h' at their intersection point, L_a, or *Laue* point (Figure 4). *Bragg*'s condition is exactly satisfied according to the geometrical theory of diffraction when M lies at L_a. The departure $\Delta\theta$ from *Bragg*'s incidence of an incident wave will be defined as the angle between the corresponding wavevectors $\mathbf{OM}$ and $\mathbf{OL_a}$. As $\Delta\theta$ is very small as compared to the *Bragg* angle in the general case of X-rays or neutrons, one may write

$$\left.\begin{aligned} \mathbf{K_o^{(a)}} &= \mathbf{OM} = \mathbf{OL_a} + \mathbf{L_aM} \\ \Delta\theta &= \overline{L_aM}\,/k \end{aligned}\right\}$$

The tangent T_o' is oriented in such a way that $\Delta\theta$ is negative when the angle of incidence is smaller than *Bragg*'s angle.

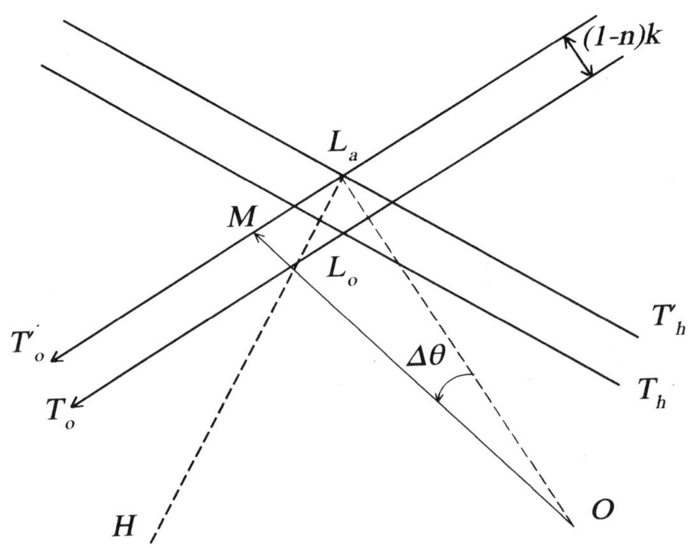

Figure 4. Departure from Bragg's law of an incident wave.

3.2. Transmission and Reflection Geometries

The boundary condition for the continuity of the tangential component of the wavevectors is applied by drawing from M a line, $\mathbf{Mz}$, parallel to the normal, $\mathbf{n}$, to the crystal surface. The tiepoints of the wavefields excited in the crystal by the incident wave are at the intersections of this line with the dispersion surface. Two different situations may occur:

• **Transmission, or Laue Case** (Figure 5). The normal to the crystal surface drawn from M intersects *both* branches of the dispersion surface (at points P_1 and P_2). The reflected wave is directed towards the *inside* of the crystal. Let γ_o and γ_h be the cosines of the angles between the normal to the crystal surface, **n**, and the incident and reflected directions, respectively, and γ their ratio, called *asymmetry ratio* :

$$\gamma_o = \cos\,(\mathbf{n}, \mathbf{s_o}) \;\; ; \;\; \gamma_h = \cos\,(\mathbf{n}, \mathbf{s_h}) \;\; ; \;\; \gamma = \gamma_h/\gamma_o \tag{24}$$

It will be noted that γ_o, γ_h and γ are all positive in the Laue case.

• **Reflection, or Bragg Case** (Figure 6). There are in this case three possible situations: the normal to the crystal surface drawn from M intersects either branch *1* (at P' and P'' in the example of Figure 6) or branch *2* of the dispersion surface, or the intersection points are imaginary. The reflected wave is directed towards the *outside* of the crystal. The cosine γ_h and the asymmetry ratio γ are now negative.

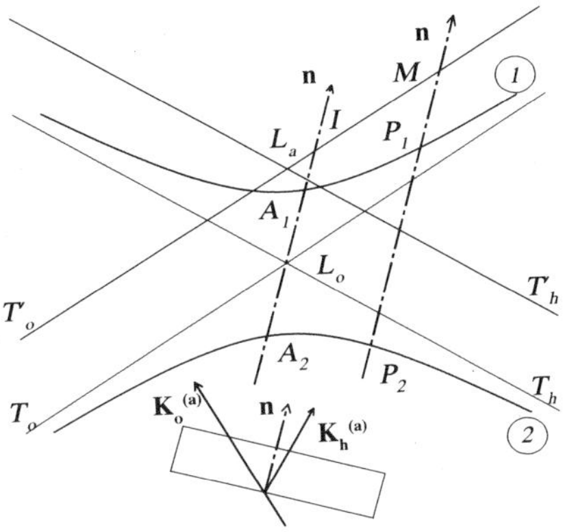

Figure 5. Transmission, or *Laue*, geometry. Boundary conditions at the entrance surface

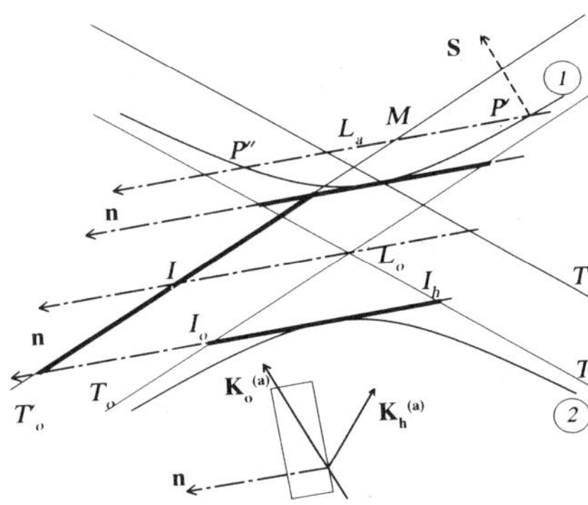

Figure 6. Reflection, or *Bragg*, geometry. Boundary conditions at the entrance surface

3.3. Middle of the Reflection Domain

The incident wavevector corresponding to the middle of the reflection domain is, in both cases, **OI** where I is the intersection of the normal to the crystal surface drawn from the Lorentz point, L_o, with T'_o (Figures 5 and 6), while, according to *Bragg*'s law, it should be $\mathbf{OL_a}$. The angle, $\Delta\theta$, between the incident wavevectors, $\mathbf{OL_a}$ and **OI**, corresponding to the middle of the reflecting domain according to the geometrical and dynamical theories, respectively, is

$$\Delta\theta_o = \overline{L_a I}/k = -\chi_o(1 - \gamma)/(2 \sin 2\theta)$$

In the *Bragg case*, the asymmetry ratio, γ, is negative and $\Delta\theta_o$ is never equal to zero. This difference in Bragg angle between the two theories is due to the refraction effect which is neglected in the geometrical theory. In the *Laue case*, $\Delta\theta_o$ is equal to zero for symmetric reflections ($\gamma = 1$).

3.4. Deviation Parameter

The solutions of dynamical theory are best described by introducing a reduced parameter called the *deviation parameter*

$$\eta = (\Delta\theta - \Delta\theta_o)/\delta \tag{25}$$

where

$$\delta = |C|\sqrt{|\gamma|\chi_h\chi_{\bar{h}}}/\sin 2\theta = R\,\lambda^2|C|\sqrt{|\gamma|F_hF_{\bar{h}}}/[\pi\, V \sin 2\theta] \tag{26}$$

In an absorbing crystal, η, $\Delta\theta_o$ and δ are complex, and the real part of δ is half the width $2\delta_r$ of the rocking curve, called the *Darwin width* (see § **4** and § **5**).

3.5. Pendellösung and Extinction Distances

The length Λ_o defined by

$$\Lambda_o = \frac{\sqrt{\gamma_o|\gamma_h|}}{|C|\sqrt{\chi_h\chi_{\bar{h}}}} = \frac{\lambda|\gamma_h|}{\delta \sin 2\theta} \tag{27}$$

plays a very important role in the dynamical theory of diffraction, both by perfect and deformed crystals.

• In the *transmission* geometry, it gives the period of the interference between the two excited wavefields. This interference constitutes the *Pendellösung* effect first described by Ewald (1917) - see § **4.3** - Λ_o is in that case called *Pendellösung* distance, denoted Λ_L hereafter. Its geometrical interpretation, in the zero-absorption case, is the inverse of a diameter $\overline{A_2A_1}$ of the dispersion surface in a direction defined by the cosines γ_h and γ_o with respect to the reflected and incident directions, respectively (Figure 5). It reduces to expression (20) in the symmetric case.

• In the *reflection* geometry, it gives the absorption distance in the total-reflection domain and is called the *extinction* distance, denoted Λ_B - see §5.2. Its geometrical interpretation in the zero-absorption case, is the inverse of the length $\overline{I_hI_o}$, (Figure 6).

3.6. Solution of the Dynamical Theory

The coordinates of the tiepoints excited by the incident wave are obtained by looking for the intersection of the dispersion surface, (19), with the normal **n** to the crystal surface drawn from M (Figures 5 and 6). The ratio ξ of the amplitudes of the waves of the corresponding wavefields is related to these coordinates by (21) and is found to be

$$\left.\begin{aligned}\xi_j &= D_{hj}/D_{oj}\\ &= -\,S(C)S(\gamma_h)\left[\sqrt{F_hF_{\bar{h}}}/F_{\bar{h}}\right]\left[\eta \pm \sqrt{\eta^2 + S(\gamma_h)}\right]/\sqrt{|\gamma|}\end{aligned}\right\} \tag{28}$$

where the *plus* sign corresponds to a tiepoint on branch *1* ($j = 1$) and the *minus* sign to a tiepoint on branch *2* ($j = 2$) and $S(\gamma_h)$ means sign of γ_h (*+1* in transmission geometry, *-1* in reflection geometry).

3.7. Standing Waves

The various waves in a wavefield are coherent and interfere. In the two-beam case, the intensity of the wavefield is, using (11) and (21)

$$|D|^2 = |D_o|^2 \exp(4\pi \, \mathbf{K_{oi}} \cdot \mathbf{r}) \left[1 + |\xi|^2 + 2C|\xi| \cos 2\pi(\mathbf{h} \cdot \mathbf{r} + \Psi)\right] \quad (29)$$

where Ψ is the phase of ξ, ($\xi = |\xi| \exp i\Psi$). Equation (29) shows that the interference between the two waves is at the origin of *standing waves*. The corresponding nodes lie on planes such that $\mathbf{h} \cdot \mathbf{r} = constant$. These planes are therefore parallel to the diffraction planes and their periodicity is $d_{hkl} = \lambda/2\sin\theta$. Their position within the unit cell is given by the value of Ψ.

- In the *Laue* case, Ψ is equal to $\pi + \varphi_h$ for branch *1* and to φ_h for branch *2*, where φ_h is the phase of the structure factor (9). This means that the *nodes* of standing waves lie on the maxima of the *hkℓ Fourier* component of the electronic density for branch *1* while it is the *anti-nodes* for branch *2*.

- In the *Bragg* case, Ψ varies continuously from $\pi + \varphi_h$ to φ_h, when the angle of incidence is varied from the low angle to the high angle side of the reflection domain by rocking the crystal. The *nodes* lie on the maxima of the *hkℓ Fourier* components of electronic density on the low angle side of the rocking curve. As the crystal is rocked, they are progressively shifted by half a lattice spacing until the *anti-nodes* lie on the maxima of electronic density on the high angle side of the rocking curve.

Standing waves are at the origin of the phenomenon of anomalous absorption (§ **3.8.** and § **4.**) and are used for the location of atoms in the unit cell at the vicinity of the crystal surface: when X-rays are absorbed, fluorescent radiation and photoelectrons are emitted. Detection of this emission for a known angular position of the crystal with respect to the rocking curve and therefore for a known value of the phase Ψ enables to localize the position of the emitting atom within the unit cell (see the Chapters by Patel and Lagomarsino, this Volume).

3.8. Anomalous Absorption

It has been shown in § **2.2** that the wavevectors of a given wavefield all have the same imaginary part and therefore the same absorption coefficient μ, (15). Borrmann (1950, 1954) has shown that this coefficient is much smaller than the normal one, μ_o, for wavefields whose tiepoints lie on branch *1* of the dispersion surface, and much larger for wavefields whose tiepoints lie on branch *2*. The former case corresponds to the *anomalous transmission* effect, or *Borrmann* effect. In favourable cases the minimum absorption coefficient may be as low as a few percent of the normal one. The physical interpretation of the *Borrmann* effect is to be found in the standing waves described in § **4**. When the nodes of electric field lie on the planes corresponding to the maxima of the *hkℓ* component of electron density, the wavefields are absorbed less than when there is no diffraction. It is just opposite for branch *2* wavefields whose *anti-nodes* lie on the maxima of electron density, and which are absorbed more than normal. Using (15), (18), (21) and (28), the effective absorption coefficient, μ_j, is shown to be:

$$\mu_j = -4\pi \, \gamma_o K_{oi} = \mu_o - 4\pi X_{oi} = \mu_o - 2\pi\lambda \, C \, \mathcal{I}(F_{\bar{h}}\xi_j) \quad (30)$$

4. INTENSITIES OF PLANE WAVES IN THE TRANSMISSION GEOMETRY

4.1. Boundary Conditions for the Amplitudes at the Entrance Surface - Intensities of the Reflected and Refracted Waves

Let us consider an infinite plane wave incident on a crystal plane surface of infinite lateral extension. As has been shown in § **3**, two wavefields are excited in the crystal, with tiepoints P_1 and P_2 (Figure 5), and amplitudes D_{o1}, D_{h1}, D_{o2}, D_{h2}, respectively. Maxwell's boundary conditions imply continuity of the tangential component of the electric field and of the normal component of the electric displacement across the boundary. Because the index of refraction is so close to unity, and using the same approximation as for reducing the fundamental equations of the dynamical theory, (16), to system (17), one can assume with a very good approximation that there is continuity of the three components of both the electric field and the electric displacement. As a consequence, it can easily be shown that one has, *along the entrance surface*, for all components of the electric displacement:

$$\left.\begin{aligned} D_o^{(a)} &= D_{o1} + D_{o2} \\ 0 &= D_{h1} + D_{h2} \end{aligned}\right\}$$

where $D_o^{(a)}$is the amplitude of the incident wave.

The amplitudes D_{hj} and D_{oj} at the entrance surface can be expressed by means of their ratio ξ_j (28):

$$\left.\begin{aligned} D_{h1} &= D_o^{(a)}\frac{\xi_1\xi_2}{\xi_2 - \xi_1}; \quad D_{h2} = D_o^{(a)}\frac{-\xi_1\xi_2}{\xi_2 - \xi_1} \\ D_{o1} &= D_o^{(a)}\frac{\xi_2}{\xi_2 - \xi_1}; \quad D_{02} = D_o^{(a)}\frac{-\xi_1}{\xi_2 - \xi_1} \end{aligned}\right\} \tag{31}$$

The amplitudes are multipled by their respective phase factors which, in the absorbing case, have an imaginary part inside the crystal. Using (28), (30) and (31), it can be shown that the *intensities* of the four waves are

$$\left.\begin{aligned} |D_{oj}|^2 &= \left|D_o^{(a)}\right|^2 \exp\left(-\mu_j z/\gamma_o\right)\frac{\left[\sqrt{1+\eta_r^2} \mp \eta_r\right]^2}{4\left(1+\eta_r^2\right)} \\ |D_{hj}|^2 &= \left|D_o^{(a)}\right|^2 \exp\left(-\mu_j z/\gamma_o\right)\left|\frac{F_h}{F_{\bar{h}}}\right|\frac{\gamma^{-1}}{4\left(1+\eta_r^2\right)} \end{aligned}\right\} \tag{32}$$

with

$$\mu_j = \mu_o\left[1/2\left(1+\gamma^{-1}\right) \mp \frac{1/2\eta_r\left(1-\gamma^{-1}\right) + |C|\sqrt{\gamma^{-1}}\,|F_{ih}/F_{io}|\cos\varphi}{\sqrt{\eta_r^2+1}}\right]$$

where $\varphi = \varphi_{rh} - \varphi_{ih}$ is the phase difference between F_{rh} and F_{ih}, the upper sign corresponds to branch *1* and the lower sign to branch *2* of the dispersion surface.

Figure 7 represents the variations of the intensities $|D_{o1}|^2 + |D_{h1}|^2$ and $|D_{o2}|^2 + |D_{h2}|^2$ of the two wavefields with the deviation parameter. It can be seen that far from the reflection domain on the small angle side ($\eta_r \Rightarrow -\infty$)

wavefield *1* is the strongest while on the large angle side ($\eta_r \Rightarrow +\infty$) it is wavefield *2*. One says that the wavefield of highest intensity "jumps" from one branch of the dispersion surface to the other one across the reflection domain. This is an important property of dynamical theory which also holds in the Bragg case, and when a wavefield crosses a highly distorted region in a deformed crystal (the so called *interbranch scattering* - see for instance Authier and Balibar, (1970).

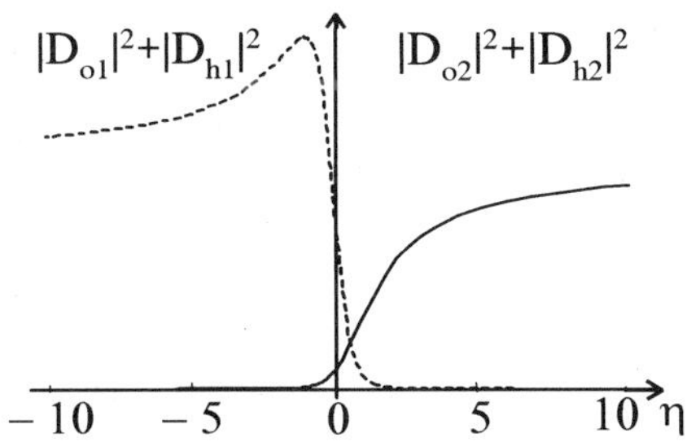

Figure 7. Variations of the intensities of the two wavefields with the deviation parameter (arbitrary parameters).

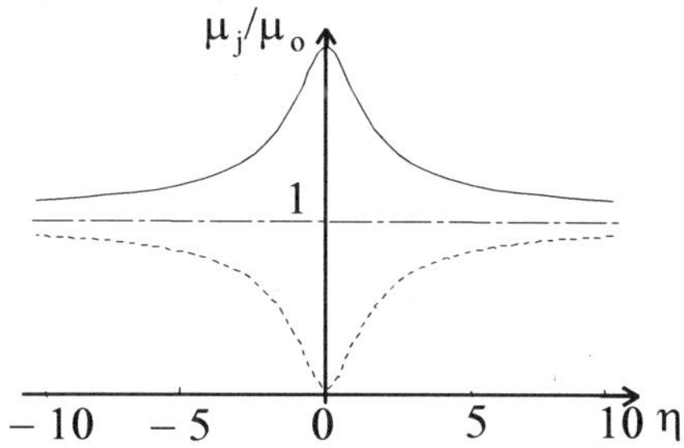

Figure 8. Variation of the effective absorption coefficient with the deviation parameter

Figure 8 shows the variations of the effective absorption coefficient μ_j with η_r for wavefields belonging to branches *1* and *2* respectively, in the case of GaAs, a symmetric 400 reflection and $Cu\ K_\alpha$. It can be seen that for $\eta_r = 0$ the absorption coefficient for branch *1* becomes significantly smaller than the normal absorption coefficient, μ_o. The minimum absorption coefficient, $\mu_o(1 - |CF_{ih}/F_{io}|\cos\varphi)$, depends on the nature of the reflection through the structure factor and on the temperature through the Debye-Waller factor which is included in F_{ih} (9).

4.2. Boundary Conditions at the Exit Surface

4.2.1. Wave vectors. When a wavefield reaches the exit surface, it breaks up into its two constituent waves. Their wavevectors are obtained by applying again the condition of the continuity of their tangential components along the crystal surface. The extremities, M_j and N_j, of these wavevectors

$$\mathbf{OM_j} = \mathbf{K_{oj}^{(d)}} \ ; \quad \mathbf{HN_j} = \mathbf{K_{hj}^{(d)}}$$

lie at the intersections of the spheres of radius k and centred at O and H, respectively, with the normal **n'** to the crystal exit surface, drawn from P_j (j = 1 and 2) (Figure 9).

4.2.2. Reflecting Power - Pendellösung. Two wavefields arrive at any point of the exit surface. Their constituent waves interfere and generate emerging waves in the refracted and reflected directions (Figure 10). Their respective amplitudes are given by the boundary conditions:

$$\left.\begin{aligned} D_o^{(d)} \exp\left(-2\pi i\ \mathbf{K_o^{(d)}}\cdot\mathbf{r}\right) &= D_{o1}\exp\left(-2\pi i\ \mathbf{K_{o1}}\cdot\mathbf{r}\right) + D_{o2}\exp\left(-2\pi i\ \mathbf{K_{o2}}\cdot\mathbf{r}\right) \\ D_h^{(d)} \exp\left(-2\pi i\ \mathbf{K_h^{(d)}}\cdot\mathbf{r}\right) &= D_{h1}\exp\left(-2\pi i\ \mathbf{K_{h1}}\cdot\mathbf{r}\right) + D_{h2}\exp\left(-2\pi i\ \mathbf{K_{h2}}\cdot\mathbf{r}\right) \end{aligned}\right\}$$

where **r** is the position vector of a point on the exit surface, the origin of phases being taken on the entrance surface.

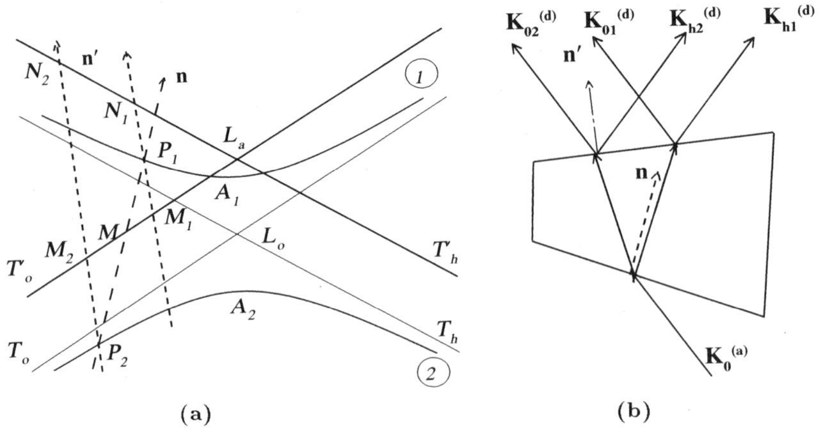

Figure 9. Boundary condition for the wave vectors at the exit surface. **n**: normal to the entrance surface; **n'**: normal to the exit surface; **(a)**: reciprocal space - **(b)**: direct space

In a plane parallel crystal, these expressions reduce to

$$D_o^{(d)} = D_{o1} \exp\left(-2\pi i\ \overline{MP_1}\,.t\right) + D_{o2} \exp\left(-2\pi i\ \overline{MP_2}\,.t\right)$$

$$D_h^{(d)} = D_{h1} \exp\left(-2\pi i\ \overline{MP_1}\,.t\right) + D_{h2} \exp\left(-2\pi i\ \overline{MP_2}\,.t\right)$$

where t is the crystal thickness.

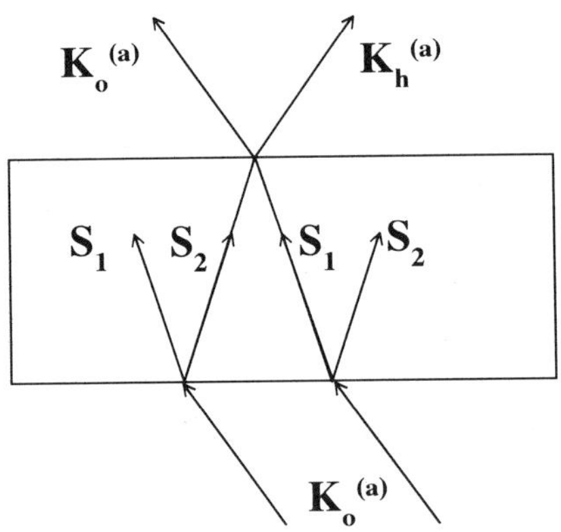

Figure 10. Interference of wavefields belonging to branches *1* and *2* of the dispersion surface, respectively, when they reach the exit surface.

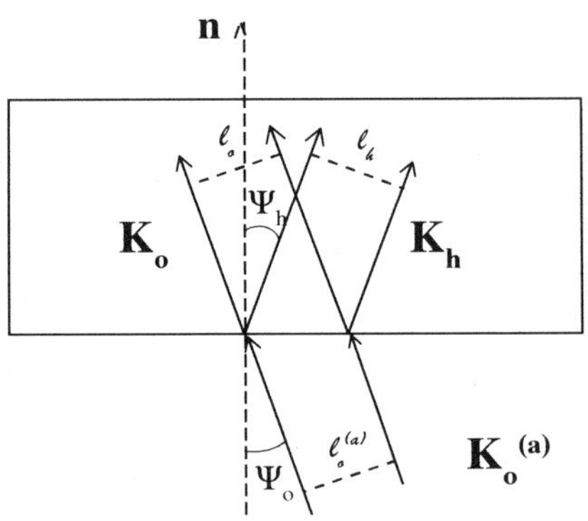

Figure 11. Cross sections of the incident, refracted and reflected waves.

What is actually measured in a counter receiving the reflected or the refracted beam is the *reflecting power*, namely the ratio of the *energy* of the reflected or refracted beam on one hand and the energy of the incident beam on the other. The energy of a beam is obtained by multiplying its intensity by its cross-section.

If ℓ is the width of the trace of the beam on the crystal surface, the cross-sections of the incident (or refracted) and reflected beams are respectively proportional to $\ell_o = \ell\gamma_o$ and $\ell_h = \ell\gamma_h$ (Figure 11). The reflecting powers are therefore respectively:

$$\left.\begin{aligned} refracted\ beam:\quad I_o &= \frac{\ell_o\left|D_o^{(d)}\right|^2}{\ell_o\left|D_o^{(a)}\right|^2} = \frac{\left|D_o^{(d)}\right|^2}{\left|D_o^{(a)}\right|^2} \\ reflected\ beam:\quad I_h &= \frac{\ell_h\left|D_h^{(d)}\right|^2}{\ell_o\left|D_o^{(a)}\right|^2} = \gamma\,\frac{\left|D_h^{(d)}\right|^2}{\left|D_o^{(a)}\right|^2} \end{aligned}\right\} \tag{33}$$

$$\left.\begin{aligned} I_o &= \exp\left[-\frac{\mu_o}{2}\left(\frac{1}{\gamma_o} + \frac{1}{\gamma_h}\right)t\right]\frac{\left|\eta\left[\mathcal{E}_1 + \mathcal{E}_2\right] + \sqrt{1+\eta^2}\left[\mathcal{E}_1 - \mathcal{E}_2\right]\right|^2}{4\left|1+\eta^2\right|} \\ I_h &= \left|\frac{\chi_h}{\chi_{\bar{h}}}\right|\exp\left[-\frac{\mu_o}{2}\left(\frac{1}{\gamma_o} + \frac{1}{\gamma_h}\right)t\right]\frac{\left|\mathcal{E}_1 - \mathcal{E}_2\right|^2}{4\left|1+\eta^2\right|} \end{aligned}\right\} \tag{34}$$

with $\mathcal{E}_1 = \exp(\pi i t/\Lambda)$; $\mathcal{E}_2 = \exp(-\pi i t/\Lambda)$; $\Lambda = \Lambda_L/\sqrt{1+\eta^2}$.

The variations of I_o and I_h with the deviation parameter present oscillations which are due to Pendellösung and have more or less contrast, depending on the absorption coefficient. The oscillations in the rocking curve were first observed by Lefeld-Sosnowska and Malgrange (1969). Figure 12 a and b show the shapes of the reflected and refracted beams rocking curves for a thin crystal and Figure 13 for a thick absorbing crystal. The *width at half-height of the rocking curve*, averaged over the Pendellösung oscillations, corresponds, in the non-absorbing case, to $\Delta\eta = 2$, that is to $\Delta\theta = 2\,\delta$, where δ is given by (26).

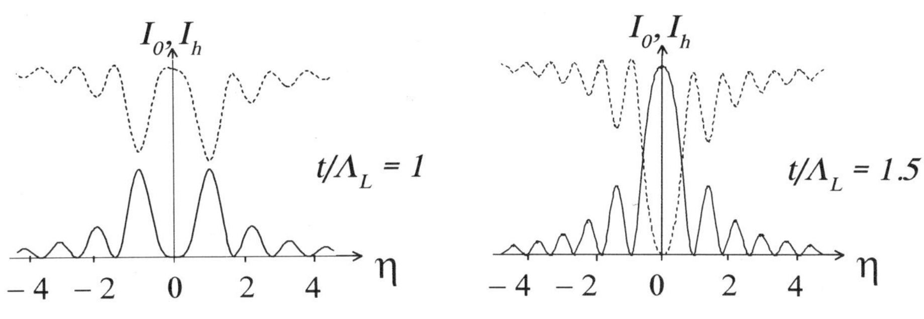

Figure 12. Rocking curves in the transmission case, thin crystals (arbitrary units). Silicon, MoKα, 220; *Solid curve*: reflected beam; *dashed curve*: refracted beam (**a**): $t/\Lambda = 1.0$; (**b**): $t/\Lambda = 1.5$.

4.3. Integrated Intensity

The integrated intensity is the ratio of total energy recorded in the counter (when the crystal is rocked) to the intensity of the incident beam. It is proportional to the area under the line profile:

$$I_{hi} = \int_{-\infty}^{+\infty} I_h d(\Delta\theta) \tag{35}$$

The integration has been performed by von Laue (1960) in the zero-absorption case. Using (25) and (34), the result is:

$$I_{hi} = \frac{\pi \,|C\chi_h|\sqrt{\gamma}}{2 \sin 2\theta} \int_0^{2\pi t \Lambda_L^{-1}} J_0(z)dz$$

where $J_0(z)$ is the zeroth order Bessel function

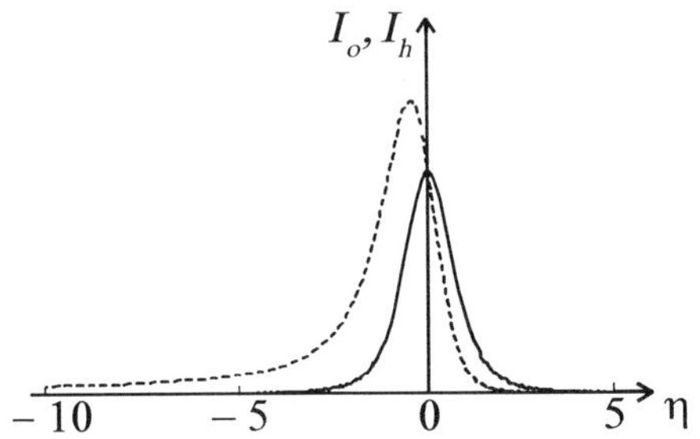

Figure 13. Rocking curves in the transmission case, thin crystals (arbitrary units); GaAs, MoKα, 400, $\mu t = 5.8$.

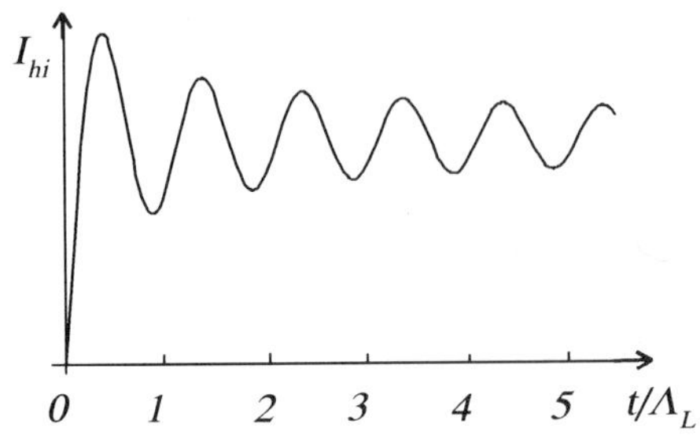

Figure 14. Integrated Intensity in the transmission case for non-absorbing crystals (arbirtrary units)

Figure 14 represents the variations of the integrated intensity with t/Λ_L. The integration has been performed for absorbing crystals by Kato (1955). The integrated intensity is given is this case by:

$$I_{hi} = \frac{\pi \,|C\chi_h|\sqrt{\gamma}}{2 \sin 2\theta} \left|\frac{F_h}{F_{\bar{h}}}\right| \exp\left[-\frac{\mu_o t}{2}\left(\frac{1}{\gamma_o} + \frac{1}{\gamma_h}\right)\right] \left[\int_0^{\frac{2\pi t}{\Lambda_L}} J_0(z)dz - 1 + I_0(\zeta)\right]$$

where

$$\zeta = \mu_o\, t \left\{ \left[|C|^2 \left|\frac{F_{ih}}{F_{io}}\right|^2 \cos^2\varphi + \frac{\gamma_h - \gamma_o}{4\gamma_o\gamma_h} \right] / (\gamma_o\gamma_h) \right\}^{1/2}$$

and $I_0(\zeta)$ is the modified Bessel function of zeroth-order.

4.4. Thin Crystals - Comparison with Geometrical Theory

The reflecting power (34) of the reflected beam may also be written, in the non-absorbing case:

$$I_h = \pi^2 t^2 \Lambda_L^{-2} f(\eta)$$

with:

$$f(\eta) = \left[\frac{\sin U\sqrt{1+\eta^2}}{U\sqrt{1+\eta^2}}\right]^2 \; ; \; U = \pi\, t\, \Lambda_L^{-1}$$

When $t\,\Lambda_L^{-1}$ is very small, $f(\eta)$ tends asymptotically towards the function

$$f_1(\eta) \;=\; \left[\frac{\sin\, U\,\eta}{U\,\eta}\right]^2$$

and I_h towards the value given by the geometrical theory. The condition for the geometrical theory to apply is therefore that the crystal thickness be much smaller than the Pendellösung distance.

5. INTENSITY OF PLANE WAVES IN THE REFLECTION GEOMETRY

5.1. Thick Crystals

The geometrical construction in Figure 6 shows that, in the Bragg case, the normal to the crystal surface drawn from the extremity of the incident wavevector intersects the dispersion surface either at two points of the same branch, P' and P'', or at imaginary points. It has been shown in § **2.7** that the propagation of the wavefields inside the crystal is along the normal to the dispersion surface at the corresponding tiepoints. Figure 6 shows that this direction is oriented towards the outside of the crystal for one only of two intersection points (P'). In a very thick crystal, these wavefields cannot exist because there is always a small amount of absorption. One concludes that in the thick crystal case and in the reflection geometry, only one wavefield is excited inside the crystal. It corresponds to branch *1* on the low angle side of the rocking curve and to branch *2* on the high angle side. Using the same approximations as in § **6.2**, the amplitude, $\mathbf{D}_{\mathbf{h}}^{(\mathbf{a})}$, of the wave reflected at the crystal surface is obtained by applying the boundary conditions which are particularly simple in this case:

$$\mathbf{D_o} \;=\; \mathbf{D}_{\mathbf{o}}^{(\mathbf{a})};\;\; \mathbf{D}_{\mathbf{h}}^{(\mathbf{a})} \;=\; \mathbf{D_h}$$

The reflecting power is given by an expression similar to (33):

$$I_h \;=\; |\gamma|\,|\xi_j|^2$$

where the expression of ξ_j is given by (28), with $S(\gamma_h) \;<\; 0$, and the $\pm$ sign is chosen so that the solution converges:

$$I_h \;=\; \left|\eta \;-\; S(\eta_r)\sqrt{\eta^2 - 1}\right|^2$$

where η_r is the real part of η.

In the zero-absorption case, the dispersion surface and η are real. When the normal to the entrance surface does not intersect any of the branches of the dispersion surface, that is when $-\;1 \;<\; \eta \;<\; +\;1$,

$$|\xi|^2 \;=\; |\gamma|^{-1},\; I_h \;=\; 1$$

and there is *total reflection*. The rocking curve has the well known top-hat shape (Figure 15). Far from the total-reflection domain, the curve can be approximated by the function $I_h(appr.) \;=1/(4\eta^2)$

The width of the total-reflection domain is $\Delta\eta = 2$ and its angular width is therefore equal, using (25), to $2\,\delta$, where δ is given by (26). It increases with increasing wavelength and it is proportional to the structure factor, the polarization factor C, and to the square root of the asymmetry factor $|\gamma|$. Using an asymmetric reflection, it is therefore possible to decrease the width at will. This property is used in the design of monochromators.

The integrated intensity is defined by (35) and is equal in the zero-absorption case to

$$I_{hi} = 8\,\delta/3$$

Within the total-reflection domain, there are two wavefields propagating inside the crystal, with imaginary wavevectors, one towards the inside of the crystal and the other one in the opposite direction, so that they cancel out and, globally, no energy penetrates the crystal. The absorption coefficient of the waves penetrating in the crystal is

$$\mu = -\,4\pi K_{oi}\gamma_o = 2\pi\gamma_o\sqrt{1-\eta^2}/\Lambda_B$$

where Λ_B is the value taken by Λ_o (27) in the Bragg case.

The *penetration depth* is a minimum at the middle of the reflection domain and is then $\Lambda_B/2\pi$. This attenuation effect is called *extinction*, and Λ_B is called the *extinction distance*. It is a specific property due to the existence of wavefields. The resulting propagation direction of energy is parallel to the crystal surface, but with a cross-section equal to zero: it is an *evanescent wave* (see, for instance, Cowan *et al.*, 1986).

When the crystal is absorbing, there is no longer a total-reflection domain and energy penetrates inside the crystal at all incidence angles although with a very high absorption coefficient within the domain $|\eta_r| \leq 1$. Figure 15 gives an example of rocking curve for a thick absorbing crystal. It was first observed by Renninger (1955). The shape is asymmetric and is due to the anomalous-absorption effect: it is lower than normal on the low angle side which is associated with wavefields belonging to branch *1* of the dispersion surface and larger than normal on the high angle side, which is associated with branch *2* wavefields.

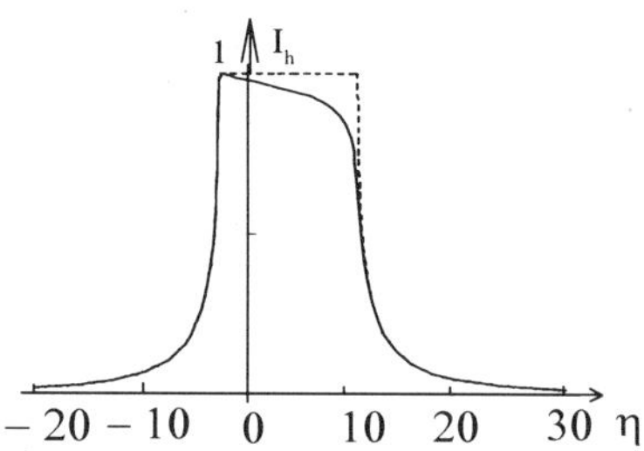

Figure 15. Theoretical rocking curve in the reflection case for a thick crystal. *Solid curve*: GaAs, 400, CuKα; *dashed curve*: non-absorbing crystal.

5.2. Thin Crystals

5.2.1. Boundary conditions. If the crystal is thin, the wavefield created at the reflecting surface at A and penetrating inside can reach the back surface at B (Figure 16 a). The *incident* direction there points towards the *outside* of the crystal, while the *reflected* direction points towards the inside. The wavefield propagating along AB will therefore generate at B:

- a *partially transmitted wave* outside the crystal, $\mathbf{D_o^{(d)}} \exp\left[-2\pi i\ \mathbf{K_o^{(d)}} \cdot \mathbf{r}\right]$

- a *partially reflected wavefield* inside the crystal. The corresponding tiepoints are obtained by applying the usual condition of the continuity of the tangential components of wavevectors (Figure 16 b). If the crystal is a plane-parallel slab, these points are M and P_2 , respectively and $\mathbf{K_o^{(d)}} = \mathbf{K_o^{(a)}}$

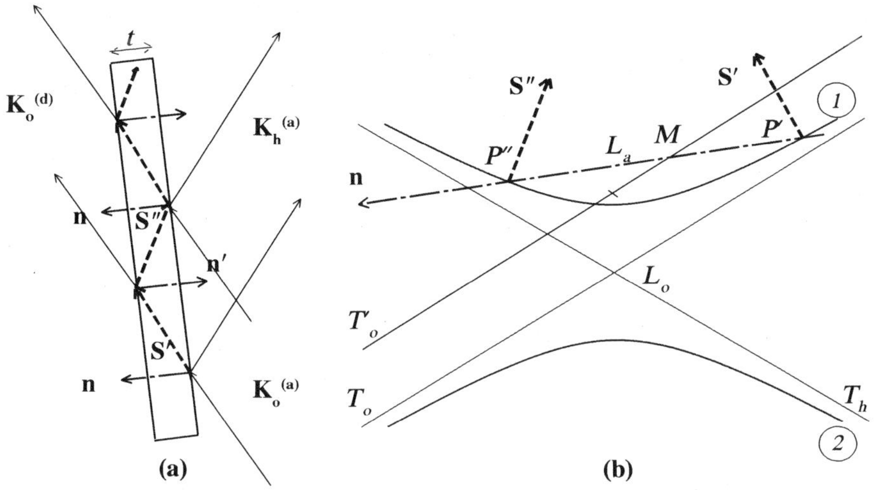

Figure 16. Bragg case - Thin crystals. Two wavefields propagate in the crystal. **(a)** direct space; **(b)**: reciprocal space.

The boundary condition are then written:

- *entrance surface*:

$$D_{o1} + D_{o2} = D_o^{(a)} \; ; \; D_{h1} + D_{h2} = D_h^{(a)}$$

- *back surface*:

$$D_{o1} \exp\left[-2\pi i\ \mathbf{K_{o1}} \cdot \mathbf{r}\right] + D_{o2} \exp\left[-2\pi i\ \mathbf{K_{o2}} \cdot \mathbf{r}\right] = D_o^{(a)} \exp\left[-2\pi i\ \mathbf{K_o^{(d)}} \cdot \mathbf{r}\right]$$

$$D_{h1} \exp\left[-2\pi i\ \mathbf{K_{h1}} \cdot \mathbf{r}\right] + D_{h2} \exp\left[-2\pi i\ \mathbf{K_{h2}} \cdot \mathbf{r}\right] = 0$$

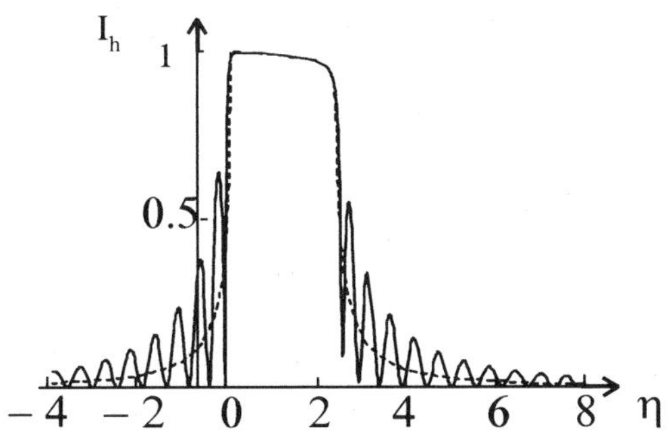

Figure 17. Rocking curve for a thin crystal in the Bragg case ($t/\Lambda_B = 2$). *Dashed curve*: rocking curve for a thick crystal

5.2.2. Rocking curve. Using (28), it can be shown that the expressions of the intensity reflected at the entrance surface is

$$I_h = |\gamma| \left| \frac{D_h^{(a)}}{D_o^{(a)}} \right|^2 = \frac{|\mathcal{E}_1 - \mathcal{E}_2|^2}{\left| \eta[\mathcal{E}_1 - \mathcal{E}_2] + \sqrt{\eta^2 - 1}\ [\mathcal{E}_1 + \mathcal{E}_2] \right|^2} \tag{36}$$

where $\mathcal{E}_2 = \exp(-\pi i t/\Lambda)$, $\mathcal{E}_1 = \exp(\pi i t/\Lambda)$ and $\Lambda = \Lambda_B/\sqrt{\eta^2 - 1}$. Figure 17 represents the variations of the reflected intensity for a thin silicon crystal, a 220 reflection, MoKα and $t/\Lambda_B = 0 = 2$ ($t = 13.5\mu m$).

5.2.3. Comparison with geometrical theory. When t/Λ_L decreases towards zero, expression (36) tends towards

$$\left[\frac{\sin\ [\pi\ t\eta/\Lambda_L]}{\eta} \right]^2$$

Using (25) and (27), it can be shown that, in the non-absorbing symmetric case:

$$I_h = \frac{R^2 \lambda^2 C^2 |F_h|^2 t^2}{V^2 \sin^2\ \theta} \left[\frac{\sin\ (2\pi\ k\ \cos\ \theta\ t\ \Delta\theta)}{(2\pi\ k\ \cos\ \theta\ t\ \Delta\theta)} \right]^2$$

where $\Delta\theta$ is the difference between the angle of incidence and the middle of the reflection domain. This expression is the classical expression given by geometrical theory (see, for instance, James, 1950).

5.2.4. Integrated intensity. The integrated intensity is

$$I_{hi} = \pi\ \delta\ \tanh\ [\pi t/\Lambda_B] \tag{37}$$

where t is the crystal thickness. When t/Λ_B is very small (thin crystals or weak reflections), the integrated intensity tends towards

$$I_{hi} = R^2 \lambda^3 t\, |F_h|^2 \,/\, \left[V^2 \gamma_o \sin\ 2\theta\right] \tag{38}$$

which is the expression given by geometrical theory. If we call this intensity $I_{hi}(geom.)$, comparison of expressions (37) and (38) shows that the integrated intensity for crystals of intermediate thicknesses can be written:

$$I_{hi} = I_{hi}(geom.) \frac{\tanh\ [\pi\ t/\Lambda_B]}{[\pi\ t/\Lambda_B]}$$

which is the expression given by Darwin (1922) for primary extinction.

6. REAL WAVES

6.1. Introduction

Diffraction of a plane wave as described in the preceding sections is never encountered in practice, although with various devices, in particular using synchrotron radiation, it is possible to produce highly collimated monochromated waves which behave like pseudo plane waves. The wave emitted by an X-ray tube is best represented by a spherical wave. The first experimental proof is due to Kato and Lang (1959), in the transmission case and to Uragami (1969, 1970) in the reflection case. Using a Fourier expansion, Kato extended the dynamical theory to spherical waves for non-absorbing (1961 a and b) and absorbing crystals (1968 a and b) in the transmission case and Saka, Katagawa and Kato (1973) in the reflection case. Another method for treating the problem was used by Takagi (1962, 1969) who solved the propagation equation in a medium where the lateral extension of the incident wave is limited and where the wave amplitudes depend on the lateral coordinates (see the Chapter Dynamical Theory of X-ray Diffraction - II. Deformed Crystals, this Volume).

6.2. Borrmann Triangle

When the incident beam falling on the crystal can be considered to be a spherical wave, its divergence is much larger than the angular width of the reflection domain. Figure 18 a shows a spatially collimated beam falling on a crystal in the transmission case and Figure 18 b represents the corresponding situation in reciprocal space. The plane waves of the Fourier expansion of the spherical wave excite every point of both branches of the dispersion surface. The propagation directions of the corresponding wavefields cover the angular range between those of the incident and reflected beams (Figure 18 a) and fill up what is called the *Borrmann* triangle. The intensity distribution within this triangle has important properties. The first one is that the angular density of wavefield paths is inversely proportional to the curvature of the dispersion surface around their tiepoints.

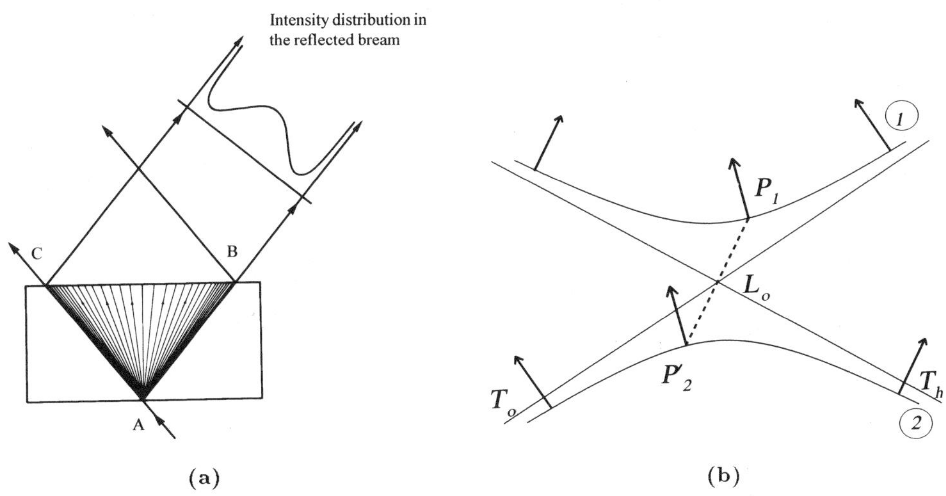

Figure 18. Borrmann triangle. **(a)** Paths of wavefields within the Borrmann triangle - intensity distribution on the exit surface. **(b)** Reciprocal space: the wavefields with conjugate tiepoints P_1 and P'_2 propagate along path Ap in direct space.

Let us consider an incident wave packet of angular width $\delta(\Delta\theta)$. It will generate a packet of wavefields propagating within the Borrmann triangle. The angular width $\delta\alpha$ between the paths of the corresponding wavefields is related to the radius of curvature $\mathcal{R}$ of the dispersion surface by

$$\mathcal{A} = \delta\alpha/\delta(\Delta\theta) = k \cos\theta/(\mathcal{R} \cos\alpha) \tag{39}$$

where α is the angle between the wavefield path and the lattice planes (23) and $\mathcal{A}$ is called the amplification ratio. In the middle of the reflecting domain, the radius of curvature of the dispersion surface is very much shorter than its value, k, far from it (about 10^4 times shorter) and the amplification ratio is therefore very large. As a consequence, the energy of a wavepacket of width $\delta(\Delta\theta)$ in reciprocal space is spread in direct space over an angle $\delta\alpha$ given by (39). The intensity distribution on the exit surface BC (Figure 18 a) is therefore proportional to $I_h/\mathcal{A}$.

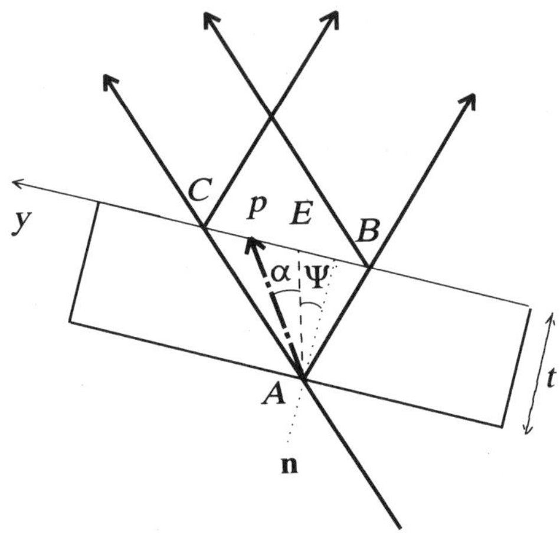

Figure 19. Angular relations within the Borrmann triangle: **n**: normal to the entrance surface, AE: trace of lattice planes.

The expression of $\mathcal{A}$ in terms of the deviation parameter, η, is given in the transmission case by:

$$\mathcal{A} = \frac{\Lambda_L}{\lambda} \frac{\sin^2 2\theta \cos^2(\alpha + \Psi)}{2\gamma_o^2\gamma_h(1 + \eta^2)^{3/2}} \tag{40}$$

where Ψ is the angle between the normal to the entrance surface and the reflecting planes (Figure 19), γ_o and γ_h are the cosines of the angles ψ_o and ψ_h between the normal to the crystal surface and the incident, AC, and reflected, AB, directions, respectively. Λ_L is the *Pendellösung* distance defined in § 3.5. The maximum value of the amplification ratio $\mathcal{A}$ is obtained at the centre of the reflecting domain ($\eta = 0$) and is of the order of the ratio Λ_L/λ of the Pendellösung distance to the wavelength. It is proportional to the structure factor and is usually very large, of the order of 10^4. Expression (40) is not valid for large values of η since the curvature of the Ewald sphere in the vicinity of the Laue and Lorentz points has been neglected. Far from the reflection domains, $\alpha \Rightarrow \theta$, the radius $\mathcal{R}$ of the dispersion surface tends towards the radius k of the Ewald sphere and $\mathcal{A}$ tends towards 1.

6.3. Intensity Distribution along the Base of the Borrmann Triangle

The intensity of the beam coming out of the crystal is the flux of energy per unit surface and unit time. Let us consider an incident beam limited by a fine vertical slit normal to the plane of incidence. We shall neglect the vertical divergence of the beam and the curvature of the dispersion surface in the direction normal to the plane of incidence.

Let E be the midpoint of the base BC of the Borrmann triangle (Figure 19) and **Ey** an oriented axis parallel to BC. The length of the base $\overline{BC}$ of the Borrmann triangle is related to the thickness t of the crystal and to the angular parameters by:

$$\overline{BC} = t \sin 2\theta \,/\, (\gamma_o \gamma_h)$$

The intensity distribution of the two wave components of each wavefield arriving at any point p of coordinate y on the exit surface ($y = \overline{Ep}$) is a function of this coordinate, $I_{mj}(y)$, ($m = 0, h; j = 1, 2$). It is given by:

$$\left.\begin{aligned} I_{hj}(y) &= \frac{\gamma_o^2 \gamma_h}{2 \sin^2 2\theta} \left| \frac{F_h}{F_{\bar{h}}} \right| \frac{\lambda}{t\Lambda_L} (1 - Y^2)^{-1/2} \exp(-\mu_j t/\gamma_o)\; I_o(\Delta\theta) \\ I_{oj}(y) &= \frac{\gamma_o^2 \gamma_h}{2 \sin^2 2\theta} \frac{\lambda}{t\Lambda_L} \frac{1 + Y}{1 - Y} (1 - Y^2)^{-1/2} \exp(-\mu_j t/\gamma_o)\; I_o(\Delta\theta) \end{aligned}\right\}$$

Where μ_j is given by (32), and

$$Y = y/y_{Max} = \overline{Ep}/\overline{EC} = \mp \eta/\sqrt{1 + \eta^2}$$

is the normalized coordinate of y. The upper sign corresponds to $j = 1$ (wavefields with tiepoints on branch 1 of the dispersion surface) and the lower one to $j = 2$ (wavefields with tiepoints on branch 2). The intensity distribution for the reflected beam is represented schematically on Figure 18 a. It increases very sharply at the edges where the density of wavefields is large, although it is the reverse for the reflecting power I_{hj}. This effect, called the *margin effect*, was predicted qualitatively by Borrmann (1959) and von Laue (1960), demonstrated experimentally by Kato and Lang (1959), and calculated by Kato (1960). The maximum in the center is due to anomalous absorption: the wavefields propagating along the edges of the Borrmann triangle undergo normal absorption, while those propagating parallel to the lattice planes (or nearly so) correspond to tiepoints in the centre of the dispersion surface and undergo anomalously low absorption. For values of μt larger than 10 or so, practically only the wavefields propagating parallel to the lattice planes go through the crystal which acts as a wave guide; this is the Borrmann effect.

7. SPHERICAL WAVE DYNAMICAL THEORY

7.1. Fourier Expansion of the Spherical Wave in Plane Waves

7.1.1. Introduction. The principle of the theory (Kato, 1961a and b; Kato, 1974) is, in a first step, to show, by means of a Fourier transform, that the spherical wave may be expanded in a distribution of plane waves having the same wavenumber. Plane-wave dynamical theory is then applied to each plane-wave component of the spherical wave. The amplitudes of the reflected and refracted spherical waves are obtained by an inverse Fourier transform performed by means of the stationary phase method.

7.1.2. The incident wave is a scalar wave. A scalar wave emitted by a point-source is written:

$$\phi(\mathbf{r}) = \phi_o \frac{\exp[2\pi i(\nu t - kr)]}{4\pi r}$$

The Fourier transform of $\phi(\mathbf{r})$ is:

$$F(\mathbf{K}) = \int_{-\infty}^{+\infty}\int_{-\infty}^{+\infty}\int_{-\infty}^{+\infty} \phi(\mathbf{r}) \exp[2\pi i\ \mathbf{K}.\mathbf{r}]\ d\tau$$

Taking the inverse transform, $\phi(\mathbf{r})$ may be written:

$$\phi(\mathbf{r}) = \int_{-\infty}^{+\infty}\int_{-\infty}^{+\infty}\int_{-\infty}^{+\infty} F(\mathbf{K}) \exp[-2\pi i\ \mathbf{K}.\mathbf{r}] d^3 K$$

The spherical wave may thus be considered as a distribution of plane waves having **K** as wavevector and $F(\mathbf{K})$ as amplitude. Its propagation in a given direction **Oz** will be studied and the value of its amplitude in the plane $z = z_o$ tangential to the wave front will be evaluated. Taking the directions of reciprocal axes OK_x, OK_y and OK_z to coincide with the axes, *Ox, Oy* and *Oz*, the expression of the spherical wave may be written:

$$\phi(\mathbf{r}) = \int_{-\infty}^{+\infty}\int_{-\infty}^{+\infty} \Phi(\mathbf{K}) \exp\left[-2\pi i\ (K_x x + K_y y)\right] dK_x dK_y$$

with

$$\Phi(\mathbf{K}) = \int_{-\infty}^{+\infty} F(\mathbf{K}) \exp\left[-2\pi i\ K_z z_o\right]\ dK_z$$

Using the properties of Fourier transforms and noticing that $F(\mathbf{K})$ has spherical symmetry, it is possible to show that

$$\phi(\mathbf{r}) = i\frac{\phi_o}{4\pi}\int_{-\infty}^{+\infty}\int_{-\infty}^{+\infty} \frac{\exp[-2\pi i\mathbf{K}.\mathbf{r}]}{K_z} dK_x dK_y \tag{41}$$

where **K** is a wavevector of length k and components

$$K_x, K_y, \text{ and } K_z = \sqrt{k^2 - [K_x^2 + K_y^2]} \tag{42}$$

Expression (41) can be shown to be a surface integral on a sphere of radius k and may therefore be interpreted as an *expansion in plane waves having all the same wavenumber* k.

7.1.3. The incident wave is a vector wave. Electromagnetic waves are actually vector waves. The divergence of the incident beam is usually much larger than the width of the dispersion surface, so that the wave can be considered as a spherical wave and the whole dispersion surface is excited. The geometrical situation is represented on Figure 20 (a: direct space, b: reciprocal space). F is a point source, **Fz** the propagation direction. The point of impact of the beam on the crystal in the incidence plane, A, may also be assimilated to a point source; ABC is the Borrmann triangle. The **Ax** axis is normal to the incident direction, and **Ay** is normal to the plane of incidence (and the plane of the figure). The coreesponding axes in reciprocal space (Figure 20 b), are, respectively: $\mathbf{L_aK_x}$, $\mathbf{L_aK_y}$, $\mathbf{L_aK_z}$; $\mathbf{L_aK_x}$ is along the tangential plane to the Ewald sphere, $\mathbf{L_aK_y}$ along the normal to the plane of incidence, $\mathbf{L_aK_z}$ along the propagation direction.

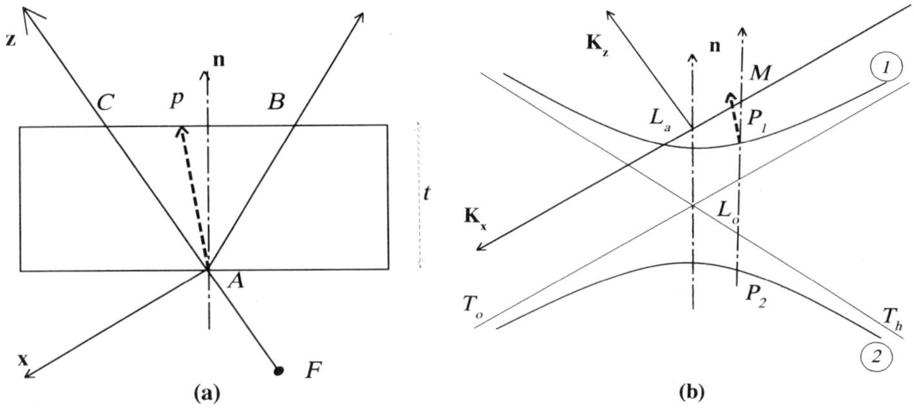

Figure 20. Spherical wave dynamical theory: axes used for the integration of the Fourier expansion. **(a)** Direct space; **(b)** Reciprocal space

Kato has shown that the reflected wave can be expanded inside the crystal in a similar way to (41):

$$\mathbf{D_h} = \frac{i}{4\pi} \int_{-\infty}^{+\infty} \int_{-\infty}^{+\infty} K_z^{-1} \sum_j \{\mathbf{D_{hj}} \exp -i\, \varphi_j(\mathbf{K}, \mathbf{r})\}\, dK_x dK_y \tag{43}$$

where the summation is over both branches of the dispersion surface (j = *1, 2*), $\varphi_j(\mathbf{K},\mathbf{r}) = 2\pi[\mathbf{hr} + \mathbf{K}\cdot\mathbf{r_s} + \mathbf{K_{oj}}\cdot(\mathbf{r} - \mathbf{r_s})]$ is the phase factor, $\mathbf{r} = \mathbf{Fp}$ is the position vector of a point, p, on the exit surface of the crystal (Figure 20 a), $\mathbf{r_s} = \mathbf{FA}$, and $\mathbf{r} - \mathbf{r_s} = \mathbf{Fp}$, $\mathbf{K}$ is a running wavevector in vacuum of length k and of extremity M in reciprocal space (Figure 20 b), $\mathbf{K_{o1}}$ and $\mathbf{K_{o2}}$ are the wavevectors of the two waves it generates inside the crystal, of tiepoints P_1 and P_2. To a good approximatiom, the spherical wave can be replaced by a cylindrical wave normal to the plane of incidence. The solutions are therefore independent of K_y and it suffices to study the state of the wave in this plane: the y coordinate of $\mathbf{r}$ can be taken equal to zero. It will furthermore be assumed that the normal $\mathbf{n}$ to the crystal surface lies in the $\mathbf{K_o}$, $\mathbf{K_h}$ plane.The phase factor can therefore be written:

$$\varphi_j(\mathbf{K},\mathbf{r}) = 2\pi \left[\mathbf{h}\cdot\mathbf{r} + K_x \cdot x + K_z(z - z_F) + \overline{MP_j}\, t \right]$$

where x, *0*, z are the coordinates of p and 0, 0, z_F those of F, K_z is given by (42) and the phase $2\pi\mathbf{h}\cdot\mathbf{r}$ is a constant and can be taken out of the integral.

7.2. Direct Integration

The double integral (43) may be separated into two integrals:

$$\mathbf{D_h} = \frac{i}{4\pi}\int_{-\infty}^{+\infty}\left[\sum_j\left\{\mathbf{D_{hj}}\exp -2\pi i\left[\mathbf{h.r} + K_x x + \overline{MP_j}\,t\right]\right\}\right]\mathcal{I}(K_z)dK_x$$

$$\text{with } \mathcal{I}(K_z) = \int_{-\infty}^{+\infty} K_z^{-1}\,\exp\left[-2\pi i\;K_z\,(z\;-\;z_F)\right]\,dK_y$$

The integral $\mathcal{I}(K_z)$ may be calculated using the stationary phase method. The result is:

$$\mathcal{I}(K_z)\;=\;\left[K_z(z\;-\;z_F)\right]^{-1/2}\exp\left[2\pi K_z(z\;-\;z_F)\;+\;\pi/4\right]$$

Using this expression, the reflected amplitude becomes:

$$\mathbf{D_h}\;=\;\frac{\exp(3i\pi/4)}{4\pi\sqrt{z-z_F}}\times$$

$$\int_{-\infty}^{+\infty} K_z^{\frac{-1}{2}}\sum_j \mathbf{D_{hj}}\exp\left[-2\pi i\left(K_x x\;+\;K_z z\;+\;\overline{MP_j}\,t\right)\right]\;dK_x$$

The component K_x of the wavevector is nothing else but the abscissa $\overline{L_a M}\;=\;k\Delta\theta$ of the extremity of the vacuum wavevector (Figure 20 b). It is related to the deviation parameter η by equation (25) and (26):

$$K_x\;=\;\eta\;\frac{\gamma_h}{\Lambda_L\sin 2\theta}\;+\;k\Delta\theta_o$$

The value of K_x is much smaller than that of K_z, which is of the order of k. The quantity $K_z^{-1/2}\exp\left[-2\pi i(z\;-\;z_F)\right]$ may therefore be considered as constant and taken out of the integral which is then of the form, using (29), (31) and (34):

$$D_h\;=\;iA_h\int_{-\infty}^{+\infty}\frac{\exp -\pi ita(x)\eta}{\sqrt{1+\eta^2}}\;\{\exp[-\alpha(\eta)]\;-\;\exp[\alpha(\eta)]\}\;d\eta$$

where $\alpha(\eta)\;=\;(\pi t/\Lambda_L)\sqrt{1+\eta^2}$, $a(x)\;=\;1\;+\;(2x\gamma_h/t\;\sin 2\theta)$ and A_h is a constant. The two exponentials inside the brackets correspond to branch *2* and *1*, respectively. By combining them, the expression of the reflected amplitude becomes:

$$D_h\;=\;-2A_h\int_{-\infty}^{+\infty}\frac{\exp -\pi ita(x)\eta}{\sqrt{1+\eta^2}}\;\{\sin[\alpha(\eta)]\}\;d\eta$$

The integral can be performed directly, using the properties of Bessel functions, and the reflected amplitude is proportional to:

$$D_h\;\sim\;J_{\,o}(\zeta)$$

where $\zeta\;=\;(\pi t/\Lambda_L)\sqrt{1\;-\;a(x)^2}$.

In this expresion, x is the distance of point p on the exit surface from the direct beam AC (Figure 20 a) and, remembering that $\overline{BC} = t \sin 2\theta / (\gamma_o \gamma_h)$ (§ **6.3**), it can be shown geometrically that

$$a(x) = Y = \overline{Ep} / \overline{EC}$$

and, therefore, that

$$\zeta = \frac{\pi t}{\Lambda_L} \sqrt{1 - Y^2}$$

7.3. Intensity Distribution on the Exit Surface

The calculation of the refracted amplitude is performed in a similar way to that of the reflected amplitude and is found to be proportional to the Bessel function of order *1*, $J_1(\zeta)$.

One obtains finally, for the intensity distribution of the reflected and refracted beams, respectively, at any point at the base of the Borrmann triangle:

$$\left.\begin{aligned} I_h(y) &\sim \left| J_0 \left[\pi \frac{t}{\Lambda_L} \sqrt{(1 - Y^2)} \right] \right|^2 \\ I_o(y) &\sim \frac{1 - Y}{1 + Y} \left| J_1 \left[\pi \frac{t}{\Lambda_L} \sqrt{(1 - Y^2)} \right] \right|^2 \end{aligned}\right\}$$

Bessel functions can be approximated by the following relations, which are more accurate the bigger the value of the argument:

$$J_0(\zeta) \simeq \sqrt{\frac{2}{\pi \zeta}} \cos(\zeta - \pi/4)$$

$$J_1(\zeta) \simeq \sqrt{\frac{2}{\pi \zeta}} \cos(\zeta + 3\pi/4)$$

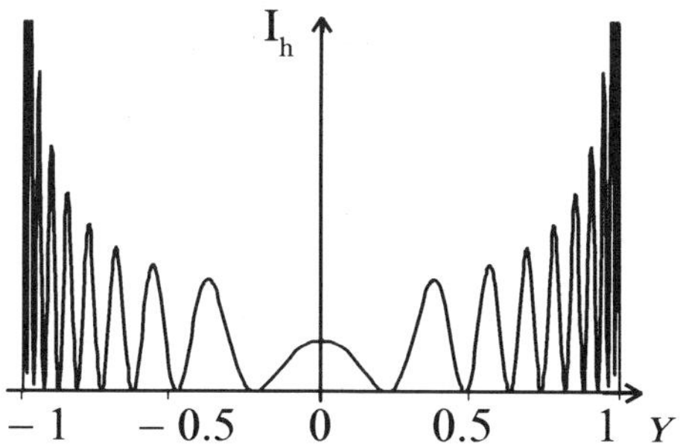

Figure 21. Intensity distribution of the reflected beam along the exit surface for an incident spherical wave in arbitrary units. Y is a normalized coordinate.

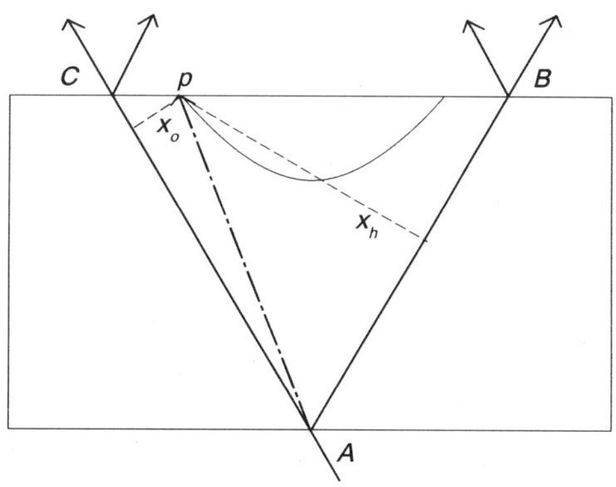

Figure 22. Pendellösung fringes for an incident spherical wave. Along any path Ap within the Borrmann propagate two wavefields with tiepoints P_1 and P'_2 (Figure 18 b).

These expressions show that the period of the oscillations of the intensity, in the centre of the Borrmann triangle, is Λ_L. Figure 21 shows the variations of the intensity of the reflected beam along the base of the Borrmann triangle. It is very similar to that represented schematically on Figure 18 a, modulated by the oscillations due to the Bessel function.

The argument of the Bessel functions may also be written:

$$\pi \frac{t}{\Lambda_L} \sqrt{(1 - Y^2)} = \frac{2\pi \sqrt{\gamma_o \gamma_h}}{\Lambda_L \sin 2\theta} \sqrt{x_o x_h}$$

where x_o and x_h are the distances of p from the sides AC and AB of the Borrmann triangle (Figure 22). The equal-intensity fringes are therefore located along the locus of the points in the triangle for which the product of the distances to the sides is constant, that is hyperbolæ having AB and AC as asymptotes (Figure 22).

REFERENCES

Authier, A., 1970, Ewald waves in theory and experiment, "Advances in Structure Research by Diffraction Methods", Ed. Brill and Mason, Vieweg und Sohn, Braunschweig, 3:1-51.

Authier, A., 1992, Dynamical theory of X-ray diffraction, "International Tables of Crystallography", Ed. U. Shmueli, Kluwer Academic Publishers, Dordrecht , Vol. B, 464-480.

Authier, A. and Balibar, F., 1970, Création de nouveaux champs d'onde généralisés dus à la présence d'un objet diffractant. II.- Cas d'un défaut isolé. *Acta Cryst.*, A26:647-654.

Batterman, B.W. and Cole, H., 1964, Dynamical diffraction of X-rays by perfect crystals, *Rev. Mod. Phys.*, 36:681-717.

Blume, M., 1985, Magnetic scattering of X-rays, *J. Appl. Phys.*, 57:3615.

Blume, M., 1994, Magnetic effects in anomalous scattering, in "Resonant Anomalous X-ray Scattering", Eds. Materlik, Sparks and Fischer, Elsevier, Amsterdam, 495.

Borrmann, G., 1950, Die Absorption von Röntgenstrahlen in Fall der Interferenz, *Z. Phys.* , 127:297-323.

Borrmann, G., 1954, Der kleinste Absorption Koeffizient interferierender Röntgenstrahlung, *Z. Krist.*, 106:109-121.

Borrmann, G., 1959, Röntgenwellenfelder, "Beit. Phys. Chem. 20 Jahrhunderts", Vieweg und Sohn, Braunschweig, 262-282.

Darwin, C.G., 1922, The reflection of X-rays from imperfect crystals, *Phil. Mag.*, 43:800-829.

Ewald, P.P., 1917, Zur Begründung der Kristalloptik. III. Röntgenstrahlen, *Ann. Physik. (Leipzig)*, 54:519-597.

Hannon, J.P., 1994, Resonant gamma-ray scatteing and coherent excitations of nuclei, in "Resonant Anomalous X-ray Scattering", Eds. Materlik, Sparks and Fischer, Elsevier, Amsterdam, 565.

Hannon, J.P., Trammell, G.T., Blume M. and Gibbs, D., 1988, X-ray resonant exchange scattering, *Phys. Rev. Letters*, 61:1245.

James, R.W., 1950, "The optical principles of the diffraction of X-rays", G. Bell and Sons Ltd., London.

James, R.W., 1963, The dynamical theory of X-ray diffraction, *Solid State Physics.*, 15:53.

Kato, N., 1955, Integrated intensities of the diffracted and transmitted X-rays due to ideally perfect crystal, *J. Phys. Soc. Japn* , 10:46-55.

Kato, N., 1960, The energy flow of X-rays in an ideally perfect crystal: comparison between theory and experiments, *Acta Cryst.* , 13:349-356.

Kato, N., 1961a, A theoretical study of Pendellösung fringes. I. General considerations, *Acta Cryst.*, 14:526-532.

Kato, N., 1961b, A theoretical study of Pendellösung fringes.Part 2 -Detailed discussion based upon a spherical wave theory, *Acta Cryst.*, 14:627-636.

Kato, N., 1968a, Spherical-wave theory of dynamical X-ray diffraction for absorbing perfect crystals : I. The crystal wave fields, *J. Appl. Phys.*, 39:2225-2230.

Kato, N., 1968b, Spherical-wave theory of dynamical X-ray diffraction for absorbing perfect crystals : II. Integrated reflection power, *J. Appl. Phys.*, 39:2231-2237.

Kato, N., 1974, X-Ray diffraction, in " X-Ray Diffraction", edited by: L.V. Azaroff, R. Kaplow, N. Kato, R.J. Weiss, A.J.C. Wilson and R.A. Young, McGraw-Hill, New York, 176-438.

Kato, N. and Lang A.R., 1959, A study of Pendellösung fringes in X-ray diffraction, *Acta Cryst.*, 12:787-794.

Laue, M. von, 1931, Die dynamische Theorie der Röntgenstrahlinterferenzen in neuer Form, *Ergeb. Exakt. Naturwiss.*, 10:133-158.

Laue, M. von, 1960, "Röntgenstrahl-Interferenzen", Akademische Verlagsgesellschaft, Frankfurt am Main.

Lefeld-Sosnowska M. and Malgrange C., 1968, Observation of oscillations in rocking curves of the Laue reflected and refracted beams from thin Si single crystals, *Phys. Stat. Sol.*, 30:K23-K25.

Pinsker, Z.G., 1978, "Dynamical Scattering of X-rays in Crystals", Springer Series in Solid-State Sciences, Springer Verlag, Berlin.

Renninger, M., 1955, Messungen zur Röntgenstrahl-Optik des Idealkristalls. I. Bestäti-gung der Darwin-Ewald-Prins-Kohler-Kurve, *Acta Cryst.*, 8:597-606.

Saka, T., Katagawa, T. and Kato, N., 1973, The theory of X-ray diffraction for finite polyhedral crystals. III The Bragg-(Bragg) m cases, *Acta Cryst. A* , 29:192-200

Uragami, T., 1969, Pendellösung fringes of X-rays in Bragg case, *J. Phys. Soc. Jap.*, 27:147-154

Zachariasen W.H., 1945, "Theory of X-ray diffraction in crystals", John Wiley, New York.

DYNAMICAL THEORY OF HIGHLY ASYMMETRIC X-RAY DIFFRACTION

Václav Holý

Dept. of Solid State Physics, Masaryk University,
Kotlářská 2, 611 37 Brno,
Czech Republic

1. SIMULTANEOUS DIFFRACTION AND REFLECTION

Usual arrangements of X-ray diffraction experiments are not sensitive to the state of the crystal surface, since the penetration depth of the incoming radiation is several μm or even higher. The penetration depth can be drastically reduced if the incident and/or diffracted beams make a very small angle with the substrate. This can be achieved in the following experimental arrangements:

X-ray reflection (XRR) – the incident *and* the scattered beams make small angles with the surface; the normal to the surface lies in the scattering plane defined by the wave vectors of the primary and scattered beams (coplanar scattering),

highly asymmetric X-ray diffraction (AXRD) – the incident *or* the scattered waves make a small angle with the surface, the arrangement being coplanar, and

grazing-incidence diffraction (GID) – the incident *and* the scattered beams make small angles with the surface, the scattering plane being nearly parallel to the surface (noncoplanar diffraction).

Since the optical Fresnel reflectivity of a surface decreases with the 4th power of the angle with the surface, X-ray reflection from the surface occurs in all these arrangements.

The conventional dynamical theory of X-ray diffraction does not properly describe these arrangements. Its validity is restricted to the close neighbourhood of the Lorentz point in reciprocal space and, in addition, it does not include X-ray reflection, since it takes into account only a simplified form of the boundary conditions.

A more general form of the dynamical diffraction theory considers both the exact form of the dispersion surface in the two–beam case and the exact boundary conditions at the interfaces. It will be demonstrated in this chapter that this theory can describe all the above experimental arrangements in a uniform way and makes it possible to calculate the diffraction and/or reflection from layered samples as well. On the basis of the

X-ray and Neutron Dynamical Diffraction: Theory and Applications
Edited by Authier *et al.*, Plenum Press, New York, 1996

generalized dynamical diffraction theory, the intensities of the diffracted and reflected beams in all mentioned experimental arrangements will be calculated in the case of a semi-infinite crystal and a pseudomorphic ideal periodical superlattice.

2. WAVEFIELDS IN THE CRYSTAL

2.1 Amplitude equations

In the generalized dynamical diffraction theory, it is advantageous to describe the wavefield propagating in the crystal by the vector $\mathbf{E}(\mathbf{r})$ instead of $\mathbf{D}(\mathbf{r})$ used in von Laue's formulation of the theory (see the book by Pinsker (1978) and the Chapter by Authier in this Volume). The overview of the generalized dynamical theory is given by Kimura et al. (1994) and in the papers cited therein. The theory of GID was published for the first time by Afanasev and Melkonyan (1983) and Aleksandrov et al. (1984).

From Maxwell equations for monochromatic radiation with frequency ω in a non-magnetic medium the wave equation for the vector $\mathbf{E}$ follows (in CGS units)

$$\operatorname{curl}\operatorname{curl}\mathbf{E}(\mathbf{r}) - K^2[1+\chi(\mathbf{r})]\mathbf{E}(\mathbf{r}) = 0, \tag{1}$$

where $\chi = \epsilon - 1$ is the crystal polarizability, $K = \omega/c = 2\pi/\lambda$, λ is the X-ray wavelength in vacuum.

In an infinite perfect crystal, the polarizability is a periodical function and it can be expanded into the Fourier series over the reciprocal lattice vectors

$$\chi(\mathbf{r}) = \sum_{\mathbf{g}} \chi_g \exp(-i\mathbf{g}\cdot\mathbf{r}).$$

Then, the solution of (1) can be found in the form of the Bloch wave

$$\mathbf{E}(\mathbf{r}) = \sum_{\mathbf{g}} \mathbf{E}_g e^{-i\mathbf{k}_g\cdot\mathbf{r}}, \tag{2}$$

where $\mathbf{k}_g = \mathbf{k}_0 + \mathbf{g}$ and $\mathbf{k}_0$ is an arbitrary reciprocal vector from the 1st Brillouin zone. Thus, the wavefield in the diffracting crystal is represented as being a superposition of the plane waves with amplitudes $\mathbf{E}_g$ and wave vectors $\mathbf{k}_g$.

Putting from (2) into (1) we obtain an infinite set of algebraic equations for the amplitudes

$$\frac{1}{K^2}\left(k_g^2\mathbf{E}_{g,g} - K^2\mathbf{E}_g\right) = \sum_{\mathbf{g}'} \chi_{g-g'}\mathbf{E}_{g'}, \tag{3}$$

where $\mathbf{E}_{g,g}$ means the component of vector $\mathbf{E}_g$ perpendicular to $\mathbf{k}_g$.

Eqs. (3) can be solved if we limit the number of the plane waves propagating through the crystals, i.e. if we restrict the number of terms in (2).

2.1.1 One–wave approximation. In this approximation, we take into account only one term in the series (2). Since the origin of the reciprocal lattice is arbitrary, we can always write the wavefield in the crystal in the form

$$\mathbf{E}(\mathbf{r}) = \mathbf{E}_0 \exp(-i\mathbf{k}_0\cdot\mathbf{r}).$$

Then, instead of the system (3), we obtain only one equation

$$[K^2(1+\chi_0) - k_0^2]\mathbf{E}_0 = 0. \tag{4}$$

The condition of the existence of a non–trivial solution of this equation is the dispersion condition

$$k_0 = nK = k = K\sqrt{1+\chi_0} \approx K(1+\chi_0/2), \tag{5}$$

since $|\chi| \ll 1$. This formula determines the refractive index n of X-rays.

From the dispersion condition (5) it follows that the refractive index does not depend on the direction of the wave vector $\mathbf{k}_0$ of the wave propagating through the crystal, and, therefore, the dispersion surface is a spherical surface with the radius $k = nK$ (see Figure 1).

In the one–wave approximation, only one plane wave propagates in the crystal. We will show later that this approximation along with the exact boundary conditions describes properly X-ray reflection from interfaces.

2.1.2 Two–wave approximation. We take two terms of (2) into account and, thus the wavefield propagating in the crystal consists of two plane waves

$$\mathbf{E}(\mathbf{r}) = \mathbf{E}_0 e^{-i\mathbf{k}_0\cdot\mathbf{r}} + \mathbf{E}_h e^{-i\mathbf{k}_h\cdot\mathbf{r}},$$

where $\mathbf{h}$ is a chosen reciprocal lattice vector (diffcration vector). In the corresponding amplitude equations (following from (3)) the components $\mathbf{E}_{0,0}$ and $\mathbf{E}_{h,h}$ occur that complicate their solution. Their values depend on the polarization of the waves. In the *σ–polarization*, the amplitude vectors $\mathbf{E}_0$ and $\mathbf{E}_h$ are perpendicular to the scattering plane. Then

$$\mathbf{E}_{0,0} = \mathbf{E}_0, \ \mathbf{E}_{h,h} = \mathbf{E}_h \tag{6}$$

and the appropriate amplitude equations can be written in the scalar form.

In *π–polarization*, $\mathbf{E}_{0,h}$ lie in the scattering plane and this simplification cannot be performed. However, the Maxwell equation can be simplified

$$0 = \text{div}\mathbf{D} = \text{div}\mathbf{E} + \text{div}(\chi\mathbf{E}) \approx \text{div}\mathbf{E},$$

since $|\chi| \ll 1$. Then, Eq. (6) is valid, too. Thus, in σ–polarization, formulas (6) *are exact*, whereas in π–polarization, they are only approximate.

Assuming (6), the scalar form of the amplitude equations in the two–beam case is

$$\frac{1}{K^2}(k_0^2 - K^2)E_0 = \chi_0 E_0 + C\chi_{\bar{h}} E_h$$

$$\frac{1}{K^2}(k_h^2 - K^2)E_h = C\chi_h E_0 + \chi_0 E_h, \tag{7}$$

where $C = \cos\phi$ is the linear polarization factor, ϕ is the angle between $\mathbf{E}_0$ and $\mathbf{E}_h$, i.e. zero in σ–polarization and $2\theta_B$ in π–polarization, $\theta_B = \text{arsin}(h/(2K))$ is the Bragg angle.

The condition of the existence of a non-trivial solution of the system (7) yields the equation of the dispersion surface (dispersion equation)

$$(k_0^2 - k^2)(k_h^2 - k^2) = K^4 C^2 \chi_h \chi_{\bar{h}} \tag{8}$$

that represents an equation of the fourth order for the unknown wave vectors $\mathbf{k}_0$ and $\mathbf{k}_h = \mathbf{k}_0 + \mathbf{h}$. The intersection of the dispersion surface with the scattering plane is depicted in Figure 2.

In the usual formulation of the dynamical diffraction theory, a wave equation for the vector $\mathbf{D}(\mathbf{r})$ is used. From this equation, the amplitude equations in the two–beam case can be derived. Comparing Eqs. (7) with the analogous set of the equations in the Chapter by Authier in this volume, we find that those systems differ only far from the Lorentz point. Close to the Lorentz point, both the sets coincide.

2.2 Boundary conditions

In the present formulation of the dynamical diffraction theory, we use the *exact* form of the boundary conditions, namely $\mathbf{E}(\mathbf{r})_{\parallel} = \text{const}$, $\mathbf{H}(\mathbf{r})_{\parallel} = \text{const}$ hold across the boundary. From these conditions, $\mathbf{D}(\mathbf{r})_{\perp} = \text{const}$, $\mathbf{B}(\mathbf{r})_{\perp} = \text{const}$ follows. Here $_{\parallel}$ and $_{\perp}$ denote the components parallel and perpendicular to the boundary, respectively.

In the following, we will assume that the incident radiation is plane. Then, all the waves propagating in the crystal as well as the waves emitted by the crystal are plane as well. The electric and magnetic components of an electromagnetic plane wave are connected by the Maxwell equation that yields the formula

$$\mathbf{H}(\mathbf{r}) = \frac{1}{K}\mathbf{k} \times \mathbf{E}(\mathbf{r}), \tag{9}$$

where $\mathbf{k}$ is the wave vector of the wave $\mathbf{E}(\mathbf{r}) = \mathbf{E}\exp(-i\mathbf{k}\cdot\mathbf{r})$.

The values of the in-plane and perpendicular components of $\mathbf{E}$ and $\mathbf{H}$ depend on the scattering geometry and on the polarization. In the σ–polarization and in the coplanar case (XRR or AXRD), the vectors $\mathbf{E}$ of all waves are parallel to the surface, while in the GID geometry the $\mathbf{H}$ vectors are parallel to the surface. In the π–polarization, the situation is opposite; in the coplanar geometry, $\mathbf{H}$ is parallel to the surface, while $\mathbf{E}$ is parallel to the surface in the GID.

In the σ–polarization of the coplanar case (or π–polarization in GID), the in-plane components of the waves are

$$\mathbf{E}_{\parallel} = E\mathbf{l},\ \mathbf{H}_{\parallel} = -\frac{k_z}{K}\mathbf{t}E, \tag{10}$$

where $\mathbf{t}$ is the unit vector parallel to the surface and to the projection of $\mathbf{k}$ into the surface, $\mathbf{l} = \mathbf{n} \times \mathbf{t}$, $\mathbf{n}$ is the unit vector of the inward surface normal (parallel to the z-axis). In the opposite case (π–polarization coplanar or σ–polarization GID) the in-plane components are

$$\mathbf{E}_{\parallel} = \frac{k_z}{K}\mathbf{t}E,\ \mathbf{H}_{\parallel} = H\mathbf{l} \approx E\mathbf{l}. \tag{11}$$

The formulas (10,11) are exact only in the σ–polarization of the coplanar case, in the other three cases, they are only approximative. The smaller χ and the smaller the angle of the scattering plane with the surface (in GID) are, the better the approximation is.

Within the approximation above, the boundary conditions at the surface $\mathcal{S}$ (assumed plane and perpendicular to the z-axis) or at an arbitrary interface in a layered sample can be written in the uniform way in all four cases

$$\left.\sum_n E_n \mathbf{l}_n e^{-i\mathbf{k}_n\cdot\mathbf{r}}\right|_{\mathbf{r}\in\mathcal{S}} = \text{const},\ \left.\sum_n E_n k_{zn}\mathbf{t}_n e^{-i\mathbf{k}_n\cdot\mathbf{r}}\right|_{\mathbf{r}\in\mathcal{S}} = \text{const}, \tag{12}$$

where the sums are calculated over all plane wave components propagating in the crystal. Since the formulas (12) must hold in every point $\mathbf{r}$ at the surface $\mathcal{S}$, the in-plane components of the wave vectors $\mathbf{k}_n$ must be same and they must equal the in-plane component of the wave vector of the primary radiation. From this condition, the well–known Ewald construction of the internal wave vectors follows. If we construct the wave vectors $\mathbf{k}_n$ so that their ending points are the points of the reciprocal lattice, their starting points must lie at the intersection of the dispersion surface (5) or (8) with the surface normal constructed from the starting point of the wave vector of the primary (vacuum) wave. The number of these intersection points (tie points) depends on the approximation used, so that the particular forms of the boundary conditions are different in the one–wave and two–wave approximations.

2.2.1 One–wave approximation. From the dispersion condition (5) it follows that the dispersion surface is spherical and the surface normal intersects this surface in two tie points. Thus, two plane waves propagate in the crystal if the primary wave is plane. These waves correspond to the transmitted (E_T) and the (optically) reflected (E_R) waves. From (12) the boundary conditions at the interface $z = 0$ follow

$$E_T + E_R = \text{const}, \ \ k_{Tz}(E_T - E_R) = \text{const}, \tag{13}$$

since the vertical component of the wave vector of the reflected wave is $k_{Rz} = -k_{Tz}$ and $\mathbf{t}_R = \mathbf{t}_T$. At the interface between the vacuum and semi-infinite crystal, the boundary conditions are

$$E_0 + E_R = E_T, \ \ K_{0z}(E_0 - E_R) = k_{Tz}E_T, \tag{14}$$

since two plane waves propagate in the vacuum above the surface (E_0 is the amplitude of the incident wave, E_R is the amplitude of the reflected wave) and only one plane wave (transmitted, with the amplitude E_T) propagates in the crystal. The other wave in the crystal (corresponding to the second tie point) is not physically relevant, since its amplitude *increases* with increasing depth below the surface. The elimination of this wave is equivalent to the boundary condition at the (distant) rear crystal surface that forbids the propagation of the waves in the crystal towards the surface. The wave vectors of the wavefields are sketched in Figure 1.

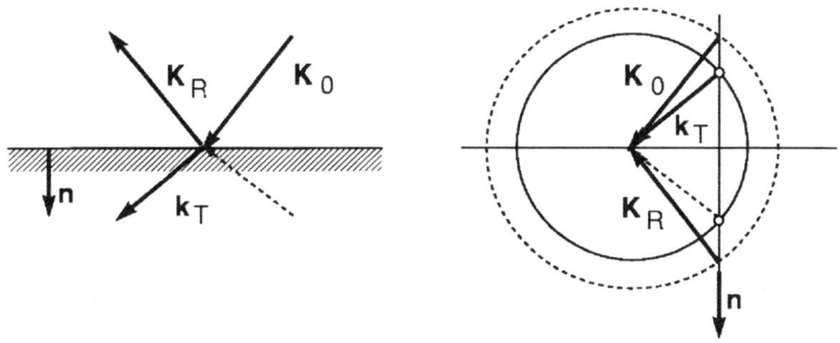

Figure 1. The wavefields in the semi-infinite crystal and in the vacuum in the one–wave approximation expressed in direct (left) and reciprocal (right) spaces. The forbidden wave is denoted by the dotted arrow. In the right panel, the intersection of the dispersion surface of the crystal with the scattering plane is plotted by the full line, that of the vacuum by the dotted line. The tie points are denoted by small empty circles.

The vertical component k_{Tz} of its wave vector $\mathbf{k}_T$ is connected with the wave vector $\mathbf{K}_0$ of the incident wave by Snell's law

$$k_{Tz} = \sqrt{k^2 - |\mathbf{K}_{0||}|^2}. \tag{15}$$

From the boundary conditions (14) and Snell's law (15) the reflectivity and transmittivity of a flat interface immediately follow, the formulas are equivalent to the Fresnel reflectivity and transmittivity coefficients in optics of visible light.

2.2.2 Two–wave approximation. The equation of an intersection of the surface normal with the dispersion surface (given by Eq. (8)) has in general 4 roots for each polarization state. Therefore, 8 plane waves propagate in the crystal in one polarization, four of them create the transmitted beam $\mathbf{E}_0$ and four the diffracted radiation $\mathbf{E}_h$. The wave vectors of those waves are $\mathbf{k}_{0n}$ and $\mathbf{k}_{hn} = \mathbf{k}_{0n} + \mathbf{h}$, where $n = 1, \ldots, 4$. Knowing these wave vectors, from the amplitude equations (7) the ratios of the amplitudes can be obtained

$$c_n = \frac{E_{hn}}{E_{0n}} = \frac{1}{C\chi_{\bar{h}}}\left[\frac{1}{K^2}(k_{0n}^2 - K^2) - \chi_0\right] = \frac{C\chi_h}{(k_{hn}^2 - K^2)/K^2 - \chi_0}, \quad n = 1, \ldots, 4. \tag{16}$$

The boundary conditions (12) provide the set of four equations (we assume the interface lying at $z = 0$)

$$\sum_{n=1,\ldots,4} E_{0n} = \text{const}, \quad \sum_{n=1,\ldots,4} k_{0nz} E_{0n} = \text{const},$$
$$\sum_{n=1,\ldots,4} c_n E_{0n} = \text{const}, \quad \sum_{n=1,\ldots,4} c_n k_{hnz} E_{0n} = \text{const}. \tag{17}$$

These formulas are valid for every polarization and for both coplanar and GID cases.

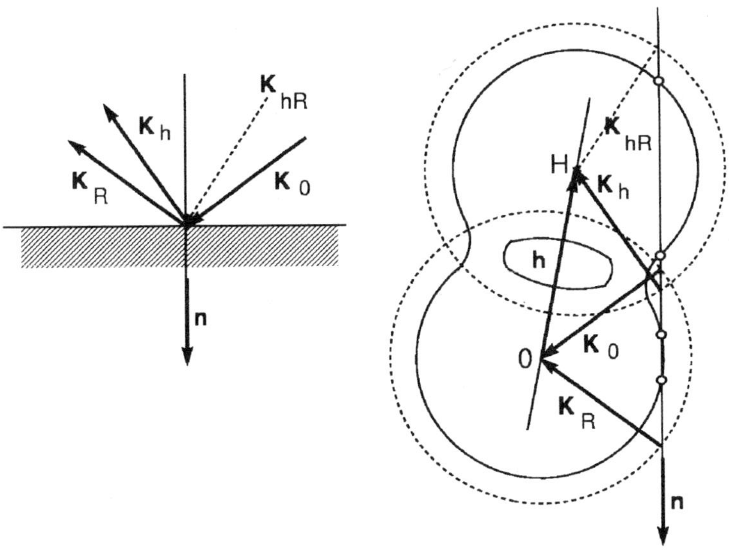

Figure 2. The wave vectors of the vacuum wavefields in the coplanar Bragg case and a semi-infinite crystal in direct space (left) and reciprocal space (right). The forbidden reflected diffracted wave $\mathbf{K}_{hR}$ is denoted by the dotted arrow. In the right panel, the dispersion surface of the crystal is plotted by the full line; the tie points are denoted by the empty circles.

In the Bragg case diffraction from a semi-infinite crystal, three waves propagate in the vacuum above the crystal surface, namely the incident wave with the amplitude E_0 and the wave vector $\mathbf{K}_0$, the diffracted wave with the amplitude E_h and the wave vector $\mathbf{K}_h$, and the (optically) reflected wave with the amplitude E_R and the wave vector $\mathbf{K}_R$. Within the crystal, only two of the four tie points are physically relevant, the others correspond to the waves with increasing amplitudes with depth. If we denote the indices of the physically relevant tie points by $n_{1,2}$, the boundary conditions for a semi-infinite crystal and Bragg case diffraction are

$$E_0 + E_R = E_{0n_1} + E_{0n_2}, \; K_{0z}(E_0 - E_R) = k_{0n_1z}E_{0n_1} + k_{0n_2z}E_{0n_2}$$

$$E_h = c_{n_1} E_{0n_1} + c_{n_2} E_{0n_2}, \; K_{hz} E_h = c_{n_1} k_{hn_1 z} E_{0n_1} + c_{n_2} k_{hn_2 z} E_{0n_2}. \tag{18}$$

These formulas represent a set of four algebraic linear equations for the unknown amplitudes E_h, E_R, E_{0n_1} and E_{0n_2} that can be simply solved. The vacuum wavefields corresponding to the coplanar Bragg case and a semi-infinite crystal are shown in Figure 2, the GID case is depicted in Figure 3.

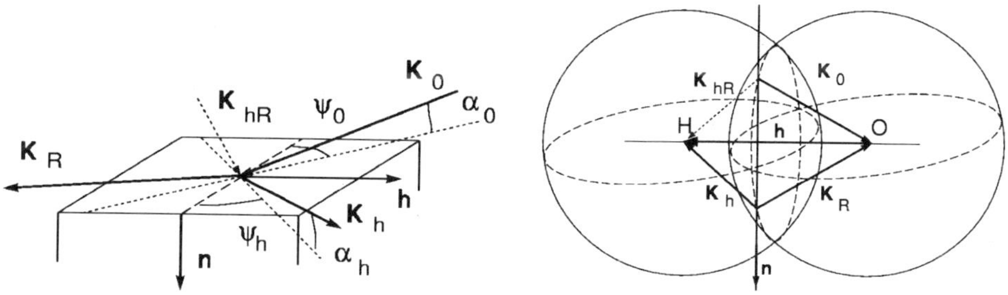

Figure 3. The wave vectors of the vacuum wavefields in GID geometry of a semi-infinite crystal in direct space (left) and reciprocal space (right). The vacuum dispersion surfaces are plotted in the right panel by full lines. The wavefield $\mathbf{K}_{hR}$ is forbidden.

Similar formulas can be obtained for the boundary conditions at the input and exit surfaces of a planeparallel crystal slab assuming the Bragg or the Laue cases. In the Laue case, two plane waves propagate in the vacuum above the input surface, namely the primary wave $\mathbf{E}_0$ and the reflected wave $\mathbf{E}_R$, and two waves propagate below the exit surface (the diffracted wave $\mathbf{E}_h$ and the optically reflected diffracted wave $\mathbf{E}_{hR}$).

3. WAVEFIELDS IN LAYERED SAMPLES

In the previous section we have derived the amplitude equations (7) and the boundary conditions (12) for the wavefields in a diffracting crystal taking into account the more exact generalized dynamical diffraction theory. From these formulas we are able to calculate the amplitudes and the wave vectors of the internal wavefields, and, using the boundary condition for the exit surface (similar to (12)), we can obtain the amplitudes of the waves emitted by a single crystal.

We have shown that the one–beam approximation leads to the description of optical reflection of X-rays and to the Fresnel reflectivity and transmittivity coefficients. The formulas of the two–beam case can be used both for strongly asymmetric coplanar diffraction and for the grazing incidence diffraction.

In this section, we find a matrix formalism for calculating the internal wavefields and the emitted wavefields for a layered perfect crystal. The formulas will be derived within the two–beam approach and we demonstrate that the one–beam case (i.e. X-ray reflection) can easily be obtained from them. The matrix approach to AXRD and GID has been formulated by Stepanov (1994,1995). The intensity of the diffracted beam can also be calculated using a semi-kinematical approach (see the paper by Rhan et al. (1993), and the citations therein).

Let us assume a sample consisting of N layers with flat interfaces deposited on a semi-infinite substrate. Each layer is characterized by its thickness d_j, polarizability coefficients $\chi_{0,h,\bar{h}j}$ and the vertical component h_{zj} of the diffraction vector. We assume a pseudomorphic structure of all layers, i.e. the in-plane components of the diffraction vectors $\mathbf{h}_j$ of all the layers are same and they equal to that of the substrate. The layer $j = 1$ lies just beneath the free surface, layer $j = N$ lies at the substrate. The interface j lies between the layers j and $j + 1$. We introduce the column vector

$$\vec{\mathbf{E}}_j = \begin{pmatrix} \mathcal{E}_{01j} \\ \mathcal{E}_{02j} \\ \mathcal{E}_{03j} \\ \mathcal{E}_{04j} \end{pmatrix}, \tag{19}$$

where $\mathcal{E}_{0nj}$ denotes the amplitude of the wave E_{0n} in the top of the layer j just below the interface $j - 1$ *including* the phase term $\exp(-ik_{0nz}^{(j)}z)$ (the origin $z = 0$ lies at the surface). The phase terms $\exp(-i\mathbf{k}_{||} \cdot \mathbf{r}_{||})$ vanish since the in-plane components of all the wave vectors are same.

The boundary conditions (12) at the interface j can be written in the matrix form as follows

$$\hat{\mathbf{C}}_j \hat{\mathbf{\Phi}}_j \vec{\mathbf{E}}_j = \hat{\mathbf{C}}_{j+1} \vec{\mathbf{E}}_{j+1}, \tag{20}$$

where

$$\hat{\mathbf{C}}_j = \begin{pmatrix} 1 & 1 & 1 & 1 \\ k_{01z}^{(j)} & k_{02z}^{(j)} & k_{03z}^{(j)} & k_{04z}^{(j)} \\ c_1^{(j)} & c_2^{(j)} & c_3^{(j)} & c_4^{(j)} \\ k_{h1z}^{(j)} c_1^{(j)} & k_{h2z}^{(j)} c_2^{(j)} & k_{h3z}^{(j)} c_3^{(j)} & k_{h4z}^{(j)} c_4^{(j)} \end{pmatrix} \tag{21}$$

and Φ_j is the diagonal matrix containing the phase shifts of the waves between the top and the bottom of the layer j

$$\hat{\mathbf{\Phi}}_j = \begin{pmatrix} e^{-ik_{01z}^{(j)} d_j} & 0 & 0 & 0 \\ 0 & e^{-ik_{02z}^{(j)} d_j} & 0 & 0 \\ 0 & 0 & e^{-ik_{03z}^{(j)} d_j} & 0 \\ 0 & 0 & 0 & e^{-ik_{04z}^{(j)} d_j} \end{pmatrix} \tag{22}$$

Deriving these shifts, we took into account the condition $\exp(-ik_{hnz}^{(j)} d_j) = \exp(-ik_{0nz}^{(j)} d_j)$ that must hold in pseudomorphic structures. The forms of the matrices $\hat{\mathbf{C}}_j$ and $\hat{\mathbf{\Phi}}_j$ have been obtained within the two–beam case. For the one–beam case (X-ray reflection), these matrices are the submatrices of (21) and (22), respectively, built from the first and the second columns and rows.

Eq. (20) represents a recurrence relation permitting the calculation of the amplitudes of the emitted waves. Using this formula N times, we obtain

$$\vec{\mathbf{E}}_0 = \hat{\mathbf{\Phi}}_0^{-1} \hat{\mathbf{C}}_0^{-1} \hat{\mathbf{C}}_1 \hat{\mathbf{\Phi}}_1^{-1} \hat{\mathbf{C}}_1^{-1} \hat{\mathbf{C}}_2 \hat{\mathbf{\Phi}}_2^{-1} \hat{\mathbf{C}}_2^{-1} \cdots \hat{\mathbf{C}}_N \hat{\mathbf{\Phi}}_N^{-1} \hat{\mathbf{C}}_N^{-1} \hat{\mathbf{C}}_{sub} \vec{\mathbf{E}}_{sub}. \tag{23}$$

Here $\hat{\mathbf{\Phi}}_0$ represents the phase shift of the wavefields between the crystal surface and the detector. The intensities of the emitted wavefields do not depend on the position of the detector, so that the distance d_0 between the detector and the crystal surface is arbitrary. The column vector $\vec{\mathbf{E}}_0$ contains the amplitudes of the incident, reflected, and diffracted waves outside the crystal, its fourth element is zero. The matrix $\hat{\mathbf{C}}_0$ is defined by

$$\hat{\mathbf{C}}_0 = \begin{pmatrix} 1 & 1 & 0 & 0 \\ K_{0z} & K_{Rz} & 0 & 0 \\ 0 & 0 & 1 & 1 \\ 0 & 0 & K_{hz} & K_{hz} \end{pmatrix}.$$

Assuming a semi-infinite substrate and the Bragg case diffraction, the column vector $\vec{\mathbf{E}}_{sub}$ will contain only two non-zero terms $\mathcal{E}_{0n_1}$ and $\mathcal{E}_{0n_2}$ corresponding to the physically relevant tie points in the substrate. Therefore, the matrix formula (23) represents a set of 4 algebraic equations for 4 unknown amplitudes E_h, E_R, E_{0n_1} and E_{0n_2}.

4. NUMERICAL EXAMPLES

In this section, we calculate the intensities of the waves diffracted and/or reflected from a semi-infinite Si crystal and from a pseudomorphic ideal AlAs/GaAs superlattice. We use the CuKα_1 radiation and diffraction 004, and we assume that the primary wave is perfectly monochromatic and plane. The superlattice consists of 10 periods, each contains an AlAs layer 150 Å thick and a 70 Å GaAs layer.

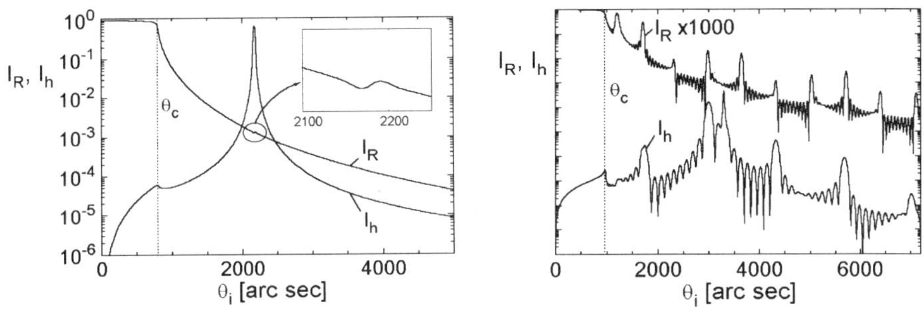

Figure 4. Diffracted I_h and reflected I_R intensities vs. the angle of incidence calculated for a semi-infinite crystal (left) and for the periodical superlattice (right). Strongly asymmetric diffraction, the diffracting planes (004) make the angle $\theta_B - 0.5°$ with the surface.

4.1 Coplanar scattering

Figure 4 shows the dependences of the diffracted (I_h) and reflected (I_R) intensities on the angle of incidence θ_i. An additional maximum can be seen on the diffraction curve $I_h(\theta_i)$ at a point where θ_i equals the critical angle of total external diffraction $\theta_c = \sqrt{\mathrm{Re}\chi_0}$. Below θ_c, the diffraction intensity is very weak, this is caused by the very small penetration depth of the radiation. Analogously, a tiny s-shaped structure can be observed on the I_R curve in the diffraction position. The experimental proof of this phenomenon can be found in the paper by Höche et al. (1988). In the case of a periodical superlattice, satellite maxima can be observed both in the diffracted and in the reflected intensities.

4.2 Grazing incidence diffraction

In the noncoplanar geometry, the direction of the primary wave vector is defined by two angles – its angle α_0 with the surface and its azimuthal angle ψ_0 (i.e. the angle with the diffracting net plane – see Figure 3). These angles determine the direction of the diffracted and reflected waves and their intensities. In Figure 5, the diffracted intensity is plotted as a function of the "take–off" angle α_h of the diffracted wave with the surface for given α_0. The azimuthal angles ψ_0, ψ_h are chosen so that they correspond to the given values of $\alpha_{0,h}$.

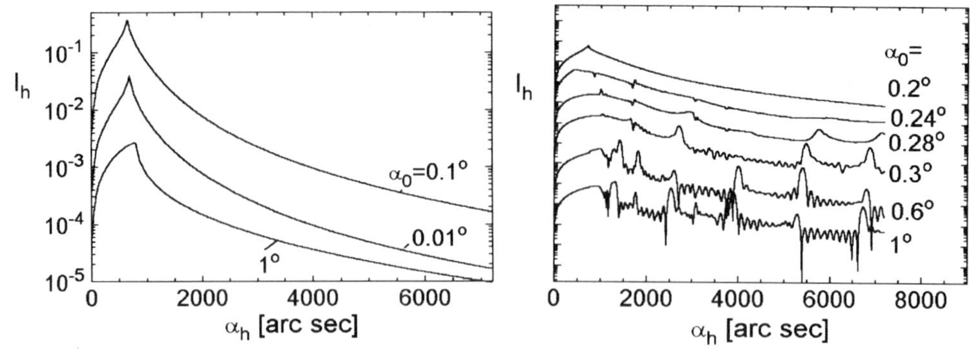

Figure 5. The intensities of GID vs. the take–off angle α_h for constant α_0 calculated for a semi-infinite crystal (left) and for a periodical superlattice (right).

It can be seen that the intensity diffracted from a semi-infinite sample has a maximum for $\alpha_h = \theta_c$, and, changing α_0 for constant α_h, we find a maximum at θ_c again. In the case of a periodical superlattice, the satellite maxima depend strongly on α_0. If $\alpha_0 < \theta_c$, the penetration depth is smaller than the superlattice period and no satellite maxima can be observed. Increasing α_0 we increase the penetration depth, and, therefore, the number of irradiated superlattice periods. Thus, the satellite width decreases with increasing α_0.

In conclusion, we have demonstrated that X-ray reflection, strongly asymmetric X-ray diffraction as well as grazing incidence diffraction can be described by the generalized dynamical diffraction theory in a uniform way. The matrix formulation of the boundary conditions facilitates the calculation of the diffracted and reflected intensities in the case of a multilayered sample.

REFERENCES

Afanasev, A.M., and Melkonyan, M.K., 1983, X-ray diffraction under specular reflection conditions. Ideal crystals, *Acta Cryst.* A39:207.

Aleksandrov, P.A., Afanasev, A.M., and Stepanov, S.A., 1984, Bragg-Laue diffraction in inclined geometry, *phys. stat. solidi (a)* 86:143.

Höche, H.-R., Nieber, J., Clausnitzer, M., and Materlik, G., 1988, Modification of specularly reflected X-ray intensity by grazing incidence coplanar Bragg-case diffraction, *phys. stat. solidi (a)* 105:53.

Kimura, S., and Harada, J., 1994, Comparison between experimental and theoretical rocking curves in extremely asymmetric Bragg cases of X-ray diffraction, *Acta Cryst.* A50:337.

Pinsker, Z.G., 1978, *Dynamical Scattering of X-rays in Crystals*, Springer, Berlin.

Rhan, H., Pietsch, U., Rugel, S., Metzger, H., and Peisl, J., 1993, Investigation of semiconductor superlattices by depth-sensitive X-ray methods, *J. Appl. Phys.* 74:146.

Stepanov, S.A., and Köhler, R., 1994, A dynamical theory of extremely asymmetric X-ray diffraction taking account of normal lattice strain, *J. Phys. D: Appl. Phys.* 27:1922.

Stepanov, S.A., Pietsch, U., and Baumbach, G.T., 1995, A matrix approach to X-ray grazing incidence diffraction in multilayers, *Z. Phys B - Cond. Matter* 96:341.

DYNAMICAL THEORY OF X-RAY DIFFRACTION - II. DEFORMED CRYSTALS

A. AUTHIER

Université P. et M. Curie
4, Place Jussieu
75252 Paris Cedex 05, France

1. PROPAGATION OF X-RAYS IN LIGHTLY DEFORMED CRYSTALS

1.1. Effective misorientation

1.1.1. Local reciprocal-lattice vector. If the crystal undergoes a deformation, any point p of position vector $\mathbf{r}$ is transformed in a point p' of position vector $\mathbf{r'}$ (Figure 1):

$$\mathbf{r}' = \mathbf{r} + \mathbf{u}(\mathbf{r})$$

This equation can also be written:

$$\mathbf{r} = \mathbf{r}' - \mathbf{u}(\mathbf{r}) \simeq \mathbf{r}' - \mathbf{u}(\mathbf{r}') \tag{1}$$

if the coefficients of the strain tensor, $\partial u_i/\partial x_i$, are small with regard to unity, which is the usual approximation in elasticity.

Consider a perfect infinite crystal asymptotic at p to the local, real crystal, and let $\mathbf{h'}$ be its reciprocal-lattice vector. It can be related to the reciprocal-lattice vector *before* deformation, $\mathbf{r}$, by means of the displacement vector, $\mathbf{u(r)}$. The equation of a plane of the family of direct lattice planes $(hk\ell)$ is:

$$f \equiv \mathbf{h}.\mathbf{r} = \mathcal{N} \tag{2}$$

The definition of the reciprocal-lattice vector $\mathbf{h}$ can be taken to be the gradient of f:

$$\mathbf{h} = \overrightarrow{\nabla}(f)$$

If one replaces in (2) **r** by its expression in the function of the position vector *after* deformation, **r'**, the equation of the lattice planes becomes:

$$f' \equiv \mathbf{h}.\mathbf{r}' - \mathbf{h}.\mathbf{u}(\mathbf{r}') = \mathcal{N}$$

This is the equation of the surfaces into which the lattice planes have been transformed by the deformation. The reciprocal-lattice vector, **h'**, of the asymptotic perfect crystal is given by:

$$\mathbf{h}' = \overrightarrow{\nabla} f' = \mathbf{h} - \overrightarrow{\nabla}(\mathbf{h}.\mathbf{u}) \quad (3)$$

The local variation $\boldsymbol{\delta}\mathbf{h}$ of the reciprocal lattice is therefore given by:

$$\boldsymbol{\delta}\mathbf{h} = \mathbf{h}' - \mathbf{h} = -\overrightarrow{\nabla}(\mathbf{h}.\mathbf{u}) \quad (4)$$

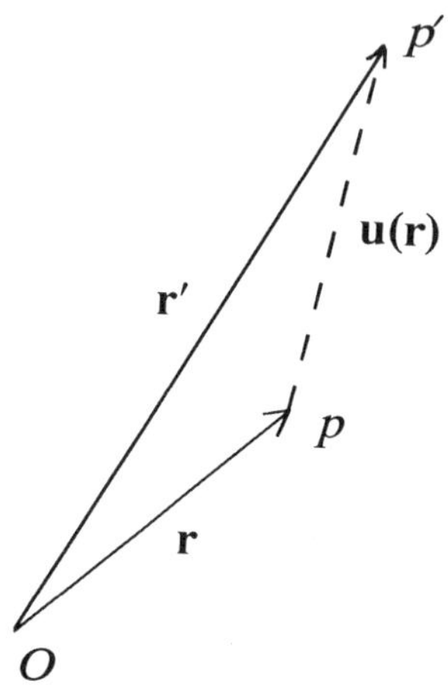

Figure 1. Deformation of a material. $\mathbf{u}(\mathbf{r})$: deformation vector.

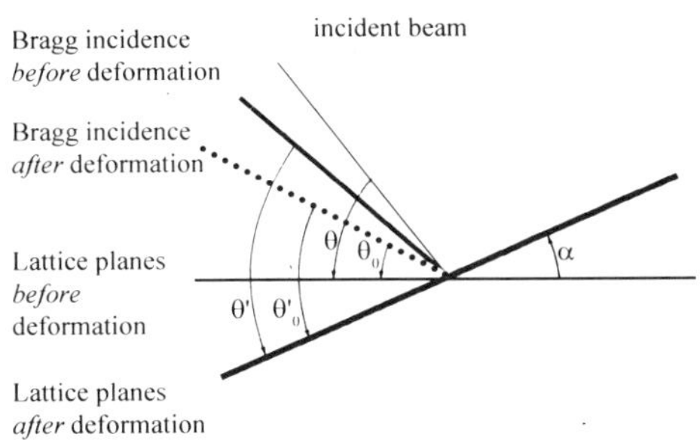

Figure 2. Variation of the departure from Bragg's incidence with deformation.

1.1.2. Effective misorientation in direct space. Let θ_o be the Bragg angle and $\Delta\theta$ the departure from Bragg's angle of an incident wave ($\Delta\theta = \theta - \theta_o$ - Figure 2). During a deformation, the lattice planes are rotated and there is a relative variation $\delta d/d$ of the lattice parameter. If α is the component of the rotation of the lattice planes around the normal to the plane of incidence, the angle of the incident beam with the lattice planes, *after* deformation, is:

$$\theta' = \theta + \alpha$$

The Bragg angle, *after* deformation, is:

$$\theta'_o = \theta_o - \frac{\delta d}{d} tan\theta_o$$

and the *new* departure from Bragg's angle of the incident beam, *after* deformation, is:

$$\Delta\theta' = \theta' - \theta'_o$$

The local variation of the departure from Bragg's law, or *effectice misorientation* is:

$$\delta\theta = \Delta\theta' - \Delta\theta = \alpha + \frac{\delta d}{d} tan\theta_o$$

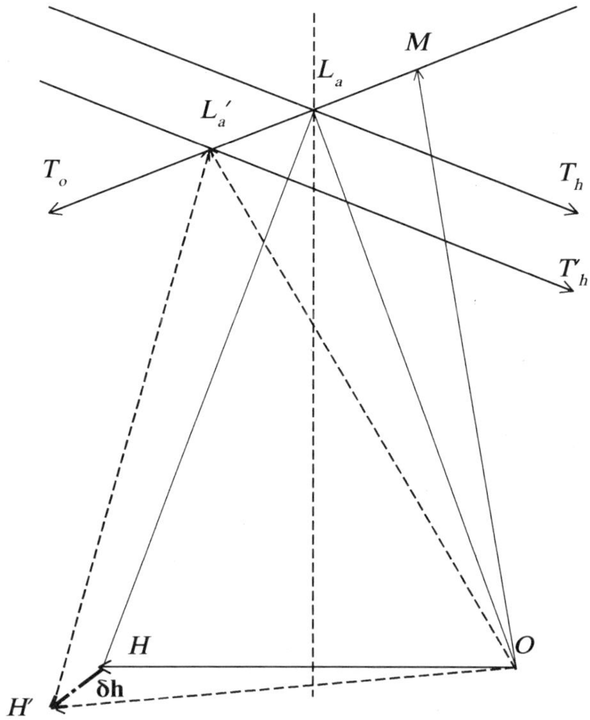

Figure 3. Modification of the reciprocal lattice due to the deformation of the crystal. L_a: Laue point *before* deformation: L'_a: Laue point *after* deformation; $\delta\mathbf{h}$: variation of the reciprocal lattice vector.

1.1.3. Effective misorientation in reciprocal space. Let **HH'** $= \boldsymbol{\delta}\mathbf{h}$ be the projection on the plane of incidence of the local variation of the reciprocal-lattice vector (Figure 3). The Laue point, L_a, lies, by definition, on the sphere of radius $k = 1/\lambda$ and centre O, and on the mediatrix of **OH**. The new, local, Laue point, L'_a, after deformation, lies on the same sphere and on the mediatrix of the new reciprocal-lattice vector, **OH'**. If **OM** $= \mathbf{K}_\mathbf{o}^{(\mathbf{a})}$ is an incident wavevector, the effective misorientation is:

$$\delta\theta = \Delta\theta' - \Delta\theta = \left(\overline{L'_aM} - \overline{L_aM}\right)/k = \overline{L'_aL_a}/k$$

Expressed in terms of the local variation of the reciprocal lattice vector, the effective misorientation is:

$$\delta\theta = \frac{\boldsymbol{\delta}\mathbf{h}\cdot\mathbf{s_h}}{k\ sin\ 2\theta} = -\frac{\mathbf{s_h}\cdot\overrightarrow{\nabla}(\mathbf{h}\cdot\mathbf{u})}{k\ sin\ 2\theta} \tag{5}$$

The effective misorientation (5) is a linear combination of components of the strain tensor $\partial u_i/\partial x_j$. It is equal to zero if $\boldsymbol{\delta}\mathbf{h}.\mathbf{s_h} = \mathbf{0}$, that is if the local variation of the reciprocal-lattice vector is normal to the reflected direction. This means that any deformation of the crystal such that

$$\mathbf{h}'.\mathbf{s_h} = constant \tag{6}$$

at every point has no influence on the propagation of X-rays.

1.1.4. Strain gradient. The derivative of the effective misorientation along the incident direction is one of the components of the strain gradient and plays an important role in the properties of the propagation of X-rays in a deformed crystal. It is usually introduced by means of a coefficient β proportional to its ratio over the angular width δ (equation 26, in Dynamical Theory of X-ray Diffraction, Part I Perfect Crystals, this Volume, called AAI thereafter) of the reflection domain:

$$\beta = \frac{1}{\delta cos\theta} \frac{\partial(\delta\theta)}{\partial s_o} = - \frac{\Lambda_o}{cos^2\theta} \frac{\partial^2(\mathbf{h}\cdot\mathbf{r})}{\partial s_o \partial s_h} \tag{7}$$

The case of a uniform strain gradient is of particular interest. It corresponds, for instance, to an homogeneous bending or to a constant thermal gradient.

1.2. Eikonal Approximation - Ray Theory

The properties of the propagation of X-rays in a slightly deformed crystal can be found by analogy with the properties of usual optics. When a wavepacket of light emitted by a source of width e propagates, it diverges slightly, the angular width being equal to $\alpha = \lambda/e$ where λ is the wavelength. One distinguishes along the path of the packet two regions; the first one, within a distance from the source of about $r_o = e/\alpha = e^2/\lambda$, called the *Fresnel region*, within which the width of the wavepacket remains roughly constant, and the second one, from the distance r_o to infinity, called the *Fraunhofer region*, where the lateral width of the wavepacket increases. If the beam propagates in a region where the index of refraction n varies, the direction of the mean wavevector of the wavepacket varies accordingly in such a way that, at every point:

$$\boldsymbol{\delta}\mathbf{k} \sim \overrightarrow{\nabla n} \tag{8}$$

so as to ensure the continuity of the tangential component of the wavevector. The path of the beam is bent and can be found by applying Fermat's principle.

In the case of X-rays propagating in a crystal while satisfying the Bragg condition, the properties of geometrical optics can be applied in the regions where the strain gradient is not too large, with certain modifications. The limits of validity of this approximation will be discussed in § **2.6.3**. The first difference with ordinary optics is that the extension of the properties of optics applies not to the individual waves, but to the *wavefields* resulting from the coupling of the reflected and refracted waves. This is the basic assumption of the *ray theory*. The angle α mentioned in the preceding paragraph is then to be multiplied by the amplification ratio $\mathcal{A}$ defined in § **8.2**, which is of the order of magnitude of Λ_o/λ. The limit between the Fresnel and Fraunhofer regions becomes:

$$r_o = e^2/\Lambda_o$$

If one takes the width of the beam to be of the order the Pendellösung distance, Λ_o, the coherence length of a wavepacket is also of the same order of magnitude.

When a crystal is deformed, plane-wave dynamical theory cannot be used any more. Penning & Polder (1961) made the hypothesis that, when the deformation is small enough, it is possible to consider at each point p of the crystal a local perfect crystal, *asymptotic* to the real, deformed, crystal, where plane-wave dynamical theory applies, where one can define a local dispersion surface and the notion of wavefields is kept. At a first approximation, the structure factor, and therefore the diameter of the dispersion surface are not affected by the deformation. As the Laue point, L_a and the Lorentz point, L_o, lie necessarily on spheres centred at O and with radii k and nk, that is, in practice, on their tangential planes T_o' and T_o, respectively, the dispersion surfaces at neighbouring points p and p' are simply translated with respect to one another, the shift being $\overrightarrow{L_oL_o'} = k\delta\theta$, where $\delta\theta$ is given by (5) (Figure 4). The local deviation parameter varies from point to point, the propagation direction of the wavefields remains at all times normal to the local dispersion surface and, by applying Fermat's principle, it is possible to find the path of the wavefields in the deformed crystal (Penning & Polder, 1961; Kato, 1963, 1964 a et b). This is the so-called *Eikonal approximation* (Kato 1963, Penning 1966), which is the X-ray analogue of the propagation of light in a medium with slowly varying index of refraction, as mentioned above. Let us consider a certain wavefield passing through a point p in direct space (Figure 1). Its tiepoint is a point P on the dispersion surface *before* deformation (Figure 4).

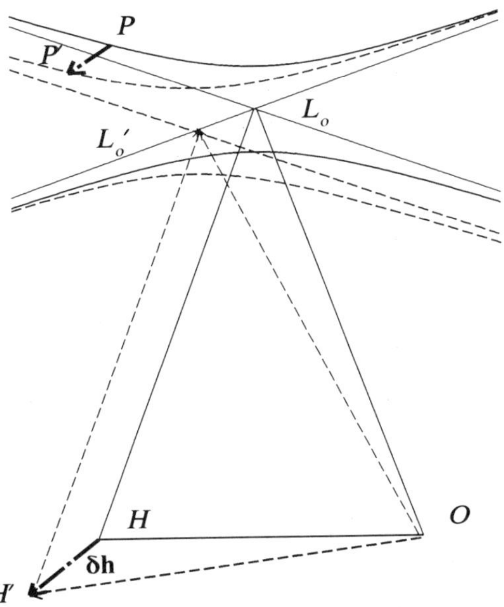

Figure 4. Translation of the dispersion surface due to deformation: L_O: Lorentz point *before* deformation: L'_O: Lorentz point *after* deformation; P: tiepoint *before* deformation; P': tiepoint *after* deformation

During the deformation, point p will be transformed into point p', and the same wavefield, now passing through p', will have, *after* deformation, as tiepoint a certain point P' on the translated dispersion surface (Figure 4). The wavevectors vary and

$$\delta \mathbf{K_o} = \overrightarrow{\mathbf{PP'}} = \overrightarrow{\mathbf{OP'}} - \overrightarrow{\mathbf{OP}}$$

The dispersion surface is invariant for a variation of the local reciprocal-lattice vector such that $\boldsymbol{\delta}\mathbf{h}.\mathbf{K_h} = 0$. Equation (6) can therefore be interpreted as the equation of constant index of refraction and, by analogy with the optical case (8), the variation of the local wavevector is given by:

$$\delta \mathbf{K_o} = \vec{\nabla} (\mathbf{h}.\mathbf{K_h}) \tag{9}$$

The deviation parameter η associated with the tiepoint P is also modified and it can be shown, using (9) and the properties of the dispersion surface, that, in the symmetric case, the local variation $\delta\eta$ at p' is given by:

$$\delta\eta = \beta\delta z$$

where β is given by (7) and δz is the difference between the depths of points p and p' from the entrance surface.

2. TAKAGI'S THEORY

2.1. Introduction

The theory developed by Takagi (1962, 1969) constitutes a generalization of the dynamical theory for any kind of incident wave, as well as for deformed crystals with any kind of deformation. In the case of perfect crystals, it reduces to the classical dynamical theory or to Kato's theory if the incident wave is a plane wave or a spherical wave. In the case of deformed crystals, it applies even in highly distorted regions. Several theories have been derived which are similar to Takagi's (Taupin, 1964, Schlangenotto, 1967). When the column approximation can be applied (Hirsch, Howie & Whelan, 1960), namely, when the Bragg angle is very small as in the case of electron diffraction, Takagi's theory reduces to Howie & Whelan's two-beam electron diffraction dynamical theory for distorted crystals (Howie & Whelan , 1960, 1961, 1962).

Plane-wave dynamical theory applies to a semi-infinite perfect crystal and its solutions do not depend on the lateral coordinates. If there is a lateral limitation of the wave, that is when there is diffraction of the incident plane wave by an object, it is obvious that the solutions of the propagation equation must depend on the lateral coordinates. This is the case for a spherical wave which can be generated, for instance, by limiting a plane wave by an infinitely thin slit. One way to solve the problem is to consider a Fourier expansion of the incident wave. The advantage is that this method describes the nature and the distribution of wavefields inside the crystal as well as the distribution of their propagation directions. This is a good way if, as in the case of Kato's spherical-wave dynamical theory, it is possible to integrate the Fourier transform, but, unfortunately, in the general case, this is not possible.

In a similar way, it has been shown in § **1.** that plane-wave dynamical theory is not valid when the crystal is deformed, and that the path of the wavefields can be obtained using ray theory. However, rigorously speaking, it is not correct anymore to speak of wavefields since these are, by definition, characterized by *individual* tiepoints in reciprocal space which are associated with waves of infinite lateral extension in direct space. One has, therefore, to consider wavebundles having a narrow width both in direct and reciprocal space. Furthermore, the theory breaks down when the strain gradient becomes too large and diffraction (in the optical sense) of the propagating wavebundles becomes predominant. The situation is anolagous to the optical case, with the complicating effect that diffraction of wavebundles is accompanied, in the X-ray case, by the generation of new wavefields whose tiepoints belong to the other branch of the dispersion surface than that of the tiepoints of the original wavebundle. This effect which was intuitively predicted by Penning (1966) and empirically suggested by Authier (1967) to interpret features of defect images on X-ray topographs was demonstrated by Balibar, Chukhovskii & Malgrange (1983).

The two preceding paragraphs have shown that, when diffraction effects occur, either in a perfect or an imperfect crystal, the notion of wavefields which have definite propagation direction in direct space and are represented by tiepoints in reciprocal space breaks down. Takagi's theory relates the total amplitude and phase of the reflected or refracted wave at any point inside the crystal to the amplitude and phase of another point in the crystal, but it does not provide any analysis of the crystal wave in terms of wavefields nor of tiepoints in reciprocal space. The relationship is given through a Green-Rieman function, or *influence* function which depends on the strain distribution in a deformed crystal. The value of the amplitude and phase at any point of the exit surface is obtained by convolution of this function with the amplitude and phase distribution on the entrance surface, expressed as a distribution of point sources. In general, it is not possible to obtain an analytical expression. However, in the case of a perfect crystal and a single point source (the incident wave is a spherical wave), the Green-Rieman function reduces to the Bessel function $J_o(\zeta)$ obtained by Kato's theory. In the case of a constant strain gradient, Chukhovskii (1974), Katagawa & Kato (1974), Litzman & Janacek (1974) have given the exact form of the Green-Rieman function and shown that the results are in agreement with those obtained with the Eikonal approximation. Except in such favourable situations, it is necessary to integrate Takagi's equations numerically. Suitable algorithms have been derived for this purpose (Authier *et al.*, 1968; Epelboin, 1983)

2.2. Generalized Fundamental Equations

2.2.1. Modulated waves. Takagi's approach consists in considering the crystal wave to be a modified Ewald wave, which can be developed as a sum of *modulated* waves associated with each reciprocal lattice which lies close to the Ewald sphere:

$$\mathbf{D} = \sum \mathbf{D_h}(\mathbf{r}) exp(-2\pi i\ \mathbf{K_h}.\mathbf{r}) \tag{10}$$

where $\mathbf{K_h} = \mathbf{K_o} - \mathbf{h}$. This development is analogous to (12 - AAI) for plane waves but, here, its coefficients $\mathbf{D_h}(\mathbf{r})$ are slowly varying functions of the position vector $\mathbf{r}$ and $\mathbf{K_o}$ has an *arbitrary* orientation and is of length nk, where n is the index of refraction of the medium and k the wavenumber in vacuum. The hypothesis that $\mathbf{D_h}(\mathbf{r})$ is a slowly varying function means that $\mathbf{\Delta}\,[\mathbf{D_h}(\mathbf{r})]$ can be neglected.

The idea underlying Takagi's theory is the following: in all the dynamical theories, for perfect or imperfect crystals, making use of the notion of wavefields, the phase of the waves can be considered as being the product of two terms, a fast oscillation and a slow modulation. The fast oscillation, which is a *microscopic* effect, is simply due to the periodicity of the crystal lattice, $1/h$, and represents a carrier wave, while the slow modulation, which is a *macroscopic* effect, is due to interferences between wavefields: such as Pendellösung interferences between wavefield associated with both branches of the dispersion surface, or interferences between the wavefields which make up a given wavepacket. The corresponding period is of the order of the Pendellösung distance Λ. If the crystal is deformed, or if there are diffraction effects due to a limitation of the incident wave by a slit, both these periods become space dependent and are no longer constant. The originality of Takagi's theory consists in decomposing the phase factor in a different, phenomenological way, namely as the product of a fast, but constant phase factor, $exp(-2\pi i\ \mathbf{K_h}.\mathbf{r})$, and a slowly varying term which can be incorporated into the amplitude $\mathbf{D_h}(\mathbf{r})$. The vectors $\mathbf{K_h}$ are constant and chosen arbitrarily, but, as indicated in the previous paragraph, in such a way that the phase factor should be very close to the real, varying one. This has the advantage of simplifying the calculations, but the drawback is that these phase factors have no physical meaning.

2.2.2. Electric susceptibility of a deformed crystal. The electric susceptibility is a triply periodic function of space coordinates in a perfect crystal. This is no longer the case in a deformed crystal. If one replaces $\mathbf{r}$ in the development (7 - AAI) of the electric susceptibility by its expression (1), we have:

$$\chi = \sum_h \chi_h exp\{-2\pi i\ \mathbf{h}[\mathbf{r}' - \mathbf{u}(\mathbf{r}')]\}$$

This expression can be rewritten, replacing **r'** by **r**,

$$\chi = \sum_h \chi'_h exp\{-2\pi i\ \mathbf{h}.\mathbf{r}\} \tag{11}$$

where the coefficients

$$\chi'_h = \chi_h exp\{-2\pi i\ \mathbf{h}.\mathbf{u}(\mathbf{r})\}$$

of the Fourier expansion of the electric susceptibility are no longer constants.

2.2.3. Takagi's equations. Let us now put the developments (10) and (11) into the propagation equation (6 - AAI). This gives a partial differential equation connecting the unknown functions $\mathbf{D_h}(\mathbf{r})$ and their derivatives. This equation is rather complicated due to the vectorial nature of $\mathbf{D}$. It is simplified by making use of the fundamental assumption of Takagi's theory, namely the macroscopic character of the variations of $\mathbf{D_h}$ as compared to the microscopic character of the variations of $exp[-2\pi i\ \mathbf{K_h}.\mathbf{r}]$. This means that the second order derivatives of $\mathbf{D_h}$ can be neglected. The details of the simplifications are rather involved and are given in Takagi (1969). The principle of the derivation is the same as that for plane waves (§ **2.4** - AAI) and the propagation equation (7 - AAI) becomes an equation with an infinite sum of terms which is shown to be equivalent to two infinite sets of equations whose unknowns are the components of $\mathbf{D_h}$. However, since the amplitudes of the modified Ewald waves are space dependent, the system is now a system of first order *partial differential* equations. In the two-beam case, there are simply two sets of two equations, one set for the components perpendicular to the $\mathbf{K_o}$, $\mathbf{K_h}$

plane (the usual polarization factor takes the value $C = 1$) and one set for the components parallel to that plane ($C = cos2\theta$):

$$\left.\begin{aligned}\frac{\partial D_o(\mathbf{r})}{\partial s_o} &= -\, i\pi k C\chi'_{\bar{h}}(\mathbf{r})D_h(\mathbf{r}) \\ \frac{\partial D_h(\mathbf{r})}{\partial s_h} &= -\, i\pi k\,[C\chi'_h(\mathbf{r})D_o(\mathbf{r}) \;-\; 2\beta_h D_h(\mathbf{r})]\end{aligned}\right\} \tag{12}$$

where s_o and s_h are the coordinates of the position vector $\mathbf{r}$ along directions parallel to the incident and reflected beams, respectively, and the parameter

$$\beta_h \;=\; \frac{|\mathbf{K_h}|^2 \;-\; k^2(1+\chi_o)}{2k^2} \;=\; \frac{|K_h|^2 - |K_o|^2}{2k^2} \;\simeq\; \frac{|K_h| - |K_o|}{k}$$

depends on the choice of the pair of vectors $\mathbf{k_o}$, $\mathbf{K_o}$. It is represented geometrically on Figure 5. Let T'_o, T_o and T_h be the traces on the plane of incidence of the tangential planes to the spheres centred at O with radius k, centred at O with radius nk, and centred at H with radius nk, respectively. Let further

$$\mathbf{OM} \;=\; \mathbf{k_o};\; \mathbf{OP} \;=\; \mathbf{K_o};\; \mathbf{HP} \;=\; \mathbf{K_h} \;=\; \mathbf{K_o} -\; \mathbf{OH}$$

where $\mathbf{PM}$ is along the normal to the crystal surface, in agreement with the boundary condition on the continuity of the tangential components of the wavevectors (see next section).

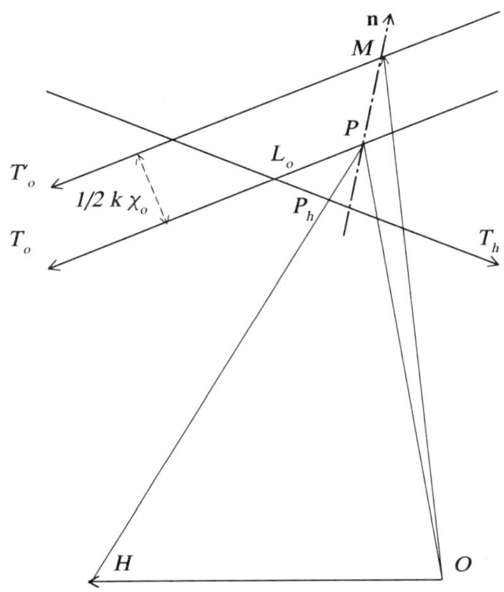

Figure 5. Takagi's theory: **OM**: incident wavevector; **OP**: refracted wavevector used in Takagi's theory; **n**: normal to the crystal surface.

It can be seen in Figure 5 that:

$$\beta_h \;=\; \overline{P_h P}\,/k \;=\; -\;\overline{L_o P}\,(sin2\theta)/k$$

where P_h is the projection of P on T_h.

2.2.4. Boundary conditions at the entrance surface. Any incident wave can be written in the form:

$$\mathbf{D_o^{(a)}} = \mathbf{D_o^{(a)}}(\mathbf{r}) exp[-2\pi i \, \mathbf{k_o}.\mathbf{r}]$$

The boundary conditions on the entrance surface are:

$$\left.\begin{aligned} \mathbf{D_o}(\mathbf{r_s}) &= \mathbf{D_o^{(a)}}(\mathbf{r_s}) \\ \mathbf{D_h}(\mathbf{r_s}) &= 0 \\ \mathbf{K_o}.\mathbf{r_s} &= \mathbf{k_o}.\mathbf{r_s} \end{aligned}\right\} \quad (13)$$

where $\mathbf{r_s}$ is a position vector on the surface.

The wavevector $\mathbf{k_o}$ can be chosen with an arbitrary orientation, but its length must be equal to the wavenumber in vacuum, k. What is important is that $\mathbf{K_o}$, $\mathbf{k_o}$ should be related by the condition of the continuity of their tangential component (the third equation in 13).

The boundary conditions for the derivatives are deduced from (12) and (13):

$$\left.\begin{aligned} \left[\frac{\partial D_o}{\partial s_o}\right]_{z=0} &= 0 \; ; & \left[\frac{\partial D_o}{\partial s_h}\right]_{z=0} &= \frac{sin\ 2\theta}{\gamma_o}\left[\frac{\partial D_o^{(a)}}{\partial y}\right]_{z=0} ; \\ \left[\frac{\partial D_h}{\partial s_o}\right]_{z=0} &= -i\pi k\frac{\gamma_o}{\gamma_h}\chi'_h D_o^{(a)} \; ; & \left[\frac{\partial D_h}{\partial s_h}\right]_{z=0} &= -i\pi k\chi'_h D_o^{(a)}(\mathbf{r_s}) \; ; \end{aligned}\right\}$$

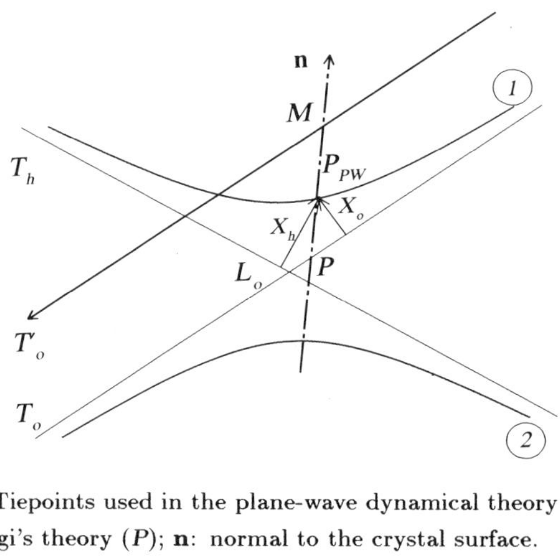

Figure 6. Tiepoints used in the plane-wave dynamical theory (P_{PW}), and in Takagi's theory (P); **n**: normal to the crystal surface.

2.2.5. Reduction of Takagi's equations in the plane-wave case. It is interesting to show that system (12) reduces to system (16 - AAI) in the case of an incident plane wave. In that case the conditions of the problem, and therefore the solutions, are not space dependent. The Ewald wave must have the form (12 - AAI):

$$\mathbf{D} = \sum_h \mathbf{D_h^{PW}} exp[-2\pi i \, \mathbf{K_h^{PW}}.\mathbf{r}]$$

where $\mathbf{D_h^{PW}}$ is constant and $\mathbf{K_h^{PW}} = \mathbf{OP_{PW}}$ is the real wavevector (Figure 6). The expression of $\mathbf{D}$ can be put in the form (10) by writing:

$$\mathbf{D} = \sum_h \{\mathbf{D_h^{PW}} exp\left[-2\pi i\, (\mathbf{K_h^{PW}} - \mathbf{K_h})\cdot\mathbf{r}\right]\} \, exp\left[-2\pi i\; \mathbf{K_{h\cdot r}}\right]$$

$$= \sum_h \mathbf{D_h} exp\left[-2\pi i\; \mathbf{K_h}.\mathbf{r}\right]$$

where $\mathbf{D_h(r)} = \mathbf{D_h^{PW}} exp\left[-2\pi i\, (\mathbf{K_h^{PW}} - \mathbf{K_h})\cdot\mathbf{r}\right]$ is the amplitude of the modulated Ewald wave in (10). If this expression is inserted in system (12), the latter becomes:

$$\left.\begin{array}{r} 2\;\overline{PP_{PW}}\;\; \gamma_o D_o = \; k\chi_{\bar{h}} C D_h \\ 2\;\overline{PP_{PW}}\;\; \gamma_h D_h = \; k\chi_h C D_o \end{array}\right\}$$

where γ_o and γ_h have the usual meanings. It can be seen on Figure 6 that this system can also be written:

$$\left.\begin{array}{r} 2X_o D_o = \; k\chi_{\bar{h}} C D_h \\ 2X_h D_h = \; k\chi_h C D_o \end{array}\right\}$$

which is equivalent to system (17 - AAI) for plane waves.

2.3. Absorbing Crystals

System (12) is still valid in the case of absorbing crystals, but, as in the classical theory, $\mathbf{K_o}$ and $\mathbf{K_h}$ have an imaginary part $\mathbf{K_{oi}} = \mathbf{K_{hi}}$ parallel to the normal to the entrance surface. The length of $\mathbf{K_o}$, which has been arbitrarily fixed, can be written:

$$\mathbf{K_o^2} = K_{or}^2 - K_{oi}^2 + 2i\mathbf{K_{or}}.\mathbf{K_{oi}} \; = \; k^2(1+\chi_o)$$

$$\simeq K_{or}^2 - K_{oi}^2 + 2i\gamma_o K_{or}.K_{oi}$$

neglecting K_{oi}^2. The value of the imaginary part of the wavevector is therefore:

$$K_{oi} \;\simeq\; k\frac{\chi_{oi}}{2\gamma_o}$$

The parameter β_h also has an imaginary part which, from this relation, is found to be:

$$\beta_{hi} \;=\; \frac{\chi_{oi}}{2}\left(\frac{\gamma_h}{\gamma_o} \;-\; 1\right)$$

2.4. Analytical Resolution of Takagi's Equations in the Perfect Crystal Case

When the crystal is perfect, and the coefficients χ_h are constant, the system (12) of two first-order partial differential equations can be replaced by a system of two *independent second*-order partial differential equations:

$$\left.\begin{array}{l} \dfrac{\partial^2 D_o}{\partial s_o \partial s_h} \; - \; 2\pi i k \beta_h \; \dfrac{\partial D_o}{\partial s_o} \; + \; \pi^2 k^2 C^2 \chi_h \chi_{\bar{h}} D_o = 0 \\[2ex] \dfrac{\partial^2 D_h}{\partial s_o \partial s_h} \; - \; 2\pi i k \beta_h \; \dfrac{\partial D_h}{\partial s_o} \; + \; \pi^2 k^2 C^2 \chi_h \chi_{\bar{h}} D_h = 0 \end{array}\right\} \qquad (14)$$

These equations are of hyperbolic form and can be solved by Riemann's method (see, for instance, Sommerfeld, 1949). Let p be the point inside the crystal where the amplitudes D_h and D_o are being calculated and A and B two points of the entrance surface such that **Ap** and **Bp** are parallel to $\mathbf{s_o}$ and $\mathbf{s_h}$, repectively (Figure 7).

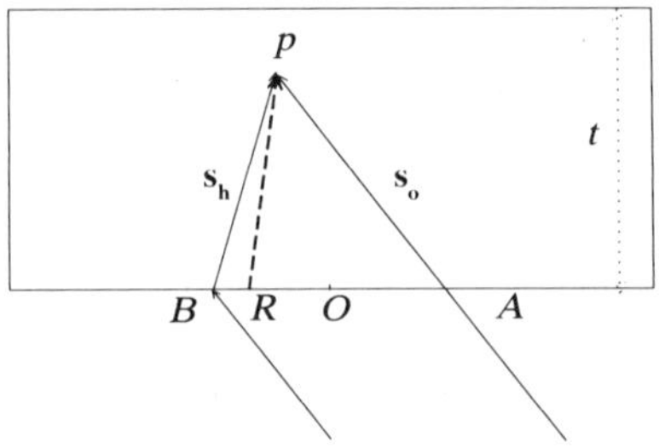

Figure 7. Inverted Borrmann triangle. *Ap* incident direction, *Bp*: reflected direction, *R*: arbitrary point source on the entrance surface.

Takagi (1969) has shown that the amplitudes of the wave at p depend only on the values of the amplitude of the incident wave and its derivatives *along BA*:

$$D_h(p) = \frac{-i\pi k C \chi_h \gamma_o}{sin\ 2\theta} \int_{BA} D_o^{(a)}(y) v(y) dy \qquad (15)$$

where y is the abscissa of a point source, R, on the entrance surface and $v(y)$is the Riemann function which is defined by:

$$\left.\begin{aligned}
\frac{\partial^2 v}{\partial s_o \partial s_h} + 2i\pi k \frac{\partial(\beta_h v)}{\partial s_o} + \pi^2 k^2 C^2 \chi_h \chi_{\bar{h}} v &= 0 \\
\frac{\partial v}{\partial s_o} = -2\pi i k \beta_h v & \quad \text{along } BP \\
\frac{\partial v}{d s_o} = 0 & \quad \text{along } AP \\
v(p) = 1 &
\end{aligned}\right\}$$

This function can be calculated analytically when β_h is equal to zero. This situation can always be achieved in the perfect crystal case by choosing the pair of vectors $\mathbf{k_o}$, $\mathbf{K_o}$ in such a way that $\mathbf{K_o} = \mathbf{OL_o}$ (P is taken to be situated at L_o, Figure 5). In this case:

$$v(y) = J_o(\zeta)$$

where ζ has the same meaning as in (**§7.2.** - AAI), but Y means here a normalized coordinate along the entrance surface:

$$Y = \overline{OR} / \overline{OA}$$

where R is a point source on the entrance surface and O is the midpoint of BA.

It is possible in a similar way to find the value of the refracted amplitude in terms of the Riemann function:

$$D_o(p) = D_o^{(a)}(A) - \frac{\pi t}{\Lambda_L} \int_{BA} D_o^{(a)}(y) \sqrt{\frac{1+Y}{1-Y}}\, J_1(\zeta) dy \qquad (16)$$

If the incident wave is a spherical wave, it can be approximated by a single point source on the entrance surface. This is equivalent to replacing $D_o^{(a)}(y)$ by a Dirac distribution. The expressions (15) and (16) of the reflected and refracted amplitudes are then identical to those given by Kato's theory.

2.5. Alternate Form of Takagi's Equations

The system (12) is not very suitable for solution because the coefficients $\chi_h(\mathbf{r})$ are space-dependent in the case of a deformed crystal. By substituting $\mathbf{r}$ by its expression (1) in the development (10) of the modulated waves, one obtains an expression for the *local* values of the modulated amplitudes:

$$D'_h(\mathbf{r}) \;=\; D_h(\mathbf{r})exp[-2\pi i\ \mathbf{h}.\mathbf{u}(\mathbf{r})] \tag{17}$$

The advantage of using expression (17) for the modulated amplitudes is that the intensities, which are the final results needed, are the same:

$$|D'_h(\mathbf{r})|^2 \;=\; |D_h(\mathbf{r})|^2$$

By inserting (17) in (10) and using (3), the development (10) can be rewritten:

$$\mathbf{D_h} \;=\; \sum_h \mathbf{D'_h} exp\,[-2\pi i\ \mathbf{K'_h}.\mathbf{r}]$$

where $\mathbf{K'_h} \;=\; \mathbf{K_o} \;-\; \mathbf{h'}$.

By substituting the expressions (17) of the modulated amplitudes into (12), the system of Takagi's equations becomes:

$$\left.\begin{aligned} \frac{\partial D'_o(\mathbf{r})}{\partial s_o} &= -\ i\pi k\,[C\chi_{\bar{h}} D'_h(\mathbf{r})] \\ \frac{\partial D'_h(\mathbf{r})}{\partial s_h} &= -\ i\pi k\,[C\chi_h D'_o(\mathbf{r}) \;-\; 2\beta'_h D'_h(\mathbf{r})] \end{aligned}\right\} \tag{18}$$

with

$$\beta'_h \;=\; \beta_h \;+\; \frac{1}{k}\ \mathbf{s_h}.\ \overrightarrow{\nabla}\ [\mathbf{h}.\mathbf{u}(\mathbf{r})] \tag{19}$$

Using the same modified waves, (17), the second order differential equations (14) are replaced by, in the deformed crystal case:

$$\left.\begin{aligned} \frac{\partial^2 D'_o}{\partial s_o \partial s_h} \;-\; 2\pi i k \beta'_h \frac{\partial D'_o}{\partial s_o} \;+\; \pi^2 k^2 C^2 \chi_h \chi_{\bar{h}} D'_o = 0 \\ \frac{\partial^2 D'_h}{\partial s_o \partial s_h} \;-\; 2\pi i k \beta'_h \frac{\partial D'_h}{\partial s_o} \;+\; \left[\pi^2 k^2 C^2 \chi_h \chi_{\bar{h}} - 2\pi i k \frac{\partial \beta'_h}{\partial s_o}\right] D'_h = 0 \end{aligned}\right\} \tag{20}$$

2.6. Solution of Takagi's Equation in the Deformed Crystal Case

2.6.1. Analytical solution. Takagi's equations have been solved analytically in the case of a constant strain gradient only. In that case, parameter β'_h (equation 19) is of the form:

$$\beta'_h = \frac{\partial(\mathbf{h}.\mathbf{u})}{\partial s_h} = \frac{-1}{2\pi i}[Ps_o + 2qs_h]$$

with $\beta_h = 0$. $P = -2\pi ik\ \partial^2(\mathbf{h}.\mathbf{u})/\partial s_o \partial s_h$ and q are constants and the phase factor $2\pi i\ \mathbf{h}.\mathbf{u}$ is of the form:

$$-2\pi i\mathbf{h}.\mathbf{u} = Ps_o s_h + qs_h^2 + rs_o^2 + Constant$$

where $r(s_o)$ is any function of s_o.

Litzman and Janacek (1974), Petrashen (1973), Chukhovskii (1974), using Riemann's function, Katagawa & Kato (1974), using Laplace transforms, have shown that the solution of (20) can be expressed in terms of confluent hypergeometric functions. They have shown that the solutions for the general case:

$$-2\pi i\ \mathbf{h}.\mathbf{u} = Ps_o s_h + q(s_h) + r(s_o)$$

where $q(s_h)$ and $r(s_o)$ are arbitrary functions, are given by:

$$D_o = D_o^1 exp[q(0) - q(s_h)]$$
$$D_h = D_h^1 exp[r(0) + r(s_o)]$$

where D_o^1 and D_h^1 are the solutions obtained for the simpler case, $-2\pi i\ \mathbf{h}.\mathbf{u} = Ps_o s_h$.

Riemanns's method of solution shows the very important fact that, as in the perfect crystal case, the solution at point p of the exit surface only depends on the distribution of the wave within the inverse Borrmann triangle BAp, and therefore that it *depends only on the strain distribution within the inverse Borrmann triangle* ABp. It can be expressed, as a generalization of the Huyghens-Fresnel principle, in terms of the amplitude distribution of the point-sources along BA and of the Green function describing the influence of the point source R at p (Chukhovskii & Petrashen, 1977):

$$D_h(p) = \int_{\overline{BA}} \mathcal{G}_{ho}(R,p)D_o(p)ds_o + \int_{\overline{BA}} \mathcal{G}_{hh}(R,p)D_h(p)ds_h$$
$$D_o(p) = \int_{\overline{BA}} \mathcal{G}_{oo}(R,p)D_o(p)ds_o + \int_{\overline{BA}} \mathcal{G}_{oh}(R,p)D_h(p)ds_h$$

The exact form of the Green functions has been given by Chukhovskii *et al.* (1978) using Laplace transforms.

2.6.2. Numerical integration. When the strain distribution is more complicated, the only way to solve of Takagi's equations is by numerical integration. Authier *et al.* (1968) have given an algorithm for integration by means of a grid illustrated on Figure 8. The principle of the calculation is as follows. Let us divide the reverse Borrmann triangle BAp into a grid whose sides Ap and Bp are parallel to the incident and reflected directions, respectively (Figure 8).

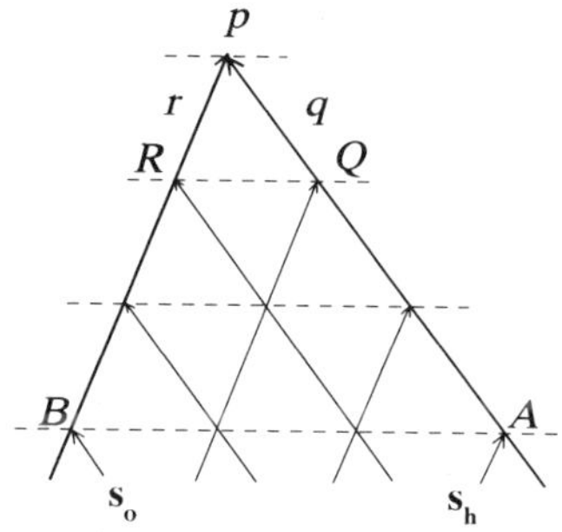

Figure 8. Numerical integration of Takagi's equations;
q: integration step in the incident direction,
r: integration step in the refracted direction .

The amplitudes D_o and D_h at point p can be expressed in terms of their values at neighbouring points Q and R on the grid by means of a Taylor series expansion. The latter involves their partial derivatives which can in turn be expressed in terms of the amplitudes by means of system *(18)*. This provides a linear system which can be calculated by the computer step by step down from the surface to point p. The value of the step is of the order of a micron and the number of steps is very large. It is therefore of paramount importance to reduce as much as possible the error committed at each step. The approximation chosen for that purpose is the so-called 'half-step' derivative approximation. Its principle can be explained using an ordinary function $f(x)$. Its first order derivative can be written:

$$\frac{df(x)}{dx} = F(f,x)$$

The Taylor expansion of $f(x+q)$ can best be approximated by considering the derivative at $x+q/2$:

$$f(x+q) \simeq f(x) + q\left(\frac{df}{dx}\right)_{x+q/2}$$

which, in turn, can be approximated by:

$$f(x+q) \simeq f(x) + qF\left[\frac{1}{2}\{f(x)+f(x+q)\}, x+q/2\right]$$

Using this approximation, the amplitudes at p can be written:

$$\left.\begin{aligned} D'_o(p) &= D'_h(Q) + \frac{\overline{Qp}}{2}\left\{\left[\frac{\partial D'_o}{\partial s_o}\right]_Q + \left[\frac{\partial D'_o}{\partial s_o}\right]_p\right\} \\ D'_h(p) &= D'_h(R) + \frac{\overline{Rp}}{2}\left\{\left[\frac{\partial D'_h}{\partial s_h}\right]_R + \left[\frac{\partial D'_h}{\partial s_h}\right]_p\right\} \end{aligned}\right\} \qquad (21)$$

These expressions can be calculated using the system (18) which, for that purpose can be written, using simplified notations:

$$\left.\begin{aligned} q\frac{\partial D'_o(p)}{\partial s_o} &= 2AD'_h(p) \\ r\frac{\partial D'_h(p)}{\partial s_h} &= 2BD'_o(p) + 2WD'_h(p) \end{aligned}\right\} \qquad (22)$$

where:

$$2A = -i\pi kC\chi_{\bar{h}}q; \quad 2B = -i\pi kC\chi_h r; \quad 2W = 2i\pi k\beta'_h r$$

and $q = \overline{Qp}$; $r = \overline{Rp}$. Using (21) and (22), one can relate the values of the amplitudes at p in terms of the values at Q and R by means of a matrix product:

$$\begin{bmatrix} D'_o(p) \\ D'_h(p) \end{bmatrix} = \frac{1}{1-W-AB}\begin{bmatrix} 1-W & A(1-W) & AB & A(1+W) \\ B & AB & B & 1+W \end{bmatrix}\begin{bmatrix} D'_o(Q) \\ D'_h(Q) \\ D'_o(R) \\ D'_h(R) \end{bmatrix}$$

Since the computing time is proportional to the square of the number of steps, there is a conflict between computing time and resolution. In order to solve this dificulty, Epelboin (1983) has introduced a varying step algorithm which uses a very small step in the regions where the strain varies rapidly, and a larger one where the strain varies more slowly. The change in the value of the step is done automatically according to the value of the strain gradient.

The numerical integration of Takagi's equations has been used for instance:

- to *simulate images of defects on X-ray topographs*:
  - *dislocations (Laue case)* (Balibar & Authier, 1967; Taupin, 1967; Authier *et al.*, 1985; Epelboin & Soyer, 1985; reviews are given in Epelboin, 1985, 1987; Chukhovskii & Petrashen, 1988)
  - *dislocations (Bragg case)* (Bubakova & Sourek, 1976, Bedynska, *et al.*, 1976; Gronkowski,1980; Riglet *et al.*, 1980; Kaganer *et al.*, 1991)
  - *precipitates* (Green *et al.*, 1990)
  - *microdefects* (Holý, 1982, Indenbom *et al.* , 1985)
  - *stacking faults* (Authier *et al.*, 1968; Wonsiewicz *et al.*, 1976; Epelboin 1985, 1987)
  - *ferromagnetic domains* (Nourtier *et al.*, 1979)
  - *antiphase domain boundaries* (Capelle *et al.*, 1982)
  - *growth striations* (Härtwig *et al.*, 1987)
- to *calculate the positions of standing-waves nodes in deformed crystals* (Authier *et al.*, 1989, Chukhovskii *et al.*, 1996)
- to *calculate rocking curve profiles*, using an appropriate model for the strain distribution, and sometimes to restore the strain profile by best fit techniques:
  - *crystals with surface layers or epilayers* (Miltat, 1984; Halliwell *et al.*, 1984; Bartels *et al.*, 1986; Bensoussan *et al.*, 1987; Faleev *et al.*; 1990; Cui *et al.*, 1990; Aristov *et al.* 1991; Fewster *et al.*,1992; Servidori *et al.* 1992)
  - *crystals with implanted layers* (Takeuchi *et al.*, 1983; Cembali *et al.*, 1985; Servidori *et al.*, 1987)

- *crystals with surface acoustic waves* (Gabrielyan *et al.*, 1988) *or excited by ultrasounds* (Iolin, 1995)
- *bent crystals* (Cembali *et al.*, 1992; Uschmann *et al.* 1993)

- to *restore the value of the strain gradient* of crystals with uniform strain (Chukhovskii *et al.*, 1978; Voronkov *et al.*, 1984; Darbinyan *et al.*, 1991)
- to *investigate the diffraction by periodic distortions at the surface of crystal* designed for X-ray optics (Aristov *et al.*, 1991).

2.6.3. Limit of the validity of the Eikonal approximation, interbranch scattering. Katagawa & Kato (1974) and Chukhovskii & Petrashen (1977) have shown that the solutions obtained in the case of a constant strain gradient by solving Takagi's equations are identical to those obtained using the Eikonal method (Penning & Polder, 1961; Kato, 1963, 1964 a and b), provided that the value of the strain gradient is not too large and the Eikonal theory can be applied. Authier & Balibar (1970) have shown that this is the case when the strain gradient $\partial^2(\mathbf{h}.\mathbf{u})/\partial s_o \partial s_h = \partial\beta'_h/\partial s_o$ is much smaller than $1/\Lambda_o^2$. This condition is equivalent to saying that the deformation of the lattice over the unit distance is much smaller than a rotation of the order of the width of the rocking curve over a distance equal to the Pendellösung distance Λ_o.

When this condition is no longer satisfied and the Eikonal theory is no longer valid, there is a diffraction, in the sense of physical optics, of the wavebundles propagating within the crystal (Balibar & Malgrange, 1975) and new wavefields are generated whose tiepoints belong to the other branch of the dispersion surface than those which are propagating (Balibar, Chukhovskii & Malgrange, 1983, Laue case; Chukhovskii & Malgrange, 1989, Bragg case). This is the so-called 'interbranch scattering' effect which was predicted by Penning (1966) and empirically used by Authier (1967) to describe some features of dislocation images on X-ray section topographs.

REFERENCES

Aristov, V.V., Goureev, T.E., Nikulin, Yu. and Snigirev, A.A., 1991, X-ray diagnosis of the elastic stress gradient in crystals, *Phys. Stat. Sol. (a)*, 127:33-42.

Aristov, V.V., Kuznetsov, S.M., Kouyumchian, A.V. and Snigirev, A.A., 1991, X-ray dynamic diffraction on multiblock interferometers in Laue and Bragg geometry., *Phys. Stat. Sol. (a)*, 125:57-66.

Authier, A., Lefeld-Sosnowska, M., Epelboin, Y. and Soyer, A., 1985, Experimental and computer simulation study of the variation with depth of the X-ray section topograph images of a dislocation., *J. Appl. Cryst.*, 18:93-105.

Authier, A., Malgrange, C., and Tournarie M., 1968, Etude théorique de la propagation des rayons X dans un cristal parfait ou légèrement déformé, *Acta Cryst. A*, 24:126-136.

Authier, A., Gronkowski, J. and Malgrange, C., 1989, Standing waves from a single heterostructure on GaAs - A computer experiment., *Acta Cryst. A*, 45:432-441.

Authier, A., 1967, Contrast of dislocation images in X-ray transmission topography, *Adv. X-ray Anal.*, 10:9-31.

Authier, A. and Simon, D., 1968, Application de la théorie dynamique de S.Takagi au contrast d'un défaut plan en topographie par RX. I. Faute d'empilement, *Acta Cryst. A*, 24:517-526.

Authier, A. and Balibar, F., 1970, Création de nouveaux champs d'onde généralisés dus à la présence d'un objet diffractant. II.- Cas d'un défaut isolé., *Acta Cryst. A*, 26:647-654.

Balibar, F., Chukhovskii, F.N. and Malgrange, C., 1983, Dynamical X-ray propagation: a theoretical approach to the creation of new wave fields, *Acta Cryst. A*, 39:387-399.

Balibar, F. and Malgrange, C., 1975, Structure of wave packets in perfect and highly distorted crystals, *Acta Cryst. A*, 31:425-434.

Balibar, F. and Authier, A., 1967, Etude théorique et expérimentale du contraste des images de dislocations, *Phys. Stat. Sol.*, 21:413-422.

Bartels, W.J., Hornstra, J. and Lobeek, D.J.W., 1986, X-ray diffraction of multilayers and superlattices, *Acta Cryst. A*, 42:539-545.

Bedynska T., Bubakova, R. and Sourek, Z., 1976, Comparison between Experimental and Theoretical Dislocation Image in the Bragg Case, *Phys. Stat. Sol. (a)*, 36:509-516.

Bensoussan, S., Malgrange, C. and Sauvage-Simkin, M., 1987, Sensitivity of X-ray diffractometry for strain depth profiling in III-V heterostructures., *J. Appl. Cryst.*, 20:222-229.

Bubakova, R. and Sourek, Z., 1976, On the Dislocation Image in the Bragg Case, *Czech. J. Phys. B*, 26:863-864.

Capelle, B. and Malgrange, C., 1982, X-ray Topographic Study of Antiphase Domain Boundaries in Ferroelectric Ferroelastic GdDy(MoO4)3 Crystals. I. Geometry of Antiphase Domain Boundaries., *J. Appl. Phys.*, 53:6762-6766.

Capelle, B., Epelboin, Y. and Malgrange, C., 1982, X-ray Topographic Study of Antiphase Domain Boundaries in Ferroelectric Ferroelastic GdDy(MoO4)3 crystals. II. Evaluation of the Additional Translation Vector., *J. Appl. Phys.*, 53:6767-6771.

Cembali, F., Servidori, M., Gabili, E. and Lotti, R., 1985, Effect of Diffuse Scattering in the Strain Profile Determination by Double Crystal X-ray Diffraction., *Phys. Stat. Sol. (a)*, 87:225-233.

Cembali, F., Fabri, R., Servidori, M., Zani, A. Basile, G., Cavegnero, G., Bergamin, A. and Zosi, G., 1992, Precise X-ray Relative Measurement of Lattice Parameters of Silicon Wafers by Multiple-crystal Bragg Case Diffractometry. Computer Simulation of the Experiment., *J. Appl. Cryst.*, 25:424-431.

Chukhovskii, F.N., 1974, Dynamic X-ray scattering on a crystal inclined to the wave-front plane. *Kristallografiya*, 19:482-488.

Chukhovskii, F.N., Gabrielyan, K.T. and Petrashen, P.V., 1978, The Dynamical Theory of X-ray Bragg Diffraction from a Crystal with a Uniform Strain Gradient. The Green-Riemann Functions., *Acta Cryst. A*, 34:610-621.

Chukhovskii, F.N. and Malgrange, C., 1989, Theoretical Study of X-ray Diffraction in Homogeneously Bent Crystals - the Bragg Case, *Acta Cryst. A*, 45:732-738.

Chukhovskii, F.N., Malgrange, C. and Gronkowski, J., 1996, X-ray Standing Waves in Crystals Distorted by a Constant Strain Gradient. A Theoretical Study., *Acta Cryst. A*, 52:47-55.

Chukhovskii, F.N. and Petrashen P.V., 1977, A general dynamical theory of the X-ray Laue diffraction from a homogeneously bent crystal, *Acta Cryst. A*, 33:311-319.

Chukhovskii, F.N. and Petrashen, P.V., 1988, X-ray Topography of Bent Crystals., *Acta Cryst. A*, 44:8-14.

Cui, S. and Mai, Z., 1990, Double-crystal X-ray Rocking Curve Peak Splitting due to Interference in Triple-layer Epitaxic Structures, *J. Appl. Cryst.*, 23:147-150.

Darbinyan, S.P., Chukhovskii, F.N. and Voronkov, S.N., 1991, Identification of Local Deformations in Anisotropic Crystals according to the X-ray Diffraction Inclination Method., *Phys. Stat. Sol. (a)*, 125:441-449.

Epelboin, Y., 1985, Simulation of X-ray Topographs., *Mater. Sci. & Eng.*, 73:1-43.

Epelboin, Y., 1983, A varying step algorithm for numerical integration of Takagi-Taupin equations, *Acta Cryst. A*, 39:761-767.

Epelboin, Y., 1987, The Simulation of X-ray Topographic Images., *Progress in Crystal Growth and Characterization*, 14:465-506.

Faleev, N.N., Flaks, L.I., Konnikov, S.G., Solomin, I.K. and Batashova S.V., 1990, The Influence of the Directed Displacement of Atomic Planes on the X-ray Diffraction Rocking Curves., *Phys. Stat. Sol. (a)*, 120:327-337.

Fewster, P.F., 1992, The Simulation and Interpretation of Diffraction Profiles from Partially Relaxed Layer Structures., *J. Appl. Cryst.*, 25:714-723.

Gabrielyan, K.T. and Aslanian, H.A., 1988, On the Theory of X-ray Diffraction by Surface Acoustic Waves., *Phys. Stat. Sol. (a)*, 108:K85-K88.

Green, G.S., Cui, S-F. and Tanner, B.K., 1990, Simulation of Images of Spherical Defects in X-ray Section Topographs, *Phil. Mag. A*, 61:23-33 Gronkowski, J., 1980, X-ray Diffraction Contrast of the Dislocation Image in the Bragg Case, *Phys. Stat. Sol. (a)*, 57:105-112.

Halliwell, M.A.G., Lyons, M.H. and Hill, M.J., 1984, The Interpretation of Rocking Curves from III-V Semiconductor Device Structures, *J. Cryst. Growth* , 68:523-531

Hirsch, P.B., Howie, A. and Whelan, M.J., 1960, A kinematical theory of diffraction contrast of electron transmission microscope images of dislocations and other defects, *Phil. Trans. Roy. Soc. A*, 252:489.

Holý V., 1982, The Coherence Description of the Dynamical X-ray Diffraction from Randomly Disordered Crystals., *Phys. Stat. Sol. (b)*, 111:341-351.

Howie, A. and Whelan, M.J., 1960, The dynamical theory of diffraction from dislocations, *Proc. Eur. Reg. Conf. on Electron Microscopy, Delft*, 1:194.

Howie, A. and Whelan, M.J., 1961, Diffraction contrast of electron microscope images of crystal lattice defects, *Proc. Roy. Soc. A*, 263:217.

Howie, A. and Whelan, M.J., 1962, Diffraction contrast of electron microscope images of crystal lattice defects. III Results and experimental confirmation of the dynamical theory of dislocation image contrast, *Proc. Roy. Soc. A*, 267:206.

Härtwig, J., Jäckel, K.H. and Lerche, V., 1987, Simulation of Contrast in X-ray Plane Wave Topographs of Quartz Crystals with Induced Growth Striations., *Crystal Res. & Technol.*, 22:951-959.

Indenbom, V.L. and Kaganer, V.M., 1985, The Formation of Plane-wave X-ray Images of Microdefects., *Phys. Stat. Sol. (a)*, 87:253-265.

Iolin, E., 1995, Rapid Suppression and Modulation of the Diffracted Beam in a Single Crystal Excited by Ultrasound., *Acta Cryst. A*, 51:897-902.

Kaganer, V.M. and Möhling, W., 1991, Characterization of Dislocations by Double Crystal X-ray Topography in Back Reflection., *Phys. Stat. Sol. (a)*, 123:379-392.

Katagawa, T. and Kato, N., 1974, The exact dynamical wave fields for a crystal with a constant strain gradient on the basis of the Takagi-Taupin equations., *Acta Cryst. A*, 30: 830-836.

Kato, N., 1963, Pendellösung fringes in distorted crystals. I. Fermat's principle for Bloch waves, *J. Phys. Soc. Jap.*, 18:1785-1791.

Kato, N., 1964 a, Pendellösung fringes in distorted crystals. II. Application to two-beam cases, *J. Phys. Soc. Jap.*, 19:67-77.

Kato, N., 1964 b, Pendellösung fringes in distorted crystals. III. Application to homogeneously bent crystals, *J. Phys. Soc. Jap.*, 19:971-985.

Litzman O. and Janacek Z., 1974, The exact solution of Takagi's equations for the dynamical X-ray diffraction in an elastically bent crystal, *Phys. Stat. Sol. (a)*, 25:663-666.

Miltat, J., 1984, *IEEE Trans. Magn.*, 20:1114-1116.

Nourtier, C., Kleman, M., Taupin, D., Miltat, J. Labrune, M. and Epelboin, Y., 1979, Simulation of X-ray topographs (by Lang method) of ferromagnetic domains in iron-silicon, especially their junctions, *J. Appl. Phys.*, 50:2143-2145.

Penning P., 1966, Theory of X-ray diffraction in unstrained and lightly strained perfect crystals, *Thesis, Technical University of Delft*.

Penning P. and Polder, D., 1961, Anomalous transmission of X-rays in elastically deformed crystals, *Philips Res. Repts*, 16:419-440.

Petrashen P.V., 1973, *Fiz. Tverd. Tela*, 15:3131-3132.

Riglet, P., Sauvage, M., Pétroff, J.F. and Epelboin, Y., 1980, Synchrotron radiation plane wave topography. II. Comparison between experiments and computer simulations for misfit dislocation images in III-V heterojunctions, *Phil. Mag. A*, 42:339-358.

Schlangenotto, H., 1967, Dynamische Theorie der Röntgenbeugung für deformierte Kristalle, *Z. Phys.*, 203:7-36.

Servidori, M., Sourek, Z. and Solmi, S., 1987, Some Aspects of Damage Annealing in Ion Implanted Silicon. Discussion in Terms of Dopant Anomalous Diffusion., *J. Appl. Phys.*, 62:723-1728.

Servidori, M., Cembali, F., Fabri, R. and Zani, A., 1992, Influence of First-order Approximations in the Incidence Parameter on the Simulations of Symmetric and Antisymmetric X-ray Rocking Curves of Heteroepitactic structures., *J. Appl. Cryst.*, 25:6-51.

Takagi, S., 1962, Dynamical theory of diffraction applicable to crystals with any kind of small distortion, *Acta Cryst.*, 15:311-1312.

Takagi, S., 1969, A dynamical theory of diffraction for a distorted crystal, *J. Phys. Soc. Jap.*, 26:239-1253.

Takeuchi, T., Ohta, N., Sugita, Y. and Fukuhara, A., 1983, Determination of strain distributions in ion-implanted magnetic bubble garnets applying X-ray dynamical theory, *J. Appl. Phys.*, 54:15-721.

Taupin, D., 1964, Théorie dynamique de la diffraction des rayons X par les cristaux déformés, *Bull. Soc. Fr. Minér. Crist.*, 87:69.

Taupin, D., 1967, Prévision de quelques images de dislocations par transmission des rayons X (cas de Laue symétrique), *Acta Cryst.*, 23:5-35.

Uschmann, I., Förster, E., Gäbel, K., Hölzer, G. and Ensslen M., 1993, X-ray Reflection Properties of Elastically Bent Perfect Crystals in Bragg Geometry., *J. Appl. Cryst.*, 26:05-412.

Voronkov, S.N., Maksimov, S.K. Chukhovskii, F.N., 1984, Investigation of Homogeneously-strained monocrystal Plates by Data of X-ray Diffraction Inclination Method., *Fiz. Tverd. Tela*, 26:019.

Wonsiewicz, B.C. and Patel, J.R., 1976, Computer simulation of X-ray diffraction topographs of stacking faults, *J. Appl. Phys.*, 47:837-1845.

DYNAMICAL THEORY OF NEUTRON SCATTERING

Michel Schlenker and Jean-Pierre Guigay

Laboratoire Louis Néel du CNRS, associé à l'UJF
B.P. 166, F-38042 Grenoble, France

1. X-RAYS vs NEUTRONS WITH SPIN NEGLECTED

The orders of magnitude for the basic quantities are similar in neutron and X-ray scattering. When the neutron's spin 1/2 is irrelevant, i.e. in diffraction by non-magnetic crystals, the dynamical theory of X-ray scattering can be very simply transferred to neutrons. Original features related to the neutron spin (§ 2) appear in diffraction by magnetic crystals (§ 3). Novel possibilities also arise because the neutrons can be manipulated from outside through applied fields (§ 4).

Neutrons sense atoms mainly through two interactions: the nuclear strong force and the effect of the magnetic field created by the atom. They are described in terms either of scattering amplitudes, needed for an adequate treatment of interference, or of scattering cross-sections. The elastic scattering amplitude for scattering vector $\mathbf{s}$, $f(\mathbf{s})$, is defined such that, when the incident plane wave is $\psi_i = D \exp [i(\mathbf{k}_o.\mathbf{r} - \omega t)]$, the asymptotic form of the wave scattered by an object placed at origin can be written as $\psi_s = D \frac{f(\mathbf{s})}{r} \exp [i\,(kr - \omega t)]$ with $k = |\mathbf{k}_o| = |\mathbf{k}_o + \mathbf{s}| = 2\pi/\lambda$. The differential scattering cross-section is $(\partial\sigma/\partial\Omega) = |f(\mathbf{s})|^2$. For the strong-force interaction, the nuclei can be considered as point scatterers because the interaction range is very small; hence the scattering amplitude is isotropic and independent of λ except in the vicinity of resonances. It is conventionally written as $(-b)$ so that most values of b, called the scattering length, be positive. The typical order of magnitude is the *fm*

X-ray and Neutron Dynamical Diffraction: Theory and Applications
Edited by Authier *et al.*, Plenum Press, New York, 1996

(femtometre, i.e. 10^{-15} *m*, or Fermi), there is no systematic variation with atomic number, and different isotopes have very different scattering lengths, including different signs.

b corrresponds to $(R.f_{at})$, the atomic scattering factor $f_{at}(s)$ of X-ray diffraction times the classical electron radius *R*. Since the usage in neutron diffraction is to have structure factors that are dimensionally lengths (they are pure numbers in the X-ray case), the translation rule, to obtain formal expressions valid for neutrons, is simply to multiply the X-ray structure factors by *R* and to set the polarization coefficient *C* to 1 (the wave-function for neutrons is not a vector). The presence of different isotopes, as well as the effect of nuclear spin (disordered except under very special efforts), give rise to incoherent elastic neutron scattering which has no equivalent in the X-ray case. The scattering length corresponding to $(R.f_{at})$ is the coherent scattering length, b_{coh}, obtained by averaging the scattering length over the nuclear spin state and isotope distribution. Scattering of neutrons by condensed matter implies the use of the "bound" scattering lengths, as tabulated in *International Tables for Crystallography*, volume C.

A description in terms of potential scattering is possible using the Fermi pseudo-potential, which in the case of a single nucleus at $\mathbf{r}_o$ is $(h^2/2\pi m)\, b\, \delta(\mathbf{r}-\mathbf{r}_o)$, where δ denotes the Dirac distribution, $m = 1.675.10^{-27}$ kg is the neutron's rest mass and $h= 2\pi\hbar$ is Planck's constant. Inside a material the spatially averaged value of this potential is $U= (h^2/2\pi mV) \sum_i b_i$, where the sum is over the nuclei contained in volume *V*. The refractive index is $n = (1- 2\lambda^2 m\, U/h^2)^{1/2} = (1 - \frac{\lambda^2}{\pi V} \sum_i b_i)^{1/2}$. It is very close to 1, with $1-n$ of the order of 10^{-5}, in the same range as for X-rays.

Neutron absorption is related to nuclear reactions in which the neutron combines with the absorbing nucleus to form a compound nucleus, usually in a metastable state which later decays. The scattering length describing this resonance scattering process depends on the neutron energy, and contains an imaginary part, associated to absorption in complete analogy with the imaginary part of the dispersion correction for the X-ray atomic scattering factors. The energies of the resonances are usually far above those of interest for crystallography, and the linear absorption coefficient then varies approximately like $\frac{1}{v}$ or λ. Except for a very few cases (notably ^{3}He, ^{6}Li, ^{10}B, In, Cd, Gd), the absorption of neutrons is very small compared to that of X-rays, and can be neglected in a first approximation.

The basic equations of dynamical theory, viz. Maxwell's equations for the X-ray case, and the time-independent Schrödinger equation in the neutron case, have exactly the same form when the effect of the neutron spin can be neglected, i.e. in situations that do not involve magnetism, and when no externally applied potential is taken into account. The physics of neutron diffraction by perfect crystals is expected to be very similar to that of X-ray diffraction, with the existence of wave-fields, Pendellösung effects, anomalous transmission, intrinsic rocking-curve shapes and reflectivity vs thickness behavior in direct correspondence. All experimental tests of these predictions confirm this view.

The main difference is in the order of magnitudes of the intensities. Neutron beams are weak in comparison with laboratory X-ray sources, and weaker by many orders of magnitude than synchrotron radiation. Also the beam sources are large in the case of neutrons since they are essentially the moderators, whereas the source is very small in the case of synchrotron radiation, and this difference again increases the ratio of the brilliances in favor of X-rays. Many experiments that are quick using X-rays become very slow, and have to accept impaired resolution, in the neutron case. State-of-the-art sources, or ample measuring time, are thus essential for dynamical neutron diffraction work.

Apart from the original aspects that will be discussed below, another difference is worth mentioning. The small velocity of neutrons, in comparison with the velocity of light, and its energy dependence, make time-of-flight measurements possible. In the only experiment on dynamical diffraction by crystals performed, to our knowledge, on a pulsed spallation source, this feature, together with the near-100% reflectivity inside the Darwin width, is being used to make a cold neutron storage device (Schuster et al., 1992). Earlier experiments on the pulsed reactor IBR-2 in Dubna (Alexandrov et al., 1988) showed encouraging preliminary results, among others on Pendellösung measurement.

Basic discussions of dynamical neutron scattering were given by Stassis and Oberteuffer (1974), Sears (1978), Rauch and Petrascheck (1978), and Squires (1978). Due to space limitation, we restrict this treatment almost entirely to perfect crystals. A more comprehensive review by the present authors will be chap. 5.3 in the 2nd edition of *International Tables for Crystallography*, vol. B.

2. NEUTRON SPIN

2.1. Polarisation of a Neutron Beam and Larmor Precession in a Uniform Magnetic Field

A polarised neutron beam is represented by a 2-component spinor $|\varphi\rangle = \begin{pmatrix} c \\ d \end{pmatrix} = c\begin{pmatrix} 1 \\ 0 \end{pmatrix} + d\begin{pmatrix} 0 \\ 1 \end{pmatrix}$, the coherent superposition of 2 states, of amplitudes c and d, polarised in opposite directions along the spin quantisation axis. The spinor components c and d are generally space- and time-dependent. We suppose that $\langle\varphi|\varphi\rangle = cc^* + dd^* = 1$. The polarisation vector **P** is defined as $\mathbf{P} = \langle\varphi|\vec{\sigma}|\varphi\rangle$, where $\vec{\sigma}$ stands for the set of Pauli matrices σ_x, σ_y and σ_z. The components of **P** are:

$$P_x = (c^* \ d^*)\,\sigma_x \begin{pmatrix} c \\ d \end{pmatrix} = (c^* \ d^*)\begin{pmatrix} 0 & 1 \\ 1 & 0 \end{pmatrix}\begin{pmatrix} c \\ d \end{pmatrix} = c^*d + cd^* \qquad (1)$$

and similarly, $P_y = i\,(cd^* - c^*d)$ and $P_z = cc^* - dd^*$, since $\sigma_y = \begin{pmatrix} 0 & -i \\ i & 0 \end{pmatrix}$ and $\sigma_z = \begin{pmatrix} 1 & 0 \\ 0 & -1 \end{pmatrix}$

Thus, unlike P_z , the polarisation components P_x and P_y depend on the phase difference between the spinor components c and d.

The magnetic moment of the neutron has magnitude μ_n = 1.913 nuclear magneton or $0.996.10^{-26}$ A.m^2 and is antiparallel to the spin angular momentum.

In a region of vacuum with a uniform magnetic field **B**, the magnetic potential energy is thus represented by the matrix operator $-\mu_n \vec{\sigma}\,.\mathbf{B} = \begin{pmatrix} -\mu_n B & 0 \\ 0 & \mu_n B \end{pmatrix}$, if **B** is along the spin quantisation axis z. Consequently, different indices of refraction $n = 1 \pm (\mu_n B\, \lambda^2 m\,/h^2)$ are associated to the spinor components c and d respectively; this induces between them a phase difference which is linear in time (or, equivalently, in the distance travelled by the neutrons), hence a rotation around **B** of the component of the neutron polarisation perpendicular to this magnetic field. The temporal frequency of this Larmor precession is $2\mu_n B/h$.

2.2. Magnetic Scattering by a Single Ion with Unpaired Electrons

The spin and orbital motion of unpaired electrons in an atom or ion give rise to a surrounding magnetic field **B**(**r**) which acts on the neutron via the potential energy $-\vec{\mu}_n.\mathbf{B}(\mathbf{r})$, where $\vec{\mu}_n$ is the neutron magnetic moment. The magnetic scattering length p, proportional to the Fourier transform of $-\vec{\mu}_n.\mathbf{B}(\mathbf{r})$, is not isotropic, unlike the nuclear scattering length.

The relation div **B**(**r**) = 0 shows that **B**(**s**), the Fourier transform of **B**(**r**), is perpendicular to the reciprocal-space vector **s**. If **B**(**r**) is due to a point-like magnetic moment μ at position **r** = 0, we get $\mathbf{B}(\mathbf{r}) = \frac{\mu_o}{4\pi}\,\mathbf{curl}\,\frac{\vec{\mu} \wedge \mathbf{r}}{r^3}$, hence

$$\mathbf{B}(\mathbf{s}) = \mu_o\, \mathbf{s} \wedge \frac{\vec{\mu} \wedge \mathbf{s}}{s^2} = \mu_o\, \vec{\mu}_\perp(\mathbf{s}) \qquad (2)$$

where $\mu_o = 4\pi\ 10^{-7}$ Hm^{-1} is the permittivity of vacuum, $\wedge$ denotes the cross product, and $\vec{\mu}_\perp(\mathbf{s})$ is the projection of $\vec{\mu}$ on the planes perpendicular to **s** (reflecting planes).

This result is generalized by volume integration to a spatially extended magnetization distribution, the atomic shell of the unpaired electrons. It is thus shown that the magnetic scattering length p is proportional to $\vec{\mu}_n.\vec{\mu}_{i\perp}$, where $\vec{\mu}_{i\perp}$ is the projection of the magnetic moment of the considered ion on the reflecting planes.

For a complete description of magnetic scattering, which involves the spin polarization properties of the scattered beam, it is necessary to represent the neutron wave-function in the form of a two-component spinor and the neutron magnetic moment as the spin operator $-\mu_n \vec{\sigma}$. The magnetic scattering length is therefore itself a (2x2) matrix:

$$(\mathbf{p}) = \frac{2\pi m}{h^2} \mu_n \vec{\sigma}.\mathbf{B}(\mathbf{s}) = \mu_o \frac{2\pi m}{h^2} \mu_n \vec{\sigma}.\vec{\mu}_{i\perp}(\mathbf{s}) f_i(\sin\theta/\lambda) \tag{3}$$

where $f_i(\sin\theta/\lambda)$ is the dimensionless magnetic form factor of the ion and tends towards a maximum value of 1 when the scattering angle 2θ tends towards 0 (forward scattering). $p_1 = \mu_o \frac{2\pi m}{h^2} \mu_n \mu_i$ is equal to 2.70 femtometers for $\mu_i = 1$ Bohr magneton.

According to (2) or (3), there is no magnetic scattering for **s** along the ion's magnetic moment $\vec{\mu}_i$. Magnetic scattering effects are maximum when **s** and $\vec{\mu}_i$ are perpendicular.

The matrix (**p**) is diagonal if the direction of $\vec{\mu}_{i\perp}(\mathbf{s})$ is chosen as the spin quantization axis. Therefore, there is no spin-flip scattering if the incident polarisation is parallel or antiparallel to $\vec{\mu}_{i\perp}(\mathbf{s})$.

It is more usual to choose the spin quantisation axis (Oz) along μ_i. Let β be the angle between the vectors $\vec{\mu}_i$ and **s**; the (xyz)-components of $\vec{\mu}_{i\perp}(\mathbf{s})$ are then $(-\mu_i \sin\beta \cos\beta, 0, \mu_i \sin^2\beta)$, if the y-axis is chosen along $\vec{\mu}_i \wedge \mathbf{s}$. The total scattering length, the sum of the nuclear and the magnetic scattering lengths, is then represented by the matrix

$$(\mathbf{q}) = \begin{pmatrix} b + p \sin^2\beta & -p \sin\beta \cos\beta \\ -p \sin\beta \cos\beta & b - p \sin^2\beta \end{pmatrix} \tag{4}$$

where $p = \mu_o \frac{2\pi.m}{h^2} \mu_n \mu_i f_i(\frac{\sin\theta}{\lambda}) = p_1 \mu_i f_i(\frac{\sin\theta}{\lambda})$, with μ_i in Bohr magnetons.

The diagonal and nondiagonal elements of matrix (**q**) are respectively the non-spin-flip and spin-flip scattering lengths. The measurable quantities are the cross sections. For neutrons polarised parallel or antiparallel to the ion's magnetic moment they are:

$$(d\sigma/d\Omega)_\pm = b^2 \pm 2b\,p \sin^2\beta + (p \sin\beta)^2 \tag{5}$$

i.e. the sum of the non-spin-flip and spin-flip cross-sections $(b \pm p \sin^2\beta)^2$ and $(p \sin\beta \cos\beta)^2$ respectively. In the case of unpolarised neutrons, the cross-section is:

$$(d\sigma/d\Omega) = b^2 + (p \sin\beta)^2 \tag{6}$$

3. DIFFRACTION BY PERFECT MAGNETIC CRYSTALS

We first deal with perfect ferro- or colinear ferrimagnetic crystals. The most direct way to develop the dynamical theory in the 2-beam case, which involves a single Bragg diffracted beam, is to consider spinor wave-functions of the form

$$\Psi(\mathbf{r}) = \exp(i\mathbf{K_o}.\mathbf{r}) \begin{pmatrix} D_o \\ E_o \end{pmatrix} + \exp[i(\mathbf{K_o}+\mathbf{h})\mathbf{r}] \begin{pmatrix} D_h \\ E_h \end{pmatrix} \tag{7}$$

as approximate solutions of the wave-equation inside the crystal,

$$\Delta\Psi(\mathbf{r}) + k^2\Psi(\mathbf{r}) = \{u(\mathbf{r}) - \vec{\sigma}.\mathbf{Q}(\mathbf{r})\}\Psi(\mathbf{r}) \tag{8}$$

where $u(\mathbf{r})$ and $-\vec{\sigma}.\mathbf{Q}(\mathbf{r})$ are respectively the nuclear and magnetic potential energies multiplied by $\frac{2m}{\hbar^2}$. In the calculation of $\varphi(\mathbf{r})$ in the 2-beam case, we need only 3 terms in the expansions of the functions $u(\mathbf{r})$ and $\mathbf{Q}(\mathbf{r})$ into Fourier series:

$$u(\mathbf{r}) = u_o + u_h \exp(i\mathbf{h}.\mathbf{r}) + u_{-h} \exp(-i\mathbf{h}.\mathbf{r}) + \ldots\ldots\ldots$$

$$\mathbf{Q}(\mathbf{r}) = \mathbf{Q_o} + \mathbf{Q_h} \exp(i\mathbf{h}.\mathbf{r}) + \mathbf{Q_{-h}} \exp(-i\mathbf{h}.\mathbf{r}) + \ldots\ldots\ldots$$

We suppose that the crystal is magnetically saturated by an applied magnetic field $\mathbf{H_a}$. $\mathbf{Q_o}$ is equal to $\mu_n \frac{2m}{\hbar^2}\mathbf{B}$, where $\mathbf{B}$ is the macroscopic mean magnetic field $\mathbf{B} = \mu_o(\mathbf{M} + \mathbf{H_a} + \mathbf{H_d})$, with $\mathbf{M}$ the magnetisation and $\mathbf{H_d}$ the demagnetising field. Then $\mathbf{Q_h}$ and $\mathbf{Q_{-h}}$ are proportional to the projection of $\mathbf{M}$ on the reflecting planes.

The 4 coefficients D_o, D_h, E_o and E_h of (7) are solutions of a system of 4 homogeneous linear equations. The condition that the associated determinant vanish defines a dispersion surface of order 4, with 4 branches. An incident plane wave thus excites 4 wave-fields of the form of (7), generally polarised in various directions. An example of such a dispersion surface is shown on Figure 1 b. This is much more complicated than for non-magnetic crystals, where the scalar wave-functions involve 2 coefficients D_o and D_h and are related to hyperbolic dispersion surfaces of order 2.

In fact, all neutron experiments related to dynamical effects in diffraction by magnetic crystals have been performed in such conditions that the magnetisation is perpendicular to the diffraction vector $\mathbf{h}$. The vectors $\mathbf{Q_h}$ and $\mathbf{Q_{-h}}$ are then parallel or antiparallel to $\mathbf{Q_o}$ which is chosen as the spin quantisation axis. The matrices $\vec{\sigma}.\mathbf{Q_o}$, $\vec{\sigma}.\mathbf{Q_h}$ and $\vec{\sigma}.\mathbf{Q_{-h}}$ are all diagonal and we get for the two spin states ($\pm$) separated dynamical equations similar to those for the scalar case, but with different structure factors :

$$F_+ = F_N + F_M \quad \text{and} \quad F_- = F_N - F_M \tag{9}$$

the nuclear and magnetic structure factors F_N and F_M being related to the scattering lengths of the ions in the unit cell of volume V_c :

$$F_N = V_c\, u_{\mathbf{h}} = \Sigma_i\, b_i \exp(-i\mathbf{h}.\mathbf{r}_i)$$

$$F_M = V_c\, |Q_{\mathbf{h}}| = -\frac{\mu_o m}{2h^2}\, \mu_n\, \vec{\sigma}\, \Sigma_i\, \mu_{i\perp}(\mathbf{h})\, f_i(\frac{\sin\theta}{\lambda}) \exp(-i\mathbf{h}.\mathbf{r}_i)$$

The dispersion surface of order 4 degenerates into 2 hyperbolic surfaces, each of them corresponding to one of the polarisation states (±) (Figure 1 a). The asymptotes are different because so are the refractive indices for neutron polarisation parallel or antiparallel to $\mathbf{Q_o}$. The dynamical theory is then again similar to the X-ray case.

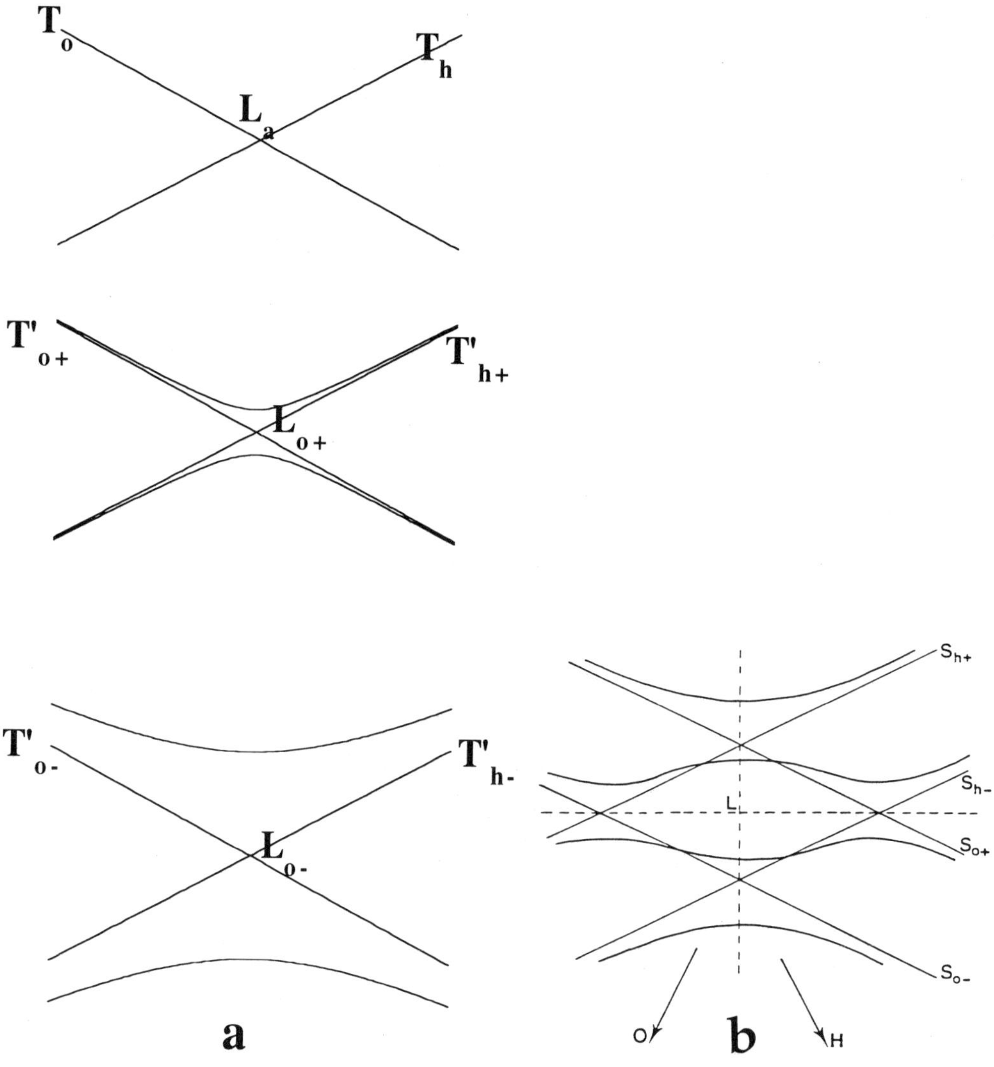

Figure 1. Schematic plot of the 2-beam dispersion surfaces in the case a) of a simple situation: ferrimagnetic yttrium iron garnet (YIG) with B = 0.18 T perpendicular to scattering vector 440 (a mixed reflection), λ = 1 Å; b) of a purely magnetic reflection such that $Q_{\mathbf{h}} = Q_{-\mathbf{h}} = Q_{\mathbf{o}}$, with angle $\frac{\pi}{4}$ between $\mathbf{Q}_o$ and $\mathbf{Q_h}$.

If the scattering vector **h** is along the magnetisation, the reflection is purely nuclear (no magnetic contribution), since $F_M = 0$. Pure magnetic reflections (without nuclear contribution) also exist if the magnetic structure involves several sublattices.

In the case of perfect colinear antiferromagnetic crystals, there is no average magnetisation ($\mathbf{Q_o}= 0$). It is then convenient to choose the quantisation axis in the direction of $\mathbf{Q_h}$ and $\mathbf{Q_{-h}}$. The dispersion surface degenerates into two hyperbolic surfaces corresponding to each polarisation state along this direction, for any orientation of **h** relative to the direction of the magnetic moments of the sublattices. These two hyperbolic surfaces have the same asymptotes. In the case of a purely magnetic reflection, they are identical.

In polarised neutron diffraction by a magnetically saturated magnetic sample, it is usual to measure the ratio $R = \frac{I_+}{I_-}$ of the intensities I_+ and I_- reflected when the incident beam is polarised parallel or antiparallel to the magnetisation. R is an experimentally well-defined quantity, independent of incident beam intensity, temperature factor or absorption.

An ideally imperfect crystal has the wavelength-independent flipping ratio

$$R_{\mathrm{kin}}\,(\mathbf{h}) = (I_+/I_-)_{\mathrm{kin}} = \left\{ \frac{|F_N + F_M|}{|F_N - F_M|} \right\}^2 \tag{10}$$

In the case of an ideally perfect very thick crystal, the limiting value is simply

$$R_{\mathrm{dyn}}\,(\mathbf{h}) = (I_+/I_-)_{\mathrm{dyn}} = \frac{|F_N + F_M|}{|F_N - F_M|} \tag{11}$$

In general, R_{dyn} depends on wavelength and on the crystal thickness; these dependences disappear, in (11), if the path length in the crystal is much larger than the extinction distances for both polarisation states. The determination of R_{kin} or R_{dyn} allows the determination of the ratio $\frac{F_M}{F_N}$, hence of F_M if F_N is known. In fact, real crystals are neither ideally imperfect nor ideally perfect, and one usually introduces an extinction factor y, the ratio of the real reflectivity over that of the ideally imperfect crystal. Different extinction coefficients y_+ and y_- are expected for the two polarisation states, which complicates the determination of $\frac{F_M}{F_N}$. The λ-dependence of R is a good test for extinction.

4. EFFECT OF EXTERNAL FIELDS ON NEUTRON SCATTERING BY PERFECT CRYSTALS

Because neutrons have mass and a magnetic moment, they can be affected by external fields, such as gravity and magnetic fields, both during their propagation in air or vacuum and while being diffracted in crystals. Experiments completely different from the X-ray case can thus be performed with perfect crystals and with neutron interferometers.

The theory was given by Werner (1980), using the approaches (migration of tie-points, and Takagi-Taupin equations) that are customary in the treatment of imperfect crystals. Zeilinger, Shull, Horne and Finkelstein (1986) pointed out that the effective-mass concept, familiar in describing electrons in solid-state physics, can shed new light on this behavior: because of the curvature of the dispersion surface at near-exact Bragg setting, effective masses 5 orders of magnitude smaller than the rest mass of the neutron in vacuum can be obtained. Elegant experiments were performed using the forces due to magnetic fields, rotation (Coriolis force) and gravity.

5. APPLICATIONS OF DYNAMICAL NEUTRON SCATTERING

The applications of dynamical neutron scattering include four areas: neutron optics, the measurement of scattering lengths by Pendellösung effects, neutron interferometry, and imaging (topography). The two last items are covered in separate chapters of this book.

Most experiments in standard neutron scattering require an intensity-effective use of the available beams, at the cost of relatively high divergence and wavelength spread. The monochromators must then be imperfect ("mosaic") crystals. In some cases, however, it is important to have a small divergence and wavelength band. One example is the search for small variations in neutron energy in inelastic scattering, without neutron spin-echo. Perfect crystals must then be used as monochromators or analysers, and dynamical diffraction is directly involved. As in the X-ray case, special designs can lead to strong decrease in the intensity of harmonics, i.e. of contributions of $\frac{\lambda}{2}$ or $\frac{\lambda}{3}$ (Hart and Rodrigues, 1978). The possibility of focusing neutron beams by the use of perfect crystals with the incident beam spatially modulated in amplitude through an absorber, or in phase through an appropriate patterning of the surface, in analogy with the Bragg-Fresnel lenses developed for X-rays, was suggested by Indenbom (1979).

The use of two identical perfect crystals in non-dispersive (+, -, //) setting provides the way of measuring the very narrow intrinsic rocking-curves expected from dynamical theory. Any divergence added between the two crystals can sensitively be measured. Thus perfect crystals provide interesting possibilities for measuring very-small-angle neutron scattering. Imaging applications are described in another chapter of this book. Curved almost-perfect crystals, or crystals with a gradient in the lattice spacing, can provide focusing [Albertini et al. (1976)], and vibrating crystals give the possibility of tailoring the reflectivity of crystals, as well as of modulating beams in time (Michalec et al., 1988). A double-crystal arrangement with bent crystals was shown by Eichhorn (1988) to be a flexible small-angle neutron scattering device.

As with X-rays, Pendellösung oscillations provide an accurate way of measuring structure factors, hence coherent neutron scattering lengths. Three kinds of measurements were made. Sippel et al. (1965) measured the integrated reflectivity from a perfect crystal of silicon the thickness of which they varied by polishing after each measurement. Shull (1968) restricted the measurement to wave-fields that propagated along the reflecting planes, hence at exact Bragg incidence, by setting fine slits on the entrance and exit faces of 3 to 10 mm thick silicon crystals, and measured the oscillation in diffracted intensity as he varied the wavelength of the neutrons used by rotating the crystal. Shull and Oberteuffer (1972) showed that a better interpretation of the data corresponds, when the beam is restricted to a fine slit, to the spherical wave approach (actually cylindrical wave), and the boundary conditions were discussed more generally by Arthur and Horne (1985). The inclination method, in which the integrated reflectivity is measured as the effective crystal thickness is varied non-destructively, by rotating the crystal around the diffraction vector, is discussed by Belova et al. (1983).

REFERENCES

Albertini, G., Boeuf, A., Lagomarsino, S., Mazkedian, S., Melone, S., Rustichelli, F., 1976, Neutron properties of curved monochromators. *Proc. Conf. on Neutron Scattering, Gatlinburg, TN,* ORNL, USERDA CONF 760601-P2: 1151-1158

Alexandrov, Yu.A., Chalupa, B., Eichhorn, F., Kulda, J., Lukáš, P., Machekhina, T.A., Michalec, R., Mikula, P., Sedláková, L.,N., and Vrána, M., 1988, Neutron optical experiments at the IBR-2 pulsed reactor. *Physica B,* 151: 108-112

Arthur, J., Horne, M.A., 1985, Boundary conditions in dynamical neutron diffraction. *Phys. Rev. B,* **32**: 5747-5752

Belova, N.E., Eichhorn, F., Somenkov, V.A., Utemisov, K., Shil'shtein, S.Sh., 1983, Analyse der Neigungsmethode zur Untersuchung von Pendellösungsinterferenzen von Neutronen und Röntgenstrahlen. *Phys. Stat. Sol. (a),* **76**: 257-265

Eichhorn, F., 1988, Perfect crystal neutron optics. *Physica B,* **151**: 140-146

Hart, M., Rodrigues, A.R.D., 1978, Harmonic-free single-crystal monochromators for neutrons and X-rays. *J. Appl. Cryst,* **11**: 248-253

Indenbom, V.L., 1979, Diffraction focusing of neutrons. *JETP Lett.,* **29**: 5-8

International Tables for Crystallography. Kluwer Academic Publishers. Dordrecht

Michalec, R., Mikula, P., Vrána, M., Kulda, J., Chalupa, B., Sedláková, L., 1988, Neutron diffraction by perfect crystals excited into mechanical resonance vibrations. *Physica B,* **151**: 113-121

Rauch, H., Petrascheck, D., 1978, Dynamical neutron diffraction and its application. in *Neutron Diffraction,* H. Dachs, ed., vol 6 of Topics in Current Physics, pp. 305-351; Springer, Berlin

Schuster, M., Jericha, E., Carlile, C.J., and Rauch, H., 1992, A cold neutron storage device and a neutron resonator. *Physica B* **180-181**: 997-999

Sears, V.F., 1978, Dynamical theory of neutron diffraction. *Canad. J. of Phys. / Journal canadien de physique,* **56**: 1261-1288

Shull, C.G., 1968, Observation of Pendellösung fringe structure in neutron diffraction. *Phys. Rev. Lett.,* **21**: 1585-1589

Shull, C.G., Oberteuffer, J.A. , 1972, Spherical wave neutron propagation and Pendellösung fringe structure in silicon. *Phys. Rev. Lett.,* **29**: 871-874.

Sippel, D., Kleinstück, K., Schulze, G.E.R. , 1965, Pendellösungs-Interferenzen mit thermischen Neutronen an Si-Einkristallen, *Phys. Letters,* **14**: 174-175

Squires, G.L., 1978, *Introduction to the Theory of Thermal Neutron Scattering. Cambridge University Press*

Stassis, C., Oberteuffer, J.A., 1974, Neutron diffraction by perfect crystals. *Phys. Rev. B,* **10**: 5192-5202

Werner, S.A., 1980, Gravitational and magnetic field effects on the dynamical diffraction of neutrons. *Phys. Rev. B,* **21**: 1774-1789

Zeilinger, A., Shull, C.G., Horne, M.A., Finkelstein, K.D., 1986, Effective mass of neutrons diffracting in crystals. *Phys. Rev. Lett.,* **57**: 3089-3092

X-RAY OPTICAL BEAMLINE DESIGN PRINCIPLES

Michael Hart

National Synchrotron Light Source
Brookhaven National Laboratory
Upton
Long Island
New York 11973

1. INTRODUCTION

Beamline optics, as in the familiar visible part of the electromagnetic spectrum, have the dual functions of conditioning the beam quality and delivering photons with the necessary divergence and spatial extent at the sample from the source. In practice, the source characteristics are usually given and the specimen requirements are a variable from one experiment to another. Often, geometrical beamline optimization is simply not possible for a range of sample geometries so that specialized designs must evolve. Apart from the obvious parameter, *flux*, photon beams have a number of electromagnetic attributes such as energy (or wavelength), energy bandwidth (or monochromaticity), polarization state and time structure. All must be controlled to a degree determined by the experimental needs in the beamline.

Other Chapters in this volume outline the dynamical theory of X-ray diffraction and also describe the main properties of synchrotron radiation sources.

2. WHITE X-RAY BEAMLINE

The concept of a beamline for white radiation is apparently so simple that there are few detailed designs in existence! Even today, many beamlines, worldwide, "get by" with bits of lead tacked here and there near and around the specimen region and especially between the specimen and the detector. In the beamline permanent slits are often of the four-jaw type edged with heavy metal absorbers. Beam defining apertures very close to the front end of the beamline are extremely specialized devices whose design and construction is dominated by considerations of the thermal flux and are therefore not of primary interest in this X-ray physics course. To some extent the lessons which were learnt about the X-ray

X-ray and Neutron Dynamical Diffraction: Theory and Applications
Edited by Authier *et al.*, Plenum Press, New York, 1996

physics of slit design for laboratory apparatus, published in classic works (Cullity, 1978, International Tables for X-ray Analysis, 1968), have been forgotten.

2.1. Black Velvet For X-rays?

Black paint and black velvet are familiar concepts in the visible part of the electromagnetic spectrum; optical instruments cannot work without them! Some means of absorbing unwanted "stray" radiation is essential to limit background radiation and to preserve signal-to-noise in optical instruments. In the infrared, where *any* body at a finite temperature is a source of radiation even when the design calls for the object to be an absorber, the problem of absorbing unwanted radiation is well known and understood. Thus, a hot slit is not a beam limiting device but is a new source of radiation which, potentially, corrupts the purpose of the infrared optical system. Although the physics is quite different a similar situation pertains too in the X-ray part of the spectrum.

A beam of white X-rays impinging on a solid material is absorbed by two main processes; photoelectric absorption and inelastic scattering. In principle the attenuation ratio required in synchrotron radiation experiments may be very high indeed. For example, an undulator source or a focused wiggler beam might deliver about 10^{13} photons per second per square millimeter of beam at the sample. A germanium solid state detector can have a background counting rate for monochromatic radiation of less than 10^{-3} photon per second. "Black velvet" as a sample would need to attenuate the beam by a factor of 10^{16} ! Let us use the example of tungsten, a widely used absorber to illustrate the challenges of slit and beam stop design.

Table 1. The attenuation length for tungsten (Hubbell, 1969, Henke et al. 1993) at several selected X-ray energies.

Energy / keV	1	10	10.4	20	30	69	70	100	1000
Length / μ m	0.13	5.6	2.4	8.8	25	234	51	131	9015

The sudden jump in absorption between 10 keV and 10.4 keV is due to the L_{III} absorption edge of the tungsten at 10.2 keV. Similarly, the jump between 69 keV and 70 keV is due to the K absorption edge of the tungsten at 69.5 keV. As laboratory experience tells us, at 10 keV just 0.1 mm of tungsten is totally absorbing; $\exp(-\mu\ t) = 10^{--18}$. However, in a microbeam fluorescence experiment or in experiments on powder diffraction in diamond anvil high pressure cells a 10 μm pinhole might be required to define the X-ray beam. Since it is difficult to make small pinholes with very high aspect ratios it may not be possible to make the pinhole in foil thicker than say 25 μm. The attenuation is then only by a factor of 2.9×10^3 - if the primary beam is 1 mm in diameter and the tungsten pinhole is used to reduce it to a 10 μm spot then three times as much background would pass through the 25 μm foil as useful intensity would pass through the pinhole! A signal to background ratio of less than 1 would result. A solution in this case would be to raise the beam energy to 10.4 keV so as to increase the absorption coefficient of the pinhole material (Nelmes and McMahon, 1995). The foil attenuation coefficient would be increased to 2.6×10^{10}

Whereas the maximum energy of photons from a laboratory source is typically 50 keV the situation at a storage ring is quite different with some flux at all energies up to the particle beam energy which may be 1 GeV at a small storage ring or up to 8 GeV at SPring-8. In practice, depending on the source energy, the direct beam can only be stopped with substantial amounts of material, for example 100 mm of tungsten which has an attenuation of 10^{11} at 1 Mev photon energy. With such thick "slits" and beamstops it is feasible to work in the forward beam direction even on high field wiggler beamlines at high energy sources such as the European Synchrotron Radiation Facility which operates at 6 GeV.

It was long ago realized that scattering and fluorescence from slit materials must be controlled. Tungsten, our example of a typical slit material, is usually in the form of polycrystalline metal. It will fluoresce at all of tungsten's characteristic energies [59.32, 57.98, 67.24, 8.40, 8.36, 9.67, 9.96, 11.29 keV for the $K\alpha_1$,$K\alpha_2$,$K\beta_1$,$L\alpha_1$, $L\alpha_2$, $L\beta_1$, $L\beta_2$, $L\gamma_1$ respectively] if the incident beam has sufficient energy. In addition a monochromatic X-ray beam will produce a complete powder diffraction pattern from the tungsten - a set of diffracted beams lying on concentric cones (the Debye-Scherrer pattern). Perhaps the most troublesome of these cones is the one with the narrowest angle in the forward direction since it is both the strongest and the closest to the defined beam. The Bragg angle is of course energy dependent and easily calculated from Bragg's Law. Each monochromatic component of a white X-ray beam produces it's own Debye-Scherrer pattern so that the space is filled with scattered radiation. In any particular direction a detector would record a complete energy-dispersed polycrystalline diffraction pattern. Finally, close to the primary beam from this slit the tungsten would produce a small angle diffraction pattern and, at angles smaller than the critical angle, very intense specular reflection.

2.2. Clean-up After The First Slit

At first sight the appearance of so many sources of background seems daunting. However, the low divergence of synchrotron radiation sources and the consequence that large distances between optical components carry little penalty other than cost, conspire to provide increased signal to noise with distance. The collimated beam has a low divergence while most of the background which is produced results in additions to the intrinsic beam divergence.

Fluorescent radiation is divergent over 4π steradians. Consider a 1 mm diameter pinhole at 10m from the source. After travelling a further metre the collimated beam is just 1.1 mm in diameter but the fluorescent background is spread over 1.3×10^7 mm^2. If we assume that the fluorescence yield (efficiency) is 10% then the effective attenuation factor of background to collimated beam in the same area is 1.6×10^8.

Bragg reflections have an integrated reflectivity of typically 10^{-5} (Authier, this Volume). However, this is not the relevant attenuation parameter from the present point of view because Bragg reflections occur at discrete angles. In practice, at the energies of interest, the Bragg angles for tungsten would be at least several degrees, providing tens of millimetres of displacement after a flight path of 1 m and so missing the next optical component or sample completely. Thus, in practice, Bragg reflections from slit and beam-stop materials are not a problem.

Small angle and especially specular reflection is troublesome when beams with sharp edges must be defined. Near the critical angle, perhaps 0.2 degrees for tungsten at the photon energy required for diffraction experiments, the reflectivity is almost unity (Fewster, this Volume) and the beam deviation small. When the slit is followed by a perfect crystal monochromator or mirror, which is often the case, then even the small deviation of the scattered beam is usually enough for the scattered radiation to be rejected by the optical component.

3. UNFOCUSSED MONOCHROMATIC BEAMLINE

The earliest beamlines constructed for spectroscopy used double crystal monochromators with no provision for focusing. For the purposes of this section we will only need to use the characteristics of synchrotron radiation produced by simple dipole magnets; the detailed differences between that and the radiation fields from more sophisticated wigglers and undulators do not influence the design motivation and principles.

3.1. Synchrotron Source Characteristics

The universal spectrum of synchrotron radiation is characterized by the photon energy at the centroid of the power distribution E_C. At higher energies the intensity falls rapidly, to about 10^{-3} of maximum at a photon energy of 10 times E_C while at lower energies the intensity is almost constant, decreasing by only one decade over an energy range of photon energies down to E_C / 1000. In practical units the following results are useful (Kirz et al., 1986, Murphy, 1993).

$$E_C\ [\text{keV}] = 0.665\ E^2\ [\text{GeV}]\ B\ [\text{T}]$$

$$d^2N/d\theta d\phi = 1.327 \times 10^{13}\ E^2\ [\text{GeV}]\ I\ [\text{A}]\ H_2(y)\ \text{photons s}^{-1}\ \text{mr}^{-2}\ (0.1\%\ \text{bandwidth})^{-1}$$

$$dN/d\theta = 2.457 \times 10^{13}\ E\ [\text{GeV}]\ I\ [\text{A}]\ G_1(y)\ \text{photons s}^{-1}\ \text{mr}^{-1}\ (0.1\%\ \text{bandwidth})^{-1}$$

$H_2(y)$ and $G_1(y)$ are universal spectral functions with a value unity at the characteristic photon energy E_C and N is the number of photons per second. The total thermal power radiated is given by;

$$P[\text{kW}] = 0.633\ E^2\ [\text{GeV}]\ B^2\ [\text{T}]\ I\ [\text{A}]\ L\ [\text{m}]$$

where the length of the magnet is L. At the characteristic photon energy in the vertical plane the beam divergence is $1/\gamma$ or, in familiar units, m_0c^2/E. For 2.5 GeV electrons this is just 2 x 10^{-4} radians or 100 seconds of arc. At higher energies the vertical emittance decreases, approximately with the square root of the photon energy. At lower photon energies the limiting emittance function increases roughly with the cube root of energy.

For dipole magnets with 1.2 Tesla fields we find that the photon characteristic energy is 5.0 keV (2.48Å) and that the total power radiated is 1.71 kW per metre. Integrated over the vertical emittance the horizontal intensity density $dN/d\theta$ is about 1.84 x 10^{13} photons s^{-1} mr^{-1} (0.1% bandwidth)$^{-1}$ if the machine current is 300mA.

The above formulae give convenient scaling laws for other storage ring conditions and further summaries for wigglers and undulators can be found in standard reference hand books.

3.2. Perfect Crystal Dynamical Diffraction

For the purposes of preliminary design discussions we look first at the reflectivity of perfect crystals such as silicon which are commonly used as monochromators. The Darwin width for total reflection in the Bragg case is given by

$$\Delta\omega = 2\,C\,/\,\sin 2\theta|\chi_h|\ \ (|\gamma_h|\,/\,\gamma_0)^{1/2}\ =\ 2\,C\,/\,\sin 2\theta|\chi_h|\ \gamma^{-1/2}$$

where the symbols have their usual meaning (see, for example, the Chapter by Authier, this Volume). At the characteristic energy calculated above (5.0 keV) we calculate the Darwin widths for several common monochromator reflections in Table 2 below.

Table 2. Darwin widths for selected crystals at 5.0 keV, 2.48 Å in symmetric reflection.

Crystal / hkl	Ge 111	Ge 220	Si 111	Si 220	C 111	C 220
$\Delta\omega$ Darwin	133×10^{-6}	115×10^{-6}	56×10^{-6}	47×10^{-6}	44×10^{-6}	37×10^{-6}

Note that the Darwin widths are similar to the angular range of vertical emittance from the synchrotron radiation source (200×10^{-6}). At lower energies the match is closer but at higher energies, such as those preferred for structure analysis and often required for spectroscopy, the match becomes poorer.

According to Bragg's Law the energy resolution corresponding to the Darwin width is given by;

$$\Delta E/E = \Delta\omega\ \cot\theta = A\ d^2\ F_{hkl} \text{ where A is a constant.}$$

In normal circumstances, where the synchrotron radiation beam has a significant divergence we can estimate the energy resolution by combining the effects of Darwin resolution and divergence in quadrature as;

$$\Delta E/E = (\Delta\omega^2 + \Delta\theta^2)^{1/2}\ \cot\theta > A\ d^2\ F_{hkl}$$

3.3. Typical beamline optical layout

Early beamlines essentially depend on a slit to determine the beam divergence and energy resolution and a channel cut or non-dispersive double crystal spectrometer such as that illustrated in Figure 1. Such arrangements have the advantage of simplicity and stability and for those reasons are still used even though not truly optimized.

With a single Bragg reflection monochromator, such as might be used in a laboratory set-up the monochromatic beam would be deviated through the Bragg angle. While that may be acceptable in some situations, for example on fixed wavelength experimental stations used in diffraction and scattering experiments, it is not useable in spectroscopy because large beam motions occur at the sample position. By using the now standard non dispersive, or parallel, double crystal setting the monochromatic beam is returned to the forward direction and is therefore fixed in direction and essentially stationary on the sample. For efficiency the two parts of the monochromator must be stable and oriented one with respect to the other within a small fraction of the Darwin width; in practice to within one micro-radian or so! While such precision is achievable, though costly, channel cut crystals are extremely simple, stable and cost effective (Bonse and Hart, 1965, Cernik and Hart, 1989). Compared with optical spectrometers these spectrometers have very wide operating ranges, for example 3 keV to 30 keV in a single instrument.

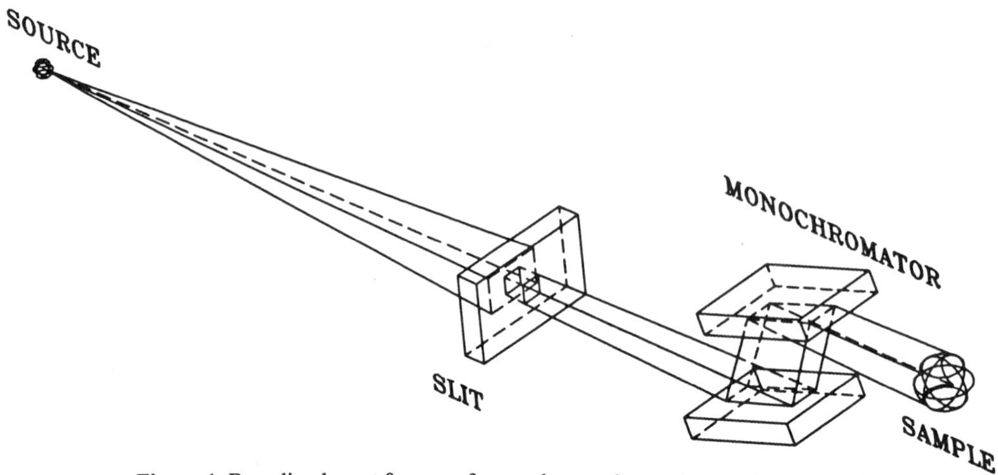

Figure 1. Beamline layout for an unfocussed monochromatic experimental station.

Almost all storage rings have source dimensions which are elliptical, typically ten times wider than they are high. To achieve maximum energy resolution and flux most beamlines diffract in the vertical plane. This also has the incidental advantage of minimizing the intensity loss which would accrue from the fact that the beams from dipole magnets are linearly polarized with the electric vector in the horizontal plane. With a vertical emittance of 2×10^{-4} radians the X-ray beam is about 2 mm high at the monochromator if it is 10 m from the source and if the emergent beam aperture is 10 milli-radians it will by then be 100 mm wide. These figures indicate that the natural geometry and scale is such that available experimental beams are not ideal since many samples are very much smaller than this and reducing the beam size with slits simply causes loss of intensity. As experience and the above calculations show, high resolution in energy, space or angle all benefit from or in some cases require fine beams.

4. HORIZONTAL FOCUSING MONOCHROMATIC BEAMLINE

Much of the loss in brightness in the synchrotron radiation beam comes from the natural horizontal divergence of the source. Moderately tunable beamlines have been made which simply use a conventional singly bent crystal monochromator focusing in the horizontal plane.

In typical situations, illustrated in Figure 2, 10-30 mm of horizontal beam can be refocused into about 1-2 mm giving a practical flux gain through a millimeter sized pinhole or sample of ten to twenty times. In laboratory applications the design usually assumes quasi-monochromatic characteristic radiation. For use with synchrotron radiation these designs simply require re-scaling to handle the much larger beam dimensions. Many crystal designs are described in detail in the International Tables for X-ray Crystallography (1968) and in the reference list published by Ice and Sparks (1993). However, optimization for white radiation at a storage ring source requires compromise between spatial and spectral parameters arising from the fact that the focusing Bragg reflecting crystal is also a dispersing element.

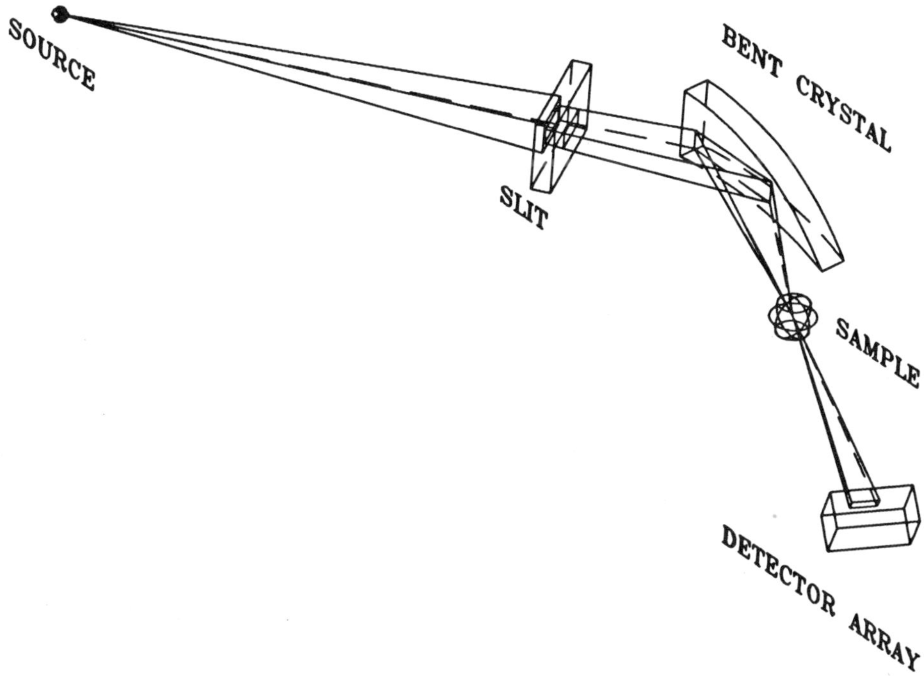

Figure 2. A single bent crystal focusing beamline providing flux gains of about ten or twenty times in typical samples. Note that changes in energy require large scale movement of the sample position and diffractometer.

Although simple this design has a number of convenient features when used as a general purpose monochromatic beamline. For example, the sample can be placed about 1 m before the focus and an area detector can be placed in the focal plane to produce a small angle scattering camera. Alternatively, the crystal can be overbent so that the focus becomes polychromatic. Then spectroscopy can be done on samples placed at the focal position with an area detector some way behind the focus.

5. MIRROR FOCUSSED MONOCHROMATIC BEAMLINE

Just as in the visible part of the electromagnetic spectrum mirrors can be used to focus beams. However, since the refractive index is so close to 1 total external reflection is only possible in the X-ray case with very small glancing angles of incidence, typically less than 0.01 radians. The available aperture is thus restricted.

5.1. Specular reflection at mirrors

The complex refractive index in the X-ray energy range can be written in terms of the X-ray scattering amplitudes as;

$$n = 1 - \alpha - i\beta$$

The critical angle of grazing incidence θ_c below which total external reflection occurs is given by ;

$$\cos^2 \theta_c = 2\,\alpha$$

$$\text{or,}\quad \theta_c = 0.00234\,\lambda\,(\,\rho\,Z\,/\,M\,)^{1/2}$$

Some values are given in Table 4 for several materials which are commonly used for X-ray mirrors.

Table 3. Critical angles at 50% reflectivity for several mirror materials at 10 keV.

Material	Silicon	Silica	Copper	Gold	Platinum
θ_c	0.1817°	0.1812°	0.3149°	0.4407°	0.4623°
θ_c /radians	3.17×10^{-3}	3.16×10^{-3}	5.50×10^{-3}	7.69×10^{-3}	8.07×10^{-3}

Table 4. Critical angle for silicon at various X-ray energies.

Silicon	10 keV	15 keV	20 keV	25 keV	30 keV
θ_c	0.1817°	0.1211°	0.0909°	0.0727°	0.0606°
θ_c /radians	3.17×10^{-3}	2.11×10^{-3}	1.59×10^{-3}	1.27×10^{-3}	1.06×10^{-3}

Since the maximum aperture of the mirror cannot exceed θ_c, there is a strong incentive to make mirrors with heavy elements so as to maximize the critical angle. Unfortunately most of the materials which can be worked to the necessary figure are composed of light elements or have a low density - for example, copper, silicon carbide, silica and various glasses. The solution is to coat the mirror with a heavy element since the skin depth for the evanescent mode which enables total external reflection is only 100 Å or so at most energies of interest. Unfortunately the L absorption edges of heavy elements such as platinum or gold are in the region of 10 keV, in the middle of the spectral band where many spectroscopic measurements are made. At high energies the critical angle becomes very small with the consequence that the precision of finish required becomes prohibitively expensive.

An essential feature of focusing systems is that the surface roughness must not be so large as to direct stray radiation outside of the focus. In a simple way we can say that the allowed roughness is less than a few percent of the penetration depth of the radiation field, that is about 1 Å! That this estimate is essentially correct is borne out by detailed calculations such as those described elsewhere in this Volume to determine the structure of thin layers by X-ray scattering (Fewster, Servidori, Patel, this Volume).

5.2. Toroidal focusing Silicon mirrors at NSLS

Figure 3 shows the calculated reflectivity of a silicon mirror at 10 keV, assuming a roughness of 5 Å rms. Notice that the critical angle is about 0.18 degrees and that below that the mirror is almost totally reflecting.

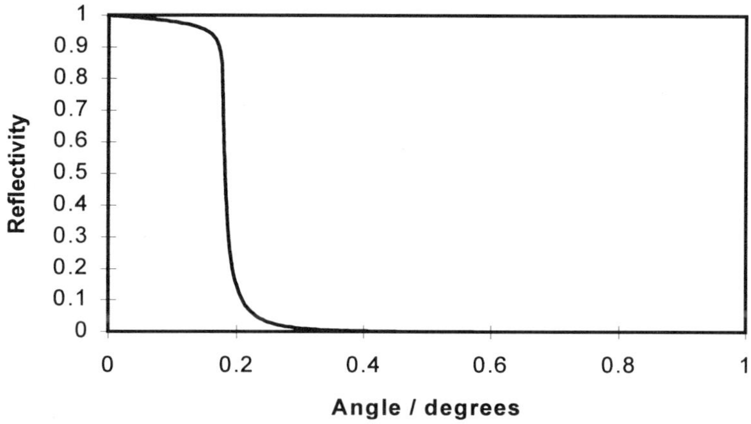

Figure 3. Angular reflectivity for silicon at 10 keV X-ray energy.

The corresponding result in Figure 4, showing reflectivity as a function of X-ray energy at an angle of incidence of 0.18 degrees, is more complicated. As expected, the critical energy below which total reflection might be expected is 10 keV but the reflectivity is low at 1.84 keV where the silicon K absorption edge occurs. In this spectral region the mirror absorption is high.

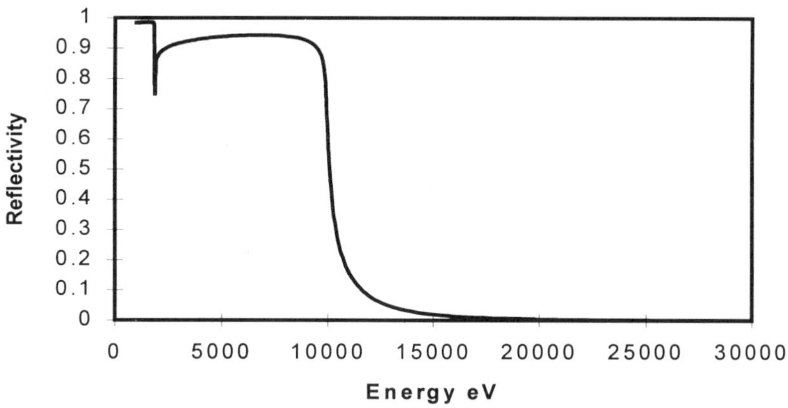

Figure 4. Reflectivity of a Silicon mirror at 0.18 degrees incidence vs X-ray energy.

To increase the critical angle and to thereby increase the angular aperture of the focusing mirror it can be coated with a heavy element, for example platinum as shown in Figures 5 and 6. Here the platinum thickness is 900 Å which is thick enough to exceed the penetration depth of X-rays, even at energies up to 30 keV and also sufficiently thick that the interference fringes (Kissing fringes) do not substantially influence the reflectivity variation with X-ray energy.

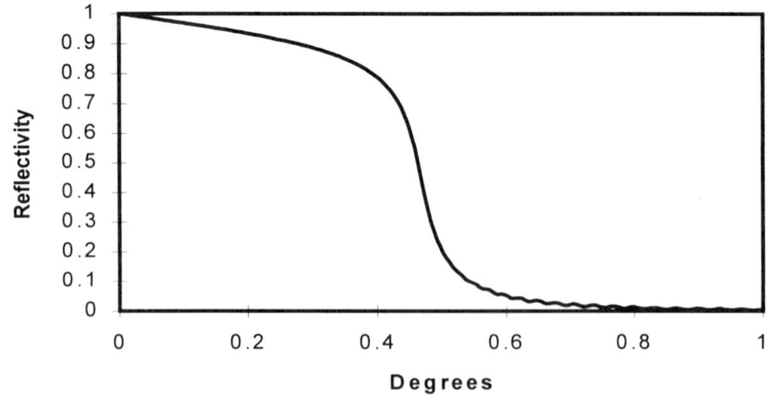

Figure 5. Reflectivity vs angle for a platinum coated silicon mirror.

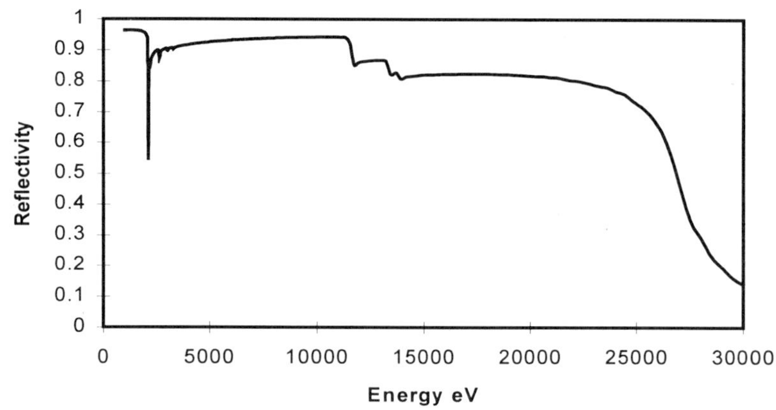

Figure 6. Reflectivity vs X-ray energy for the platinum coated silicon mirror at 0.18 degrees incidence.

At the original silicon critical angle of 0.18 degrees the mirror reflectivity is over 90% and at this same angle the mirror is now useful up to 26 keV. By comparison with Figure 4 the influence of the platinum L edges can be clearly seen; a reflectivity of above 80% in the 10 to 25 keV X-ray energy range where the reflectivity of the uncoated mirror was very low before.

6. SAGITTAL FOCUSING WITH A DOUBLE CRYSTAL MONOCHROMATOR

Well established principles of geometrical optics for visible light are applicable in the X-ray case; provided that the angle of incidence is below the critical angle for total external reflection. Otherwise little or no intensity is available. It follows that the angular aperture of any focusing system of mirrors cannot exceed the critical angle - 0.18 degrees (approximately 3 milli radians) for silicon at 10 keV (Figure 4) or for platinum at 27 keV (Figure 6). One way to obtain perfect reflection at higher angles is by Bragg reflection from perfect crystals and this is commonly done so as to achieve larger numerical aperture optical focusing. Unfortunately, only certain simple curvatures can be achieved by elastic bending so that it is generally not possible to make doubly bent Bragg reflecting mirrors but singly bent crystals for either sagittal or meridional focusing have been extensively used.

6.1. Historical development survey

The first publications by Sparks et al. (1980, 1982) laid down the principles of sagittal focusing and found the 3:1 focusing condition which minimized the aberrations. One of the first uses of these design principles by Batterman and Berman (1986) used an 11-rib triangle crystal was bent with a four-rod bender. The ribs were on a 3 mm pitch and effectively brought 11 Bragg reflections into a single 3.4 mm wide spot. The authors described the crystal as a polygon with 11 facets joined by weak links. These tests were done on a CHESS bending magnet beamline. Focal distances were 12 m from the source and 4 m to the sample.

In the next step Mills et al. (1986) realized that the reflecting facets could be rectangular even though the bending regions had to be triangular; and this was achieved by suitable design. In these experiments the crystal had 21 facets on 0.1 inch centers and focal distances of 11.3 m and 3.75 m. Again, tests were done on a bending magnet at CHESS. The triangle base was supported by flexure pivots and bending was done by a cam driven at the triangle vertex. Whereas a perfectly bent crystal would provide a focal width of 0.7 mm, the polygonal crystal gave a focus only 2.3 mm wide corresponding to its facet width. A very similar design reported by Iwazumi (1992) had been tested on a multi-pole elliptical wiggler at the accumulator ring in Tsukuba. With 2.7 kW heat-load from a high energy insertion device it was necessary to provide some cooling for the sagittal second crystal. This was done by water cooling a baseplate which was thermally coupled to the bent crystal through a gallium-indium eutectic layer. The rhombohedral or double triangle ribbed design of Matsushita et al. (1986) was the forerunner of most of the present designs used at the Photon Factory in Tsukuba. The bending mechanism was cam driven and also provided for fixed exit conditions. The overall width was 125 mm and the facets were on 3.5 mm centers.

At this stage interest in multifaceted sagittally bent crystals was "turned over" so that X-rays were diffracted from the smooth side of the crystal. Now the ribs are intended to reduce the anticlastic bending which causes severe X-ray optical aberrations. Knapp et al. (1992) tested a new four-rod bending design on the bending magnet beamline X-6B at NSLS. It had two important new features; it is 125 mm long in the beam direction and, following those tests the design was optimized by ANSYS finite element analysis. The dimensions allow energy scanning over several hundred eV without the need to translate the sagittal crystal. It is clear that FEA investigation of any proposed design is important. In the special (but common) case of silicon 111 it is worth noting that Poisson's ratio is isotropic normal to the [111] direction.

Kushnir et al. (1993) have shown by Finite Element Analysis that an anisotropic rectangular lamella bent by couples applied along two parallel edges suffers no anticlastic curvature if the length-to-width ratio is correctly chosen. Thus, complicated rib structures are not necessary.

Yet more sophisticated is the design concept published by Zontone and Comin (1992). They propose bending both crystals of a non-dispersive double crystal pair, the first in the meridional direction and the second in the sagittal geometry while simultaneously taking account of the beam heating. While this paper evaluates a design concept, at this stage they reported only Finite Element calculations, it gives an indication of the shape of things to come.

Sagittal focusing by a conically bent crystal has been explored in experiments at NSLS by Ice and Sparks (1993, 1994) and by Habenshuss et al. (1988).] Extensive ray tracing using "Shadow" has been done. The geometry is well suited to synchrotron radiation sources with up to 15 mrad of horizontal divergence, with magnifications between 0.2 and 2 and energies between 3 keV and 40 keV. On beamline X-14 at NSLS the conically bent crystal collects ~5 mrad of horizontal divergence from 3 keV to 30 keV and focuses to 0.36 mm horizontal width. A vertical focusing cylindrical mirror focuses the beam to 0.24 mm. These dimensions are compared with aberration free calculations which indicate 0.28 mm x 0.16 mm. The measured transmission is within 20% of theory. Calculations for undulators indicate that demagnification of x20 or more should be easily obtained with sagittal focusing because the beams are so small that aberrations become negligible.

6.2. Universal Sagittal Focusing Crystals

From the above summary it is clear that different source types will require different sagittal focusing crystals.

Remarkably, crystals for undulator beamlines seem to be the simplest of all because the required apertures are so very small. A generic design might be only 5-10 mm wide with a rib along each edge and could be at least 100 mm long. No scanning stage would be necessary for the sagittally bent crystal as the energy is tuned (Ice and Sparks,1993). Since the requirements for bending magnets and for wigglers are geometrically similar, if translation stages are to be avoided, then the length along the beam direction will be at least 100 mm as in the undulator case. Across the beam the dimension must cover the desired angular aperture. Third generation sources so far planned provide only 6 mr available aperture from the bending magnets, but at 30 m from the source even that requires 180 mm *useful aperture* on the crystal. Similar dimensions apply to wigglers on these sources. On second generation sources the situation is eased because, in some cases, the first optical element may be 10 m or even less from the tangent point. Thus 15 mr amounts to only 150 mm on the crystal so much larger angular collection apertures are foreseen with conveniently sized available crystals.

Whereas focal spots in the region of 1-3 mm in size have been acceptable in the past on second generation synchrotron radiation sources, the specification for new X-ray optical systems seems to be moving towards the 0.2 mm size range. That is to image the source without loss of brilliance.

7. HIGH ENERGY RESOLUTION

Based on the dynamical theory of X-ray diffraction by perfect crystals there are two distinct approaches to providing high energy resolution; in the first the Bragg angle is set close to 90 degrees so that the chromatic dispersion is maximized while the second seeks to minimize the Darwin width by utilizing high order Bragg reflections. To some extent both can be combined.

Where the synchrotron radiation beam has a significant divergence we can estimate the energy resolution by combining the effects of Darwin resolution and divergence in quadrature as;

$$\Delta E/E = (\Delta\omega^2 + \Delta\theta^2)^{1/2} \cot\theta$$

According to Bragg's Law the intrinsic energy resolution corresponding to the Darwin width is given by;

$$\Delta E/E = \Delta\omega \cot\theta = A\, d^2\, F_{hkl} \text{ where A is a constant.}$$

In dispersive double crystal spectrometers where the energy resolution is independent of the source divergence the energy resolution is determined by the Darwin width of the chosen Bragg reflection. Values for selected silicon reflections are given in Table 6.

Table 5. Energy resolution for silicon monochromators.

Si / hkl	111	220	311	222	400	331	511
$\Delta E / E$	133×10^{-6}	$55.7 \times 10^{-}$	$26.3 \times 10^{-}$	$0.80 \times 10^{-}$	$23.2 \times 10^{-}$	$13.3 \times 10^{-}$	8.1×10^{-6}

Owing to thermal expansion Bragg angles are of course temperature dependent. Because the intrinsic energy resolution for a given Bragg reflection is of fundamental importance in designs and because it is independent of energy it is useful to evaluate the Darwin resolution as an equivalent temperature change. Since $\Delta d/d = \alpha\Delta T$ and $\Delta\theta = -\Delta d/d \cot\theta$ the Darwin or energy widths given in Table 5 can be represented as an equivalent "temperature width". In effect this is the change of temperature which would be required to change the Bragg angle, through thermal expansion, by one Darwin width. Values are calculated in Table 6.

Table 6. Darwin widths for Silicon Bragg reflections.

Si / hkl	111	220	311	222	400	331	511
ΔT / °C	52	21.8	10.3	0.3	9.1	5.2	3.2

Another way to interpret these results is to note that if the temperature of the crystal within the beam footprint varies by more than a small fraction of the Darwin "temperature width" then the crystal is effectively inhomogeneous and imperfect. When synchrotron radiation sources can deliver kilowatts of radiation to a beamline this requirement is seen to be very severe in practice.

8. TYPICAL BEAMLINE LAYOUT

A very common situation requires doubly focusing the X-ray beam to an image of the source at the sample. This can be achieved in many ways, and one solution which combines a flat crystal monochromator with a toroidal mirror is shown in Figure 7. In principle the mirror or the crystal can be the first optical element in the beamline and that is the arrangement on beamline X-25 at NSLS. In practice the choice depends on such factors as the heat load on the first optical element and whether or not the beamline is also used without the focusing mirror or with a number of different monochromators.

For comparison a slightly more sophisticated beamline design is shown in Figure 8. Here, the first mirror provides some defocusing of the slightly divergent incident beam so as to match the beam divergence to the Darwin width of the first crystal of the monochromator. The beam is then focused horizontally by the sagittal bent first crystal and refocused vertically onto the sample by the second mirror. Such a system is employed on the General Purpose Italian Line for Diffraction and Absorption (GILDA) at the ESRF.

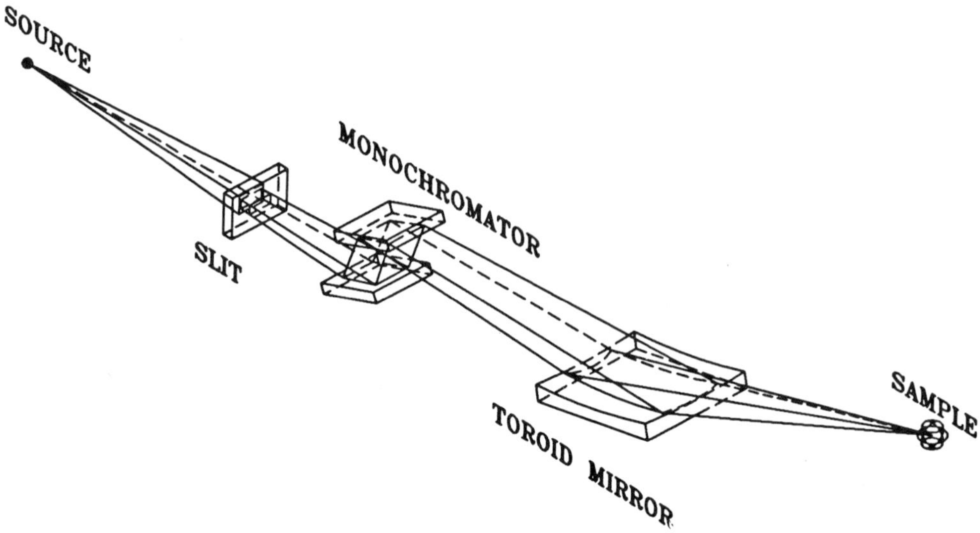

Figure 7. Typical focused monochromatic beamline optics.

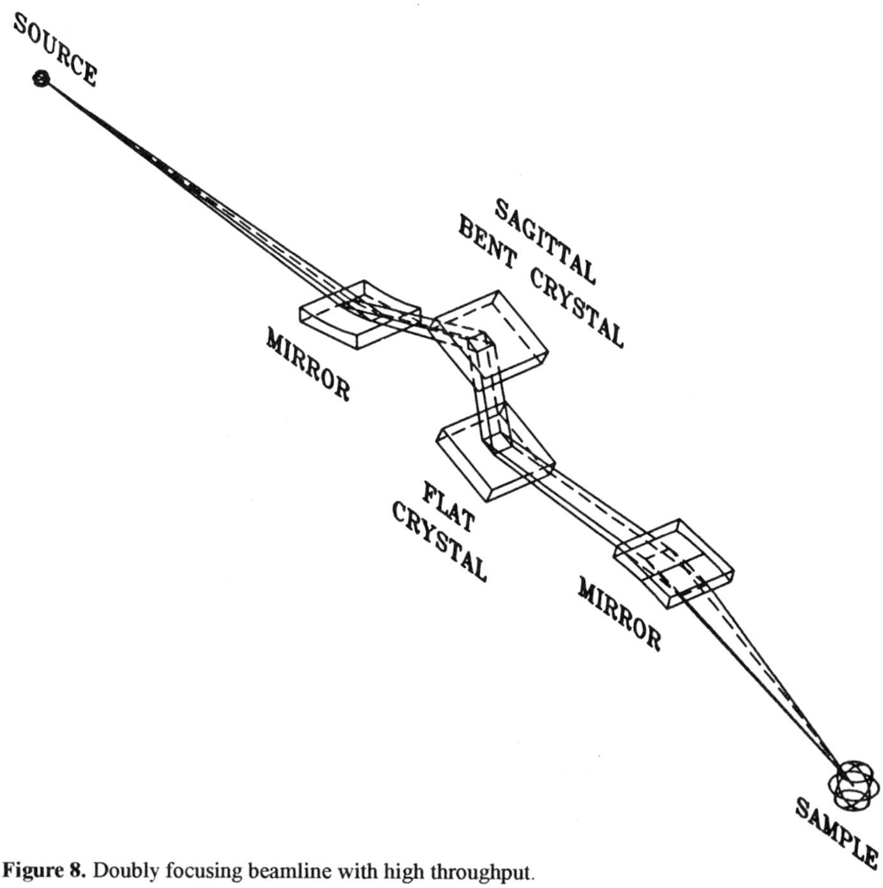

Figure 8. Doubly focusing beamline with high throughput.

9. HIGH POWER X-RAY OPTICS

We indicated in the previous section that the thermal power in the photon beam can be a cause of temperature changes and inhomogeneity in the monochromator itself. Mirrors too may suffer sufficiently large heat loads to compromise their optical performance. Sample calculations show that the total photon power produced by wigglers can be in the kW range while undulators and focused white radiation beams from bending magnets and wigglers can result in power densities in the 100 W/mm^2 range. Bearing in mind that the surface figure of mirrors is critical to their performance and that the preservation of perfection and homogeneity in crystals is crucial to delivering the Darwin efficiency it is clear that the thermal response of these optical components is very important in practice.

For the present purpose two simple thermal calculations suffice to illustrate the problems which arise in high power X-ray beams.

9.1 One dimensional heat flow

The steady state thermal diffusion equation can be solved exactly in this case. With a power input Q / A per unit area we require that

$$Q / A = k \nabla T = k (T_H - T_C) / t$$

where t is the crystal or mirror thickness. In either case the parallel plate becomes spherically bent to a radius R which is given by $R = t / \alpha \ (T_H - T_C)$. The resulting curvature is independent of the thickness of the material since we can combine these last two results to give $Q / A = k / \alpha R$. In the dynamical diffraction regime sensitivity to curvature is extremely high, for example 1km bend radii are easily detected. With R = 1 km in silicon (for which k = 160 W / mK and $\alpha = 2.33 \times 10^{-6}$) we find Q / A = 68 mW / mm^2; very low power density indeed compared with that routinely available at synchrotron radiation sources.

9.2 Three dimensional heat flow with cylindrical symmetry

The beam which emerges from the storage ring dipole magnets and from most wigglers is a horizontal swath of radiation . From the thermal viewpoint the footprint of these beams is therefore a line source of heat on a wafer or more particularly a semi-infinite conducting medium. In this case too the thermal diffusion equation has simple analytical solutions. The steady state solution in cylindrical polar coordinates is

$$Q / L = 2\pi k (T_H - T_C) / \ln(R / r)$$

where Q / L is the heat input per unit length. The important part of the solution is the ln R dependence of the heat loading on the distance to the distant cylindrical boundary at which the cooling heat-sink is provided. For the case R>>r this result is exactly the same as the one dimensional result, as expected. If R = 0.1m, corresponding to a thick single crystal block, then the hot beam footprint reaches the melting point of silicon when the heat flux exceeds 270 W / mm of input thermal power if the cold boundary is at room temperature. Such linear power densities are frequently encountered at synchrotron radiation sources.

9.3 Choice of materials for monochromators and mirrors

At the highest thermal beam powers encountered there are clearly problems with silicon monochromators and with normal mirror materials from the point of view of X-ray optical performance.

In the case of mirrors, as outlined in sections **5.1** and **5.2** the mechanical properties of the mirror material can be chosen almost independently of the optical properties of the mirror surface coating because the reflecting layer needs to be only a few hundred Å thick from the optical point of view. In crystal diffraction the perfect crystal layer too only needs to be several extinction distances thick, that is several hundred μm thick, but it must be ideally perfect. Such perfect epitaxy is simply not possible in practice so that the monochromator must be a single perfect crystal (Fewster, Servidori, Patel, this Volume).

To solve the crystal thermal load problem two solutions have been explored; to use different materials (Van Zuylen et al., 1986) or to adapt the crystal shape so as to compensate for the thermal deformation (Hart, 1990, Berman and Hart, 1991, Berman et al., 1992, Oyanagi, 1992, Berman, 1994, Quintana et al. 1995, Quintana and Hart 1995).

Two different materials have emerged offering excellent solutions to the high heat load monochromator problem; silicon used at about 120 K where the thermal expansion coefficient is zero (Bilderback, 1986) and the thermal conductivity is almost ten times higher than at room temperature and diamond used at room temperature (Freund, 1995). From the point of view of thermal design the figure of merit for a mirror or monochromator material is just k / α. Materials with high thermal conductivity and low thermal expansion are clearly advantageous in X-ray optics. However, the power absorbed in thin slices is determined by the X-ray linear absorption coefficient μ , but the meaning of the description "thin" is dependent on the X-ray energy. Silicon wafers are almost totally absorbing at 5 keV but almost totally transparent at 100 keV so that the incorporation of X-ray parameters such as absorption or scattering power into X-ray-optical figures of merit is complicated in practice. Some values are given in Table 8.

Table 8. Figures of merit for mirror and monochromator materials

Material	Diamond	Silicon	Germanium
μ (10 keV) / mm^{-1}	0.38	0.72	20.6
k at 300K / $Wm^{-1}K^{-1}$	3500	160	64
k at 77K / $Wm^{-1}K^{-1}$		1330	360
$10^6\ \alpha$ / C at 300K	0.8	2.33	5.8
$10^6\ \alpha$ / C at 77K		-0.5	1.4
$k / 10^6\ \alpha$ / Wm^{-1} at 300K	4375	68.7	11.0
$k / 10^6\ \alpha$ / Wm^{-1} at 77K		2660	257
$k / 10^9\ \alpha\ \mu$ / Wm^{-1} at 300K	11510	95.4	0.53
$k / 10^9\ \alpha\ \mu$ / Wm^{-1} at 77K		3694	4.0

The figures of merit show the advantages of diamond and cryogenically cooled silicon over other possible materials. Details of mirror and crystal designs and measurements of optical performance in high power beams from insertion devices have been published by many authors. The figure of merit $k/\alpha\mu$ which includes the X-ray absorption coefficient μ is particularly relevant in the Laue case of transmission whereas the value of k/α is relevant to the Bragg case of reflection and to mirrors.

This work was supported under Contract No. DE-AC02-76CH00016 with the U.S. Dept. of Energy.

REFERENCES

Batterman, B.W. and Berman, L. E., 1983, Sagittal focusing of sychrotron radiation, *Nucl. Instrum. Methods Phys. Res*,. 208: 327-331

Berman, L.E., Preserving the high source brightness with X-ray beamline optics, *Rev. Sci. Instrum.*, 66:2041-2047 (1994)

Berman, L.E. and Hart, M., 1991, Adaptive crystal optics for high power synchrotron sources, *Nucl. Instrum. Methods Phys. Res.*, A302: 558-562

Berman, L.E. and M. Hart, 1991, Performance of water jet cooled silicon monochromators on a multipole wiggler beamline at NSLS, *Nucl. Instrum. Methods Phys. Res.*, A300: 415-421

Berman, L.E., Hart, M. and Sharma, S., 1992, Adaptive crystal optics for undulator beamlines, *Nucl. Instrum. Methods Phys. Res.*A321: 617-628

U. Bilderback , D.H., 1986, The potential of cryogenic silicon and germanium X-ray monochromators for use with large synchrotron heat loads, *Nucl. Instrum. Methods Phys. Res*,.A246: 434-436

Bonse, U. and Hart, M., 1965, Tailless X-ray single crystal reflection curves obtained by multiple reflection, *Appl. Phys. Lett.* 7: 238-240

Cernik, R. and Hart, M., 1989, Medium power X-ray crystal optics for synchrotron radiation sources, *Nucl. Instrum. Meth.* A281: 403

Cullity, B.D., 1978, *Elements of X-ray Diffraction*, Addison-Wesley Publishing Company, Inc. Reading, Massachusetts, Second Edition,

Freund, A.K., 1995, Diamond single crystals: the ultimate monochromator material for high -power X-ray beams, *Optical Engineering,* 34: 432-440

Habenschuss, A., Ice, G.E., Sparks, C.J. and Neiser, R. A., 1988, The ORNL beamline at the National Synchrotron Light Source: a Guide for Users, *Nucl. Instrum. Methods Phys. Res*,. A266: 215-219

Hart, M., 1990, X-ray monochromators for high power synchrotron radiation sources, *Nucl. Instrum. Methods Phys. Res.,* A297: 306-311

Henke B.L., Gullikson, E.M. and Davis, J.C., 1993, X-ray interactions: photoabsorption, scattering, transmissionand reflection at E=50 -30000 eV, Z=1-92, *Atomic Data and Nuclear Data Tables*, 54: 181-342

Henke, B.L., Gullikson, E.M. and Davis, J.C. at the following site on the world wide web:http://www-cxro.lbl.gov/optical_constants/

Hubbell, J.H., 1969, Photon Cross Sections, Attenuation Coefficients, and Energy Absorption Coefficients From 10 keV to 100 GeV, *Nat.Stand.Ref.Data.Ser., Nat. Bur. Stand. (U.S.),* 29: 1-80

Ice, G.E. and Sparks, C.J., 1993, Conical geometry for sagittal focusing as applied to X-rays from synchrotrons, *Oak Ridge National Laboratory Report ORNL/TM-12327 June 1993*

Ice, G.E. and Sparks, C.J., 1994, Conical geometry for sagittal focusing as applied to X-rays from synchrotrons, *J. Opt. Soc Amer.* 11: 1265-1271

International Tables for X-ray Crystallography, 1968, pp 73-88, The Kynoch Press, Birmingham

Iwazumi, T., Sato, M. and Kawata, H., 1992, Double-crystal monochromator for high-power and high-energy synchrotron radiation, *Rev. Sci. Instrum.,* 63: 419-422

Kirz, J., Attwood, D.I., Henke, B.L., Howells, M.R., Kennedy, K.D., Kim, Kwang-Je , Kortright, J.B., Perera, R.C., Pianetta, P., Riordian, J.C., Scofield, J.H. , Stradling, G.L., Thompson, A.C., Underwood, J.H., Vaughan,D. , Williams, G.P. and Winick , H., 1986 *X-ray Data Book* PUB-490 Rev. from Lawrence Berkeley Laboratory

Knapp, G. S., Ramanathan, M., Nian, H.L., Macrander, A. T. and Mills, D. M., 1992, A simple sagittal focusing crystal which utilises a bimetallic strip, *Rev. Sci. Instrum*,. 63: 465-467

Kushnir, I., Quintana, J. P. and Georgopoulos, P., 1993, On the sagittal focusing of synchrotron radiation with double crystal monochromator, *Nucl. Instrum. Methods Phys. Res.*, A328: 588-591

Matsushita,T, Ishikawa , T. and Oyanagi, H., 1986, Sagittally focusing double-crystal monochromator with constant exit beam height at the Photon Factory, *Nucl. Instrum. Methods Phys. Res.*, A246: 377-379

Mills, D.M., Henderson, C. and Batterman, B. W., 1986, A fixed exit sagittal focusing monochromator utilizing bent single crystals, *Nucl. Instrum. Methods Phys. Res.,* A246: 356-359

Murphy, J., 1993, *Synchrotron Light source Data Book*, BNL informal report #42333

Nelmes, R.J. and McMahon , M.I., 1995, Ordered superstructure of InSb-IV, *Phys. Rev.Lett.*, 74: 106-109

Oyanagi, H., 1992, Adaptive silicon monochromators for high heat-load insertion devices, *S. P. I. E.* 19-24 July 1992, San Diego.

Quintana, J.P. and Hart, M., 1995, Adaptive silicon monochromators for high power wigglers;Design, Finite Element analysis and Laboratory tests, *J. Synchrotron Rad.*, 2: 119-123

Quintana, J.P., Hart, M., Bilderback, D., Henderson, C., Richter, D., Setterson, T., White, J., Hausermann, D., Krumrey, M. and Schulte-Schrepping, H., 1995, Adaptive silicon monochromators for high power insertion devices; tests at CHESS, ESRF and HASYLAB, *J.Synchrotron Rad.* 2: 1-5

Sparks, C.J., Jr., Borie, B.S. and Hastings, J. B., 1980, X-ray monochromator for focusing synchrotron above 10keV, *Nucl. Instrum. Methods Phys. Res.* 172: 237-42

Sparks, J., Jr., Ice, G. E., Wong J. and Batterman, B. W., 1982, Sagittal focusing of synchrotron x-radiation with curved crystals, *Nucl. Instrum. Methods Phys. Res.*, 194: 73-78

Van Zuylen, P., Lemaire, A.D. and Wijsman, A.J.Th.M.A., 1988, Cooling of silicon crystals for X-ray monochromators, *Report to ESRF*

Zontone, F. and Comin, F., 1992, Heat load and anticlastic effect compensation on an ESRF monochromator: An exhaustive ray-tracing study for a meridional-sagittal geometry, *Rev. Sci. Instrum.* 63:501-504

X-RAY POLARIZATION AND APPLICATIONS

Cécile Malgrange

Laboratoire de Minéralogie-Cristallographie, Universités Paris 6 et 7,
associé au CNRS, case 115, 4 place Jussieu,
75252 Paris Cedex 05, France

1. INTRODUCTION

X-ray polarization phenomena have only been studied by few people until the development of synchrotron sources which present three advantages in this respect :

- polarization (§.2) whereas X-ray tubes emit unpolarized radiation,
- high flux which permits to modify polarization, keeping enough photons to make experiments,
- white spectrum which allows to select wavelengths near absorption edges where many polarization-dependent phenomena are enhanced.

Linearly polarized X-rays but also elliptically polarized X-rays with a high rate of circular polarization are now available and polarized X-rays form now a challenging new experimental probe useful for a large range of studies. Furthermore, when an experiment is perfomed at an absorption edge it becomes selective in chemical species and sometimes even site dependent.

Polarized X-rays are fundamental to study magnetic materials either through magnetic diffraction (de Bergevin and Brunel, 1972, 1981, 1991) which brings information complementary to those given by neutron diffraction, or Compton magnetic scattering (Sakai and Ono, 1976) or X-ray magnetic circular dichroism (Schütz et al., 1987). The last two techniques require circularly polarized photons whereas in magnetic Bragg diffraction playing with the incident polarization and analysis of the diffracted polarization permits to obtain different types of information.

X-ray and Neutron Dynamical Diffraction: Theory and Applications
Edited by Authier *et al.*, Plenum Press, New York, 1996

2. POLARIZATION OF X-RAY SOURCES

Before the use of synchrotron sources any experiment concerning polarization had to start with unpolarized X-rays since X-ray tubes deliver unpolarized radiation.

In the first synchrotron rings the charge particles were accelerated only by bending magnets and today many synchrotron beam lines still use the radiation delivered by electrons or positrons accelerated by bending magnets.

Let us first analyse the radiation emitted by a charged particle whose energy E is γ times its rest mass m_0c^2, $E=\gamma\ m_0c^2$. This radiation is highly

- directional and concentrated in a narrow cone issued from the particle with an apex angle of the order of γ and axis tangent to the orbit,

- polarized: in the orbit plane the X-ray electric vector is linearly polarized and parallel to the orbit plane. For emission at an angle ψ with respect to the orbital plane the electric field is elliptical. The excentricity of the ellipse increases with the angle ψ but at the same time the intensity decreases. Figure 1 shows the dependence on the vertical angle ψ of the intensities of the electric field components respectively parallel and perpendicular to the orbit plane. It is then possible to get linearly polarized X-rays by selecting with an horizontal slit the beam in the orbital plane. In the same manner one obtains elliptically polarized photons, left- or right-handed, by selecting the beam at a given vertical angle above or below the orbital plane. For real sources the curves presented in Figure 1 have to be convoluted by the position and momentum gaussian distributions of the charged particules. As a result of this convolution the beam in the orbit plane for example is not totally linearly polarized.

Insertion devices consist of special arrays of magnets with different polarizations which give special trajectories to the charged particles (a good brief review is given in Elleaume

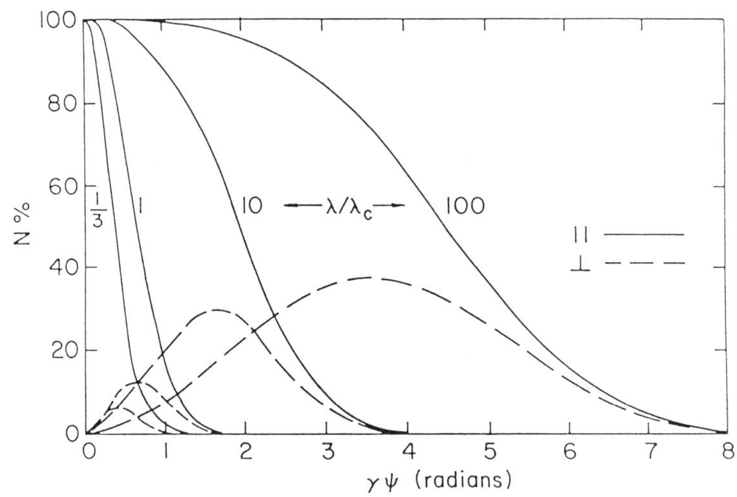

Figure 1. Intensity of the component linearly polarized parallel (full line) and perpendicular (dashed line) to the orbit plane as a function of the vertical angle of emission ψ for a beam emitted by an electron whose energy is γ times its rest mass. λ_c is the characteristic wavelength.

1992). The transverse wigglers and undulators are designed so that the magnetic field has a sinusoidal dependence leading to sinusoidal trajectories in the horizontal plane. A given direction of observation **n** is tangent to the trajectory in different points (in general the number of points is equal to the number of magnetic poles) which are all sources of radiation for an observer at infinity in the **n** direction. Such devices deliver linear polarization in the orbit plane ($\psi=0$) but also outside the orbit plane ($\psi\neq0$) because the total radiation is a sum of alternatively right- and left-circularly polarized radiation due to the alternate signs of trajectory curvature.

In order to associate the advantages of wigglers and undulators without losing the possibility to obtain circularly polarized photons from such sources, special insertion devices have been realized such as

- asymmetric wigglers where the alternate magnetic fields have different values leading to a non-sinusoidal trajectory and different values of the curvatures of opposite signs ; there is then a residual elliptical polarization above and below the orbit plane (Goulon, Elleaume & Raoux, 1987).

- elliptical wigglers derived from a conventional wiggler by adding a small horizontal magnetic field of identical period but shifted by $\pi/2$ with respect to the main verical field (Yamamoto et al., 1989).

- crossed undulators made of two successive linear undulators with magnetic fields in perpendicular planes (Kim, 1984).

- helical undulators where the magnetic field is helical ; they can deliver high circular polarization rates on the axis (Elleaume, 1994).

3. POLARIZATION PHENOMENA IN THE X-RAY RANGE

3.1. Fundamentals

In the range of optical wavelengths transformation of the polarization state is easy since linear polarizers and phase retarders are available. Their principle is based on the anisotropy of the dielectric constant in the visible light range. The dielectric properties of materials at X-ray wavelengths are very different because X-ray energies are close to the energies of K or L electrons which are not sensitive to crystalline fields. This problem has been considered theoretically by Molière as early as 1939 (Molière, 1939). He concluded that birefringence is expected to be nearly zero at X-ray wavelengths except very close to an absorption edge.

The situation is different when Bragg diffraction occurs. The Thomson electron scattering depends on the geometrical relation between the electric field and the incident and scattering directions. The scattering amplitude is proportional to the polarization factor equal to 1 and cos 2θ for the σ and π polarization respectively (electric field perpendicular and parallel to the scattering plane respectively). Then if the Bragg angle is equal to 45°, σ polarization only is diffracted and the diffracting crystal acts as a linear polarizer or analyser.

This holds for a powder as well ; the diffracting plane has then to be defined by a thin slit in front of the detector. In the case of perfect or very good crystals, X-ray diffraction is governed by dynamical theory leading to very interesting specific X-ray optics properties and particularly, for our concern here, birefringence properties which will be presented in the next paragraph.

3.2. Dynamical theory and polarization

As has been mentioned in the first paragraph, one can get elliptically polarized X-rays by selecting the radiation emitted above or below the orbit plane by charged particles deviated by a bending magnet in a synchrotron ring. Exotic insertion devices can also deliver elliptically polarized radiation. Another way is to transform a linearly polarized beam (delivered either in the orbit plane in a bending magnet section or along its axis by a planar undulator) into circular polarized radiation with a quarter-wave plate. As was seen in the previous paragraph and as is detailed in Authier (this volume), birefringence occurs at Bragg diffraction. Let us consider an incident plane wave linearly polarized with a wave vector $\mathbf{K_o}=\mathbf{OM}$ corresponding to a departure from Bragg angle equal to $\Delta\theta$ and for example a Laue symmetric geometry. It gives rise inside the crystal to four Bloch waves or "wavefields" with tie-points $P_{1\sigma}$, $P_{1\pi}$, $P_{2\sigma}$, $P_{2\pi}$ (Figure 2) and respective wave vectors in the forward diffracted direction $\mathbf{k}^{o}_{ij}= \mathbf{OP}_{ij}$ (i=1,2 and j=σ,π) and in the diffracted direction $\mathbf{k}^{h}_{ij}= \mathbf{HP}_{ij}$. Going through the crystal each forward-diffracted (resp. diffracted) wave undergoes a phase-shift Φ^{o}_{ij} equal to $2\pi\,(\mathbf{k}^{o}_{ij}.\mathbf{r})$ (resp. $\Phi^{h}_{ij} = 2\pi\,(\mathbf{k}^{h}_{ij}.\mathbf{r})$).

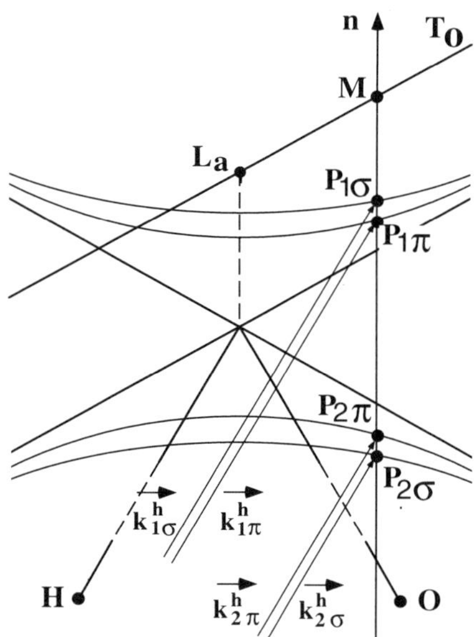

Figure 2. Dispersion surfaces for σ and π polarizations. Mn is normal to the entrance surface and L_aM is equal to $k\Delta\theta$.

If the incident wave is unpolarized (case of X-ray tubes), σ and π components are incoherent and their intensities I_j ($j=\sigma, \pi$) add together. Each intensity I_j is the result of interferences between wavefields 1 and 2. It is worthwile pointing out that, after the crystal, the phase-shift Φ_j is the same for the diffracted and the forward-diffracted waves since it is given by

$$\Phi_j = 2\pi\,[(\mathbf{k}^{\mathbf{h}}_{1j} - \mathbf{k}^{\mathbf{h}}_{2j}).\mathbf{r}] = 2\pi\,[(\mathbf{k}^{\mathbf{o}}_{1j} - \mathbf{k}^{\mathbf{o}}_{2j}).\mathbf{r}] = \mathbf{P_{2j}\,P_{1j}.r}$$

$$= 2\pi\,(P_{2j}\,P_{1j})\,z = 2\pi\,z\,\frac{\sqrt{1+\eta_j^2}}{\Lambda_j} \tag{1}$$

where z is the thickness of the crystal and $1/\Lambda_j$ is the distance between the apices of the dispersion hyperbola. This phenomenon gives rise to oscillations in the rocking curve of thin crystals (Lefeld-Sosnowska and Malgrange, 1969). When scanning a rocking curve, the point M moves along the line T_o and the phase-shift oscillates between 0 and 2π giving rise to oscillations in the rocking curve with beats between σ and π components. Such a beating between σ and π interferences was also observed with an incident spherical wave and wedge shaped crystals (Hart and Lang, 1965 ; Hattori, Kuriyama and Kato, 1965).

If the incident wave is linearly polarized, its σ and π components are coherent. Their propagation inside the crystal induces a phase-shift between them which results in an elliptically polarized wave for each wavefield 1 and 2. If the two wavefields overlap inside the crystals the two elliptical waves interfere. However, let us first consider wavefields 1 and 2 separately. The nature of the ellipse depends on the phase-shift between the two components σ_i and π_i and on the ratio of their amplitudes.

The phase-shifts for the forward-diffracted $\Phi^{\mathbf{o}}_i$ (i=1,2) and diffracted $\Phi^{\mathbf{h}}_i$ amplitudes are equal and

$$\Phi^{\mathbf{o}}_i = \Phi^{\mathbf{h}}_i = 2\pi\,(P_{i\sigma}P_{i\pi})\,z$$

The ratio of the amplitudes is mainly governed by the Borrmann absorption (see Authier I, this volume, eq.32). The absorption coefficients for the four waves (in the symmetrical case which is considered here) are

$$\mu_{1\sigma} = (\mu_o/\cos\theta)\,(1 - (1+\eta^2)^{-1/2}\chi_{ih}/\chi_{io})$$
$$\mu_{1\pi} = (\mu_o/\cos\theta)\,(1 - (1+\eta^2)^{-1/2}|\cos 2\theta|\,\chi_{ih}/\chi_{io})$$
$$\mu_{2\sigma} = (\mu_o/\cos\theta)\,(1 + (1+\eta^2)^{-1/2}\chi_{ih}/\chi_{io})$$
$$\mu_{2\pi} = (\mu_o/\cos\theta)\,(1+(1+\eta^2)^{-1/2}\,|\cos 2\theta|\,\chi_{ih}/\chi_{io}). \tag{2}$$

For silicon-like crystals (i.e. germanium and diamond) the value of χ_{ih}/χ_{io} is equal to the Debye-Waller factor exp-M for even reflections and equal to exp-M$\sqrt{2}$ /2 for odd reflections. For an even reflection with small h,k,l, exp-M is close to unity.

3.3. Linear polarizers

3.3.1. Borrmann effect and forward-diffracted beam. For high values of the product $\mu_o t$, the only wave which is not much absorbed is the σ_1 wave and its amplitude is noticeable only very close to exact Bragg incidence ($|\eta|<<1$). This is the principle of an X-ray polarizer first proposed by Cole, Chambers and Wood (1961). Whatever the polarization of the incident beam, the diffracted and the forward-diffracted wave after such a crystal at Bragg incidence is polarized perpendicularly to the diffraction plane and only waves with angles of incidence close to the exact Bragg incidence propagate through the crystal. The forward-diffracted beam is then linearly polarized *and* monochromatized. Furthermore, the direction of linear polarization can be rotated at will by a rotation of the crystal around the direction of the incident beam.

3.3.2. Reflected beam. The Darwin width δ, angular range of the total reflection domain for a non-absorbing crystal, is given by

$$\delta = |C| \frac{\sqrt{|\gamma|\, \chi_h \chi_{\bar{h}}}}{\sin 2\theta} \tag{3}$$

and is proportional to the polarization factor. The integrated reflecting power for the π component tends then to zero when the Bragg angle tends towards 45° and, as already mentioned, a crystal with a 45° Bragg angle is a linear polarizer (or analyser). Such a condition is hardly realized with X-ray tubes (a good combination occurs with a Ge 333 reflection and CuKα radiation). With synchrotron radiation it is easy to get a 45° reflection but for a given wavelength only. Hart and Rodrigues (1979) proposed to use a channel-cut crystal and multiple Bragg diffractions. If the ratio of the σ and π reflecting powers is a for one reflection, it will be a^n after a channel-cut operating with n reflections. The device can still be improved if the two sides of the multiply reflecting grooved crystal are offset by a small angle α from the parallel position. The angle α has only to be larger than the width of the π rocking curve and smaller than the σ one.

4. X-RAY PHASE PLATES

4.1. Phase plates adjusted inside the rocking curve

Let us go back to cases where $\mu_{1\sigma}$ (resp. $\mu_{2\sigma}$) is not too different from $\mu_{1\pi}$ (resp. $\mu_{2\pi}$) and let us even assume, in order to simplify, that the amplitudes of the σ and π waves are equal (polarization vector making a 45° angle with respect to the diffraction plane). Let us consider an incident plane wave and a perfect crystal adjusted inside the rocking curve in order to have a high birefringence. The phase-shift Φ_1 between $\sigma 1$ and $\pi 1$ waves is opposite to the one, Φ_2, between $\sigma 2$ and $\pi 2$ waves and equal to

$$\Phi_1 = 2\pi\,(P_{1\pi}P_{1\sigma})\,z = -2\pi\,(P_{2\pi}P_{2\sigma})\,z = -\Phi_2 \qquad (4)$$

$$\Phi_1 = \pi \frac{1}{\sqrt{1+\eta^2}}\left(\frac{1}{\Lambda_\sigma} - \frac{1}{\Lambda_\pi}\right) z \text{ where } \Lambda = \lambda \frac{\cos\theta}{\sqrt{\chi_h \chi_{\bar{h}}}} \quad \text{and } \Lambda_\pi = \Lambda_\sigma / |\cos 2\theta| \qquad (5)$$

For a phase-shift Φ_1 equal to $\pi/2$, the wave becomes right-handed circular, for a π phase-shift it becomes linear and turned 90° with respect to the incident wave, for a $3\pi/2$ left-handed circular etc... If the wavefields do not overlap (which could be obtained with a narrow incident beam and a value of $|\eta| \geq 1$) one could get circularly polarized X-rays with a crystal of the ad-hoc thickness as was first suggested by Skalicky and Malgrange (1972). In fact, it would be elliptically polarized because of the difference between amplitudes due to inequal values of the absorption coefficients.

If the wavefields overlap, the two elliptical waves 1 and 2 which have opposite signs of helicity (see Figure 2 showing that $k^h_{2\sigma} < k^h_{2\pi} < k^h_{1\pi} < k^h_{1\sigma}$) but different amplitudes because wavefield 2 is more absorbed than wavefield 1, interfere. This phenomenon was experimentally shown with an X-ray tube as soon as 1972 by Skalicky and Malgrange. Then Hart (1978) suggested to prepare elliptically polarized X-rays by using a three-quarter wave plate silicon crystal with CuKα wavelength and at Bragg incidence. Wavefields 1 and 2 with their opposite helicity coexist but with different intensities resulting in an elliptically polarized wave. Later, Golovchenko et al. (1986) and Mills (1988), using synchrotron radiation, evidenced the efficiency of a silicon quarter-wave plate based on the same principle.

A different kind of phase plates based on the existence of a phase-shift between σ and π reflected beam in Bragg geometry has been developed by Batterman (1992).

4.2. Phase plates far from the reflection domain

In the preceeding section we considered the angular domain inside the rocking curve and more especially the exact Bragg incidence where the birefringence is the highest. Another solution, (first suggested theoretically by Dmitrienko and Belyakov 1980, 1989) and experimentally evidenced with thin silicon plates and a well collimated beam by Hirano et al., (1991) is to adjust the crystal outside the domain of reflection where the birefringence is much lower but still exists and naturally to use the forward-diffracted beam. Then there are two advantages : i) only one wavefield has a noticeable intensity (wavefield one for $\Delta\theta<0$ and wavefield 2 for $\Delta\theta>0$) ii) the amplitudes of σ and π waves become quite equal (see equation (2) for $|\eta|>>1$). We shall see that there is still an other and big advantage in the case of crystals with a low absorption coefficient like diamond, as first evidenced by Giles et al.(1994a) : a low sensitivity to the angular divergence of the incident beam.

Let us consider an incident plane wave with a departure from Bragg angle $\Delta\theta$ large compared to the Darwin width ($|\eta|>>1$). Dynamical theory shows (eq. 32 Authier I, this volume) that the only waves which go through the crystal are forward-diffracted waves : $\sigma 1$

and $\pi 1$ waves if $\Delta\theta<0$ and $\sigma 2$ and $\pi 2$ waves if $\Delta\theta >0$. Let us point out that this is true also for Bragg geometry. The phase-shift Φ between π and σ waves is given by

$\Phi = 2\pi\ (X_{0\pi} - X_{0\sigma})\ \mathbf{s_0.r}$ where $\mathbf{s_0}$ is a unit vector in the incident direction.

$\mathbf{s_0.r} = z/\gamma_0 = t$ where t is the beam-path since for large values of $|\eta|$ the Poynting vector is along the incident direction.

From equation (21) of Authier I, this volume, $X_0 = kC\chi_{\bar{h}}\ \xi/2$ (6)

and $\xi\chi_{\bar{h}}$ is given by Authier I (28)

$$\xi\chi_{\bar{h}} = -S(C)\ S(\gamma_h) \sqrt{\chi_h\chi_{\bar{h}}}\ \left(\eta \pm \sqrt{\eta^2 + S(\gamma_h)}\right) / \sqrt{|\gamma|} \tag{7}$$

where
$$\eta = \frac{\Delta\theta - \Delta\theta_0}{\delta} \tag{8}$$

with the Darwin width δ given by equation (3) ; the upper sign holds for wavefield 1 and the lower for wavefield 2. For a given incident plane wave characterized by $\Delta\theta$

$\eta_\pi = \eta_\sigma / |\cos 2\theta|$ and

$$X_{0\sigma} - X_{0\pi} = \frac{k\sqrt{\chi_h\chi_{\bar{h}}}\ S(\gamma_h)}{\sqrt{|\gamma|}} \left(|\cos 2\theta| (\eta_\pi \pm \sqrt{\eta_\pi^2 + S(\gamma_h)} - (\eta_\sigma \pm \sqrt{\eta_\sigma^2 + S(\gamma_h)} \right);$$

for $|\eta_\sigma| >> 1$ this formula, transformed and developed to first order, gives

$$X_{0\sigma} - X_{0\pi} = (\pm) \frac{k\sqrt{\chi_h\chi_{\bar{h}}}}{4\sqrt{|\gamma|}} \frac{\sin^2 2\theta}{|\eta_\sigma|} = (\pm) \frac{\chi_h\chi_{\bar{h}}\ k \sin 2\theta}{4\ |\ \Delta\theta - \Delta\theta_0\ |} \tag{9}$$

$\Delta\theta_0$ is the departure from Bragg angle corresponding to the middle of the rocking curve (in a symmetrical Laue case $\Delta\theta_0 = 0$) and it is simpler to take it as a reference for the angle of incidence and define then the offset $\Delta\Theta$ as

$$\Delta\Theta = \Delta\theta - \Delta\theta_0.$$

Since in formula (9) the upper (resp. lower) sign corresponds to wavefield 1 (resp. 2) for which the offset is negative (resp. positive), then one gets $\pm\ |\Delta\theta - \Delta\theta_0\ | = \Delta\Theta$

Then $\phi = -\ 2\pi\ (k/4) \dfrac{\chi_h\chi_{\bar{h}} \sin 2\theta}{\Delta\Theta}\ t$

which is the usual formula for a birefringent crystal whose birefringence is

$$n_\sigma - n_\pi = -\frac{\chi_h \chi_{\bar{h}} \sin 2\theta}{4\,\Delta\Theta} \tag{10}$$

Then $$\phi = -\frac{\pi}{2}\frac{\chi_h \chi_{\bar{h}} \sin 2\theta}{\lambda\,\Delta\Theta}\,t = -\frac{\pi}{2}\frac{r_e^2\,\lambda^3\,F_h F_{\bar{h}} \sin 2\theta}{(\pi V)^2\,\Delta\Theta}\,t \tag{11}$$

This simple formula shows that the phase-shift is inversely proportional to the offset and that it changes sign with the offset. It is then possible to obtain from linear polarization either right- or left-handed circular polarization by going from positive to negative offsets. For a given crystal the efficiency is larger for stronger reflections. The phase-shift can be rewritten in a simple form

$$\phi = -\frac{\pi}{2}\,A\,\frac{t}{\Delta\Theta} \tag{12}$$

5. QUARTER-WAVE PLATE AND CIRCULAR POLARIZATION

5.1. Interest of low absorption crystals

A linearly polarized wave whose polarization vector is inclined by an angle ψ equal to 45° with respect to the plane of diffraction of a perfect crystal has equal σ and π components. If the crystal introduces a phase-shift ϕ equal to $\pm\pi/2$ (quarter-wave plate) between these components the wave after the crystal is circularly polarized (right or left). More generally, it is easily shown that the circular polarization rate τ is given by

$$\tau = \frac{I_R - I_L}{I_R + I_L} = \sin 2\psi \sin\phi \tag{13}$$

where I_R and I_L are the right and left polarized intensities respectively.

For given crystal and wavelength, a $\pi/2$ phase-shift is obtained for a given value of $t/\Delta\Theta$. Since one uses the forward-diffracted beam rather far from Bragg incidence, the absorption coefficient is practically not affected by the Borrmann effect and the transmission coefficient can be considered as equal to exp-$\mu_0 t$. The larger is the transmission coefficient, the more efficient is the phase plate.

In order to give an order of magnitudes, Figure 3 shows the variation of the phase-shift introduced by different phase plates between σ and π components for λ=1.712 Å. The dotted line corresponds to a silicon phase plate with a beam path of 139 μm for 220 reflection whereas the full line corresponds to thicker crystals : silicon and diamond. These curves show that to obtain a $\pi/2$ phase-shift with the thin silicon crystal, one has to work at an offset of about 16 arcsec with a very collimated beam (a few arcsec) to preserve a nearly constant phase-shift of $\pi/2$ (Hirano et al., 1991,1992a) whereas with the thicker crystals one has to

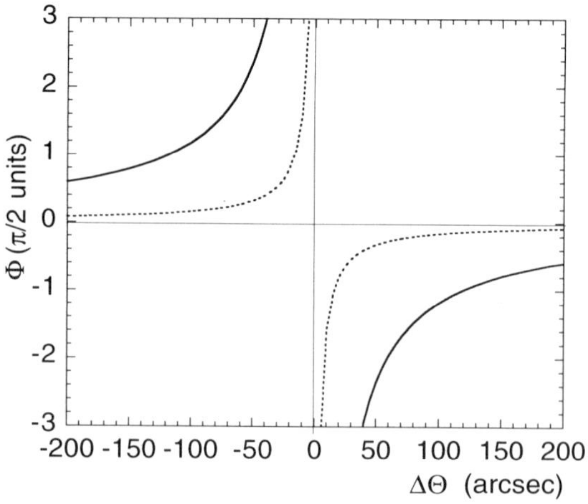

Figure 3. Variation of the phase-shift between σ and π polarizations at λ=1.712 Å for three different crystals; dotted line : Si crystal, reflection 220, beam path 139 μm ; full line: Si crystal, reflection 220, beam path 1023 μm and diamond crystal, reflection 220, beam path 1290 μm.

work at an offset of the order of 100 arcsec with a low sensitivity to the divergence of the incident beam (a few tens arcsec). In fact, the thick silicon crystal cannot be used because of its prohibitive attenuation factor (Table 1). Diamond crystals because of their small absorption coefficient and the possibility to get good quality samples are, for the moment, the best candidates for transmission X-ray phase plates (Giles et al., 1994a,b, 1995a,b,d).

crystal	t (μm)	ΔΘ(π/2) arcsec.	T=exp-μt
Si 220	139	16	0.06
C 220	1290	118	0.06
Si 220	1023	118	$\approx 10^{-9}$

Table 1. Beam path t, offset ΔΘ giving a π/2 phase-shift and transmission T for the three different crystals considered in Figure 3.

There are two means to prove the efficiency of a given quarter-wave plate :

i) To analyse the vibration after the phase plate, which needs a linear polarization analyser and a quarter-wave plate in order to distinguish between circularly polarized and unpolarized radiation. Such a complete analysis gives a measurement of the polarization rates (Giles et al., 1995a).

ii) To measure a physical phenomenon proportional to the circular polarization rate. Most often, this does not give a measure of the polarization rate but it allows to evidence the presence of circular polarization and to compare quantitatively polarization rates. Magnetic Compton scattering has been used by Golovchenko et al. (1986), Mills (1988). We have used X-ray magnetic circular dichroism (XMCD) on DCI-11 station (dispersive absorption spectrometer) at LURE (Orsay-France).

5.2. XMCD experiments to test the efficiency of diamond phase plates in forward-diffracted beam geometry

It is necessary here to present briefly the principle of the energy-dispersive spectrometer. A detailed description is given in Tolentino et al. 1988 and Dartyge et al., 1992. A curved crystal focusses at a point F a polychromatic beam issued from a source far away from its Rowland circle (Figure 4). A linear array of small diodes placed on the Rowland circle analyses at once the whole spectrum transmitted by a sample placed at the focus. The detector is positioned on the Rowland circle because, due to the non-zero size of the source, each wavelength is diffracted by the curved crystal (polychromator) with a non-zero divergence. As is well-known each divergent monochromatic beam converges on the Rowland circle.

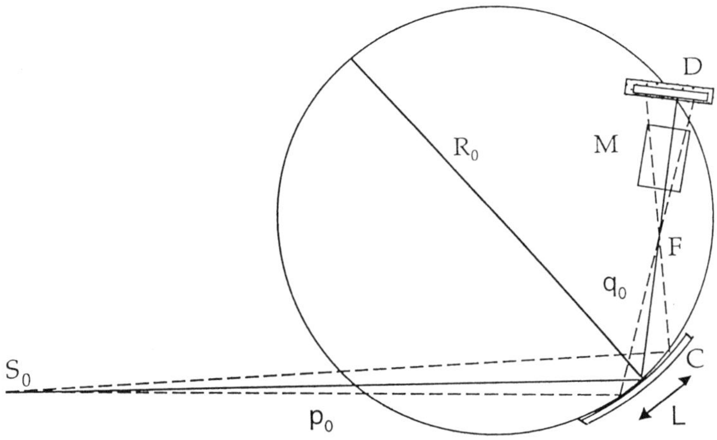

Figure 4. Schematic drawing of an energy-dispersive setup. R_0 is the radius of curvature of the bent crystal C and is equal to the diameter of the Rowland circle. The detector (array of diodes) is placed on the Rowland circle. A mirror M eliminates the harmonics.

Most magnetic samples present different absorption coefficients (or dichroism) for right- and left-circularly polarized X-rays around their absorption edge. The analysis of this magnetic circular dichroism gives information on the spin-dependent transitions with the advantage of a chemical selectivity (Schütz et al., 1989, Chen et al., 1990, Dartyge et al., 1992).

Practically, XMCD experiments are not performed by measuring the differences between absorption coefficients for right- and left-circularly polarized photons but by using right- *or* left-circularly polarized photons and measuring the difference between the absorption coefficients for a magnetic field parallel and antiparallel to the photon wave vector.

XMCD experiments on the DCI-11 station at LURE are usually done by what will be called in the following the standard technique *i.e.* by selecting with a slit the beam below the orbit plane (for X-rays emitted by positrons deviated by a bending magnet) in order to get elliptically polarized photons. In order to test diamond crystals as quarter-wave plates (QWP) the slit is adjusted to select the linearly polarized beam in the orbit plane. The diamond plate is placed between the polychromator and the sample on a θ–2θ goniometer which can be inclined at a 45° angle with respect to the horizontal plane thanks to a rotation axis parallel to the incident mean wave vector. The linearly polarized beam incident on the phase plate has then equal σ and π components with respect to the diamond diffraction plane. The offset is adjusted in order to get a π/2 phase-shift between these two components. The difference between absorption spectra with parallel and antiparallel magnetic fields is recorded and compared to the same experiment performed with the standard technique. Figure 5 shows such a comparison of the XMCD signals obtained with a ferrimagnetic $HoFe_2$ sample at the Ho L_{III} absorption edge (8071 eV and 1.534 Å). The QWP was a (111) diamond crystal 740 μm thick, adjusted in an asymmetric transmission geometry for a $11\bar{1}$ reflection. The beam path is nearly normal to the plate, the π/2 offset is equal to about 50 arcsec. The signal obtained with the QWP is higher which proves that the circular polarization rate obtained with the QWP is higher than with the standard technique. Furthermore, the intensity is also higher because the intensity lost by the absorption of the QWP is lower than lost by going away from the orbit plane.

It should be pointed out at this stage that, with the energy dispersive spectrometer, it is not possible to choose the reflecting plane on the QWP at will : in order to get the same phase-shift for the whole spectral range (equation 11), the offset ΔΘ has to be the same for all wavelengths (here one neglects the smaller variation due to the variation of λ in (11). It means that all the wavelengths should be together at Bragg reflection for a given position of the QWP. Then the right offset ΔΘ for all wavelengths is obtained by a ΔΘ rotation of the QWP. It means that the crystals should be in a non-dispersive geometry.

In the case of interest here one crystal is curved and the other one is flat and their diffraction planes make an angle ψ. The non-dispersive condition is given by a relation between the lattice spacings d_{hkl} of both crystals which includes the angle ψ (Giles et al., 1994a). Thanks to this degree of freedom and to a low sensitivity of the circular polarization rate to variations of ψ around 45° (equation 13), several couple of reflections satisfying the

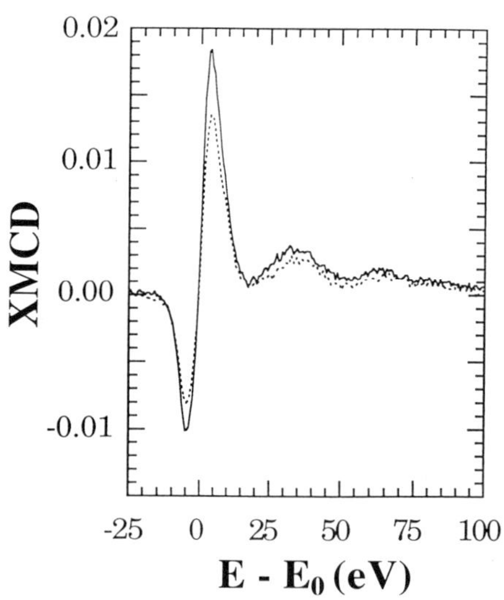

Figure 5. XMCD spectra of $HoFe_2$ alloy at the HoL_{III} edge whose energy is E_0=8071 eV. Full line : QWP technique ; dotted line : standard technique.

non-dispersive condition for a bent silicon crystal and a diamond QWP have been found (Si 311 and C 220, Si 111 and C 111). The wavelength tunability was obtained by varying ψ and evidenced with XMCD spectra at PrL_{II}, GdL_{III}, HoL_{III} and TmL_{III} absorption edges whose energies are 6440eV (1.925 Å), 7243 eV (1.712 Å), 8071 eV (1.536 Å) and 8648 eV (1.434 Å°) respectively (Giles et al., 1994b).

Different XMCD spectra have been recorded on the same $HoFe_2$ sample at the HoL_{III} absorption edge for different offsets. The height of the signals is proportional to the polarization rate. From the study of the variation of this height as a function of the offset it has been possible to evaluate the rate of circular polarization which could be obtained in such a set-up where the divergence of the beam is about 80 arcsec. The value was 70% if the incident beam is assumed to be completely linearly polarized. In fact, because of the vertical size of the source and of the finite size of the slit, the initial linear polarization rate was measured to be about 90% and then the circular polarization rate to be 63%.

It is worth pointing out that the polarization can be easily switched from right- to left-handed by a simple rotation of the plate (Hirano et al., 1992b ; Giles et al.,1994b,1995b).

5.3. Mosaic crystals

It has been shown in the previous paragraphs that perfect crystals are efficient phase plates for the forward-diffracted beam rather far from Bragg incidence even for incident beams with a rather large divergence (order of 100 arcsec). Therefore it seems probable that the crystal needs not to be totally perfect but that its mosaic spread can reach something like the divergence of the beam. Experiments similar to those described above have been performed with a beryllium crystal whose mosaic spread was of the order of 80 arcsec and results similar to those obtained with diamond crystals have been obtained (Giles et al., 1995c).

6. X-RAY POLARIMETRY

In the previous experiments, the circular polarization rate could not be very close to 1 because of the large angular divergence of the incident beam due to the size of the source (σ_x= 2.7 mm; σ_z=1.5 mm). With a third generation source (ESRF Grenoble, France) one can expect to obtain very high polarization rates and even to use the phase plate as a half-wave plate which requires offset two times smaller.

The experimental set-up which has been used is schematically described in Figure 6. A 9.1 keV beam was monochromatized by a 111 reflection on a 0.4 mm thick diamond crystal in an asymmetric transmission geometry, resulting in an outcoming beam with an angular Lorentzian profile of about 15 arcsec FWHM. In order to reduce the angular and spectral tails of the monochromatic beam, a Si (111) channel-cut crystal was set with a diffraction plane inclined at 45° from the horizontal plane. Two diamond phase plate crystals were independently inserted in the monochromatic beam. The first phase plate (beam-path of 0.99mm resulting from an asymmetric $11\bar{1}$ reflection in a (111) diamond crystal 0.74mm thick with a negative asymmetry) was also set with a diffraction plane at 45° while the second one (beam path of 1 mm resulting from an asymmetric $11\bar{1}$ reflection in a (111) diamond crystal 1 mm thick with a positive symmetry) mounted on a circle could be set with a diffraction plane at different orientations.

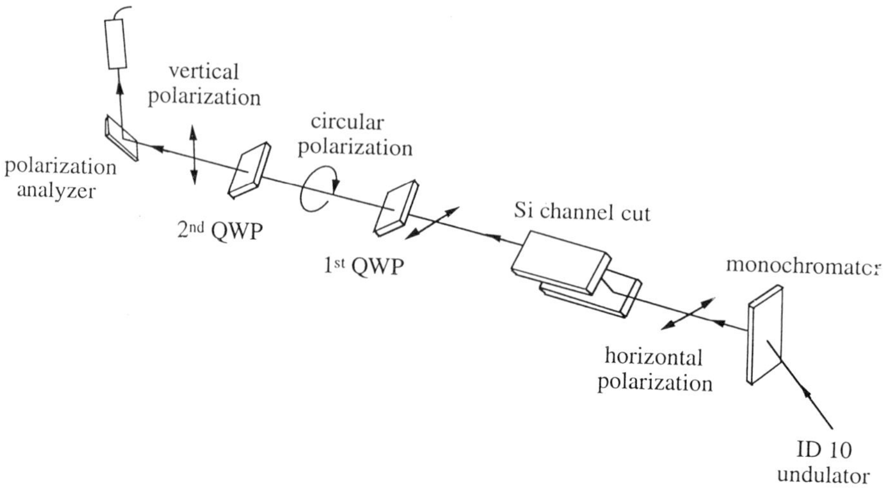

Figure 6. Schematic drawing of the setup used to measure the polarization.

The last optical element was a linear polarimeter based on integrated intensity $I(\alpha)$ measurements of powder lines at Bragg angles θ_B close to 45° as a function of the azimuthal angle α whose principle is described by Suortti and Materlik (1984). With this polarimeter it was possible to measure the linear polarization rates P_1 and P_2 (see appendix) through the relation

$$I(\alpha)=I_0\,(1+\cos^2 2\theta_B+\cos 2\alpha\,\sin^2 2\theta_B\,P_1+\sin 2\alpha\,\sin^2 2\theta_B\,P_2)$$

where I_0 is the incident intensity (Vaillant 1977). Several measurements have been made : i) after the diamond monochromator ($P_1 = 0.998$ and $P_2 = -0.02$), ii) after the Si channel-cut crystal (Figure 7 dashed line, $P_1 = 0.97$, $P_2 = -0.09$), iii) after the first phase plate set at a QWP offset ($\Delta\Theta=42$ arcsec) giving $P_1 = 0.00\pm0.01$and $P_2 = 0.08\pm0.01$ (dotted line in Fig.7), iv) after the first phase plate set at a half-wave plate (HWP) offset ($\Delta\Theta= -21$ arcsec) giving $P_1 = -0.95\pm0.01$ and $P_2 = -0.05\pm0.02$. The results are very good since we obtain with the HWP a highly linearly polarized beam in the vertical plane giving an efficiency of 98% in the rotation of the polarization plane. At this stage we can infer that a very high circular polarization rate is obtained in case iii) but in order to distinguish between a circularly polarized and an unpolarized beam it is necessary to use another QWP which transforms the circularly polarized radiation into a linear polarization and has no effect on the unpolarized radiation. If the wave after the first QWP is circularly polarized, the second QWP (adjusted at

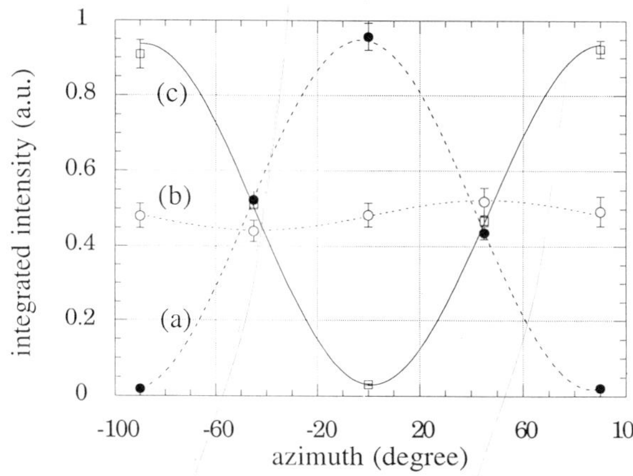

Figure 7. Intensity measured by the linear polarimeter as a function of the azimuthal angle α. For $\alpha = 0$ the diffraction by the powder occurs in the vertical plane and the polarimeter measures the horizontal component of the electric field. (a) Dashed line: after the Si channel-cut. (b) Dotted line : after the first phase plate set at QWP offset. (c) after the phase plate set at HWP offset.

the same offset as the first one) with a plane of diffraction parallel to the first one ($\chi = 45°$) transforms it into a linearly polarized wave perpendicular to the initial linear polarization as shown in Figure 8 (full line) where the measured value of P_1 is - 0.97. It proves that the combination of the two QWP equivalent to a single HWP completely rotates the incoming horizontal linear polarization into the vertical plane and then that the efficiency of each QWP is close to 100%. If the plane of diffraction of the second QWP makes a 45° angle with the first one ($\chi = 0°$) then the linear polarization is inclined by 45° with respect to the first one. The measured value of P_2 was found to be equal to 0.94 (dotted line in Figure 8). The same principle has also been used with a silicon QWP to characterize the circular polarization from an insertion device (elliptical multipole wiggler) (Ishikawa et al., 1991, 1992).

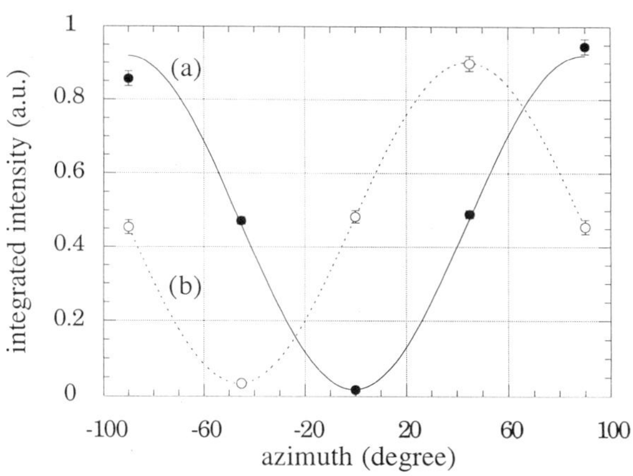

Figure 8. Polarization measurements as in Figure 7 after the two QWPs. (a) Full line : both QWPs in the same setting leading to vertical polarization. (b) Dotted line : the second QWP is set at χ=0 instead of χ=45° leading to linear polarization at 45°.

7. Some concluding remarks

Polarized X-ray research is now in full development and polarization becomes one of the characteristics of X-ray photons one wants to know and to be able to analyse and modify. In this respect one essential optical device is the phase plate. Different types of phase plates have been imagined and tested, all based on X-ray birefringence of perfect crystals at Bragg diffraction.

The most convenient device is certainly the non-absorbing phase plate (diamond for example) in forward-diffracted beam (FDB) mode which stands rather large divergences compared to the other modes but what is the wavelength range where FDB mode phase plates can be used efficiently ? In order to have a high transmission rate one should use preferentially diamond crystals. Unfortunately because of its small lattice parameter (3.567 Å) diamond does not diffract any longer for wavelengths larger than 4.1 Å. A thin silicon crystals (about 16 μm) has been used at 4.38 Å or 2.8 KeV (Goulon et al., 1996) but, naturally, the transmission was quite low (0.2 %). Towards high energies one has to consider the efficiency of QWP which decreases nearly as λ^4 (equation 11) and the absorption which decreases also but more slowly. One could say that the limiting energy is around 17 to 20 keV taking also into account that the beam path in diamond plates cannot exceed something like 5 to 10 mm. It seems that towards higher energies one should go back to phase plates using the reflected beam at Bragg incidence in Laue geometry and crystals with high structure factors like germanium but then a highly collimated beam is required and there is no wavelength tunability (Yahnke et al., 1994).

It can be considered as rather surprising that birefringence which is explained by dynamical theory of diffraction still exists rather far from the diffraction profile and that non-perfect crystals (as for example a beryllium crystal with a "mosaicity" of about 80 arcsec) still present a birefringence. In fact, the birefringence given by formula (10) can be retreived (de Bergevin, 1994) by considering all possible two successive scatterings on coherent lattice planes, each scattering being then polarization dependent ; the consideration of one scattering only (a forward scattering independent of polarization) would give the ordinary index of refraction. Dynamical theory leading to all specific properties of perfect crystals inside the domain of reflection is valid if the number of coherent lattice planes is high. Far from this domain, the required number of coherent planes is smaller and then the birefringence becomes less sensitive to imperfections in the crystal.

Finally, a few words should be said about X-ray rotary power which can be evidenced between crossed linear polarizers. The first attempts to evidence X-ray rotary power were not convincing (Hart & Rodrigues, 1981, Sauvage et al., 1983). The reason was that it can be observed only very near an absorption edge as was evidenced later (Siddons et al., 1990). In fact, rotary power and circular dichroism are the same phenomenon *i.e.* different indices of refraction for right- and left-handed circularly polarized photons (real part for rotary power and imaginary part for dichroism) and the two effects are related through a Kramers-Kronig transformation.

REFERENCES

Batterman W., 1992, X-ray phase plates, *Phys. Rev., B*45, 12677-12681.

Belyakov V.A. and Dmitrienko V.E., 1989, Polarization phenomena in X-ray optics, *Sov. Phys. Usp.*, 32, 697-719.

Born and Wolf, 1959, "Principles of Optics," Pergamon press, London.

Chen C.T., Sette F., Ma Y., Modesti S., Smith N.V., 1990, *Phys.Rev., B*42, 7262-7265.

Cole H., Chambers F.W. and Wood C.G., 1961, X-ray polarizer, *Journal Appl.Phys.*, 32, 1942-1945.

Dartyge E., Fontaine A., Baudelet F., Giorgetti C., Pizzini S. and Tolentino H., 1992, An essential property of synchrotron radiation : linear and circular polarization for X-ray absorption spectroscopy, *J.Phys. I France,* 2, 1233-1255.

De Bergevin, 1994, Private communication.

De Bergevin F. and Brunel M., 1972, Observation of magnetic superlattice peaks by X-ray diffraction on an antiferromagnetic NiO crystal, *Phys. Lett., A*39, 141-142.

De Bergevin F. and Brunel M., 1981, Diffraction of X-rays by magnetic materials I. General formulae and measurements on ferro- and ferrimagnetic compounds, *Acta Cryst., A*37, 314-324.

De Bergevin F.and Brunel M., 1991, Magnetic scattering, in "Handbook of Synchrotron Radiation", vol. 3, 535-564, North Holland, Amsterdam.

Dmitrienko V.E. and Belyakov V.A., 1980, Polarization conversion of X-rays in single crystals, *Sov. Tech. Phys. Lett.*, 6, 621-622.

Elleaume P., 1992, Insertion devices for the new generation of synchrotron sources : a review, *Rev. Sci. Instrum.*, 63, 321-326.

Elleaume P., 1994, Helios : a new type of linear/helical undulator, *J. Synchrotron Rad.*, 1, 19-26.

Giles C., Malgrange C., Goulon J., de Bergevin F., Vettier C., Dartyge E., Fontaine A., Giorgetti C. and Pizzini S., 1994a, Energy-dispersive phase plate for magnetic circular dichroism experiments in the X-ray range, *J.Appl.Cryst.*, 27, 232-240.

Giles C., Malgrange C., Goulon J., de Bergevin F., Vettier C., Fontaine A., Dartyge E., and Pizzini S., 1994b, Energy and polarization-tunable X-ray quarter-wave plates for energy-dispersive absorption spectrometer, *Nucl. Instr. and Meth., A*349, 622-625.

Giles C., Vettier C., de Bergevin F., Malgrange C., Grübel G. and Grossi F., 1995a, X-ray polarimetry with phase plates, *Rev. Sci. Instrum.*, 66, 1518-1521.

Giles C., Malgrange C., Goulon J., de Bergevin F., Vettier C., Dartyge E., Fontaine A., Pizzini S., Baudelet F. and Freund A., 1995b, Perfect crystal and mosaic crystal quarter-wave plates for circular magnetic X-ray dichroism experiments, *Rev. Sci. Instrum.,* 66, 1549-1553.

Giles C., Malgrange C., de Bergevin F., Goulon J., Baudelet F., Fontaine A., Vettier C. and Freund A., 1995c, Mosaic crystals as X-ray phase plates, *Nucl. Instrum. and Meth., A*361, 354-357.

Giles C., Malgrange C., Goulon J., de Bergevin F., Vettier C., Dartyge E., Fontaine A., Pizzini S., Baudelet F. and Freund A., 1995d, Tunable X-ray quarter-wave plates for X-ray magnetic circular dichroism experiments with the energy-dispersive absorption spectrometer, *Physica, B*208 & 209, 784-786.

Golovchenko J.A., Kincaid B.M., Levesque R.A., Meixner A.E. and Kaplan D.R., 1986, Polarization Pendellösung and the generation of circularly polarized X-rays with a quarter-wave plate, *Phys. Rev. Lett.*, 57, 202-205.

Goulon J., Elleaume P. and Raoux D., 1987, Special multiple wiggler design producing circularly polarized synchrotron radiation, *Nucl. Instr. and Meth., A*254, 192.

Goulon J., Malgrange C., Giles C., Neumann C., Rogalev A. Moguiline E., de Bergevin F., Vettier C., 1996, Design of an X-ray phase plate analyser to measure the circular polarization rate of a helical undulator source, to appear in *J. Synchrotron Rad.*, 3.

Hart M. and Lang A.R., 1965, The influence of X-ray polarization on the visibility of Pendöllusung fringes in X-ray diffraction topographs, *Acta Cryst.*, 19, 73.

Hart M., 1978, X-ray polarization phenomena, *Phil. Mag.*, 38, 41-56.

Hart M. and Rodrigues A.R.D., 1979, Tuneable polarizers for X-rays and neutrons, *Phil. Mag.*, 40, 149-157.

Hart M. and Rodrigues A.R.D., 1981, Optical activity and the Faraday effect at X-ray frequencies, *Phil. Mag. B*43, 321-332.

Hattori H., Kuriyama H. and Kato N., 1965, Effects of X-ray polarization on Pendellösung fringes, *J. Phys. Soc. Japan*, 20, 1047-1050.

Hirano K., Izumi K., Ishikawa T., Annaka S. and Kikuta S., 1991, An X-ray phase plate using Bragg-case diffraction, *Jpn. J. Appl. Phys. Lett.* , 30, L407-L410.

Hirano K., Kanzaki K., Mikami M., Miura M., Tamasaku K., Ishikawa T. and Kikuta S., 1992a, Tunable-wavelength production of circularly polarized X-rays with a perfect crystal quarter-wave plate, *J. Appl. Cryst.*, 25, 531-535.

Hirano H., Ishikawa T., Koreeda S., Fuchigami K., Kanzaki K., and Kikuta S., 1992b, Switching of photon helicities in the hard X-ray region with a perfect crystal phase retarder, *Jpn. J. Appl. Phys.*, 31, L1209-L1211.

Hirano K., Ishikawa T. and Kikuta S., 1995, Development and application of X-ray phase retarders, *Rev. Sci. Instrum., 66*, 1604-1609.

Ishikawa T., Hirano K. and Kikuta S., 1991, Complete determination of polarization state in the hard X-ray region, *J. Appl. Cryst.*, 24, 982-986.

Ishikawa T., Hirano K., Kanzaki K. and Kikuta S., 1992, A multiple crystal diffractometer for generation and characterization of circularly polarized X-rays at the Photon Factory, *Rev. Sci. Instrum.,* 63, 1098-1103.

Kim K.J. 1984, A synchrotron radiation source with arbitrarily adjustable elliptical polarization, *Nucl. Instr. and Meth.*, 219, 425.

Lefeld-Sosnowska M. and Malgrange C., 1969, Experimental evidence of plane wave rocking curve oscillations, *Phys. Stat. Sol.*, 34, 635-647.

Mills D.M., 1988, Phase-plate performance for the production of circularly polarized X-rays, *Nucl. Instr. and Meth., A*266, 531-537.

Molière G., 1939, Quantenmechanische Theorie der Röntgenstrahlinterferenzen in Kristallen, *Ann. Phys.* 3, 272-313.

Sakai N., and Ono K., 1976, Compton profile due to magnetic electrons in ferromagnetic iron measured with circularly polarized γ rays, *Phys. Rev. Lett.* 37, 351-353.

Sauvage M., Malgrange C. and Petroff J.F., 1983, Rotatory power measurements in the X-ray range with synchrotron radiation : experimental set-up and preliminary results for $NaBrO_3$ single crystals, *J. Appl. Cryst.*, 16, 14-20.

Schutz G., Frahm R., Mautner P. Wienke R., Wagner W. Wilheim W. and Kienle P., 1989, Spin-dependent extended X-ray absorption fine structure : probing magnetic short-range order, *Phys. Rev. Lett.*, 62, 2620-2623.

Siddons D.P., Hart M., Amemiya Y. and Hastings J.B., 1990, X-ray optical activity and the Faraday effect in cobalt and its compounds, *Phys. Rev. Lett.*, 64, 1967-1970.

Skalicky P. and Malgrange C., 1972, Polarization phenomena in X-ray diffraction, *Acta Cryst.*, A28, 501-507.

Suortti P. and Materlik G., 1984, Measurement of the polarization of X-rays from a synchrotron source, *J.Appl. Cryst.*, 17, 7-12.

Tolentino H., Dartyge E., Fontaine A. and Tourillon G., 1988, X-ray absorption spectroscopy in the dispersive mode with synchrotron radiation : optical considerations, J. *Appl. Cryst.*, 21, 15-21.

Vaillant F., 1977, Utilisation des paramètres de Stokes dans le calcul de l'état de polarisation d'un faisceau de rayons X après diffraction par un cristal mosaïque, *Acta Cryst., A*33, 967-970.

Yahnke C.J., Srajer G., Haeffner D.R., Mills D.M., Assoufid L., 1994, Germanium X-ray phase plates for the production of circularly polarized X-rays, *Nucl. Instr. and Meth.*, A347, 128-133.

Yamamoto S., Kawata H., Kitamura H., Ando M., Saki N., Shiotami N., 1989, First production of intense circularly polarized hard X-rays from a novel multipole wiggler in an accumulation ring, *Phys. Rev. Lett.*, 62, 2672-2675.

APPENDICES

A1. Some general results of optics

We follow here the notations, in Chapter 10 of Born and Wolf (1959). A plane wave $\boldsymbol{E}$ in vacuum whose wave vector is $\mathbf{k_0}$ ($|\mathbf{k_0}| = k_0 = 1/\lambda$) is written as

$$\boldsymbol{E} = \mathbf{E_0} \cos 2\pi(\mathbf{k_0}.\mathbf{r}-\nu t) = \mathrm{Re}\, \mathbf{E_0} \exp i2\pi(\mathbf{k_0}.\mathbf{r}-\nu t).$$

We choose an orthogonal coordinate system with **Oz** axis parallel to the wave vector $\mathbf{k_0}$. Then $E_x = E_{0x} \mathrm{Re}[\exp i2\pi(k_0 z-\nu t)]$, $E_y = E_{0y} \mathrm{Re}\, [\exp i2\pi(k_0 z-\nu t)]$

leading to the Jones vector notations

$$E_x = E_{0x} \exp i2\pi k_0 z, \quad E_y = E_{0y} \exp i2\pi k_0 z \quad \text{or} \quad E_x = E_{0x}, \quad E_y = E_{0y}$$

since a common phase change in both components can be done at will : it is equivalent to a change in time origin.

If such a plane wave goes through a crystal plate whose thickness is z and which presents different indices of refraction n_x and n_y for E_x and E_y components, the wave after the plate becomes

$$E_x = E_{0x} \exp i2\pi k_0 n_x z, \quad E_y = E_{0y} \exp i2\pi k_0 n_y z$$

$$\text{or } E_x = E_{0x}, \; E_y = E_{0y} \exp i2\pi k_0 (n_y - n_x) z = E_{0y} \exp i\phi$$

$$\text{with } \phi = \frac{2\pi}{\lambda}(n_y - n_x) z$$

If $\phi = \pm\pi/2$ and $E_{0x} = E_{0y} = E_0$ one gets after the crystal

$$E_x = E_0 \cos 2\pi(k_0 z - \omega t), \; E_y = -(\pm) E_0 \sin 2\pi(k_0 z - \omega t).$$

At a given time the extremity of the electric vector **E** describes a right-handed helix (resp. left-handed) for the lower (resp. upper) sign and will be said to be right-handed (resp. left-handed) circularly polarized. This definition has the advantage not to depend on the direction one looks at, since the right or left character of an helix is intrinsic.

Finally, with our conventions on the form of the wave ($\exp i2\pi(\mathbf{k.r}-\omega t)$), the wave is right-handed (resp. left-handed) circularly polarized for $\phi = -\pi/2$ (resp. $+\pi/2$) and has a Jones vector $(E_0, -iE_0)$ (resp. (E_0, iE_0)).

A2. Stokes parameters

In general the electric field is a superposition of different waves emitted by different charged particles which leads to define the following average values : $\langle E_x E_x^* \rangle$, $\langle E_y E_y^* \rangle$, $\langle E_x E_y^* \rangle$ and $\langle E_x^* E_y \rangle$ generally presented as a matrix J. **Oz** is the direction of propagation of the waves and **Ox** and **Oy** are two orthogonal directions in the plane perpendicular to **Oz**.

The four Stokes parameters s_0, s_1, s_2 and s_3 are defined as follows:

$$s_0 = \langle E_x E_x^* \rangle + \langle E_y E_y^* \rangle = J_{xx} + J_{yy},$$

$$s_1 = \langle E_x E_x^* \rangle - \langle E_y E_y^* \rangle = J_{xx} - J_{yy},$$

$$s_2 = 2 \,\mathrm{Re} \langle E_x E_y^* \rangle = 2 \,\mathrm{Re}\, J_{xy},$$

$$s_3 = 2 \,\mathrm{Im} \langle E_x E_y^* \rangle = 2 \,\mathrm{Im}\, J_{xy},$$

s_0 is the total intensity,

$P_1 = s_1/s_0$ is the linear polarization rate in the two orthogonal directions **Ox** and **Oy** ,

$P_2 = s_2/s_0$ is the linear polarization rate for components inclined 45° with respect to **Ox** and **Oy**,

$P_3 = s_3/s_0$ is the circular polarization rate defined as $P_3 = (I_R - I_L) / (I_R + I_L)$.

P_1, P_2 and P_3 are called Poincaré components.

A totally polarized beam is such that $I_{total} = s_0 = (s_1^2 + s_2^2 + s_3^2)^{1/2}$

The degree of polarization R is defined as $R = I_{polarized} / I_{total} = \dfrac{(s_1^2 + s_2^2 + s_3^2)^{1/2}}{s_0}$.

STATISTICAL THEORY OF DYNAMICAL DIFFRACTION IN CRYSTALS

Norio Kato

S 512, Hoshigaoka Iris, Meito-ku
Meito-Honmachi, Nagoya, Japan

1. INTRODUCTION

Originally, the theory to be described below was initiated for a better understanding of extinction phenomena (Kato, 1976a, b). In the second stage (Kato, 1980a, b), it was aimed in addition at interpreting observations on the contrast and spacing of Pendellösung fringes in nearly perfect crystals. In order to explain the underlying physics, a brief historical review on classical extinction theories and on the fringe behaviour in homogeneously bent crystals will be given in the next section. Next, the basic concepts (coherence and incoherence of crystal waves) used in the present theory are explained and an outline of the theory is given. Sections 4 and 5 are devoted to more technical details.

The theory is developed by using two approaches which are different from the methodological viewpoint, viz. path-integral (constructive) and differential approaches. The former is adequate to understand the global picture of intensities in crystals, whereas the latter gives the local relations among the various intensities in the form of differential equations.

In the final section, a few unsettled problems are discussed in connection with recent work of Al Haddad & Becker (1988), Becker & Al Haddad (1990, 1992), Guigay (1989) and Guigay & Chukhovskii (1992). Also, a few comments will be added on relevant statistical theories of dynamical diffraction having different styles from the present one.

2. HISTORICAL REVIEW

2.1. Extinction

In good crystals, the integrated intensity of strong Bragg reflections is often less than the value expected by kinematical theory, which assumes only single scattering. This phenomenon, called "extinction", may cause serious problems in the practice of accurate structure determination. Obviously, the true integrated intensity must be evaluated by dynamical theory, which takes multiple scattering into account.

In the early period of crystal diffraction, eminent pioneers like Darwin, Ewald and Bragg, already obtained clear physical insights into this problem. They classified the phenomena into primary and secondary extinctions (PE and SE) starting from the wave equations for PE and the energy transfer equations for SE. In other words, they proposed two independent theories for the two extreme cases, in which the crystal waves are either perfectly coherent or perfectly incoherent. The perfect incoherence can be justified to some extent by the intuitive model of "mosaic crystals".

Around 1960, were derived the following wave equations and energy transfer equations, which are adequate to treat two-dimensional problems in a diffraction (reflection) plane:

The wave equations read (Takagi, 1962; Taupin, 1964):

$$(\partial/\partial s_o)\, d_o(s_o,s_g) = i\kappa_o\, d_o + i\kappa_{-g}\, d_g \qquad (2.1a)$$

$$(\partial/\partial s_g)\, d_g(s_o,s_g) = i\kappa_o\, d_g + i\kappa_g\, d_o \ , \qquad (2.1b)$$

whereas the energy transfer equations (Hamilton, 1957) can be written as:

$$(\partial/\partial s_o)\, P_o(s_o,s_g) = -(\mu_o + \sigma)\, P_o + \sigma\, P_g \qquad (2.2a)$$

$$(\partial/\partial s_g)\, P_g(s_o,s_g) = -(\mu_o + \sigma)\, P_g + \sigma\, P_o \ . \qquad (2.2b)$$

In these equations, d_o and d_g denote direct (O) and Bragg-reflected (G) waves respectively, P_o and P_g denote the power of the O and G beams per unit area, and μ_o is the linear absorption coefficient. In addition, (s_o, s_g) are the oblique coordinates along the O and G directions, and the coupling constants (κ_o, $\kappa_{\pm g}$ and σ) represent the strength of diffraction per unit length of (s_o, s_g). The details will be discussed in Subsec. 2.3.

Although the two theories could provide some practical recipes to get rid of extinction effects, most people were not satisfied with the ad hoc separation of PE and SE, because in principle they should be unified. One motivation of the present theory is to attempt this unification on the sound basis of optics.

2.2. Perfect crystals

For simplicity, let us consider non-absorbing crystals under the symmetrical Laue conditions. Also, it is assumed that the incident wave is monochromatic and sufficiently narrow so that the expression $A_e\delta(x_o) = (A_e/\sin 2\theta_B)\delta(s_g)$, where $x_o = s_g \sin 2\theta_B$ is the coordinate perpendicular to the O direction, may be adopted. This treatment is called spherical wave theory (Kato, 1961a, b; 1968a, b). The solution of Eq.(2.1) is given by

$$d_g(s_o, s_g) = A_e\,(i\,\kappa_g/\sin 2\theta_B)\,J_0(\,2\kappa\sqrt{s_o s_g}\,)\,\exp[\,i\,\kappa_o\,(s_o + s_g)] \qquad s_o, s_g > 0$$

$$= 0 \qquad s_o, s_g < 0 \qquad (2.3)$$

where J_0 is the first kind of Bessel function of 0-th order and $\kappa^2 = \kappa_g\,\kappa_{-g}$. The intensity field is schematically drawn in Figure 1. It is confined within the Borrmann fan. The thick contours indicate the Pendellösung fringes.

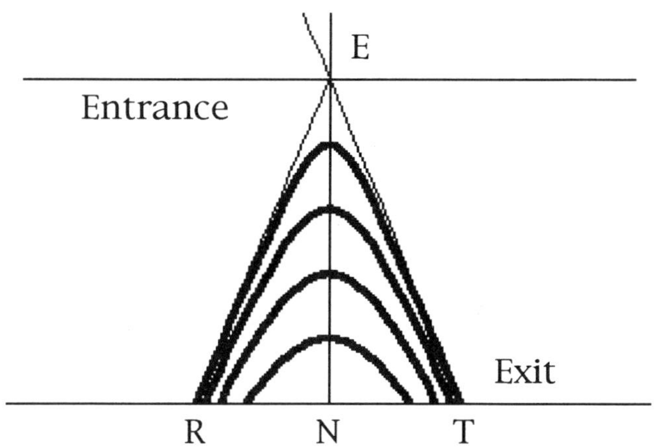

Figure 1. Schematic behaviour of the intensity field (G-wave) for ideally perfect crystals. E : The entrance point of a narrow wave. The triangle ETR : The Borrmann fan. EN: The bisector of the angle TER.

The normalized amplitude $|\,d_g\,|$ along the net plane is shown in Figure 2. The integrated intensity R_g is given by the integral of $|\,d_g\,|^2$ over the exit surface of the crystal. It turns out to be a function of the crystal thickness T, which has the form shown in Figure 3 (Kato, 1968b). Incidentally, the plane wave theory gives the same result. On the other hand, it is well known that the kinematical value R_g^K represented by the curve K is proportional to T. The deviation of R_g from this linearity implies PE.

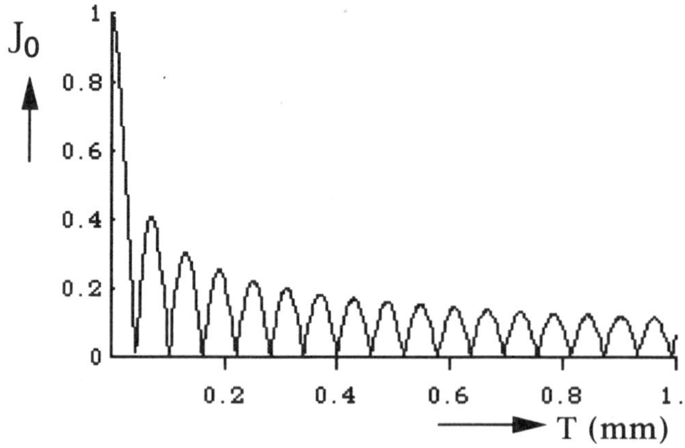

Figure 2. The normalized amplitude of the G-wave [J_0 in Eq.(2.3)] for perfect crystals along EN of Figure1. (Λ = 60 μm, θ_B = 10 degree, which will be used also in Figures 3 and 4).

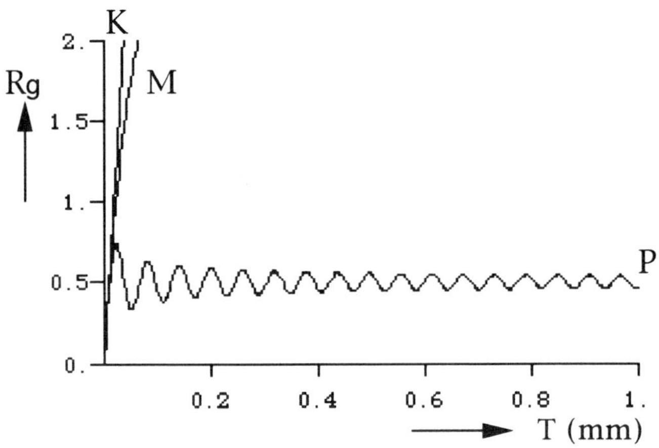

Figure 3. The integrated intensities for the ideally perfect crystal (P), a mosaic crystal (M) and the ideally mosaic crystal (K).

2.3. Mosaic crystals

Traditionally, Eqs.(2.2) have been solved under the boundary conditions that an homogeneously wide beam impinges on the crystal. Here, however, in order to compare with the results of Sec. 2.2, the similar conditions are used; i.e. the incident beam has the form of $P_e\delta(x_o)$. Because Eqs.(2.1) and (2.2) have the similar mathematical structure, one can easily see that the solution has the form,

$$P_g = (P_e\, \sigma/\sin 2\theta_B)\; I_0(\,2\sigma \sqrt{s_o s_g}\,)\; \exp[-\,\sigma\,(s_o + s_g)]\,, \qquad s_o, s_g > 0 \qquad (2.4)$$

where I_0 is the modified Bessel function [$I_0(x) = J_0(ix)$]. Here, we shall present the expressions for κ_g and σ.

$$\kappa_g = (\lambda\, r_c/v)\; C\; |F_g| \equiv \pi/\Lambda\;, \qquad (2.5)$$

$$\sigma = QW, \qquad Q = \lambda\kappa^2/\sin 2\theta_B\;, \qquad (2.6a, b)$$

where r_c is the classical radius of electron, v the volume of a unit cell, C the polarization factor and F_g is the structure factor. Λ, defined in Eq.(2.5), is called the extinction length.

The expression (2.5) is derived from the basic Maxwell equations so that there is no ambiguity. Q in Eqs.(2.6) is the power reflectivity of the net plane per unit path length in single crystals. It is also well known in the kinematical theory. However, in mosaic crystals, the beams may not always encounter mosaic blocks which create the Bragg reflection. Therefore, the effective reflectivity must be modified by a factor W. In the classical treatment of SE the factor is assumed to be the angular distribution $W(\varphi)$ of the blocks. This intuitive assumption is not straightforward, so that some modifications are proposed to improve the arguments (for examples, Zachariasen, 1945; Becker & Coppens, 1974a, b; the review article of Werner et al., 1986). In any case, a solution like Eq.(2.4) is angular-dependent through $W(\varphi)$ so that the integrated intensity R_g must be a double (spatial and angular) integral. When the distribution $W(\varphi)$ is sufficiently broad, viz. QW(0) is sufficiently small, P_g can be approximated by the first factor in Eq.(2.4). Then, the integrated intensity R_g amounts to the kinematical value R_g^K. This case is called "ideally mosaic". In general, as shown by the curve M of Figure 3, R_g deviates from K. This behaviour implies nothing else but SE.

2.4. Crystals with constant strain gradient

In considering the connection between PE and SE, it seems worthwhile to discuss the case in which the crystal perfection can be characterized by a single parameter. In general, the lattice distortion can be represented by the displacement field $\mathbf{u}(\mathbf{r})$ of the lattice points from their position in the perfect state. Then, the coupling constants of Eq.(2.1) are modified as $\kappa_g = \kappa_g\, \Phi$ and $\kappa_{-g} = \kappa_{-g}\, \Phi^*$, where

$$\Phi = \exp[-2\pi\, (\mathbf{g}.\mathbf{u})] \qquad (2.7)$$

is called the *lattice phase factor*, which obviously is a function of position through $\mathbf{u}(\mathbf{r})$.

When the crystal has a constant strain gradient (as example, in homogeneously bent crystals), the lattice phase has the quadratic form

$$\phi = 2\pi\,(\mathbf{g}.\mathbf{u}) = \alpha s_o s_g + \beta(s_o)^2 + \gamma(s_g)^2 \,. \tag{2.8}$$

Since, however, the last two terms can be eliminated by a suitable unitary transformation of the wave fields, only the first term on the r.h.s. is retained in the following treatment.

In this particular case, the solution of Eq.(2.1) can be written in the form (Katagawa and Kato, 1974),

$$d_g(s_o, s_g) = A_e\,(i\,\kappa_g/\sin 2\theta_B)\;{}_1F_1(\,f;\,1;\,-\,\phi)\,\exp[\,i\,\kappa_o\,(s_o + s_g)] \qquad s_o,\,s_g > 0 \tag{2.9}$$

$$f = -\,i\,\kappa^2/\alpha \tag{2.10}$$

where ${}_1F_1(\,a;\,c;\,z)$ is Kummer's confluent hypergeometric function.

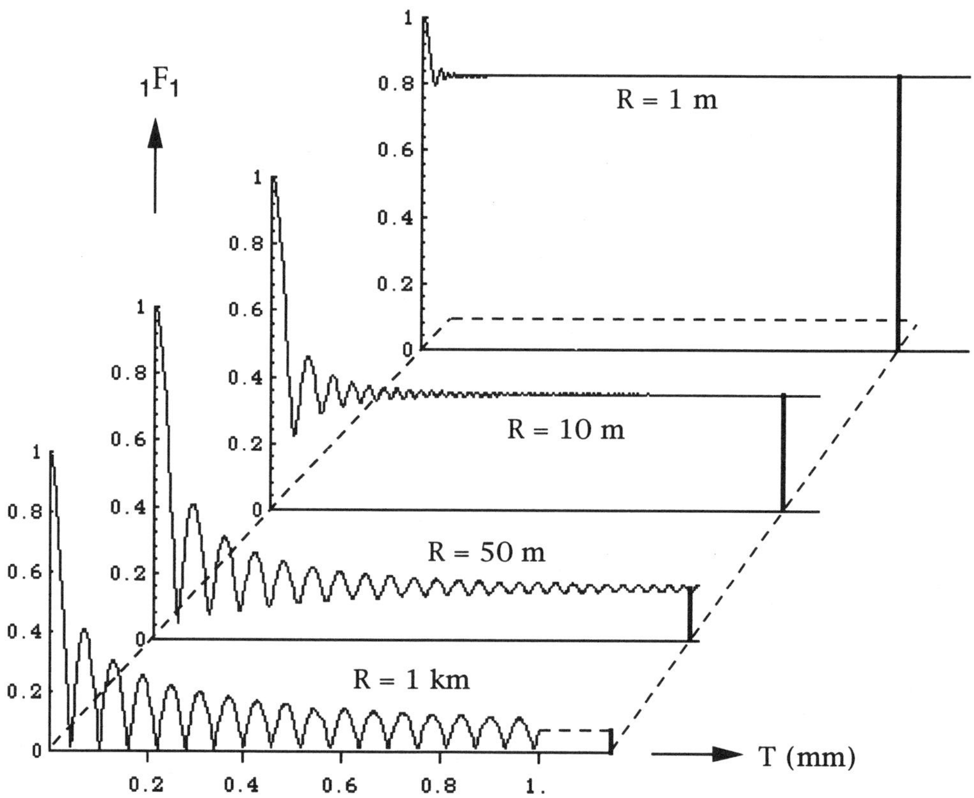

Figure 4 The normalized amplitudes [${}_1F_1$ in Eq.(2.9)] for homogeneously bent crystals. R : The radius of curvature, which is proportional to α^{-1}.

Figure 4 illustrates the amplitude distributions along the net plane. As the bending (α) increases, the background of the fringes is enhanced and the fringe amplitude decreases. In the following the behaviour is referred to as the rule of contrast. We notice also that the (local) fringe spacing decreases with increasing α and the depth from the entrance surface (the rule of spacing). The distribution for a large radius of curvature R = 1 km is nearly the same as for the ideally perfect case shown in Figure 2.

Although the "rule of contrast" represents apparently the transition from PE to SE in a qualitative sense, the theory is not very satisfactory. First, the model of the lattice distortion is too specific to apply the results to real crystals in general. Second, even when Si crystals are handled carefully to avoid lattice bending, an appreciable background is detected in real experiments (Wada & Kato, 1977). Third, contradicting to the above rule of spacing, recent observations of Si crystals show often that the fringe spacing is larger than the theoretical value of the perfect crystal (see the review article by Schneider et al., 1992). This is particularly true in crystals grown with the use of quartz crucibles, for which there is a lot of evidence to believe that the crystal includes tiny Si-oxide clusters.

3. THE OUTLINE OF THE STATISTICAL THEORY

The considerations at the end of the previous section suggest that one needs a new statistical treatment for a better understanding of the diffraction phenomena, although the classical SE theories include implicitly statistical considerations. When, as in the cases of perfect and bent crystals, the lattice phase factor Φ is specified uniquely as a function of position, Eq.(2.1) can be solved in principle, either analytically or numerically. In mosaic crystals and crystals including tiny defects, however, one can not use a similar procedure, because Φ is known only in a statistical sense. In other words, Φ is a statistical quantity rather than an ordinary function of position. Consequently, the wave fields also are statistical quantities. Mathematically, they are a "functional" of the variable Φ. Under these circumstances, the best way is to start by postulating for the observed intensities

$$I^{ob} \equiv \langle d^* d \rangle = \langle d^* \rangle \langle d \rangle + \langle \delta d^* \delta d \rangle \tag{3.1}$$

where the symbols < > and δ indicate the ensemble average and the fluctuations from the average, respectively. Here, the suffices o and g are suppressed in the relevant quantities. Henceforth, this convention will often be used.

3.1. Coherence and incoherence of the crystal wave

The first term on the r.h.s.of Eq.(3.1) is called the *coherent intensity* , I^{co}, and the second is called the *incoherent intensity*, I^{in}. This is nothing else but the definition of coherence and incoherence. Since, however, there are good reasons for this usage, this terminology is widely employed in optics.

In the ideally perfect crystal, Φ is a constant so that the statistical ensemble consists of a single entity. In this pedagogical example, I^{ob} is simply I^{co} because $< d > = d$ and δd does not exist. If the crystal is randomly and highly distorted, as will be discussed in Sec. 4, $< d >$ is nearly zero but $< \delta d^* \ \delta d >$ may still be appreciable. In this case, I^{ob} must be described by I^{in}. In general, I^{co} and I^{in} are superposed so that, hopefully, one can connect PE and SE on the sound basis of optics.

3.2. Statistical properties of the random variable Φ

The simplest one is the ensemble average

$$<\Phi> = E \quad , \tag{3.2}$$

which is called the static Debye-Waller factor in an analogy with the case of thermal vibrations. We assume that E is a real constant (i.e. spatial homogeneity and isotropy).

The next simplest one is the spatial correlation function defined by

$$G(z) \equiv <\Phi^*(\mathbf{r})\ \Phi(\mathbf{r} + z)> = <\Phi^*><\Phi> + <\delta\Phi^*(\mathbf{r})\ \delta\Phi(\mathbf{r} + z)> \tag{3.3a}$$

$$= E^2 + (1 - E^2)\ g(z) \tag{3.3b}$$

where $\mathbf{r}$ indicates an arbitrary position, and the function g(z) is called the intrinsic correlation function, with the normalization g(0)=1. The factor $(1 - E^2)$ is required in Eq.(3.3b), because $< \Phi^*(\mathbf{r})\ \Phi(\mathbf{r}\ ) >$ is always unity. Here, homogeneity and isotropy are assumed also for $g(\mathbf{z})$ so that it is simply a function of the separation distance z. In this context, the correlation lengths are defined by

$$\tau_n = \int_0^\infty \{g(z)\}^n \, dz \ . \tag{3.4}$$

In Figure 5, the functions G(z) is schematically illustrated. Higher order correlations are conceivable, but they are disregarded in the rest of this article.

The present theory attempts to obtain I^{co} and I^{in} in terms of $\{ E, \tau_n \}$, which characterize the statistical nature of the crystalline media. The theory is not concerned directly with any specific model of the lattice distortion.

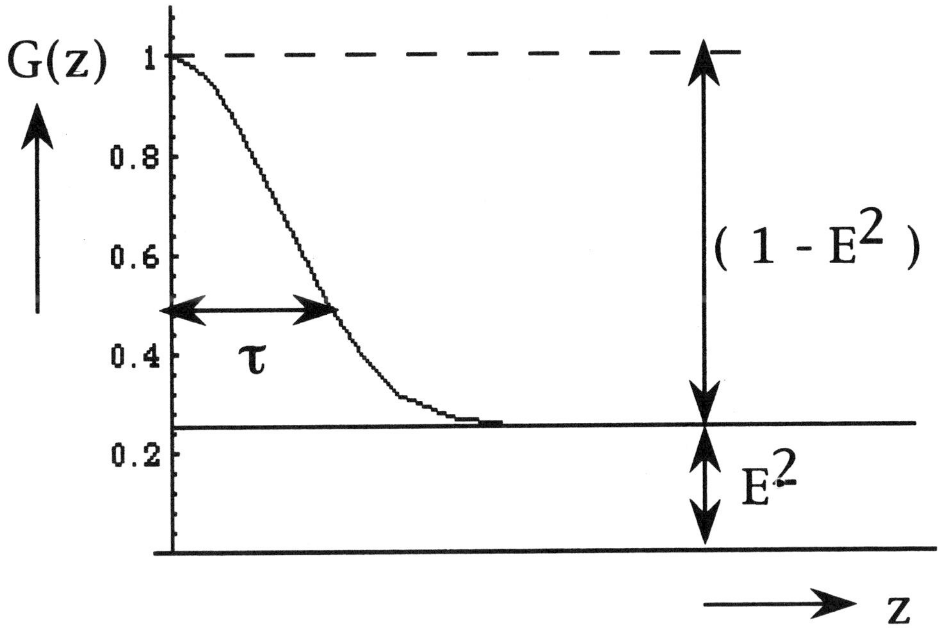

Figure 5. The schematic behaviour of the correlation function G(z) defined by Eq.(3.3). τ : The correlation length. E : The static Debye-Waller factor.

4. CONSTRUCTIVE (OR PATH-INTEGRAL) APPROACH

The incident wave is assumed to be $(A_e/\sin 2\theta_B)\ \delta(s_g)$ as in Subsecs. 2.2 and 2.4. However, the amplitude factor is omitted in the following treatments, for convenience.

4.1. The wave fields, d_o and d_g

First, we are concerned with a single wavelet specified by a zigzag path, which starts from the entrance point E and arrives at an observation point P (Figure 6). The number of kinks must be even and odd for the O and G waves, respectively. At the kinks, **g** and - **g** reflections take place alternately.

With these wavelets, one can construct the O-wave field at $P(s_o, s_g)$ in the form

$$d_o(s_o, s_g) = \delta(s_g) + \Sigma_{r=1}\, S_{2r}\, (i\kappa_g)\ \Phi_1\ (i\kappa_{-g})\ \Phi^*_2 \cdots (i\kappa_{-g})\ \Phi^*_{2r}\ , \tag{4.1a}$$

where the suffices of $\{\Phi, \Phi^*\}$ indicate primarily the sequence order of kinks and implicitly also the location of the kinks. The first term on the r.h.s. implies the non-diffracted wave corresponding to r = 0. The rest is the O-wave due to even diffraction events. Notice that it exists only for $s_g > 0$. The symbol S_{2r} implies the sum (integral) of the wavelets over all possible paths having 2r kinks within the parallelogram specified by (s_o, s_g).

Similarly, the G-wave can be written in the form

$$d_g(s_o, s_g) = \Sigma_{r=0} S_{2r+1} (i\kappa_g) \Phi_1 (i\kappa_{-g}) \Phi^*_2 \cdots (i\kappa_g) \Phi_{2r+1} . \qquad (4.1b)$$

Again, the wave field can exist only for $s_g > 0$. Incidentally, in the kinematical case (r = 0), we have

$$d_g(s_o, s_g) = (i\kappa_g) \Phi(s_o, 0) . \qquad (4.2)$$

4.2. The coherent waves < d >

In the following, it is assumed that $\kappa^2 = \kappa_g\kappa_{-g}$ is the same as in the case of perfect crystals and only $\{\Phi\}$ are random variables. Then, the averaged fields have the forms,

$$<d_o>(s_o,s_g) = \delta(s_g) + \Sigma_{r=1} S_{2r} (-\kappa^2)^r <\Phi_1 \Phi^*_2 \cdots \Phi^*_{2r}> \qquad (4.3a)$$

$$<d_g>(s_o,s_g) = i\kappa_g \Sigma_{r=0} S_{2r+1} (-\kappa^2)^r <\Phi_1 \Phi^*_2 \cdots \Phi_{2r+1}> . \qquad (4.3b)$$

For simplicity, we shall consider the correlation of Φ's up to second order. Then, as shown in Figure 6, each wavelet consists of isolated kinks and kink pairs in an arbitrary configuration. If any one of the isolated kinks is associated with the fluctuation $\delta\Phi$, the wavelet does not contribute to the average field. After factorizing < phase factors >, all possible configurations of the isolated kinks and kink pairs must be summed up.

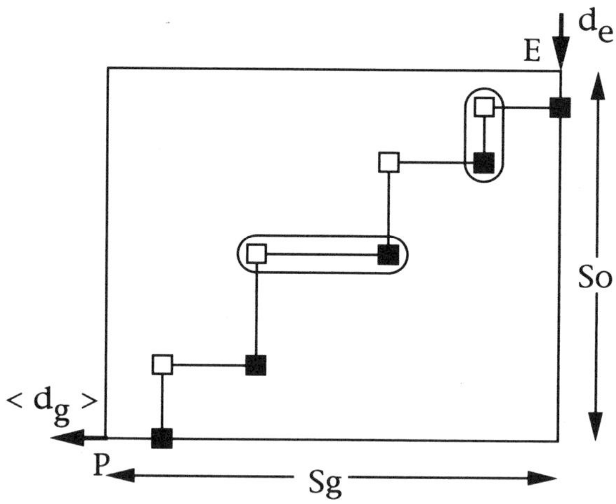

Figure 6. A topological illustration of a wavelet (G-wave) and the factorization into isolated kinks and kink pairs. Black and white squares show the kinks associated with $\kappa_g\Phi$ and $\kappa_{-g}\Phi^*$, respectively. Elliptically enclosed ones are the kink pairs.

In order to perform this additional summation, as well as Σ and S in Eq.(4.3), a mathematical technique using the Laplace transform is very powerful (Kato, 1994). Here, the final results are presented; viz.

$$<d_o>(p,q) = G_g\,[\,G_o G_g + (\kappa\, E)^2\,]^{-1} \tag{4.4a}$$

$$<d_g>(p,q) = i\kappa_g\, E\,[\,G_o G_g + (\kappa\, E)^2\,]^{-1}\ , \tag{4.4b}$$

where (p, q) are the conjugate variables associated with (s_o, s_g), respectively, in the two-dimensional Laplace transform, and

$$G_o(p,q) = p + (1 - E^2)\,\kappa^2\, g(q)\ , \tag{4.5a}$$

$$G_g(p,q) = q + (1 - E^2)\,\kappa^2\, g(p)\ , \tag{4.5b}$$

$$g(p) = \int_0^\infty g(z_o)\, \exp(-\,p\, z_o)\, dz_o\,,\quad g(q) = \int_0^\infty g(z_g)\, \exp(-\,q\, z_g)\, dz_g\ . \tag{4.6a, b}$$

The coherent waves are given by the inverse Laplace transform of $<d_o>(p,q)$ and $<d_g>(p,q)$ given by Eqs.(4.4).

For ideally perfect crystals (E = 1), Eqs.(4.4) are reduced to

$$<d_o>(p,q) = q\,/\,[\,p\,q + \kappa^{\,2}\,]\ , \tag{4.7a}$$

$$<d_g>(p,q) = i\kappa_g\,/\,[\,p\,q + \kappa^{\,2}\,]\ . \tag{4.7b}$$

The inverse Laplace transforms are easily obtained and the results are identical to those of the spherical wave theory [Eq.(2.3) for $<d_g>(s_o, s_g)$].

4.3. The expressions for $(\partial/\partial s_o)<d_o>$ and $(\partial/\partial s_g)<d_g>$

In order to obtain these expressions, we calculate the Laplace transforms, p $<d_o>(p,q)$ and q $<d_g>(p,q)$ from Eqs.(4.4). After some algebraic manipulations, they can be found to be,

$$p<d_o> = (i\,\kappa_{-g}E)<d_g> - (1 - E^2)\,\kappa^2\, g(q) <d_o> + 1\ , \tag{4.8a}$$

$$q<d_g> = (i\,\kappa_g E) <d_o> - (1 - E^2)\,\kappa^2\, g(p) <d_g>\ . \tag{4.8b}$$

Here, the arguments (p, q) are suppressed in $<d_o>$ and $<d_g>$. Then, the inverse Laplace transforms can be written as,

$$(\partial/\partial s_o)\langle d_o\rangle(s_o,s_g) = \delta(s_o)\delta(s_g) + (i\,\kappa_{-g}E)\,\langle d_g\rangle(s_o,s_g)$$

$$- (1 - E^2)\,\kappa^2 \int_0^{s_g} g(z_g)\,\langle d_o\rangle(s_o,\, s_g - z_g)\, dz_g \tag{4.9a}$$

$$(\partial/\partial s_g)\langle d_g\rangle(s_o,s_g) = (i\,\kappa_g E)\,\langle d_o\rangle(s_o,s_g)$$

$$- (1 - E^2)\,\kappa^2 \int_0^{s_o} g(z_o)\langle d_g\rangle(s_o - z_o,\, s_g)\, dz_o\ . \tag{4.9b}$$

The δ-function in Eq.(4.9a) implies the incident pulse-like wave and as long as we are interested only in the wave fields in s_o, $s_g > 0$, this term can be omitted. The details of deriving Eqs.(4.9) and the implications are described by Kato (1994).

4.4. Approximate treatments

So far everything is exact. For practical purposes, however, we need some approximations. In most problems, the effective range of the correlation function g(z) is the correlation length τ (see Figure 5) in the order of magnitude. On the other hand, the variation of the coherent fields < d > is gentle over the extinction distance Λ (≡ π / κ). Therefore, at least in the case of τ / Λ << 1, the coherent fields < d > can be taken out of the integrals in Eqs.(4.9) and the integrals render simply the correlation length τ_1 defined by Eq.(3.4). This treatment is called the approximation of gentle variation (AGV) of the fields. Then, the integro-differential equations (4.9) can be reduced to

$$(\partial/\partial s_o)\langle d_o\rangle = \delta(s_o)\delta(s_g) +$$

$$i\,\kappa_{-g}\,E\langle d_g\rangle - (1 - E^2)\,\kappa^2\,\tau_1\langle d_o\rangle \tag{4.10a}$$

$$(\partial/\partial s_g)\langle d_g\rangle = i\,\kappa_g\,E\langle d_o\rangle - (1 - E^2)\,\kappa^2\,\tau_1\langle d_g\rangle\ . \tag{4.10b}$$

The formulae are similar to the original wave equations (2.1) with modified coupling constants. They are easily integrated and the result for the G-wave is obtained in the form,

$$\langle d_g\rangle(s_o,\, s_g) = i(A_e\,\kappa_g E/\sin 2\theta_B)\, J_0(\, 2\kappa E\sqrt{s_o s_g}\,)$$

$$\times \exp[\, -(1-E^2)\,\kappa^2\,\tau_1\,(s_o + s_g)] \qquad s_o,\, s_g > 0\ . \tag{4.11}$$

The coherent intensity I_g^{∞} is simply given by multiplying this expression with its complex conjugate . Important is the introduction of the apparent absorption factor and the reduction of $\kappa_{\pm g}$ by E, as compared to the case of ideally perfect crystals, Eq.(2.3). Thus, an elongation of the spacing and a damping of the amplitude of the Pendellösung fringes result.

Returning to the expressions (4.8) of the Laplace transforms, one can see that the AGV used for $< d >(s_o, s_g)$ is equivalent to assuming $g(0) = \tau_1$ for g(p) and g(q). With these approximations, the same result as Eq.(4.11) can be derived through the inverse Laplace transform.

5. INCOHERENT INTENSITIES, I_o^{in} and I_g^{in}

The incoherent intensities are defined in the context of Eq.(3.1) so that one needs the expressions for the fluctuation δd of the wave fields, from which we can obtain the intensity fields I^{in} in a rather formal way. This is similar to the constructive approach described in Sec. 4. It is valuable to understand the mathematical structure of I^{in} but not necessarily convenient to obtain practical formulae. For this reason, in the section 6, we shall describe the conventional approach in which an attempt is made to derive differential relations for I^{in} directly from the differential equations (2.1).

5.1. The fluctuation fields, δd

As in the case of Eqs.(4.1), we shall consider a wavelet, first. If any one of the phase factors $\{\Phi_j\}$ is $\delta\Phi_j$, the wavelet contributes to the fluctuation fields $\delta d(s_o, s_g)$ for $s_o, s_g > 0$. Such a kink point is called a "bridge kink" for the reason explained in Subsec. 5.2. Next, we carry out the summation over all possible wavelets while fixing the bridge kinks and the entrance and exit points. Then, two of them are connected by a coherent wave field which occupies a block of the size (ρ, σ) as illustrated in Figure 7. This coherent field is called the "block field" and denoted by $C_i(\rho, \sigma)$. The suffix i indicates primarily the sequential order of the block fields. However, depending on the type of the bridge kinks at the top and the bottom of the block, C_i is either one of C_{oo}, C_{og}, C_{go} and C_{gg} in which the first and second subscripts indicate the directions of the incident and exit wavelets of each block field, respectively. Therefore, the suffix i indicates implicitly also the type of $\{C_i\}$.

The types of $\{C_i\}$ are constrained by a diffraction rule. For example, if the top bridge kink is associated with $(i\kappa_{-g}\delta\Phi^*)$ and the bottom one is $(i\kappa_g\delta\Phi)$, C_i must be C_{oo}. Also, If both of the bridge kinks are associated with $(i\kappa_{-g}\delta\Phi^*)$, C_i must be C_{og}. Thus, it turns out that when the sequence of the bridge kinks is given, the types of $\{C_i\}$ are automatically fixed and *vice versa*. Here, we define an "extended wavelet (EWt)" which consists of the bridge kinks $(i\kappa_g\delta\Phi)$, $(i\kappa_{-g}\delta\Phi^*)$ and the block fields $\{C_i\}$. The expression has the form

$$(\text{EWt}) = C_1 \, (i\kappa_g \, \delta\Phi)_1 \, C_2 \, (i\kappa_g \, \delta\Phi)_2 \cdots C_{f-1} \, (i\kappa_g \, \delta\Phi)_{f-1} \, C_f \quad , \tag{5.1}$$

where $(i\kappa_g\delta\Phi)_i$ stands for either $(i\kappa_g\delta\Phi)$ or $(i\kappa_{-g}\delta\Phi^*)$, and the minimum number of f is 2 because at least one bridge kink must be included.

5.2. The expression for $I^{in} = < \delta d^* \delta d >$

First, a product of $(EWt)^*$ and (EWt) is considered. Obviously, the kinks associated with $\{\delta\Phi\}$ in (EWt) must have a one-to-one correspondence with those of $\{\delta\Phi^*\}$ in $(EWt)^*$. They must be of the same type, and locate close enough so that the correlation $< \delta\Phi^* \delta\Phi >_i$ has an appreciable value. Otherwise, the product vanishes on average. Consequently, the number f of (EWt) is identical to f of $(EWt)^*$. This kind of pairing among conjugate waves is called bridging, which explains the term "bridge kink" introduced above. Now, we have

$$I^{in} = \Sigma_f \, S_{W'(f)} \, S_{W(f)} \, |\kappa_g|^{2m} \, |\kappa_{-g}|^{2n}$$

$$I_1 G_1 \, I_2 G_2 \cdot \cdot \cdot \, I_{f-1} G_{f-1} \, I_f \quad , \tag{5.2}$$

where m and n are the number of bridges of the respective types (m + n = f -1), and

$$I_i = C_i^*(\rho', \sigma') \, C_i(\rho, \sigma) \quad , \tag{5.3a}$$

$$G_i = [\, \delta\Phi^*(z_o, z_g) \, \delta\Phi\,(0,0)\,]_i \quad . \tag{5.3b}$$

In addition, W(f) and W'(f) imply (EWt) and $(EWt)^*$ having f - 1 bridge kinks, respectively, (ρ, σ) and (ρ', σ') are the block size, and (z_o, z_g) indicate the separation of the bridge kinks, the suffix i being suppressed. The symbol S implies the sum over the possible configurations with respect to the positions of the bridge kinks and also the types of G_i and I_i. Figure 7 illustrates schematically the expression of Eq.(5.2) in the case of G-beam.

Here, a few remarks should be made. First, as in the case of Eq.(5.1), if the sequence of the bridge kinks is given, the types of $\{I_i\}$ are automatically fixed. Next, the boundary conditions for I^{in} should be taken carefully; namely, (a) I_1 must be either I_{oo} or I_{og}, (b) I_f is either I_{oo} or I_{go} if we are concerned with I_o^{in}, and similarly (c) I_f is either I_{og} or I_{gg} in the case of I_g^{in}. Therefore, each of I_o^{in} and I_g^{in} is classified into four categories, depending on the types of I_1 and I_f.

5.3. Correlation factors

One can see the relation

$$(z_o, z_g)_i = \Sigma_k^{\,i} \, \{(\rho', \sigma')_k - (\rho, \sigma)_k\} \tag{5.4}$$

where $\Sigma_k^{\,i}$ implies the summation over the blocks preceding the i-th kink bridge. Thus, $\{z_o, z_g\}$ depend on $\{\rho', \sigma'\}$ and $\{\rho, \sigma\}$. For avoiding the complexity which arises by taking the two summations S in Eq.(5.2), henceforth, $\{z_o, z_g\}$ are dealt with as if they were independent of the variables of the block size in most parts. This treatment is equivalent to assuming that $\{\rho', \sigma'\}$ are nearly equal to $\{\rho, \sigma\}$. Since the effective range of G_i is of the order of τ, the approximation may not be very serious except for G_1 and G_f. In particular, if the block size is of the order of the extinction distance Λ or larger, the approximation is similar to the AGV used in subsection 4.4.

With this approximation, the summation of possible kink bridges renders simply the factor

$$\overline{G}_i = (1 - E^2) \int_-^+ \int_-^+ g(r)\, dz_o\, dz_g = (1 - E^2)\, 2\pi \int_0^+ r\, g(r)\, dr/\sin 2\theta_B \quad (5.5)$$

where r is the separation distance of the bridge kinks and the integral limits ± are assumed sufficiently larger than $\pm\,\tau$. $\overline{G}_i$ is called the correlation factor of the bridge. Notice that it is λ-dependent through the Bragg angle.

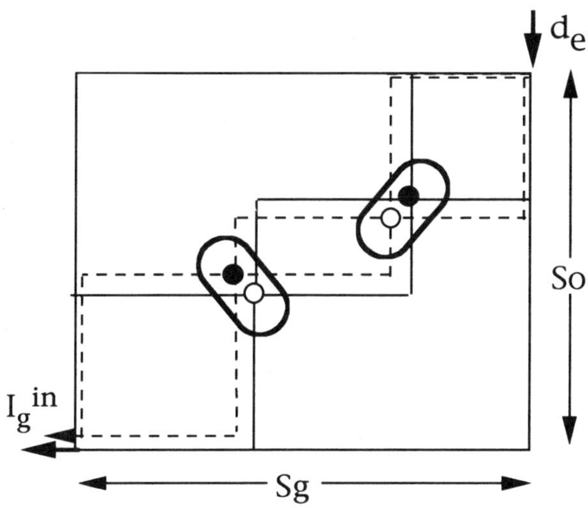

Figure 7. A topological illustration of an extended wavelet (G-wave). The dashed one is the conjugate wave. Rectangular blocks are the coherent wave fields of the block size (ρ, σ) and (ρ', σ'). Thick elliptic loops indicate the bridging kinks.

It is necessary to take a special care in treating the final correlation function G_{f-1}. If I_f happens to be unity, namely the final bridge kinks align on the edge of the parallelogram which is effective for the intensity at the observation point P, the correlation factor is

$$\overline{G}_{f-1} = (1 - E^2) \int_-^+ g(z)\, dz = 2(1 - E^2)\, \tau_1 \quad (\text{ if } I_f \text{ is unity}) \qquad (5.6a)$$

due to spatial degeneracy. The correlation length τ_1 is defined by Eq.(3.4). The same care should be observed in the first correlation function G_1; namely

$$\overline{G}_1 = 2(1 - E^2)\, \tau_1 \qquad (\text{ if } I_1 \text{ is unity}) \qquad (5.6b)$$

5.4. An approximate expression of I^{in}

Using the correlation factors instead of the correlation functions, we obtain an approximate formula from Eq.(5.2)

$$I^{in} = \Sigma_f\, S_{W(f)}\, |\kappa_g|^{2m}\, |\kappa_{-g}|^{2n}\, \bar{I}_1\, \bar{G}_1\, \bar{I}_2\, \bar{G}_2 \cdot\cdot\cdot\, \bar{I}_{f-1}\, \bar{G}_{f-1}\, \bar{I}_f \ , \tag{5.7}$$

in which the single summation S is sufficient, and

$$\bar{I}_i = C_i^*(\rho, \sigma)\, C_i(\rho, \sigma) \tag{5.8}$$

implies the coherent intensity of the block size (ρ, σ). The essence of the approximation is the assumption $(\rho', \sigma') \approx (\rho, \sigma)$.

The summations in Eq.(5.7) can be performed by use of the Laplace transform, as in the case of Eq.(4.3). However, not only I_o^{in} and I_g^{in} but also the four categories depending on the type of $(\bar{I}_1, \bar{I}_f)$ must be separately treated (see the last paragraph of Subsec. 5.2). The calculations are rather complicated and we do not treat the technical details here.

6. AN APPROACH BASED ON DIFFERENTIAL EQUATIONS

In this section, an alternative to the constructive approach, which has been developed earlier (Kato, 1980a), will be explained.We start from Eqs.(2.1), but κ_0-terms are omitted because they can be eliminated by a suitable unitary transformation. First, the differential equations for the coherent fields <d> are discussed. Then, the coherent intensities I^{co} are derived straightforwardly. Next, the incoherent intensity fields I^{in} and their differential equations are discussed. This treatment is called the "differential approach".

6.1. The coherent fields

If the lattice phase factors are written explicitly, the average of Eqs.(2.1) gives the relations:

$$(\partial/\partial s_o)\langle d_o\rangle = i\,\kappa_{-g} \langle \Phi^* d_g\rangle \ , \tag{6.1a}$$

$$(\partial/\partial s_g)\langle d_g\rangle = i\,\kappa_g \langle \Phi\, d_o\rangle \ . \tag{6.1b}$$

On the other hand, the integral form of Eq.(2.1b) can be written as

$$d_g(s_o, s_g) = i\,\kappa_g \int_0^{s_g} \Phi(s_o, s_g - z_g)\, d_o(s_o, s_g - z_g)\, dz_g \ . \tag{6.2}$$

Inserting Eq.(6.2) into Eq.(6.1a) and taking the average, we have

$$(\partial/\partial s_o)\langle d_o\rangle = i\,\kappa_{-g}\, E \langle d_g\rangle - (1 - E^2)\,\kappa^2 \int_0^{s_g} g(z_g)\,\langle d_o\rangle(s_o, s_g - z_g)\, dz_g \ . \tag{6.3}$$

In this manipulation, only the correlation of $\{\Phi\}$ between neighbouring kinks is taken into account. Figure 8(a) illustrates the two terms of the r.h.s.. A similar integro-differential equation is obtained for $(\partial/\partial s_g)\langle d_g\rangle$ from Eq.(6.1b) and the integral form of Eq.(2.1a) [see Figure 8 (b)].

If the AGV explained in Subsec. 4.4 is used, these integro-differential equations are reduced to the differential equations,

$$(\partial/\partial s_o)\langle d_o\rangle = i\,\kappa_{-g}\, E \langle d_g\rangle - (1 - E^2)\,\kappa^2\,\tau_1\langle d_o\rangle \tag{6.4a}$$

$$(\partial/\partial s_g)\langle d_g\rangle = i\,\kappa_g\, E \langle d_o\rangle - (1 - E^2)\,\kappa^2\,\tau_1\langle d_g\rangle \ . \tag{6.4b}$$

These are the same as Eqs.(4.10) except for the inhomogeneous term $\delta(s_o)\ \delta(s_g)$ which represents the incident wave on the entrance surface. Therefore, in order to solve Eqs.(6.4), one needs boundary conditions for $\langle d_o\rangle$, which fit to any incident wave. More will be discussed in Subsec. 6.4.

It is straightforward to obtain I^{co} from $\langle d\rangle$. Nevertheless, as a preparation to the next subsection, the differential relations for I^{co} are given here. By the manipulation of $[\langle d_o^*\rangle \times (6.4a) + \langle d_o\rangle \times (6.4a)^*]$, we obtain

$$(\partial/\partial s_o)\, I_o^{co} = i\,\kappa_{-g}\, E \langle d_o^*\rangle\langle d_g\rangle - (1 - E^2)\,\kappa^2\,\tau_1\, I_o^{co} + \text{c.c.} \ , \tag{6.5a}$$

where c.c. stands for the complex conjugate of the preceding expressions. Similarly, we have

$$(\partial/\partial s_g)\, I_g^{co} = i\,\kappa_g\, E \langle d_g^*\rangle\langle d_o\rangle - (1 - E^2)\,\kappa^2\,\tau_1\, I_g^{co} + \text{c.c.} \ . \tag{6.5b}$$

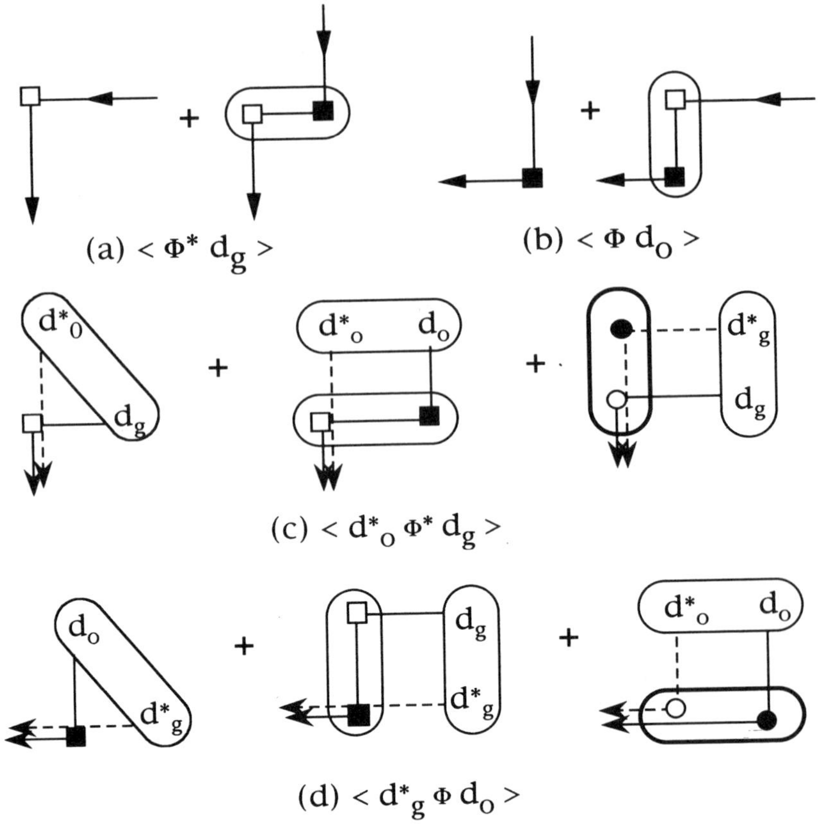

Figure 8. The correlations taken into account in the differential approach. (See Sec.6.)

6.2. The averaged intensity fields

Similarly to Eqs.(6.5), manipulations like [$d_o^* \times$ (2.1a) + $d_o \times$ (2.1a)*] and taking the average give the following relations:

$$(\partial/\partial s_o)<I_o> = (i\,\kappa_{-g})<d_o^*\,\Phi^*\,d_g> + \text{c.c.} \tag{6.6a}$$

$$(\partial/\partial s_g)<I_g> = (i\,\kappa_g)<d_g^*\,\Phi\,d_o> + \text{c.c.}\,. \tag{6.6b}$$

Each of the averages on the r.h.s. consists of three terms, which are schematically illustrated in Figure 8(c) and (d).

Again, by use of the AGV, we obtain the differential equations,

$$(\partial/\partial s_o)<I_o> = i\,\kappa_{-g}E<d_o^*\,d_g>$$
$$-\,(1 - E^2)\,\kappa^2\,\tau_1<I_o> + (1 - E^2)\,|\kappa_{-g}|^2\,\tau_1<I_g> + \text{c.c.} \tag{6.7a}$$

$$(\partial/\partial s_g)<I_g> = i\,\kappa_g E<d_g^*\,d_o>$$

$$- (1 - E^2)\, \kappa^2\, \tau_1 < I_g > + (1 - E^2)\, |\kappa_g|^2\, \tau_1 < I_o > + \text{c.c.}\ . \qquad (6.7b)$$

These are the relations for the observed intensities, I_o^{ob} and I_g^{ob}, by the postulate (3.1).

In the case of non-absorbing crystals [$\kappa_g = \kappa_{-g} = \kappa$], one can show the conservation law among I_o^{ob} and I_g^{ob}, i.e.

$$(\partial/\partial s_o)\, I_o^{ob} + (\partial/\partial s_g)\, I_g^{ob} = 0 \qquad (6.8)$$

6.3. The incoherent intensity fields

By subtracting each of Eqs.(6.5) from Eqs.(6.7) and using the relation, $< I > = I^{co} + I^{in}$, one can obtain the relations:

$$(\partial/\partial s_o)\, I_o^{in} = i\, \kappa_{-g} E < \delta d_o^* \, \delta d_g > + (1 - E^2)\, |\kappa_{-g}|^2\, \tau_1\, I_g^{co}$$

$$- (1 - E^2)\, \kappa^2\, \tau_1\, I_o^{in} + (1 - E^2)\, |\kappa_{-g}|^2\, \tau_1\, I_g^{in} + \text{c.c.} \qquad (6.9a)$$

$$(\partial/\partial s_g)\, I_g^{in} = i\, \kappa_g E < \delta d_g^* \, \delta d_o > + (1 - E^2)\, |\kappa_g|^2\, \tau_1\, I_o^{co}$$

$$- (1 - E^2)\, \kappa^2\, \tau_1\, I_g^{in} + (1 - E^2)\, |\kappa_g|^2\, \tau_1\, I_o^{in} + \text{c.c.}\ . \qquad (6.9b)$$

The third and fourth terms on the r.h.s. of each equation represent the energy balance of I^{in} in the crystal, which was intuitively postulated in the classical SE theory [cf. Eqs.(2.2)]. The physical meaning of the second terms is the source of I^{in}, which are transferred from I^{co}. The first terms also are a source of I^{in} due to the fluctuations of the wave fields.

Using Eq.(6.2) and the similar one for d_o and $d = <d> + \delta d$, one can write as

$$<\delta d_o^* \, \delta d_g > = (i\, \kappa_g)\, E \int_0^{s_g} <\delta d_o^*(s_o, s_g)\, \delta d_o(s_o, s_g - z_g)> dz_g$$

$$+ (i\, \kappa_{-g})^*\, E \int_0^{s_o} <\delta d_g^*(s_o - z_o, s_g)\, \delta d_g(s_o, s_g)> dz_o + \text{c.c.} \qquad (6.10a)$$

$$= (i\, \kappa_g)\, E\, \Gamma\, I_o^{in} + (i\, \kappa_{-g})^*\, E\, \Gamma\, I_g^{in}\ , \qquad (6.10b)$$

where ΓI_o^{in} and ΓI_g^{in} correspond to the two integrals in Eq.(6.10a), respectively. At present, unfortunately, it is not easy to evaluate Γ precisely. We shall discuss this point later on. Nevertheless, if we use the expression (6.10b) for $< \delta d_o^* \, \delta d_g >$ and its complex conjugate, one can rewrite Eqs.(6.9) in the concise forms,

$$(\partial/\partial s_o) I_o^{in} = (1 - E^2)\, \sigma_{-g}\, I_g^{co} - \tilde{\sigma} I_o^{in} + \tilde{\sigma}_{-g}\, I_g^{in} \tag{6.11a}$$

$$(\partial/\partial s_g) I_g^{in} = (1 - E^2)\, \sigma_g\, I_o^{co} - \tilde{\sigma} I_g^{in} + \tilde{\sigma}_g\, I_o^{in} \quad , \tag{6.11b}$$

where

$$\sigma_g = 2\,|\kappa_g|^2\, \tau_1 \qquad \sigma_{-g} = 2\,|\kappa_{-g}|^2\, \tau_1 \tag{6.12a,b}$$

$$\tilde{\sigma}_g = 2\,|\kappa_g|^2\, \tau_e \qquad \tilde{\sigma}_{-g} = 2\,|\kappa_{-g}|^2\, \tau_e \tag{6.12c,d}$$

$$\tilde{\sigma} = 2\mathrm{Re}(\kappa^2)\, \tau_e \ , \tag{6.12e}$$

$$\tau_e = (1 - E^2)\, \tau_1 + E^2\, \Gamma \quad . \tag{6.13}$$

6.4. The boundary conditions for I^{co} and I^{in}

First of all, in considering the boundary conditions, it is important to note that Eqs.(6.5) and (6.11) are derived by using the AGV. Precisely speaking, therefore, they are non-local equations in the sense that the coordinates s_o and s_g are ambiguous over the correlation length τ. The situation is similar to that in the fluid-dynamical equations which are not valid on a molecular scale. Since, however, we are interested in the case that τ is much smaller than the spatially resolvable distance (not only less than Λ), Eqs.(6.5) and (6.11) are dealt with as if they are ordinary differential equations. The opposite limit is out of the scope of the present theory. In such cases, individual lattice defects would be observable by diffraction topography. These subtle problems are pointed out in (Kato, 1980a).

In the case of a spherical wave, it is convenient to impose boundary conditions near the edges of the Borrmann fan. Also, it is recommended that the constructive approach be consulted in which the boundary conditions are exactly taken into account. Inside the edge ET of Figure 1, where the kinematical theory is valid, the wave field is given by Eq.(4.2) and we have

$$\langle d_g^*(s_o, s_g)\, d_g(s_o+\varepsilon, s_g)\rangle = |\kappa_g|^2 \{ E^2 + (1-E^2)\, g(\varepsilon) \} \qquad \Lambda \gg s_g > 0 \ . \tag{6.14}$$

In the expression on r.h.s. the two terms are deliberately separated by introducing an infinitesimal quantity ε. They can be interpreted to be the coherent and incoherent intensities, respectively. Thus, the boundary conditions inside the line ET of Figure 1 are

$$I_g^{co} = |\kappa_g|^2\, E^2 \ , \qquad I_g^{in} = |\kappa_g|^2\, (1-E^2) \ . \tag{6.15a,b}$$

The additional condition that no incoherent beam arrives from the outside of the Borrmann fan is required. Explicitly, it is stated as

$$I_o^{in} = 0 \qquad \text{(outside of ER)} \qquad (6.15c)$$

In this respect, the present theory is conceptually different from the old SE theory, in which an incoherent beam is assumed on the crystal surface. Incoherent beams may not cause any Bragg reflection!! All intensity fields, I^{co} and I^{in}, are created by the incident *wave field.*

Finally, it is worth mentioning that the condition (6.15a) is redundant, because it is already taken into account in Eq.(6.11a) through its first term.

7. DISCUSSION AND CONCLUSIONS

7.1. General scheme of the present theory

Figure 9 illustrates a flowchart for calculating the wave and intensity fields in the differential approach (Kato, 1981). The coherent and incoherent intensities (I^{co}, I^{in}) are defined by Eq.(3.1). They are calculated by the standard methods for the differential equations (4.10) and (6.11). The solution (4.11) gives an elongation of the fringe spacing and a damping of the fringe amplitude. The background is enhanced by the presence of I^{in}.

Incidentally, if the long range order E is zero, the coherent wave is quenched, but unless the short range order τ_1 is very small the incoherent intensity still exists in the crystal. On the contrary, if E = 1 (perfect crystals), only the coherent wave exists but does not create incoherent beams, because Eqs.(6.11) gives the null solution owing to the boundary conditions (6.15b and c).

In some cases, in particular for the extinction problems, we need the integrated intensities (R^{co}, R^{in}). They are obtained by integrating (I^{co}, I^{in}) over the exit surface. The calculations are performed by Kato (1980b), Al Haddad & Becker (1988) and Guigay (1989). The latter two references point out careless mistakes in the first paper. The method of the last paper is neat, although it seems applicable only to the symmetrical Laue case. In general, the detailed information on diffraction is obtained by examining (I^{co}, I^{in}), provided that the Bragg angle is sufficiently large.

7.2. Unsettled problems

a) Γ problems: Unfortunately, Eqs.(6.9) is not closed with respect to the independent variables (I_o^{in}, I_g^{in}). For this reason, and partly in order to fit the present formulation to the old SE theory, a kind of correlation length Γ is introduced for $\langle \delta d^*_o\, \delta d_g \rangle$ [cf. Eqs.(6.10)] in a rather intuitive way. Kato (1980a) suggested $\Gamma \approx \Lambda/E$. A few authors attempt to find the better expression in terms of parameters which characterize the crystalline media (Becker & Al Haddad, 1990; 1992; Polyakov et al., 1991). They suggested $\Gamma \approx \tau\ (\ll \Lambda)$.

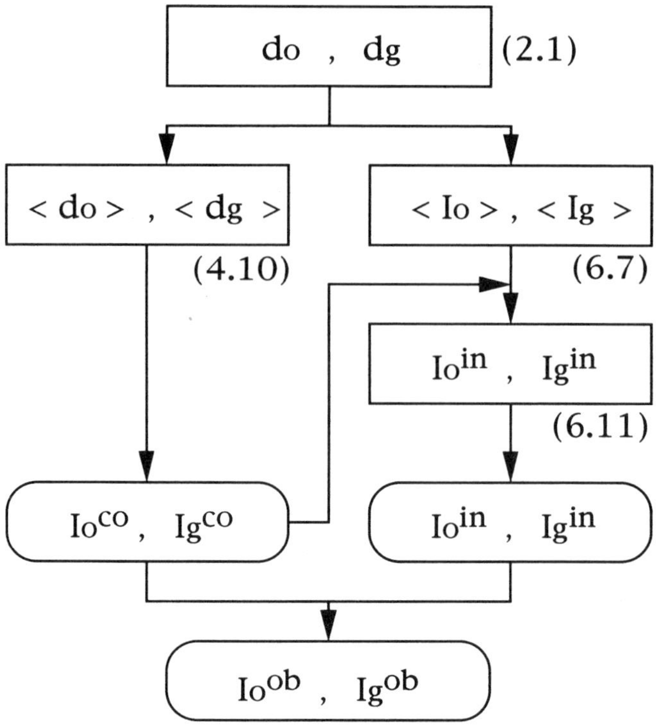

Figure 9. A flowchart for calculating the observed intensity I^{ob} in the differential approach. The rectangular blocks show the relevant differential equations, the associated (.) being the equation number in the text. Elliptic loops show their solutions.

b) τ problems: The correlation length τ appearing in Eqs.(6.4) for the coherent waves $< d >$ is τ_l. This is justified both by the constructive and differential approaches. Within the differential scheme, it turns out that τ is also τ_l in Eqs.(6.9) for I^{in}. However, Becker and Al Haddad (1990; 1992) reformulated the theory and claimed that τ_l should be replaced by τ_2. In this theory, $<d^*_o \, \Phi \, d_g>$ and $<d^*_g \, \Phi^* \, d_o>$ in Eqs.(6.6) are calculated up to fourth order in κ, and $g(z_o, z_g) = g(z_o)\, g(z_g)$ with the special form of $g(z) = \exp(- z/\tau)$ is assumed. Since the present author cannot follow their procedures, in particular the last assumption (see Subsec. 5.3), this problem is not discussed here any further.

Since Eqs.(6.11) involve these unsettled problems, the solutions of (I_o^{in}, I_g^{in}) may not be strictly justified for testing experimental results. In fact, the Laplace transforms of (I_o^{in}, I_g^{in}) based on Eqs.(6.11) do not agree with the theoretically better expressions based on the constructive approach. For this reason, it is desirable to obtain (I_o^{in}, I_g^{in}) directly from the constructive approach through the inverse Laplace transform. When Eqs.(6.11) is used for convenience, it is recommended that τ_l and τ_e be regarded as parameters to be determined experimentally rather than the theoretically defined ones [cf. Eqs.(3.4) and (6.13)].

7.3. Other relevant works

a) Diffuse scattering: The theory described in this article is based on the wave equations of the Takagi-Taupin type, which belong to the category of hyperbolic partial differential equations (e.g. Sommerfeld, 1949). The characteristic lines (physically speaking, optical paths) are, therefore, confined to zigzag paths. This simplifies the optical picture considerably but results in unrealistic situations. As an example, let us consider the kinematical solution (6.14 and 15). Then, the two terms I_g^{co} and I_g^{in} are nothing else but the Bragg reflection and diffuse scattering, respectively, in the standard kinematical theory. However, in the present theoretical framework, they propagate along the G-direction. This result is due to omitting the second derivatives in the original equations. The present author (Kato, 1991) and Polyakov et al (1991) have attempted to overcome this theoretical defect. However, the theories are still in an immature stage so that the future developments are waited. Holy & Gabrielyan (1987) also shed physical insight on this problem .

b) The cases of Bragg geometry: The present article dealt deliberately with the simplest case; a spherical wave theory for the Laue geometry. Holy (1984), Holy & Gabrielyan (1987), Bushuev (1989) and Punegov (1990) present plane wave theories for the Bragg geometry, keeping applications to some surface problems in mind. In these cases, the displacement $\mathbf{u}(\mathbf{r})$ of the lattice points has to be generalized as $\{<\mathbf{u}(\mathbf{r})> + \delta\mathbf{u}\}$ because the crystal surface may be deformed relative to the interior. Thus, the lattice phase factor must have the form as $\Phi(\mathbf{r})\ \Phi$, in which only the second factor contains random variables discussed in the present theory. In other words, the homogeneity and isotropy of the lattice phase factor is no longer valid. Since the theories are rather complicated, it is recommended to refer to the original papers.

Extension along this line is required also in the Laue cases if the crystal is elastically deformed over a wide range. In such cases, Eqs.(4.10) for $< d >$ must be modified in such a way that the coupling constant κ_g is multiplied by the factor $\Phi(\mathbf{r})$. Then, as in the case of Figure 4, the background of the Pendellösung fringes may be enhanced in addition to the presence of I^{in}. In this context, the elastic deformation must be eliminated carefully when the present simple theory is checked with the experiments.

c) The works of Kulda (1987, 1988a, b) and of Davis (1992, 1994): Finally, it seems worth mentioning two theories which are of different styles from the present one. Kulda presents a specific theory for crystal models with random elastic distortions. Davis' theory is based on the equation of Riccati's type derived by Taupin (1964) for the reflectance, R(t), where t is the depth from the entrance surface. As random variables, he adopted the resonance factor (the deviation from the Bragg condition), $\beta(t)$, instead of Φ (see Subsec.3.2). Differential equations are derived for statistical quantities such as $< R(t) >$,

< $\delta R(t)\ \delta R(t-t')\ \beta$ > and so on. It is interesting that the theory is analogous to those of Brownian motion if t is interpreted as time variable. Although this theory is a kind of setback to Darwin's one-dimensional formalism, it has the advantage of obtaining directly the rocking curves rather than the topographic behaviour. In this sense, Davis' theory is complementary to the theory discussed in this article.

Unfortunately, because of page limitations, many experimental works cannot be reviewed. The extensive review article of Schneider et al. (1992) and the recent paper of Takama & Harima (1994) are worthy of reference.

REFERENCES

Al Haddad, M. & Becker, P.J., 1988, On the statistical dynamical theory of diffraction: Application to silicon, *Acta Cryst. A*, 44:262-270.

Becker, P.J. & Al Haddad, M., 1990, Diffraction by a randomly diffracted crystal. I. The case of short range order, *Acta Cryst. A*, 46:123-129.

Becker, P.J. & Al Haddad, M., 1992, Diffraction by a randomly diffracted crystal. II. General theory, *Acta Cryst. A*, 48:121-134.

Becker, P.J. & Coppens, P., 1974a, b,. Extinction within the limit of validity of the Darwin transfer equations. I. II. , *Acta Cryst. A*, 30:129-147; 148-153.

Bushuev, V.A.,1989, Statistical-dynamical theory of diffraction of X rays in imperfect crystals with allowance for angular distribution of intensities, *Sov. Phys. Crystallogr.*, 34(2):163-167.

Davis, T.J., 1992, A stochastic model for X-ray diffraction from imperfect crystals, *Acta Cryst. A*, 48:872-879.

Davis, T.J., 1994, Dynamical X-ray diffraction from imperfect crystals: A solution based on the Fokker-Planck equation, *Acta Cryst. A*, 50:224-231.

Hamilton, W.C., 1957, The effect of crystal shape and setting on secondary extinction, *Acta Cryst.* , 10: 224-231.

Holy,V., 1984, Dynamical X-ray diffraction from crystals with precipitates. I. Theory of the Bragg case, *Acta Cryst. A*, 40:675-679.

Holy,V. & Gabrielyan, K.T., 1987, Dyson and Bethe-Salpern equations for dynamical X-ray diffraction in crysstals with randomly placed defects, *Phys. Stat. Sol. (b)*, 140:39-50.

Guigay, J.P., 1989, On integrated intensities in Kato's statistical diffraction theory, *Acta Cryst.A*, 45:241-244.

Guigay, J.P. & Chukhovskii, F.N., 1992, Reformulation of the dynamical theory of coherent wave propagation by randomly distorted crystals, *Acta Cryst. A*, 48:819-826.

Katagawa, T. & Kato, N., 1974, The exact dynamical wave fields for a crystal with a constant strain gradient on the basis of the Takagi-Taupin equations, *Acta Cryst. A*, 30:830-836.

Kato, N., 1961a, b, A theoretical study of Pendellösung fringes. Part I. II., *Acta Cryst.*, 14:526-532; 627-636.

Kato, N., 1968a, b, Spherical-wave theory of dynamical X-ray diffraction for absorbing perfect crystals. I. II, *J. Appl. Phys.*, 39:2225-2230; 2231-2237.

Kato, N., 1976a, b, On extinction. I. II. , *Acta Cryst. A*, 32, 453-457; 458-466.

Kato, N., 1980a, b, Statistical dynamical theory of crystal diffractin. I. II. , *Acta Cryst. A*, 36:763-769; 770-778.

Kato, N., 1981, A new approach to extinction problems, *Sov. Phys. Crystallogr.*, 26(5):536-539.

Kato, N., 1991, A foundation for the statistical dynamical theory of diffraction, *Acta Cryst. A*, 471-11.

Kato, N., 1994, Mathematical structure of the coherent wave field in the statistical theory of dynamical diffraction, *Acta Cryst. A*, 50:17-22.

Kulda, J., 1987, Random elastic deformation (RED) - An alternative model for extinction treatment in real crystal, *Acta Cryst. A*, 43:167-173.

Kulda, J.,1988a, b, The RED extinction model. I. II. , *Acta Cryst. A*, 44,:283-285; 286-290.

Polyakov, A.M., Chukhovskii, F.N. & Piskunov, D.I., 1991, Dynamic scattering of X rays in real crystal, *Sov. Phys JETP*, 72(2):330-340.

Punegov, V.I., 1990, Statistical-dynamic theory of diffraction of X-rays by crystals with lattice parameters which continuously change along the thickness, *Sov. Phys. Crystallogr.*, 35(3):336-340.

Sommerfeld, A. ,1949, "Partial Differential Equation in Physics ", Academic Press, New York.

Schneider, J.R., Bouchard, R., Graf, H.A. & Nagasawa, H., 1992, Experimental tests of the statistical dynamical theory, *Acta Cryst. A*, 48:804-819.

Takagi, S.,1962, Dynamical theory of diffraction applicable to crystals with any kind of small distortion, *Acta Cryst.*, 15:1311-1312.

Takama, T. & Harima, H., 1994, Experimental verification of the statistical dynamical theories of diffraction, *Acta Cryst. A*, 50:239-246.

Taupin, D.,1964, Théorie dynamique de la diffraction des rayons X par les cristaux déformés, *Bull. Soc. fr. Minéral. Cristallogr.*, 87:469-511.

Wada, M. & Kato, N., 1977, The intensity distribution of X-ray Pendellösung fringes, *Acta Cryst. A*, 33: 161-168.

Werner, S.A., Berliner, R.R. & Arif, M., 1986, Mathematical methods in the solution of the Hamilton-Darwin and Takagi-Taupin equations, *Physica B:*137:245-255.

Zachariasen, W.H., 1945, "Theory of X-Ray Diffraction in Crystals ", John Wiley, New York.

X-RAY DIFFRACTION TOPOGRAPHY: PRINCIPLES AND TECHNIQUES

Helmut Klapper

Mineralogisch-Petrologisches Institut
Universität Bonn
D-53115 Bonn

1. INTRODUCTION

The first X-ray topographic experiment with the aim of studying the perfection of large single crystals was performed by Berg (1931). Using a simple experimental setup, he could reveal the lattice perturbations at the cleavage faces of a large sodium chloride crystal before and after plastic deformation. Berg's reflection technique (Bragg case) was refined by Barrett (1945) and applied to the transmission (Laue) case by Barth & Hosemann (1958). Another pioneering experiment was performed by Ramachandran (1945), who studied the perfection of diamond plates in transmission by applying a white-beam technique. Milestones of X-ray topography, stimulating rapid instrumental developments and wide-spread applications, were the introduction of the double-crystal method by Bond & Andrus (1952) and Bonse & Kappler (1958), and - to an even larger extent - of "projection" and "section" topography by Lang (1959). In later years further progress in X-ray topography was initiated by Tuomi et al. (1974) who showed that topographs could be recorded within a few seconds by the use of white synchrotron radiation (Laue technique, see below). The availability of synchrotron radiation sources has started numerous methodical and instrumental developments and opened new fields of application of diffraction topography (see Baruchel, this volume).

In this short survey the basic x-ray topographic methods mentioned above and some methodical variations are treated. For more extended informations the reader is referred to the textbook of Tanner (1976), to the comprehensive collection of surveys on various aspects of X-ray topography edited by Tanner & Bowen (1980) and by Weissmann et al. (1984), and to the reviews of Armstrong & Wu (1973) and Lang (1978,1995).

X-ray and Neutron Dynamical Diffraction: Theory and Applications
Edited by Authier *et al.*, Plenum Press, New York, 1996

2. PRINCIPLES

All X-ray topographic methods are based on the reflection of X-rays by a set of lattice planes (hkl) of the crystal. The relation between the reflection angle (Bragg angle), the interplanar spacing and the wavelength is given by the Bragg equation. In addition, the reflection geometry is strongly determined by the orientation of the reflecting planes with respect to the crystal surface. Two cases are distinguished: The Bragg or "reflection" case and the Laue or "transmission" case. In the Bragg case the diffracted beam leaves the crystal through the entrance surface. The reflecting lattice planes may be parallel or inclined to the entrance face (symmetric or asymmetric Bragg case). Only defects within a restricted depth below the surface are imaged. This depth is determined by the absorption and - with some restrictions - by the extinction of the diffracted X-ray beam. In the Laue case the diffracted beam penetrates the crystal. In this case all defects within the crystal volume are recorded, provided the absorption is small enough to permit sufficient transmission through the crystal.

Defects embedded in an undisturbed crystal matrix are imaged by a locally changed diffracted intensity ("contrast"). It is emphasized that in most cases it is not the defect itself, but the lattice deformations surrounding the defect, which produce the x-ray topographic contrast. For example, dislocations are imaged by local intensity changes arising from their extended strain field. Three categories of "contrast" are distinguished:

(i) Orientation contrast. Misoriented crystal regions may be fully or partially out of the reflection position leading to different reflected intensities. This, however, requires the use of collimated monochromatic radiation. For white radiation different wavelengths are diffracted in differently oriented regions, leading to a shift of their topographic images.

(ii) "Kinematical" or "extinction" contrast. It arises from inhomogeneous strain (i. e. inhomogeneous tilts of lattice planes and changes of the interplanar spacings). In the low absorption case ($\mu \cdot t \leq 2$, μ: linear absorption coefficient, t: specimen thickness) this results in a considerable local increase of the diffracted intensity.

(i i i) "Dynamical" contrast. This kind of contrast is, according to its origin and appearance, subdivided into different parts which are understood from the dynamical theory of X-ray diffraction (*cf.* Authier, this volume). A detailed treatment of contrast on X-ray topographs will be given by Tanner (this volume).

The intensity of the reflected beam is mainly determined by the structure factor modulus of the reflection used. Usually the strongest reflections are selected for X-ray topography. Besides the advantage of short exposure time, strong reflections generate narrow defect images and, consequently, permit a better spatial resolution of the defects on the topographs.

3. TECHNIQUES

3.1 White-radiation methods

These methods are very simple since the crystal to be studied can just be placed into the white beam without the need of exact adjustments. Many reflections (lattice planes) will fulfill the reflection condition in the continuous spectrum and form a Laue pattern. Each Laue spot is a topograph (Laue method, Figure 1a). Thus several topographs can simultaneously be recorded

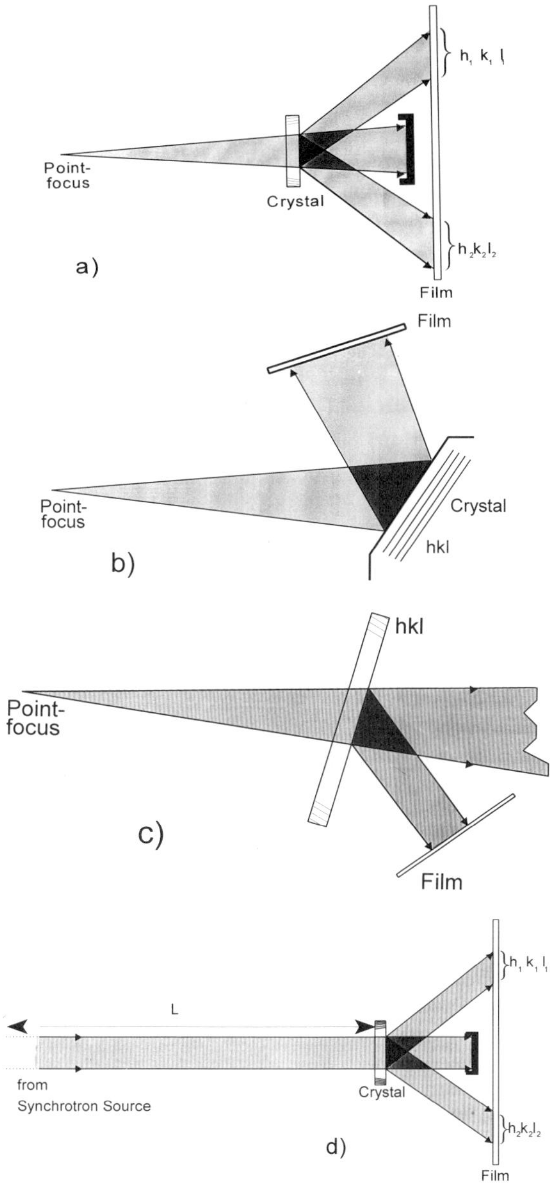

Figure 1 (a) Laue method (only two reflections of the Laue pattern are drawn).
(b),(c) Schulz method in reflection and transmission. In order to reduce image broadening and achieve good spatial resolution, the focal spot of the conventional X-ray source (sealed tube or rotating anode) should be small (a microfocus, if possible) and the crystal-to-film distance as short as possible. Due to the divergence of the incident beam, the wavelength of the diffracted beam varies slightly across the Laue topographs along a direction parallel to the incidence plane (i. e. in radial direction on the topographic Laue diagram). Different reflections with different Bragg angles are formed by different (medium) wavelengths.
(d) Laue technique with white Synchroton radiation. Due to the long distance L between source and crystal the primary beam is highly collimated (L is about 80 m in the topography line of the SRS Daresbury, England, and crystals up to 30 mm diameter can be studied). Consequently each topograph is formed by essentially one wavelength (except for higher harmonics, see text).

within one exposure on one film. In certain cases it may be useful to concentrate on one selected reflection and optimize the imaging conditions by choosing a suitable wavelength (i.e. by adjusting the Bragg angle) and placing the film normal to the diffracted beam and as close as possible to the crystal (Schulz method, 1954; Figure 1b,c).

The Laue technique is an excellent method for the application of synchrotron radiation (Figure 1d). The essential merits of this radiation compared with conventional X-ray sources (sealed tubes, rotating anode tubes) are:

(i) Short exposure times of the order of seconds (or even of fractions of a second for the new generation of SR sources, see Baruchel, this volume). With beam diameters of 20 mm and more, sufficiently large areas of the specimen can be imaged without the time-consuming translation necessary in the Lang technique (see below).

(ii) Due to the high collimation of the white beam and the large distance between the source (tangent point of the storage ring) and the crystal (about 40m at HASYLAB, Hamburg, and 80 m at the Synchrotron Radiation Source SRS, Daresbury), the divergence of the beam is extremely small (< 0.1 mrad). Thus a good resolution is obtained, even for the large crystal-to-film distances which are necessary to avoid overlapping of topographs when applying the Laue technique.

The high power of the white SR beam, however, implies an increased danger of radiation damage. Another (in general negative) feature (which is common to SR and conventional white radiation) is the inclusion of higher-order reflections with higher-harmonic wavelengths ($\lambda/2$, $\lambda/3$, etc.). The "higher-order" topographs are exactly superimposed on the required first-order topograph (Herres and Lang, 1983). Since shorter wavelengths and smaller structure factor moduli (higher-order reflections have, in general, smaller F-moduli) lead to wider kinematical images, a more diffuse contrast of defects results.

Finally an essential advantage of the white-radiation topography compared to the monochromatic-radiation methods is emphasized: even strongly distorted and warped crystals are fully imaged with the white beam, generating a more or less strongly distorted topograph, whereas they are only partially imaged along "bend contours" if monochromatic radiation is used. Thus white-radiation topography is beneficial in all studies where strong lattice distortions are involved, e.g. in plastic-deformation experiments or during in-situ topographic studies of phase transitions (e.g. Bhat et al., 1995).

3.2 Monochromatic-radiation methods

These methods require more or less exact adjustments. The crystal must be aligned in such a way, that the characteristic radiation of a conventional X-ray source (sealed or rotating anode tube) is diffracted by a chosen suitable set of lattice planes (reflection hkl) and recorded on a film. Commonly the strongest characteristic radiation of the X-ray spectrum, the $K\alpha_1$ line of the following elements is used: Cu (λ = 1.540 Å, for low absorbing crystals like organic materials), Mo (λ = 0.709 Å, for higher (but still low) absorption), and Ag (λ = 0.559 Å, for crystals of high absorption). In some special cases the $K\alpha$ radiation of other elements, e.g. of Co (λ = 1.789 Å) or even of Cr (λ = 2.290 Å, for the imaging of the Brasil twin domains in quartz making use of the anomalous scattering of Si, Lang 1965), was applied.

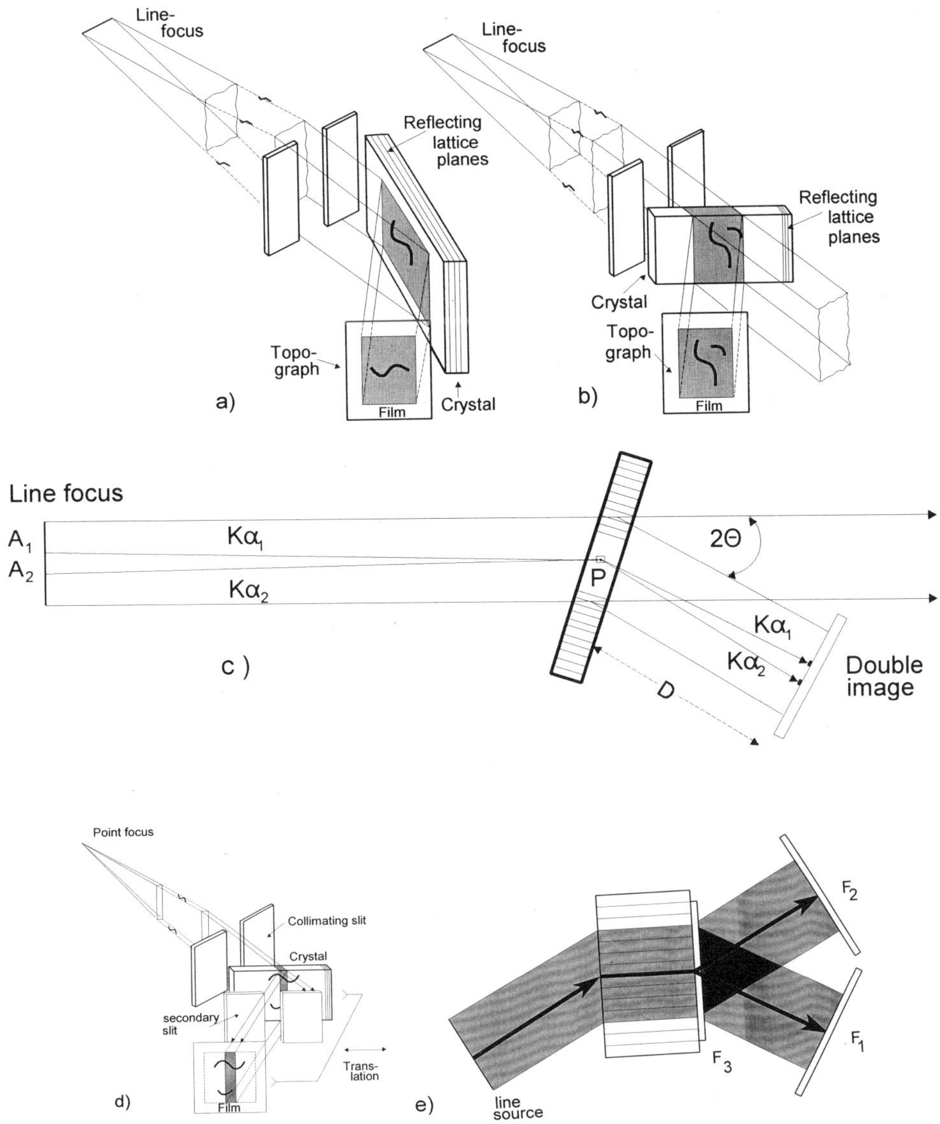

Figure 2: (a),(b) Berg-Barrett technique in reflection and transmission with film normal to the diffracted beam.

(c) X-rays originating from the loci A_1 and A_2 of the line source diffract the wavelengths λ_1 and λ_2 of the Kα-doublet in a point P of the probe, thus forming two images of P with distance $S = D\cdot(\Delta\lambda/\ \lambda)\cdot tg\theta$ on the film. The simultaneous diffraction of both doublet lines in P can be avoided by using a point source and collimation of the incident beam by a narrow slit.

(d) Lang projection technique, permitting high spatial resolution by reflection of the Ka_1-line alone. A secondary slit shields the film against the primary beam and the background radiation. With standing crystal + film a section topograph is recorded. By coupled translation of crystal and film the whole crystal is imaged.

(e) Borrmann technique for highly absorbing crystals. The thick lines and arrows show the energy flow of the waves and wave fields (inside the crystal) involved in the anomalous transmission diffraction. The film F can be alternatively placed in three positions.

3.2.1 Berg-Barrett technique (Berg, 1931; Barrett, 1945; Barth & Hosemann, 1958). These techniques are outlined in Figure 2 in self-explaining sketches. The disadvantage of this method is the simultaneous reflection of both lines of the Kα-doublet and the large crystal-film distance (except for highly asymmetric Bragg cases, *cf.* Figure 4), leading to poor resolution and blurred or double images of defects. As can be derived from Figure 2c, the simultaneous reflection of the doublet can be suppressed by using a point source with width smaller than A_1-A_2 and a strong collimation of the incident-beam divergence.

3.2.2 Lang projection topography (Lang, 1959). This method (Figure 2c) permits high resolution by collimation of the primary-beam divergence down to 1-3' (depending on the Bragg angle) and minimizing the crystal-to-film distance (typically 10 mm). This divergence is small enough to avoid simultaneous diffraction of both Kα lines, but still large enough to cover the range of lattice misorientations in the neighbourhood of defects (e.g. dislocations), thus leading to the kinematical defect contrast (extinction contrast) mentioned above.

3.2.3 Borrmann method. This technique makes use of the effect of anomalous transmission (Borrmann effect) and is applicable to highly perfect crystals in the case of high absorption (typically $\mu \cdot t > 10$). For such high absorption the incident radiation is completely absorbed except for such waves which excite wave fields with tie points close to the vertex of the dispersion surface (i. e. belonging to the centre of the perfect-crystal reflection range, see Authier, this volume). Thus, within an very narrow incidence angular range of the order 1", a transmitted ("forward diffracted") and a diffracted beam appear behind the crystal. Topographs can be recorded with the diffracted as well as with the transmitted beam (Figure 2e). The energy flow (Poynting vector) of the surviving wave field is directed parallel to the lattice planes. Since the wave field splits at the exit face of the crystal into the diffracted and transmitted waves, the film can also be placed in contact with the back side of the crystal without loss of resolution, recording both waves (Figure 2e). Lattice defects "disturb" the effect of anomalous transmission and lead to a locally reduced intensity ("Borrmann shadow"). Only defects close to the exit surface give images with strong contrast. It is pointed that this method is adequate only for highly absorbing crystals with a pronounced Borrmann effect, e.g. for crystals with diamond, ZnS and metal structures (examples: germanium, gallium arsenide, copper).

3.2.4 Section Topography. In the wide-beam topography (Figures 1 and 2) and in the Lang projection topography, the "integral reflected intensity", i.e. the reflected intensity integrated over the whole reflection range, is recorded at each point of the film. This implies the superposition of many interference and contrast effects, thus veiling contrast details and suppressing information about the defects involved. This is overcome by "section topography" which is - due to the point source - simply realized in a Lang camera by inserting a narrow collimating slit of typically 10 μm width in front of the crystal. The crystal is not moved during exposure. The X-ray wavefields excited by the incident beam "fill" the so-called Borrmann fan (Borrmann triangle) with apex angle 2θ and width $2t \cdot tg\theta$ on the exit surface of the parallel-sided crystal (t: crystal thickness). The width of the section topograph on the film (which is placed normal to the reflected beam) is $2t \cdot \sin\theta$. The intensity pattern on the topograph corresponds to the intensity distribution of the interfering reflected wavefield components at the crystal exit surface. The formation of a section topograph and its integration to a projection topograph is shown in Figure 3.

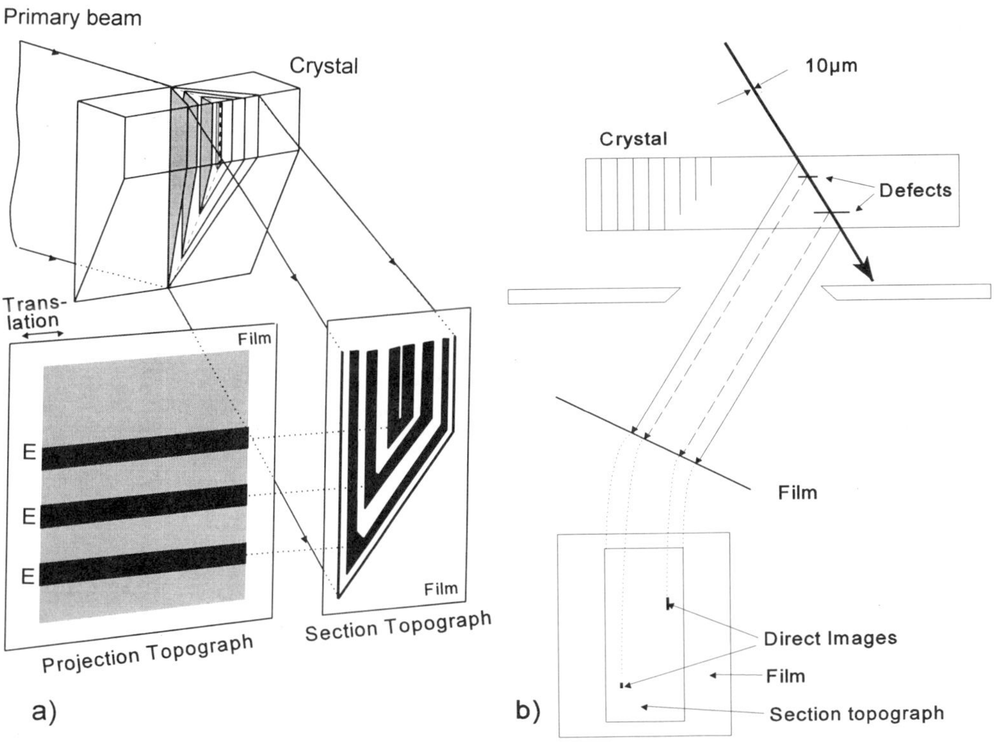

Figure 3: (a) Formation of a section and a projection topograph of a parallel-sided and a wedge-shaped perfect crystal. The primary beam of a conventional tube is very narrow (about 10 μm, thus defining a sharp apex of the Borrmann fan and yielding good spatial resolution) but divergent enough to cover (coherently) the full reflection range ("spherical waves", see Authier, this volume). In the Borrmann fan three hyperbolic cylinders (here surfaces of maximum reflected intensity) are shown. The section topograph exhibits the typical "spherical-wave" fringes ("Kato fringes"). During the coupled translation of crystal and film an intensity resulting from integration across the section topograph is recorded. The section fringe elements parallel to the translation direction produce the equal-thickness pendellösung fringes E ("Ewald fringes") of maximum (relative) intensity.
(b) Location of defects in the depths of the crystal plate, derived from the position of the direct images within the width of the section topograph. Defects close to the entrance face are close to the left, those close to the exit face close to the right-hand margin of the section (viewed against the diffracted beam).

The merits of section topography compared with projection topograph are as follows:

- Section topographs reveal pendellösung fringes ("Kato fringes") in wedge and parallel-sided crystals. Slight lattice strains result in distortion of the fringes.
- It allows the separate recording of the various images associated with defects, i.e., the "kinematical" (or "direct"), the "intermediary" and the dynamical images (Tanner, this volume);
- It reveals the location of defects in the depth of the crystal plate (Figure 3b);
- It elucidates the character of planar defects, e.g. the distinction between shift boundaries (stacking faults) and tilt boundaries;
- The theory permits computer simulation of section patterns of dislocations and planar defects (stacking faults, twin boundaries).

Apart from these "refined" applications, section topography is very beneficial for the study of thick crystals with high defect density, where any information is suppressed by the superposition of many defect images.

Section topography, however, has disadvantages too, in that it yields informations only from the restricted crystal volume intercepted by the 10 μm incident beam and, to some extent, from the volume of the Borrmann fan. In order to gain such information from a larger part of the specimen, a sequential series of section patterns with stepwise shifts of the crystal has to be recorded (e.g. on one film, if an overlap of section patterns is avoided). This procedure, however, is rather tedious. Thus section topography will not replace projection topography but rather supplement it: Section patterns will be recorded only from regions containing such defects for which more detailed information about their location and their character is required.

3.3 Plane-wave topography (double-crystal technique).

The X-ray topographic methods described in the foregoing sections apply a primary X-ray beam which contains all incidence directions covering the full reflection range of the perfect-crystal curve. For highly collimated synchroton radiation the reflection range is covered due to the continuously varying wavelength of the synchroton spectrum. This means that in such cases the local "integral reflected intensity" (which is represented by area below the perfect-crystal reflection curve) is recorded at any point on the film. For the Lang method, therefore, frequently the term "integrated-wave topography" (see Figure 3) is used.

For "plane-wave topography" a primary beam collimated down to an angular width, which is a small fraction of the perfect-crystal reflection range of the crystal to be studied, is required. Such collimation is achieved by the reflection of characteristic radiation or white synchroton radiation (which implies monochromatization of the beam) at a first crystal in a special reflection geometry. The radiation reflected from the first crystal is used for the topographic study of the second crystal. This is shown and explained in Figure 4.

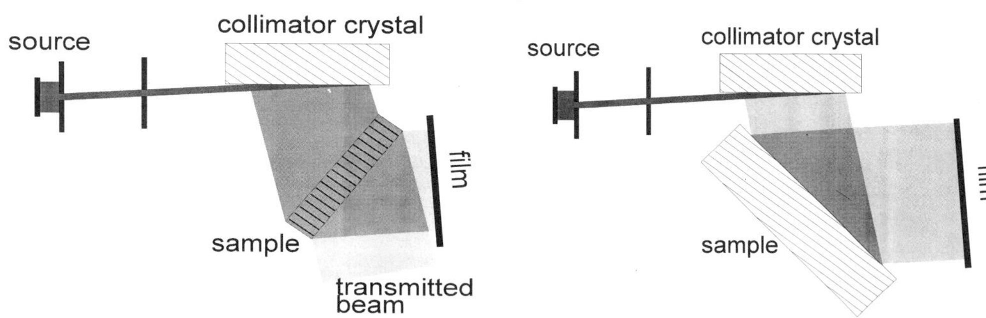

Figure 4: Double-crystal arrangement in dispersion-free (n,-n)-setting for plane-wave topography. The choice of a highly asymmetric Bragg-case reflection for the collimator crystal implies two essential merits: **(i)** A wide reflected beam is produced, permitting the imaging of a large area of the sample. **(ii)** The asymmetric reflection has a particularly narrow reflection range, leading to a considerably lower beam divergence than in the symmetric case and allowing quasi plane-wave topography. The collimator crystal should be defect-free.

Plane-wave topography is an extremely sensitive technique for the detection of minute lattice misorientations and lattice parameter changes (of order 10^{-6} to 10^{-8}). The highest sensitivity is achieved by adjusting the sample crystal on the steepest point on the flank of its rocking curve. In this case slight misorientations and d-value changes appear as black-white contrasts (e.g. Sauvage, 1980).

Plane-wave topography is also useful for achieving high spatial resolution by applying the "weak beam technique": the sample is adjusted relatively far from the reflection condition. This yields very narrow defect images (e.g. dislocation image widths down to 1 µm).

The double-crystal arrangement of Figure 4 implies that the collimator is a perfect crystal identical to the sample, in order to obtain the non-dispersive parallel (n,-n)-setting. This could be achieved for silicon, germanium and quartz crystals. For other crystals this method is not practicable except for the rare case that a quasi-parallel approximation by matched Si or Ge reflections is possible. In recent years, however, highly sophisticated harmonic-free monolithic monochromators, implying several successive asymmetric reflections, have been developed. They deliver a beam with an extremely narrow spectral window and angular divergence and permit plane-wave topography of all kinds of sufficiently perfect crystals.

4. IMAGE BROADENING

In the topographic methods using single crystals and conventional sources, "image broadening" results from:

- the finite size of the X-ray focal spot, in particular of the focus "height" H (normal to the incidence plane),
- the intrinsic angular (perfect-crystal) reflection width ω,
- the spectral width $\Delta\lambda_n$ of the $K\alpha_1$ line,
- the simultaneous reflection of both lines of the $K\alpha$ doublet.

The focus height H induces a "vertical" image broadening h (normal to the incidence plane):

$$h = (H/L)\cdot D$$

(where L is the distance focus-crystal, D the distance crystal-film). For $H/L = 2 \times 10^{-4}$ (i.e. for a 0.4 mm focus and L = 2 m or a 0.1 mm focus and L = 0.5 m) and typically D = 10 mm, a vertical broadening h = 2 µm results.

The reflection widths ω, the spectral width $\Delta\lambda_n$ of the $K\alpha_1$ line and the simultaneous reflection of both doublet lines produce a "horizontal" image streaking. The perfect-crystal reflection width ω is very small, $\omega < 5''$, and thus negligible. The broadening resulting from $\Delta\lambda_n$ is:

$$D \cdot \left(\frac{\Delta\lambda_n}{\lambda}\right) \cdot \mathrm{tg}\theta \approx 1\mu$$

for $CuK\alpha$ radiation ($\Delta\lambda_n$ is roughly 5×10^{-4} Å), a θ-value smaller than 15° and a crystal-film Distance D = 10 mm. Simultaneous reflection of the $K\alpha_1$ and $K\alpha_2$ lines, however, generates two images separated by about 7 µm (for D = 10 mm, θ = 15°, $\Delta\lambda_{1,2} = 3.8 \times 10^{-4}$ Å for the $CuK\alpha$ doublet) and should, therefore, be avoided by suitable reduction of the primary beam divergence (see Lang method). For small angles $\theta < 6°$ this broadening is small (< 2 µm), and both lines can be used without noticeable loss of resolution. This is even advantageous because the inclusion of the $K\alpha_2$ line increases the reflected intensity by 50%.

Thus the instrumental image broadening of the highly resolving Lang method is typically 2 μm, leading to a resolution limit of roughly 500 lines/mm (provided only $K\alpha_1$ is used). This broadening, however, is of minor significance compared with the rather broad widths of defect images in X-ray topography. Image widths of dislocations are usually larger that 10 μm, and for crystals with light atoms like organic materials even larger than 20 μ. In such cases even larger broadenings, (say 5 μm, e.g. due to larger focal spots or the inclusion of the $K\alpha_2$ line) may not be detrimental to a clear presentation of the defects.

5. REFERENCES

Armstrong, R.W. and Wu, C. C., 1973, X-ray diffraction microscopy, in: *Tools and Techniques for Microstructural Analysis*, p. 169, J.L. McCall and W.M. Müller, eds., Plenum Press, New York.

Authier, A., 1977, Section topography, in: *Topics Appl. Phys., Vol. 22: X-Ray Optics*, p. 145, Springer Verlag Berlin-Heidelberg-New York.

Barret, C. S., 1945, A new microscope and its potentialities, *Trans. AIME* 161:15.

Barth, H., and Hosemann, R., 1958, Anwendung der Parallelstrahlmethode im Durchstrahlungsfall zur Prüfung des Kristallinnern mit Röntgenstrahlen, *Z. Naturforsch. Teil A*, 13:792.

Berg, W.F., 1931, Über eine röntgenographische Methode zur Untersuchung von Gitterstörungen in Kristallen. *Naturwissenschaften* 19:391.

Bhat, H.L., Klapper, H. and Roberts, K.J., 1995, An X-ray topographic study of the para- to -ferroelectric phase, Transformation in nearly perfect single crystals of ammonium sulfate. *J. Appl. Cryst.* 28: 168.

Bond, W.L., and Andrus, J., 1952, Structural imperfections in quartz crystals, *American Mineralogist* 37:622.

Bonse, U., and Kappler, E., 1958, Röntgenographische Abbildung des Verzerrungsfeldes einzelner Versetzungen in Germanium-Einkristallen, *Z. Naturforsch. Teil A*, 13:348.

Herres, N. and Lang, A.R., 1983, X-ray topography of natural beryl using synchroton and conventional sources, *J. Appl. Cryst.* 16:47.

Lang, A.R., 1959, The projection topograph: A new method in X-ray diffraction microradiography, *Acta Cryst.* 12:249.

Lang, A.R. 1965, Mapping Dauphine and Brazil twins in quartz by X-Ray topography, *Appl. Phys. Letters* 7: 168.

Lang, A.R., 1978, Techniques and interpretation in X-ray topography, in: *Diffraction and Imaging Techniques in Materials Science*, p. 623, S. Amelinckx, R. Gevers and J. Van Landuyt, eds., North Holland, Amsterdam. Lang, A.R., 1995, Topography, in: *International Tables for Crystallography, Volume C*, p. 113, International Union of Crystallography, ed., Kluver Academic Publishers, Dordrecht-Boston-London.

Ramachandran, G.N., 1944, X-ray topographs of diamond, *Proc. Indian Acad. Sci., Sect. A*, 19:280.

Sauvage, M., 1980, Monochromatic Synchrotron Radiation Topography, in: *Characterization of Crystal Growth Defects by X-ray Methods*, p. 433. B.K. Tanner and D.K. Bowen, eds., Pergamon Press, New York.

Schulz, L.G., 1954, Method of using a fine-focus X-ray tube for examining the surface of a single crystal, *Trans. AIME* 200:1082.

Tanner, B.K., and Bowen, D.K., 1980, eds., *Characterization of Crystal Growth Defects by X-Ray Methods*, Plenum Press, New York.

Tanner, B.K., 1976, *X-Ray Diffraction Topography*, Pergamon Press, Oxford.

Tanner, B.K., 1977, Crystal assessment by X-ray topography using synchroton radiation, *Progress in Crystal Growth and Characterization*, Vol 1, p. 23, B.R. Pamplin, ed., Pergamon Press, Oxford.

Tuomi, T., Naukkarinen, K. and Rabe, P., 1974, Use of synchroton radiation in X-ray diffraction topography, *Phys. Status Solidi* A25:93.

Weissmann, S., Balibar, F. and Petroff, J.-F., eds., 1984, *Application of X-Ray Topographic Methods to Materials Sciences*, Plenum Press, New York-London.

CONTRAST OF DEFECTS IN X-RAY DIFFRACTION TOPOGRAPHS

B K Tanner

Department of Physics, University of Durham
South Road, Durham, DH1 3LE, U.K.

1 ORIGINS OF CONTRAST IN TOPOGRAPHS

X-ray diffraction topographs (Lang, 1978; Tanner, 1975; Tanner and Bowen, 1980) are maps of the scattering power as a function of position across the diffracted X-ray beam. The observed contrast is related therefore to the properties of both the incident beam and the specimen. Even for a very well conditioned incident beam, the task of determining the microscopic strains in the crystal from the recorded image is not a simple task of inversion. As is the case in all diffraction experiments, we are able only to measure scattered intensity, not amplitude, and therefore quantitative measurements must be undertaken by matching the experimental image to a simulated image of a model structure. In later parts of this chapter, we will see that this can now be achieved routinely for section and double crystal topographs. However, in order to obtain a starting structure for iterative simulation, a more qualitative interpretation of the contrast is essential and in many instances such analysis is sufficient. Much of the discussion in this chapter concerns this latter approach to defect analysis.

There are two fundamental mechanisms for contrast in X-ray topographs. Orientation contrast occurs where part of the crystal is misoriented in such a way that either diffraction cannot occur at the same time as that from the rest of the crystal, or that the diffracted beam from the two parts of the crystal make different directions in space. The former occurs for characteristic radiation (figure 1(a)) and contrast occurs when the effective misorientation (which may arise from dilations as well as rotations of the lattice) exceeds the divergence of the incident X-ray beam. It is clear that the zero intensity region corresponds geometrically to the misoriented region. The latter occurs for continuous radiation (figure 1(b)) and here there is a gain or loss of intensity in regions corresponding to the boundary of the misoriented region (figure 1(c)). Boundaries giving divergent beams lead to a loss of intensity and thus stronger contrast than convergent boundaries. The width of the region of intensity gain or loss is determined both by the angle of misorientation and the specimen to plate distance. As the latter is a simple relation with distance, displacement of the plate by a measured distance permits immediate measurement of the component in the incidence plane of the misorientation across the boundary.

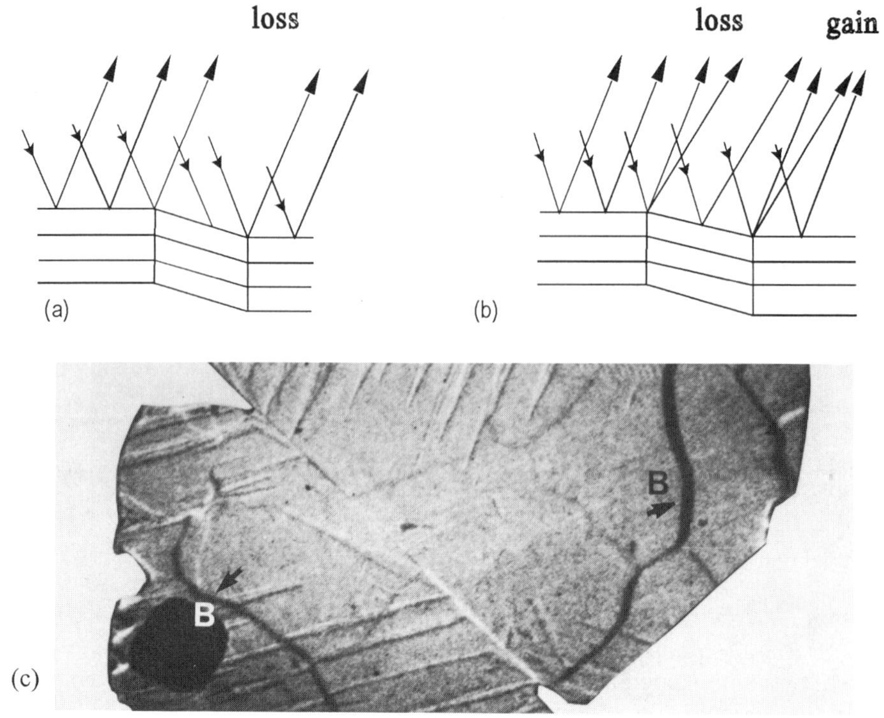

Figure 1 (a) Schematic diagram of the origin of orientation contrast with monochromatic radiation (b) Schematic diagram of the origin of orientation contrast with continuous radiation (c) Images of sub-grain boundaries (B) in Fe-3%Si using white synchrotron radiation. (courtesy D.K.Bowen)

The second type of contrast is termed extinction contrast and arises where the scattering power around the defect differs from that in the rest of the crystal. To interpret such changes, it is necessary to understand the dynamical diffraction effects occurring in thick, highly perfect crystals. There are essentially three types of extinction contrast, identified by Authier (1967), which can be seen schematically in figure 2(a). The three images which are formed are termed the direct image(1), arising from the diffraction of X-rays which do not satisfy the Bragg condition in the perfect crystal, the dynamical image (2) which is formed from changes in intensity in the Bloch wavefields propagating through the perfect crystal and the intermediary image (3), formed by interference between these wavefields and new wavefields created at the defect. These images are found in cases where the strain gradient is high, such as around a dislocation, and this leads to the region around the defect behaving as if it were the surface of the crystal. The wavefields decouple into their plane wave components and create new Bloch wavefields when the strain field becomes small again below the defect. A somewhat different type of dynamical image is formed when the strain gradients are small (figure 2(b)). Then the wavefields do not decouple, but the tie points migrate along the dispersion surface. As the propagation direction is normal to the dispersion surface, the rays paths become curved in real space. These two effects lead to redistribution of the energy between forward and diffracted beams, resulting in contrast.

Despite the section topograph being more fundamental, projection (Lang, 1959) and wide area synchrotron radiation white beam topographs (Tanner and Bowen, 1992) are much more commonly encountered. Therefore the discussion here will start with these integrated wave topographs, to provide an accessible guide to users. The subtle effects which arise in the spherical wave section topograph due to the changes in energy flow within the crystal will be examined later.

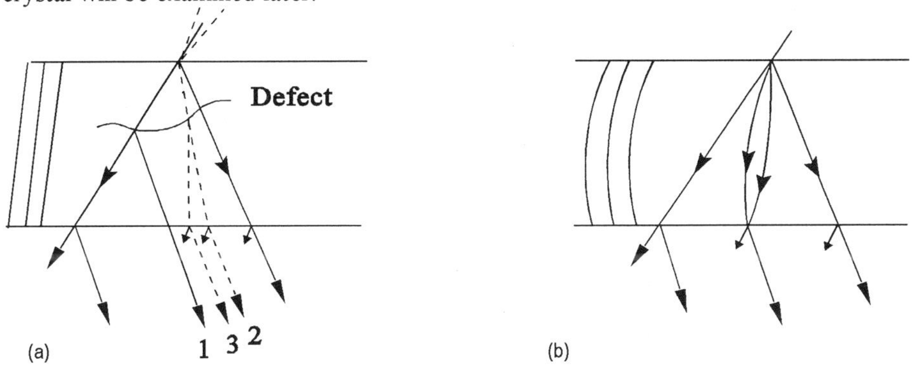

Figure 2 Formation of the types of extinction contrast image (a) Defect where the strain field gradient is high. 1 is the direct image, 2 the dynamical image and 3 the intermediary image. (b) Defect where the strain field gradient is low

2 INTEGRATED WAVE TOPOGRAPHS

We have already noted that orientation contrast is significantly different in the characteristic radiation Lang topographs and those taken with white radiation at synchrotron radiation sources. However, the extinction contrast images arise in a similar manner in the two techniques and for 1st and 2nd generation medium energy synchrotron radiation sources the differences in images are generally very small.

2.1 The Direct Image

This image is the one most commonly seen in X-ray topographs, being characterised by enhanced intensity (dark regions in positives) with respect to the perfect crystal (figure 3).

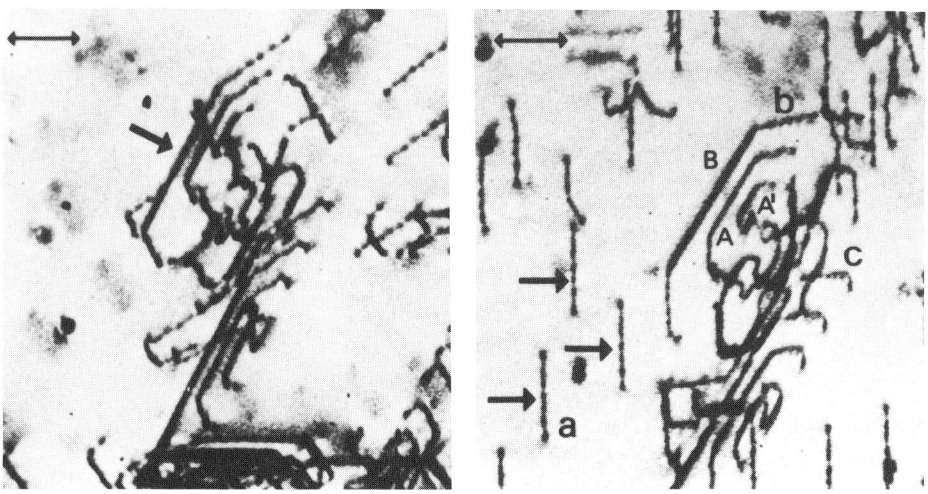

Figure 3 Direct images of dislocations in (111) silicon wafer. Lang topograph MoKα radiation ($\mu t = 0.15$) (a) $1\bar{1}\bar{1}$ reflection (b) $\bar{1}\bar{1}1$ reflection, scale mark 100 μm. From Miltat and Bowen (1975) with permission.

For a specific wavelength, the dynamical diffraction reflecting range from the perfect crystal is very small, typically a few arc seconds. This is small compared with the divergence of an unconditioned X-ray beam from a laboratory source and thus a large part of the incident beam is not diffracted strongly. However, in the deformed region around the defect, these X-rays outside the perfect crystal reflecting range can be diffracted, while those which are within the reflecting range of the perfect crystal are now not diffracted. If the crystal is very thin, the scattering from the deformed region is identical to that from equivalent undeformed material, as the intensity for kinematical scattering varies linearly with thickness. On the other hand, if the crystal is thick, the integrated intensity is no longer proportional to thickness and the scattering around the defect always leads to enhanced intensity.

Let us assume (figure 4(a)) that there exists in a crystal of thickness t, a region of thickness Δ around the defect where the effective misorientation exceeds the perfect crystal reflecting range. The effective misorientation $\delta\theta$ around a defect is the sum of the tilt component in the incidence plane $\delta\varphi$ and the change in Bragg angle θ_B due to dilation δd.

$$\delta\theta = - \tan\theta_B\,(\delta d/d) \pm \delta\varphi \qquad (1)$$

The integrated intensity for the Laue case (figure 4(b)) varies linearly with thickness for small values of thickness, but then falls off and oscillates about a constant value in the absence of absorption. [When absorption is included, there is superimposed a gradual decay of intensity with thickness.] As the gradient of the I versus t curve is everywhere is less than that at the origin, the intensity I_t due to material thickness t is always less than the sum of I_Δ and $I_{t-\Delta}$ from separate regions of thickness t and $t\text{-}\Delta$. In figure 4(b), where $t/\xi_g \approx 3$, corresponding to the third Pendellösung minimum, it is very obvious that

$$I_t < I_\Delta + I_{t-\Delta} \qquad (2)$$

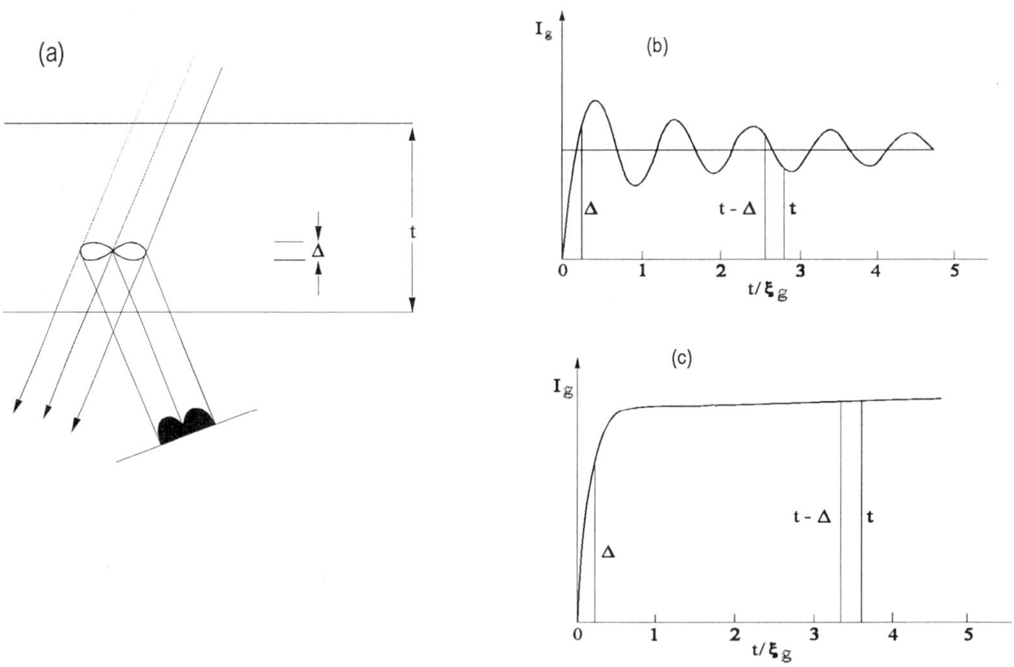

Figure 4 (a) Diagram showing the splitting of the crystal into perfect material above and below the defect and distorted material close to it. (b) Integrated intensity - Laue case (c) Integrated intensity - Bragg case.

Thus the intensity around a defect is always enhanced with maximum contrast for crystal thickness $0.88\xi_g$, i.e. at the first Pendellösung oscillation minimum. For the zero absorption Bragg case (figure. 4(c)) the intensity is independent of thickness beyond a crystal thickness of about an extinction distance ξ_g Again it is clear that the inequality of Equation 2 must hold as I_Δ is linear with thickness Δ, and $I_t = I_{t-\Delta}$. The loss of image contrast in very thin crystals where Equation 2 becomes an equality was demonstrated on wedge crystals (Tanner, 1972),the minimum thickness for image formation being between $0.15\xi_g$ and $0.4\xi_g$.

The width of the defect image can be deduced using the simple idea that the image is formed when the misorientation around the defect exceeds the perfect crystal reflecting range $\delta\omega$. For the case of a screw dislocation running normal to the Bragg planes, the line direction ***l*** coincides with the diffraction vector ***g***. Around a screw dislocation the effective misorientation $\delta\theta$ at distance r from the core is

$$\delta\theta = b / 2\pi r \tag{3}$$

The width of the dislocation image D is twice the value of r for which $\delta\theta = \delta\omega$ and is thus

$$D = b / (\pi\, \delta\omega) \tag{4}$$

As, for a symmetric reflection, $\delta\omega = 2/g\xi_g$, we have $D = gb\xi_g/2\pi$. This result can be generalised for arbitrary line direction and also for edge dislocations to

$D = \mathbf{g}.\mathbf{b}\ \xi_g/2\pi$	for screw dislocations	(5a)
$D = 0.88\ \mathbf{g}.\mathbf{b}\ \xi_g/\pi$	for edge dislocations	(5b)

Direct images of edge dislocations are approximately twice as wide as those for screw dislocations of the same ***g.b*** value. Substitution of typical values yields image widths typically a few μm. This is several orders of magnitude greater than those of transmission electron micrographs and represents a fundamental limit set by the weak scattering, and hence high strain sensitivity, of X-rays in crystals. In terms of the structure factor F_g we have

$$D \approx constant.\ [\lambda F_g]^{-1} \tag{6}$$

Thus we see that the dislocation image width goes down for increasing wavelength λ and strength of the reflection. Use of short wavelength radiation, such as available at the European Synchrotron Radiation Facility (ESRF), results in very wide dislocation images (Barrett *et al* 1995). There is thus no advantage in using very high resolution Nuclear Emulsion plates over faster, less expensive and easier-to-develop High Resolution film. It also implies that the dislocation density at which images overlap is lower than with softer radiation. As one of the advantages of using hard radiation is that thicker specimens can be examined, a requirement of higher perfection for individual defect resolution is unfortunate.

Equations 5 suggest that the image width is zero when $\mathbf{g}.\mathbf{b} = 0$. This is of course the classic criterion, originally applied to transmission electron microscopy, where the effective misorientation of the distortion around the dislocation is zero. As the Bragg planes are not distorted or tilted the dislocation is invisible in that reflection. Strictly, this criterion is that both $\mathbf{g}.\mathbf{b} = 0$ and $\mathbf{g}.\mathbf{b}\mathbf{x}\mathbf{l} = 0$ and, except when the dislocation runs parallel to a high symmetry axis, it is valid only for isotropic elasticity. Nevertheless, the contrast is often weak for just $\mathbf{g}.\mathbf{b} = 0$ and this enables ***b*** to be determined on purely geometric grounds by finding two reflections in which the dislocation is invisible or very weak. The procedure is not infallible and when in doubt the contrast in section topographs must be matched to simulation.

The above arguments follow also for synchrotron radiation, where the beam divergence is intrinsically very small, usually less than the perfect crystal reflecting range. Here the direct image is formed by X-rays of wavelength outside of the wavelength range of reflection for the perfect crystal. There is a subtle difference in that the angular directions of X-rays diffracted from the same distorted region are different, the deviation being $\delta\theta$ for monochromatic radiation and $2\delta\theta$ for continuous radiation. This divergence of the diffracted beam is not usually observed in Lang topographs, as the specimen to plate distance has to be kept short to provide good resolution, but at synchrotron radiation sources this distance can be several centimetres. Images of edge dislocations show a double lobed (bimodal) contrast when $\boldsymbol{g}.\boldsymbol{b} > 2$ and by displacing the photographic plate, this angular divergence is observed by the gradual separation of the positions of the two lobes (Midgley *et al*, 1977).

Another difference between Lang and white SR topographs is apparent from Equation 6. With the continuous synchrotron beam, diffraction for the fundamental vector $\boldsymbol{g}$ at wavelength λ occurs at the same angle as for the harmonics $2\boldsymbol{g}$ at $\lambda/2$, $3\boldsymbol{g}$ at $\lambda/3$ ....etc. For the harmonics, the image width is usually higher because of the shorter wavelength and (except for forbidden reflections), the structure factor falls with increasing scattering vector. White radiation images are thus wider than Lang topographs taken at the same wavelength and diffraction geometry. The intensity in the topograph varies with the X-ray beam power $P(\lambda)$ as

$$I = constant.\ P(\lambda)\ F_{hkl} \tag{7}$$

Thus as both $P(\lambda)$ and F_{hkl} fall with decreasing λ, the effect is small if the critical wavelength λ_c is large. For a high energy machine such as the ESRF, $P(\lambda)$ does not fall rapidly and there is very substantial intensity out to 100keV (0.13Å). Again, this is not a problem in the Bragg reflection geometry, but in the Laue transmission case it is serious for thick or high atomic number materials. The long wavelength fundamental reflection can be absorbed strongly, leaving the high energy harmonics dominating the image. Thus the image width as a function of wavelength is not a monotonic function, but shows relatively abrupt jumps as successive harmonic orders dominate the image (Zontone, 1995).

Simulation of the direct image is extremely difficult and requires much computing time. Early studies of the direct image taking the intensity of the image to be proportional to the calculated volume of the region inside the model effective misorientation contour corresponding to $C\ \delta\omega$ showed qualitative agreement (Miltat and Bowen, 1975). However the scaling constant C was not unity and varied somewhat unpredictably between reflections. Nevertheless this method is easy to implement numerically, gives a good method of determining Burgers vectors (especially when the simple $\boldsymbol{g}.\boldsymbol{b} = 0$ invisibility criterion is ambiguous) and has become a standard procedure in many places (Epelboin, 1985).

2.2 The Dynamical Image

The dynamical image arises from the change in wavefield intensity due to the defect. It is most easily observed under conditions of high absorption, where normal absorption destroys the direct image (which arises from X-rays not diffracted by the perfect crystal.) However, for X-rays within the perfect crystal reflecting range, the Borrmann anomalous transmission effect results in intensity reaching the exit surface even when the product of the linear absorption coefficient and thickness μt is considerably greater than unity. The wavefield from one branch of the dispersion surface (usually the σ polarisation branch 1) close to the exact Bragg condition shows extremely low absorption. Under these conditions (typically $\mu t > 4$) for the Laue case, the intensity in the forward and diffracted beams is equal.

Dislocations or other regions of strain destroy the conditions for wavefield propagation and the energy in this wavefield is redistributed right across the dispersion surface as new wavefields are created below the defective region. Most of these wavefields are strongly absorbed, resulting in a loss of intensity around the defect. Such dynamical images (figure. 5) show light against a dark background in the conventional positively printed topograph.

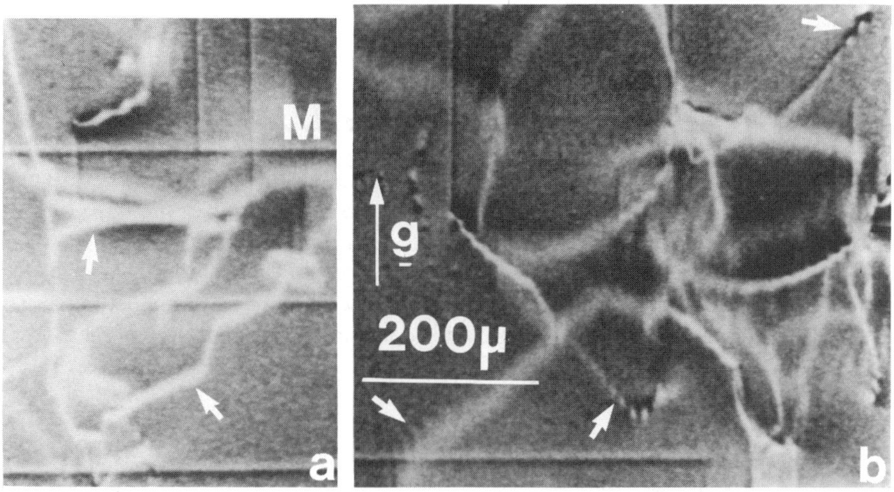

Figure 5 (a) Dynamical and (b) intermediary images in white radiation transmission topographs of GaAs wafers under conditions of high absorption. Note particularly the black-white contrast near the intersection of the threading dislocations with the exit surface. g = [220].

A second type of dynamical diffraction effect explains the reason why these dynamical images are sharp in regions corresponding to the dislocation close to the exit surface but diffuse and blurred when the dislocation is close to the entrance surface. In the region far from the dislocation, where the effective misorientation is varying slowly, the wavefields remain coupled and the phenomena can be explained by Eikonal theories (Kato, 1963; Penning and Polder, 1961). In the strained lattice, the tie points migrate along the dispersion surface and the ray paths are bent. If we consider the one σ polarisation branch 1 wavefield at the dispersion surface centre which is anomalously transmitted, we see that it moves away from this position. At the dispersion surface centre the wavefield is of the form

$$\boldsymbol{D} = \boldsymbol{D}_o \, exp\,(i\boldsymbol{K}_o.\boldsymbol{r}) + \boldsymbol{D}_g \, exp\,(i\boldsymbol{K}_g.\boldsymbol{r}) \qquad (8)$$

with $|\boldsymbol{D}_o| = |\boldsymbol{D}_g|$. As the tie point migrates D_o and D_g differ. If the defect is near the exit surface, the lattice rotates back sharply due to surface relaxation with a result that this migration is "frozen in". At the exit surface, the wavefield decouples into the diffracted direction component D_g and the forward component D_o. Depending on the sense of the effective misorientation due to the original long range distortion, D_g either increases or decreases. Thus near the intersection of the dislocation with the exit surface the contrast can be either black or white depending on the sense of the Burgers vector (see the straight misfit dislocations M in figure 5). The "rule of thumb" is that for the defect close to the exit surface, if the side corresponding to the positive sense of the $\boldsymbol{g}$ vector is enhanced, then the lattice is under compression. If it is reduced it is under tension. Note that the contrast reverses with reversal of $\boldsymbol{g}$ and of the exit to entrance surface.

Under conditions of low or moderate absorption, the dynamical image is not usually seen as it superimposes on the much stronger direct image in the integrated wave topograph. the exception is when the dislocation lies near to and parallel to the entrance surface with the dislocation line vector $\boldsymbol{l}$ perpendicular to the diffraction vector $\boldsymbol{g}$. Then the white dynamical image may be seen on the positive $\boldsymbol{g}$ side of the direct image (Tanner, 1972).

2.3 The Intermediary Image

This image arises from the interference at the exit surface of the new wavefields created below the defect with original wavefields propagating along other ray paths. Intermediary images are not normally very easy to see in integrated wave topographs and usually appear as bead-like contrast along the direct image of inclined dislocations (figure 3). The image appears to be periodically wider and narrower due to this image. In the case of moderate absorption and a small inclination of the defect to the surface projection, the scale of the image can be seen. It forms a fan within the intersections of the exit and entrance surface of the defect and has an oscillatory contrast with depth periodicity of an extinction distance ξ_g. (figure 5) The intermediary image thus provides a method of structure factor measurement.

2.4 Thickness Fringes

Another means of measuring the structure factor arises from the fringes seen in a wedge-shaped crystal. As already seen, there is a difference in wavevector in the components of wavefields associated with branch 1 and branch 2 of the dispersion surface. For a plane wave, it is easy to see that there will be interference between the two wavefields which reach the exit surface of the crystal, as there will be a varying phase difference between the waves, depending on the path length travelled through the crystal. For waves at the exact Bragg condition, i.e. with tie points at the centre of the dispersion surface, the ray paths are parallel to the Bragg planes. If the reflection is symmetric, the depth periodicity is then the reciprocal of the dispersion surface diameter, as this is the difference in the wavevectors. The reciprocal of the dispersion surface diameter is of course the extinction distance ξ_g. Thus for a plane wave, a wedge shaped crystal will show interference fringes along the constant thickness contours of the specimen.

It is not quite so obvious to see that, on integration across the whole dispersion surface, will be preserved a periodicity given by the extinction distance. However, numerical calculation reveals that the integrated wave, in the Laue geometry does show a pseudo-periodicity and such thickness (or Pendellösung) fringes are a good indicator that the crystal is highly perfect. They arise from dynamical interference, and are not found in highly imperfect crystals, which might also show uniform contrast in an integrated wave topograph. Pendellösung fringes are often seen at the edges of specimen, where the thickness grades down due to polishing or where there is an inclined cleavage edge.

3 SECTION TOPOGRAPHS

The section topograph is formed when a ribbon beam passed into the crystal and the diffracted intensity in a stripe down the crystal is recorded on the photographic plate. Whereas the intensity in the integrated wave topograph from a perfect crystal is uniform, this is not the case for the section topograph. As the divergence of the beam is large compared with the perfect crystal reflecting range, for a first approximation the wave at the entrance surface can be considered as a spherical wave. This excites the whole of the dispersion surface and wavefields propagate in all directions within the Borrmann fan (figure 6(a)).

These wavefields are coherently excited and preserve their relative phase. At a point on the exit surface P, two wavefields arrive, one from branch 1 and one from branch 2 of the dispersion surface (figure 6(b)). As these wavefields have different wavevectors there will be a phase difference between them at the exit surface and they will interfere. This phase difference clearly changes across the base of the Borrmann fan as for different directions of propagation, different parts of the dispersion surface contribute. Thus across the section topograph of a parallel sided, perfect crystal interference fringes are seen (figure 6(c)). These fringes are known as "Kato fringes" after Kato (1974) who first explained their origin.

The fringe density increases towards the margins of the section topograph image, as here the wavevector difference becomes greater for a smaller change in the normal to the dispersion surface. As the normal to the dispersion surface defines the wavefield propagation direction, the phase change between the corresponding branch 1 and branch 2 wavefields for a given angular change increases. The Kato fringes thus provide important information on the dispersion surface shape. The density of the Kato fringes is low at the centre of the pattern, resulting from the angular amplification of the angular deviation of the incident rays (Authier, 1970). With increasing absorption, the fringe contrast becomes weaker, as only the branch 1 wavefield reaches the exit surface. For very thick crystals, the section pattern becomes narrower, with only the branch 1 wavefield in the centre of the dispersion surface reaching the exit. No Kato fringes are visible under conditions of very high absorption.

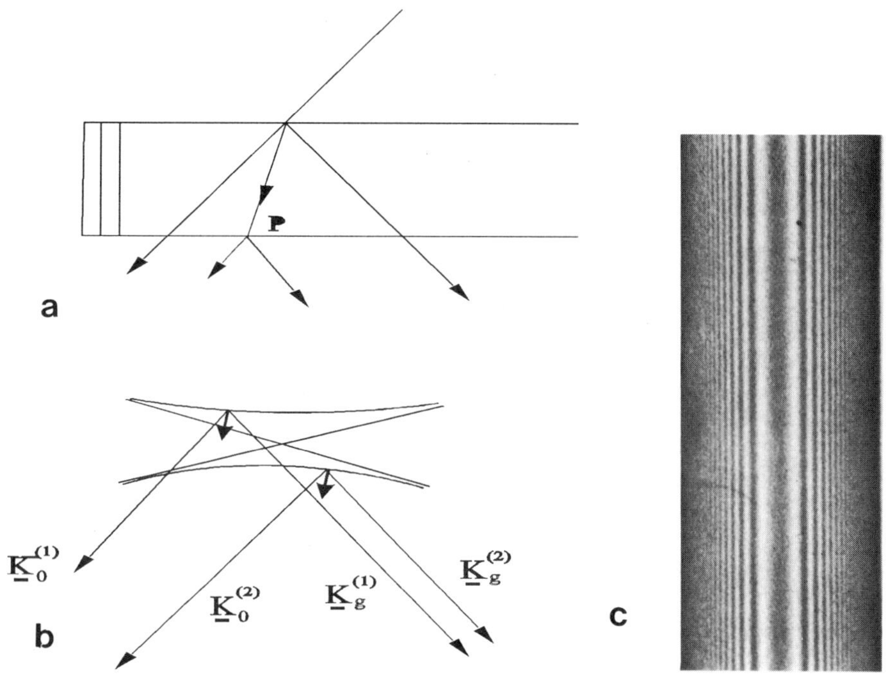

Figure 6 (a) Schematic diagram of the propagation of wavefields in the Borrmann fan. (b) Equivalent dispersion surface construction showing the tie points of the rays in (a) (c) Section topograph of a perfect parallel sided crystal showing Kato fringes.

For the case of a wedge-shaped crystal, the Kato fringes are hook shaped. Each hook corresponds to an increase in thickness of an extinction distance and this technique has been used to make very precise measurements of the structure factors of silicon and quartz.

3.1 The Direct Image

As is evident from figure 2(a), the position of the direct image in section topograph corresponds to the depth of the defect in the crystal. The direct image is formed from rays not diffracted by the perfect crystal and therefore simply project geometrically on to the image. This is a particularly powerful method of determining the defect distribution with depth and finds practical application in the semiconductor industry. Intrinsic getting is used to eliminate impurity from the near surface region of silicon used for device processing. As modern devices are shallow, only the perfection of the near surface region is important. The gettering process results in the production of a high precipitate density in the centre of the wafer, with near prefect material at the surfaces. Under conditions of low absorption, section topography can be used as a quality control tool to measure non-destructively the thickness of this denuded zone (Tuomi *et al* 1985).

3.2 The Intermediary and Dynamic Images

These are the most characteristic of the images in the section topograph. Most importantly, the intermediary image is sensitive to the sense of the lattice distortion and hence sense of Burgers vector. Typical images of dislocations are shown in figure 7.

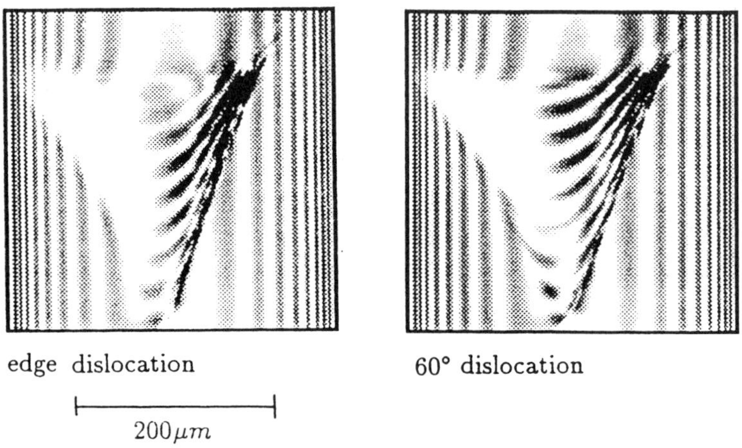

Figure 7 Simulated $MoK\alpha_1$ $3\bar{3}\bar{3}$ section topograph images of dislocation with $\boldsymbol{l}$ = [101] in (001) wafer of silicon. (a) $\boldsymbol{b}$ = ½[$\bar{1}$01] edge dislocation (b) $\boldsymbol{b}$ = ½[011] 60° dislocation (A.J.Holland, 1992)

The broad white region is the dynamical image formed by the loss of intensity due to destruction of the original wavefield at the defect and redistribution of the energy over the whole dispersion surface, and hence directions within the Borrmann fan. Linking the dynamical image and the black direct image is the intermediary image. This has characteristic interference contrast which changes dramatically with the strain field. Particularly in the case where there is significant elastic anisotropy, matching of simulation and experiment may prove to be the only method for the determination of Burgers vector (Epelboin, 1985).

Simulation of the image is achieved using the Takagi's theory of diffraction in an imperfect crystal. This theory results in a pair of coupled differential equations which describe the wave amplitude in the forward and diffracted direction at any point in the crystal (see Authier, this volume). These may be solved using numerical methods over a grid of points and using a "half-step derivative" algorithm, Authier's group succeeded in producing the first simulated section topography images (Authier, Malgrange and Tournarie, 1968).

Nowadays, on a fast personal computer, the simulation of a section topograph takes only a matter of seconds for a reasonable definition image. Simulation has become an analytical tool for the determination of the microscopic strains by matching simulated and experimental images, in much the same way as is done for high resolution diffraction rocking curve profiles. Details of the procedures may be found in the review by Epelboin (1985).

4 DOUBLE CRYSTAL TOPOGRAPHS

The image contrast in double crystal topographs can be mainly interpreted as orientation contrast in the manner originally developed by Bonse and Kappler (1958). If we make the approximation that the rocking curve profile is triangular, then the intensity varies linearly with the effective misorientation. Thus from Equation 1, we have immediately that the fractional intensity change $\Delta I/I$ is given by

$$\Delta I/I = - k\{\tan \theta_B (\delta d/d) \pm \delta\varphi\} \tag{9}$$

where k is the rocking curve gradient. This simple relation works extremely well and we see immediately that the contrast of a dislocation which is black-white on the low angle side of the rocking curve, when k is negative, will reverse to white-black on the high angle side where k is positive. Such a contrast reversal is shown in figure. 8. Kagener and Möhling (1991) have shown that such an approach works extremely well and that calculation of the misorientation contours does predict very accurately the double crystal topograph images.

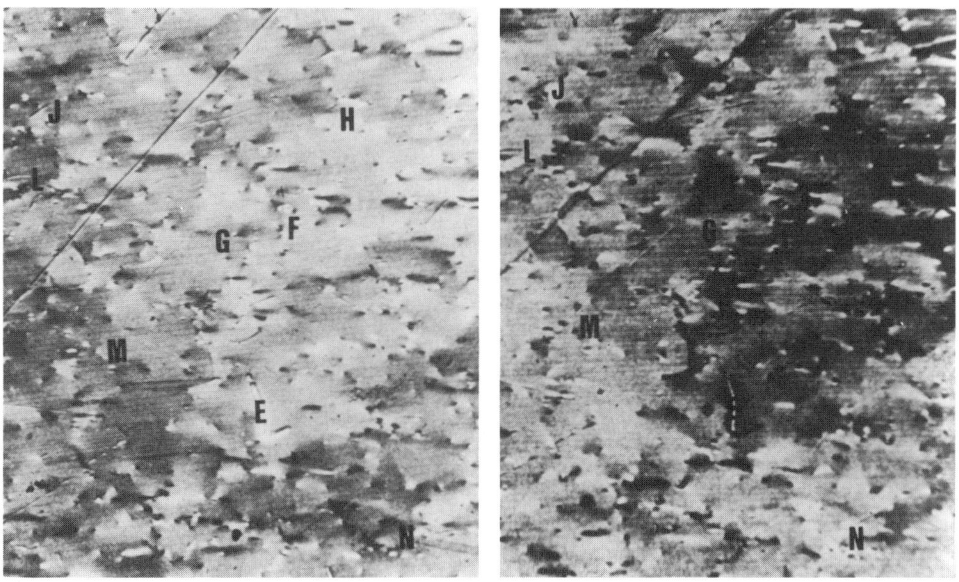

(a) (b)

Figure 8. Images of dislocations take in Bragg (reflection) geometry double crystal topographs of a bulk InP crystal. Dislocation images reverse the sense of the contrast on switching from the low angle side of the rocking curve (a) to the high angle side (b). (S.M.Abdul-Gani, Ph.D. Thesis, Durham University, 1982)

Detailed simulations based on Takagi's theory do, however, permit interference effects to be included in the topographs. Spirkl *et al* (1994) have used a variable step algorithm to simulate misfit dislocation images in (InGa)As layers grown on GaAs. When the epilayer is very thin, typically 30nm, the relaxation becomes extremely important and prevents the identification of the Burgers vector from contrast changes. Only one component of Burgers vector contributes significantly to the contrast and this is the same for all types of 60° misfit dislocation. When a capping layer is present, the interference contrast in the images can be complex and simple geometric analysis based on orientation contrast is impossible.

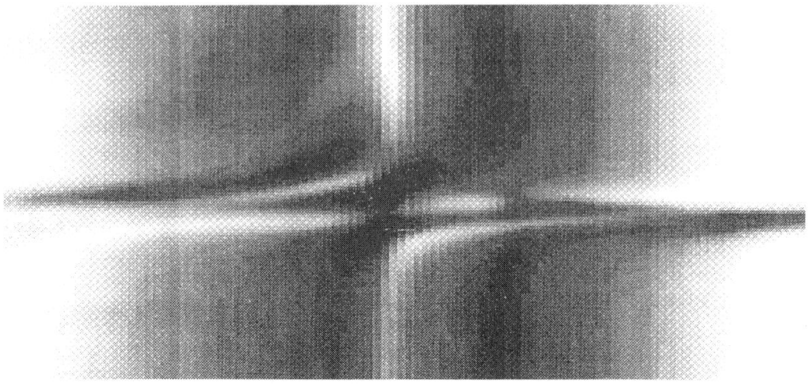

Figure 9 Simulated misfit dislocation with **b** = ½[$\bar{1}\,\bar{1}\,0$] in an InGaAs layer on (001) GaAs, capped with GaAs. The "rocking topograph" is of a distorted crystal where the left-hand side corresponds to the low angle and the right-hand part to the high angle side of the rocking curve. Note the complex contrast in the centre of the (split) rocking curve due to interference effects. 1.49Å wavelength, $\overline{22}4$ reflection

4.1 Weak Beam Topography

A very powerful technique of electron microscopy was developed in the early 1970s, known as weak beam microscopy. The technique involves setting the specimen a long way off the Bragg diffraction condition. Then very little intensity is diffracted from the perfect crystal, and only in the region near the dislocation is the effective misorientation sufficient to bring the lattice into the condition for Bragg reflection locally. Although the intensity is very low, the images are extremely narrow and use of the method enabled a number of important observations to be made for example about the structure of dissociated dislocations in silicon. In principle, such a technique should be possible using a near plane wave in X-ray topography. Until recently, the relatively large size of the sources available has meant that the technique could only be realised in a very specific geometry (Sauvage, 1980). However, the very small source size of the ESRF has meant that weak beam images can now be taken in any geometry (Barrett *et al*, 1995) and as seen in figure 10, the image width does drop very significantly. The region around the defect always shows enhanced intensity.

4.2 Zebra-Patterns

In the double crystal geometry, the angular range of Bragg reflection is extremely narrow, typically 10 arc seconds or less. Suppose the specimen contains a region which has an effective misorientation greater than the width of the rocking curve in the double axis diffraction experiment. Then, when the bulk of the crystal is set on the diffraction maximum,

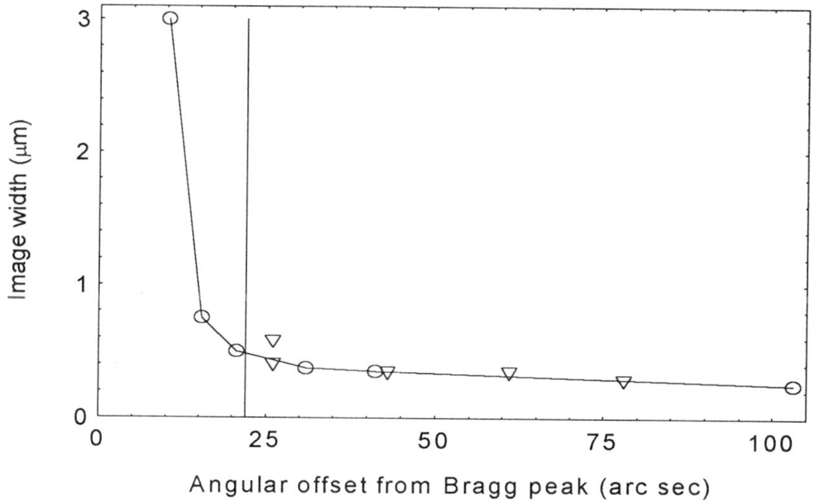

Figure 10. Simulated width of a misfit dislocation in the Bragg geometry as a function of the deviation from the exact Bragg angle. The vertical line is the rocking curve full width at half maximum. Triangles are data points scaled from transmission experiments at the ESRF, Grenoble (Zontone, 1995)

no intensity will be recorded from the misoriented region. Thus the double crystal topograph is sensitive to long range strains, unlike the single crystal arrangments. As the intensity goes to zero, the contrast is higher than for the equivalent single crystal topograph. When the crystal is continuously bent, only a small region will diffract and this effectively maps out a contour of effective misorientation. By displacing the specimen and reference crystals, a different contour can be set into the diffraction condition (Renninger, 1962). Thus by a series of topographs at different relative displacements of reference and specimen, a series of contours can be recorded on the same plate (figure 11). This simple orientation contrast provides a powerful method of strain analysis and by selecting the angular width of the double crystal rocking curve it is possible to make such measurements on such disparate materials as highly perfect crystals of silicon (Ishikawa *et al* 1985) or fatigued zinc bi-crystals (Billelo *et al*, 1989).

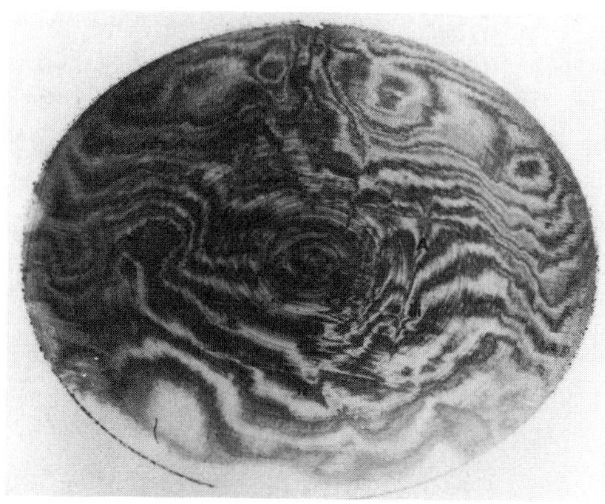

Figure 11 "Zebra pattern" images formed by superposition of double crystal topographs taken at different incidence angles from 50mm semi-insulating GaAs (S.J. Barnett, Ph.D Thesis, Durham University 1987)

5 PRECIPITATES

At an atomic scale, point defects are not visible in X-ray topographs but when such defects cluster to form a precipitate or inclusion, then characteristic contrast is observed. This is of a double "half moon" shape with a line of zero contrast perpendicular to the projection of the diffraction vector . Under conditions of low absorption, the direct image dominates and both halves of the image are dark, whereas under moderate absorption conditions the contrast is black-white if the precipitate is near to the entrance or exit surface. Using the "rule of thumb" introduced in the section on dynamical images, we see that if the precipiate is near the exit surface, if the lobe in the direction of the diffraction vector is dark, then the (coherent) precipitate has an effective volume larger than the lattice and is placing the surrounding lattice in compression.

This contrast can be explained easily in terms of a radial strain field of the form

$$\mathbf{u}(\mathbf{r}) = C\,\mathbf{r}/r^3 \qquad (10)$$

where C is a constant proportional to the effective volume of a coherent precipitate. From this, it is seen immediately that in the direction perpendicular to the diffraction vector, the Bragg planes are not distorted, giving rise to the line of null contrast.

Equation 10 can be used quantitatively to simulate the contrast of precipitates in section topographs of semiconductors such as silicon where there is only small elastic anisotropy. Considering the simplicity of the model strain field, a remarkably good agreement is found over a wide range of C values. The image varies quite markedly as a function of the position of the precipitate within the Borrmann fan and the wide variety of images shown in figure 12 are all simulated using the same value of C (Green, Cui and Tanner, 1990).

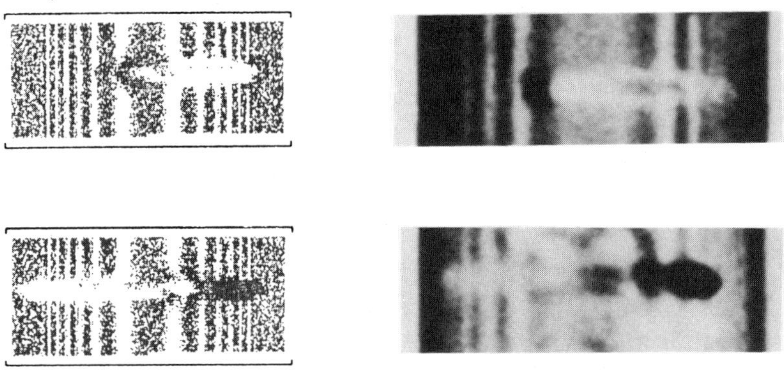

Figure 12 Experimental and simulated section topograph images of hydride precipitates in silicon

6 PLANAR DEFECTS

6.1 Stacking faults

Contrast occurs at stacking faults due to the fact that there is a change in the phase of the structure factor on crossing the fault. At the defect, the Bloch waves decouple into their component plane waves and these immediately each excite two new Bloch waves with amplitudes subject to phase matching with the new structure factor $F_g' = F_g\ exp\ (2\pi i\boldsymbol{g}.\boldsymbol{u})$ where $\boldsymbol{u}$ is the atomic displacement along the fault. The physics is best understood by consideration of a stacking fault inclined to the specimen surface in the section topograph (figure 13(a)). Consider the pair of Bloch wavefields propagating in a direction OA within the Borrmann fan. As the energy flow, or ray path, is normal to the dispersion surface, we see in figure 13(b) that these wavefields must have tie points P and Q on opposite branches of the dispersion surface. The tie points are associated with different wavevector components and so these two wavefields will be propagating with different velocities and hence a phase difference appears which is the origin of the Kato fringes.

If however, the crystal contains an inclined stacking fault XY (figure 13(a)) then at S these two wavefields decouple into their plane wave components, the wavevectors of which must lie on the normal to the stacking fault. They immediately recreate new wavefields whose wavevectors must also lie on the normal to the stacking fault, but also on the dispersion surface. Thus, new wavefields N and M are created. It is easy to see by inverting figure 13(b) and reversing the ray trajectories that, from symmetry considerations, these new wavefields propagate in the same new direction SB. Again, because these wavefields are associated with different branches of the dispersion surface, they propagate with different velocities and a phase difference exists at the exit surface. Notice that, in the crystal below the stacking fault, there are still wavefields propagating in the direction OA, reaching the exit surface at A. At the exit surface we have both old wavefields created at the top surface and new wavefields created at the stacking fault. Consider now the point B. The amplitude at B is composed of old wavefields which had propagated on path OB and new wavefields which had propagated along SB. All these wavefields have different wavevectors and interference occurs between all of them.

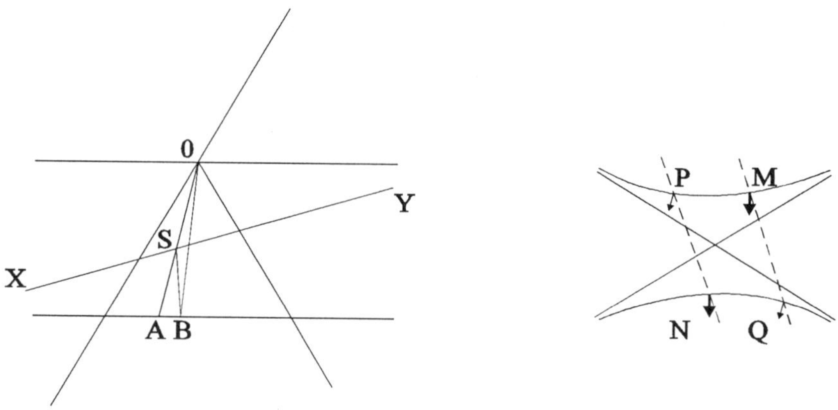

Figure 13 (a) Real space diagram of ray paths in a crystal containing a stacking fault XY (b) Reciprocal space representation showing the dispersion surface and tie points P, Q and M, N on the dispersion surface corresponding to the wavefields propagating respectively in the directions OA and SB.

As a result of this, the section topograph image of an inclined stacking fault (figure 14) is extraordinarily beautiful and resembles an hour-glass. There are in general three sets of fringes, one set associated with interference between old wavefields, one set associated with interference beyween new wavefields and a third set associated with interference between old and new wavefields. This last set has the best visibility under strong absorption conditions as interference occurs between wavefields on the same branch of the dispersion surface. The branch 1 wavefields are anomalously transmitted and in the limit are the only wavefields reaching the exit surface.

The fringe system is complicated by the presence of polarisation, which results in two branches of the dispersion surface for each polarisation state. Scattering does not occur between polarisation states and the result is simply to superimpose the two interference patterns. The result is a periodic fading of the pattern as the two sets of fringes coincide and cancel periodically. For experiments with synchrotron radiation, in which the σ polarisation state alone can be selected, such fading is not observed (Kowalski *et al* 1989). The detailed features of the interference fringe contrast can be simulated either using the Eikonal theory (Authier, 1968) or the wave theory of Takagi (Simon and Authier, 1968).

For a perfect stacking fault, as there is no strain associated with the boundary, no direct image is observed. However, it is usually the case that experimentally the "neck" of the hour-glass is thicker than predicted by the simulations. Jiang and Lang (1983) made a detailed study of the section topograph contrast of stacking faults in diamond and concluded that there existed an anomalous displacement normal to the fault. Such a displacement is consistent with the observation of birefringence contrast at the stacking fault (Jiang, Lang and Tanner, 1987). Lang (1972) has shown that under conditions of high absorption, the sense of the fault vector can be determined. The contrast of the first fringe from the exit surface is positive if $\boldsymbol{g}.\boldsymbol{u}$ is positive.

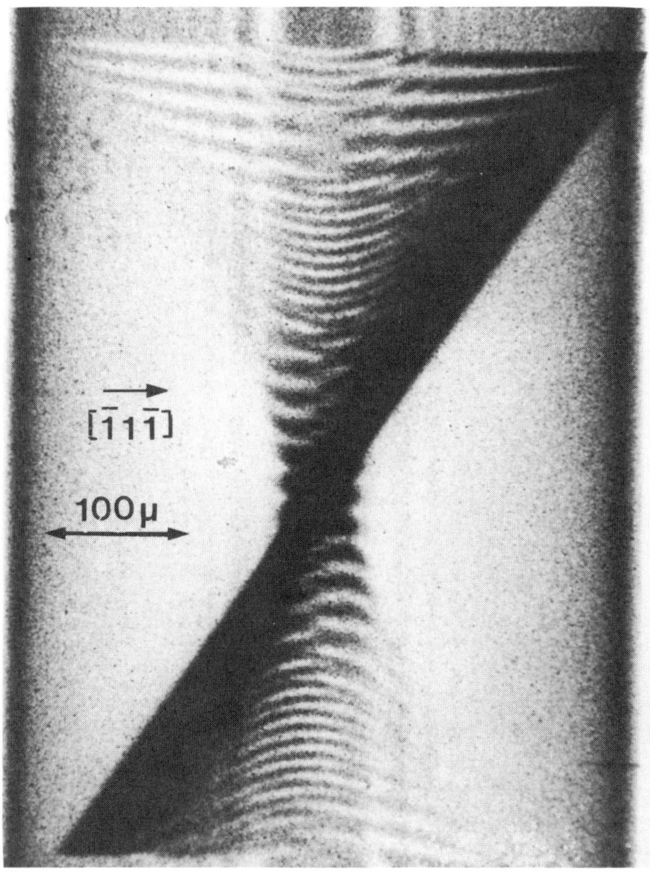

Figure 14. Section topograph of an inclined stacking fault in a crystal of diamond. (a) MoKα radiation [courtesy S.S.Jiang, Nanjing University]

The integrated wave image in the white radiation or Lang topograph is less spectacular but may well exhibit a pronounced fringe contrast if the fault is inclined to the specimen surface. The fringe spacing is such that there is one fringe for a depth of one extinction distance and this provides a measure of the structure factor, if the specimen thickness is known independently. In the case of high absorption, the fringe system is lost except in the region near to the intersection of the fault with the exit surface.

6.2 Twins

The displacement at a twin boundary is more complex. In the case of parallel-lattice twins, the magnitude as well as the phase of the structure factor changes. Such an example is the Dauphiné twin in quartz and this was studied extensively in the early years. Good agreement can be obtained between the simulated and experimental section topograph image only when a misorientation is included at the boundary (Kato, 1980). This arises from the interface being stepped on a fine scale. 180° ferroelectric domains fall into this class of defect but also may exhibit contrast from the bulk of the domain. In a non-centrosymmetric crystal the presence of anomalous dispersion results in differing intensities being diffracted from the two domains (Niizeki and Hasegawa, 1964). Use of synchrotron radiation to tune very close to the absorption edge can greatly enhance this type of contrast.

The 90° ferroelectric and ferroelastic domain boundaries, on the other hand are examples of a large class of twins where there is a misorientation of the lattice planes across the boundary. They may be treated in the same manner as the stacking fault, but with the phase shift varying continuously, being linear with depth below the defect. In the plane of the boundary, the lattice planes are coherent and thus the twin boundary vanishes when the diffraction vector is normal to the twin plane. Residual contrast is often seen however and this can be attributed to stepped interfaces (Jiang *et al* 1988). Magnetic domain contrast arises for the presence of a magnetostrictive strain in the magnetically ordered state. For the case of iron, the body centred cubic lattice becomes slightly tetragonal, and the 90° wall then behaves like a coherent twin on the $\{011\}$ planes. The weak contrast is very well described by this model (Polcarova, 1973) . Such a distortion is also present in cubic antiferromagnets and 90° antiferromagnetic domains are imaged in exactly the same manner as ferromagnetic domains, even though there is no net spontaneous magnetisation (Safa and Tanner, 1978).

It is important to note that 180° ferromagnetic domains are not normally visible as there is no magnetostrictive distortion between them. Only in special cases can the weak contrast associated with the relaxation of the strains at the free surface be imaged. Rotation and inversion twins (Tanner, Clark and Safa, 1988) also give characteristic contrast and appear most spectacularly in white radiation topographs.

6.3 Growth Bands

Another characteristic feature in topographs, particularly of crystals grown from solution are bands of contrast parallel to the growth front. These give a unique record of the growth habit of the crystal throughout its growth history. The contrast arises from the lattice dilation arising from the incorporation of impurities during growth. Changes in temperature or composition of the solution result in a different amount of impurity being included into the lattice and a region of contrast forms. As the strain is normal to the growth interface, growth bands disappear when the diffraction vector is parallel to the band, i.e. when $\boldsymbol{g}.\boldsymbol{n} = 0$, where $\boldsymbol{n}$ is the vector normal to the growth face. Boundaries between growth sectors are also observed in topographs, again arising from the lattice distortion from the impurities.

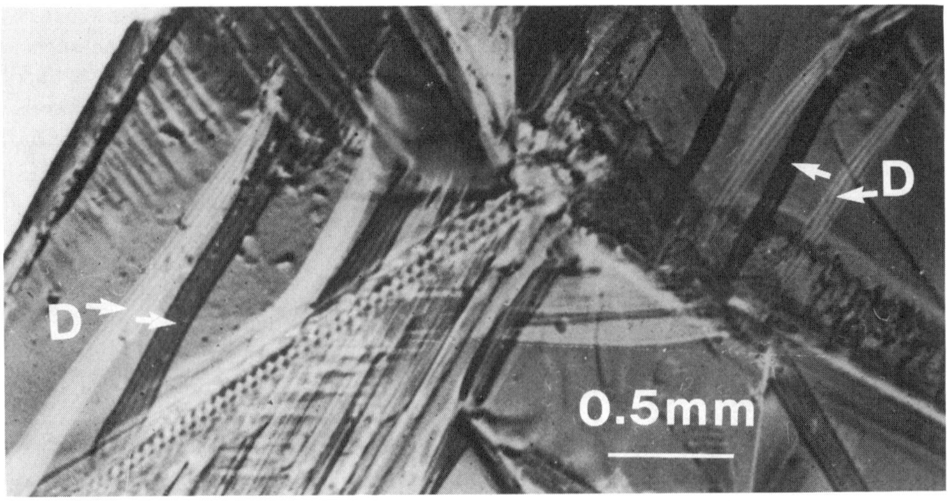

Figure 15 White radiation topograph of magnetic domains in α-Fe_2O_3. As the walls D are inclined to the crystal surface, interference fringes are seen across them. Note the black-white contrast depending on the sense of the magnetisation change across the wall.

6.4 Acoustic Waves

Whatmore *et al* (1982) and Graeff *et al* (1983)independently showed that by exciting acoustic waves in synchronisation with the electron bunches in a storage ring, high frequency stroboscopic topography could be performed. X-rays were then only diffracted at a specific point in the deformation cycle and a topograph of the strains with a time resolution of nanoseconds could be accumulated over a period of up to an hour. In the case of surface acoustic waves (figure 16), the contrast is orientation contrast arising from the misorientation of the Bragg planes as the surface wave passes. Diffracted waves propagate in different directions at different parts of the wave cycle and thus there is overlap and separation of the diffracted waves as one moves down the acoustic wave path.

Figure 16 Stroboscopic white radiation reflection topograph of surface acoustic waves in a quartz crystal. 0003 reflection from a basal plane crystal.

More recently, Capelle and co-workers have studied the contrast of bulk waves in resonators and showed that the technique is extremely good for identification of unwanted modes associated with defects (Zarka *et al*, 1988). They have also performed true section topography in stroboscopic mode to determine whether the interaction between a dislocation and the acoustic wave could be described by simple linear piezoelectric theory. Using simulation of the section topographs to analyze the data, they concluded that a non-linear interaction was present *near* to the dislocation line, linear theory working satisfactorily in the region far from the defect.

REFERENCES

Authier, A., 1967, Contrast of dislocation images in X-ray transmission topographs *Adv. X-ray Analysis* 10:9

Authier, A., 1968, Contrast of a stacking fault on X-ray topographs, *Phys. Stat. Sol.*27:77

Authier, A., 1970, Ewald waves in theory and experiment, in: *Advances in Structure Research by Diffraction Methods*, Brill and Mason, eds., Vieweg und sohn, Braundchweig 3:1

Authier, A., Malgrange, C. and Tournarie, M. 1968Etude theorique de la propagation des rayons X dans un cristal parfait ou legerement deforme, *Acta Cryst.* A24:126

Authier, A. and Simon, D., 1968, Application de la théorie dynamique de S.Takagi au contrast d'un défaut plan en topographie par RX. I. Faute d'empilement,, *Acta Cryst.* A24: 517

Barrett, R., Baruchel, J., Härtwig, J. and Zontone, F., 1995, The present status of the ESRF diffraction topography beamline: new experimental results, *J. Phys. D: Appl. Phys.* 28:A250

Bilello, J. C., Schmitz, H. A. and Dew-Hughes, D., 1989, Synchrotron radiation Bragg angle contour mapping of crack nucleation at grain-boundaries, *J. Appl. Phys.* 65:2282

Bonse, U. and Kappler, E., 1958, Röntgenographische abbildung des verzerrungsfeldes einzelner vertsetzungen in germanium-einkristallen, *Z. Naturforsch.* 13a:348

Epelboin, Y., 1985, Simulation of X-ray topographs, *Mater. Sci. Eng.* 73:1

Gluer, C-C., Graeff W. and Moller, H., 1983, Stroboscopic topography with nanoseconds time resolution, *Nucl. Inst. Meths.* 208:701

Green, G. S., Cui, S. F. and Tanner, B. K. 1990, Simulation of images of spherical strain centres in X-ray section topographs, *Phil Mag A* 61:23

Holland, A.J., 1992, Analysis of crystal defects by simulation of X-ray section topographs, Ph.D. Thesis, University of Durham

Ishikawa, T., Kitano, T. and Matsui, J., 1985, Synchrotron plane-wave X-ray topography of GaAs with a separate monochro-collimator, *Japan. J. Appl. Phys.* 24:L968

Jiang, S-S. and Lang, A. R., 1983, Stacking fault contrast in X-ray diffraction - a high-resolution experimental study, *Proc. Roy. Soc. Lond.* A388:249

Jiang, S-S., Lang, A. R. and Tanner, B. K., 1987, The contrast of stacking faults in optical birefringence micrographs of diamond, *Phil Mag A* 56:367

Jiang, S-S., Surowiec, M. R. and Tanner, B. K., 1988, Ferroelastic domain structures and phase transitions in barium sodium niobate *J. Appl. Cryst.* 21:145

Kaganer, V. M. and Möhling, W., 1991, Characterization of dislocations by double crystal X-ray topography in back reflection, *Phys. Stat. Sol* (a) 123:379

Kato, N., 1963, Pendellösung fringes in distorted crystals. I. Fermat's principle for Bloch waves, J. Phys. Soc. Japan 18:1785

Kato, N., 1974, X-ray diffraction, in: *Introduction to X-ray Diffraction*, L. G. Azaroff, R. Kaplow, N. Kato, R.J. Weiss, A.J.C. Wilson and R.A. Young, eds, Mc-Graw Hill, New York, p.425

Kato, N. 1980, Perfect and imperfect crystals, in: *Characterization of Crystal Growth Defects by Diffraction Methods,* B.K.Tanner and D.K.Bowen, eds Plenum, New York , p.264

Kowalski, G., Lang, A. R,. Makepeace A. P. W. and Moore, M. 1989, Studies of stacking-fault contrast by synchrotron X-ray section topography, *J. Appl. Cryst.* 22:410

Lang, A.R., 1959, The projection topograph: a new method in X-ray diffraction microradiography, *Acta Cryst* 12:249

Lang, A.R., 1978, Techniques and interpretation in X-ray topography, in: *Modern Diffraction and Imaging Techniques in Materials Science*, Vol II, S. Amelincx, R. Gevers, J. van Landuyt, eds., North Holland, Amsterdam, p 623

Lang, A. R., 1972, On determining the sign of fault vectors by X-ray topography, *Z. Naturforsch.* 27a:461

Midgley, D., Safa, M., and Tanner, B.K., 1977, Dislocation contrast in X-ray synchrotron topographs *J Appl Cryst*, 10:281

Miltat, J., and Bowen, D.K, 1975, On the widths of dislocation images in X-ray topography under low absorption conditions, *J. Appl. Cryst.* 8:657

Niizeki, N. and Hasegawa, M., 1964, Direct observation of antiparallel 180° domains in $BaTiO_3$ by X-ray anomalous dispersion method, *J. Phys. Soc. Japan* 19:550

Penning, P., and Polder, D., 1961, Anomalous transmission of X-rays in elastically deformed crystals, *Philips Res. Reports* 16:419

Polcarova, M., 1973, The effect of anomalous absorption on the diffraction contrast on 90° Bloch walls, *Z. Naturforsch.* 28a:639

Renninger, M., 1962, "Frozen" and reversible lattice distortions, *Phys. Lett.* 1:106

Safa, M. and Tanner, B. K., 1978, Antiferromagnetic domain wall motion in $KNiF_3$ and $KCoF_3$ observed by X-ray synchrotron topography *Phil. Mag. B* 37:739

Sauvage, M., 1980, Monochromatic synchrotron radiation topography, in: *Characterization of Crystal Growth Defects by Diffraction Methods,* B.K.Tanner and D.K.Bowen, eds., Plenum, New York, p.433

Spirkl, W., Tanner, B. K, Whitehouse, C.,. Barnett, S. J., Cullis, A. G., Johnson, A. D., Keir, A., Usher, B., Clark, G. F., Hagston, W., Hogg C. R. and Lunn, B. 1994 Dislocation contrast in X-ray reflection topographs of strained heterostructures, *Phil. Mag. A* 70: 531

Tanner, B.K., 1972, Dislocation contrast in X-ray topographs of very thin crystals *Phys. Stat. Sol.* (a) 10:381

Tanner, B.K., 1975, *X-ray Diffraction Topography,* Pergamon Press, Oxford

Tanner, B.K.and Bowen, D.K., eds. 1980, *Characterization of Crystal Growth Defects by Diffraction Methods,* Plenum, New York,

Tanner, B. K., Clark, G. F. and Safa, M., 1988, Domain structures in haematite *Phil Mag B* 57:361

Tanner, B.K., and Bowen, D.K., 1992, Synchrotron X-ray topography, *Materials Science Reports* 8:369

Tuomi, T., Tilli, M. and Anttila, O., 1985, Detection of defect-free zones in annealed Czochralski silicon with synchrotron section topography, *J. Appl. Phys.* 57:1384

Whatmore, R. W., Goddard, P. A,. Tanner B. K. and Clark, G. F, 1982, Direct imaging of travelling Rayleigh waves by stroboscopic X-ray topography *Nature* 299:44

Zarka, A., Capelle, B., Detaint J. and Schwartzel, J., 1988, Stroboscopic X-ray topography of quartz resonators J. *Appl. Cryst.* 21:967

Zontone, F., 1995, Imagerie par diffraction ("topographie') a l'installation Europeenne de rayonnement synchrotron ("ESRF"): ligne de lumiere, possibilites et premieres experiences, Thesis, University of Grenoble

X-RAY DIFFRACTION TOPOGRAPHY: APPLICATION TO CRYSTAL GROWTH AND PLASTIC DEFORMATION

Helmut Klapper

Mineralogisch-Petrologisches Institut
Universität Bonn
D-53115 Bonn, Germany

1. INTRODUCTION

X-ray diffraction topography is a non-magnifying imaging method of comparatively poor spatial resolution and therefore confined to crystals of low defect density. Despite this restricting feature it has found wide-spread applications in various fields of research, in crystal processing and in the assessment of crystals for technical applications. Classic fields of applications are crystal growth and semiconductor processing for device production. In recent times it has proved to be an excellent method also for the study of phase transitions and the formation of (twin) domains.

The particular significance of X-ray diffraction topography lies in the investigation of nearly perfect crystals, when individual defects can be resolved and characterized. In this paper such studies selected from the fields of crystal growth and dislocation glide (plastic deformation) are presented. In both cases dislocations play the essential role and are emphasized. All topographs shown in this paper were recorded on X-ray film Structurix D4 (Agfa-Gevaert) applying the technique of Lang (1959) and a conventional rotating-anode X-ray source. For the low-absorbing organic molecular crystals with oxygen as the heaviest atom CuKα radiation, for the higher (but still low) absorbing materials (e.g. sulphates of light metals) MoKα radiation was used.

For additional reading on X-ray topography and its various applications the textbook by Tanner (1976) and the reviews by Lang (1978), Authier (1977,1978), Bhat (1985) and Klapper (1980,1987,1991) are recommended.

X-ray and Neutron Dynamical Diffraction: Theory and Applications
Edited by Authier *et al.*, Plenum Press, New York, 1996

2. CRYSTAL GROWTH

The perfection of a crystal is strongly dependent on the conditions of growth. X-ray diffraction topographic studies revealing and characterizing growth defects are an excellent help in improving the growth conditions and achieving better crystals. The most important defects occurring during growth are dislocations. Other significant lattice perturbations related to crystal growth, and considered here, are inclusions, growth striations and faulted growth-sector boundaries.

2.1 Dislocations

Dislocations in crystals are generated during growth, by plastic deformation (glide) and by the condensation of self-interstitials or lattice vacancies. With respect to crystal growth, two categories of dislocations are distinguished:

- Growth dislocations, which are connected with the growth front, and
- Dislocations, which are generated "behind" the growth front. This may be still during the growth run or after growth during the period of cooling to room temperature or later.

The final dislocation configuration in a crystal at room temperature results from growth dislocations, post-growth dislocations and the movement of both after growth (e.g. during cooling). Crystals grown at low temperatures (e.g. from solutions in water) usually contain dislocations in their original "as-grown" arrangement, whereas in crystals grown at high temperatures (e.g. from the melt) the original dislocation configuration may drastically be changed by the dislocation movement and dislocation reactions after growth. The post-growth movement of dislocations is induced by thermal stress ("dislocation glide") and by the absorption of interstitials and vacancies ("dislocation climb").

2.1.1 Sources of dislocations. For topological reasons dislocation lines cannot start in the interior of a perfect crystal. They either form closed loops or they start from the (internal or external) crystal surfaces or from defects implying a "break" of the crystal lattice. Such defects are inclusions of mother liquor, impurity particles, solute precipitates (dopants) and gas (bubbles). Inclusions frequently occur in the zone of initial growth on seed crystals or in regions of unstable growth due to strong fluctuations of growth conditions. Examples showing growth dislocations starting from inclusions at the seed surface of crystals grown from supercooled melt (benzil) and from aqueous solution (sodium chlorate) are given in Figure 1 a,b. The generation of inclusions and dislocations by an accidental growth instability is shown in Figure 1c.

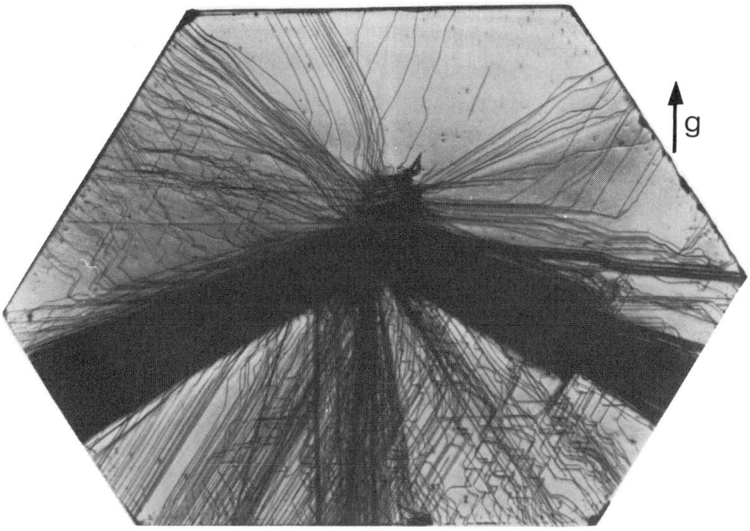

(a)

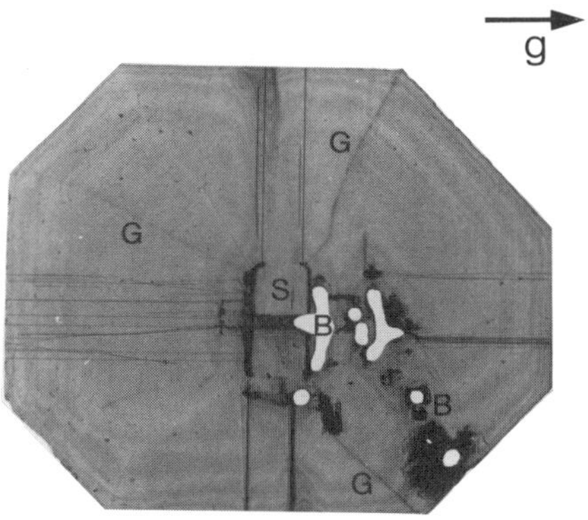

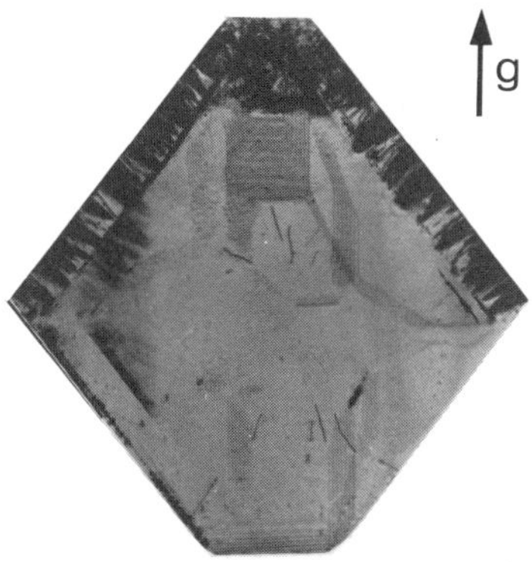

(b) (c)

Figure 1. (a) X-ray topograph (CuKα radiation, reflection 2$\bar{2}$00, **g**: diffraction vector) showing growth dislocations in a (0001) plate of trigonal benzil (C_6H_5CO-COC_6H_5) grown from supercooled melt (T_m = 96°C). The plate (about 1.2 mm thick, horizontal dimension about 35 mm) contains a part of the disturbed seed crystal. The dislocations start from tiny (non-resolved) bubbles at the seed surface. Many dislocations show post-growth movement.

(b) (001) plate of cubic sodium chlorate $NaClO_3$ (1 mm thick, 21 x 25 mm²) grown from aqueous solution by evaporation at 45°C. MoKα radiation, reflection 200. S: Seed crystal. B: Holes in the plate due to inclusions of solution opened by cutting the plate out of the bulk crystal. G: Growth-sector boundaries.

(c) (101) plate of cubic potash alum, $KAl(SO_4)_2 \cdot 12H_2O$, about 1 mm thick, 32 x 22 mm, reflection 400, MoKα radiation. Due to an unknown growth accident many inclusions and dislocations have formed on the upper octahedral faces of the nearly dislocation-free crystal.

Figure 2a is a topograph of benzophenone ($C_6H_5COC_6H_5$, melting temperature $T_m = 48°C$) grown by the Czochralski method. The crystal is highly perfect, showing faint pendellösung fringes due to a slight tapering of the plate toward its edges. Only a few dislocations have formed in the zone of initial growth on the seed crystal (*cf.* Figure 4). The salol crystal ($C_{13}H_{10}O_3$, $T_m = 42°C$) shown in Figure 2b was grown with a rather high pulling rate and poor temperature control, leading to the precipitation of small bubbles of (atmospheric) gas dissolved in the melt. The bubbles outline the growth front and are the source of new dislocations. Due to the concave shape (toward the melt) of the growth interface the dislocation lines are "focussed" toward the axis of the boule. This leads to an increasing dislocation density with proceeding growth.

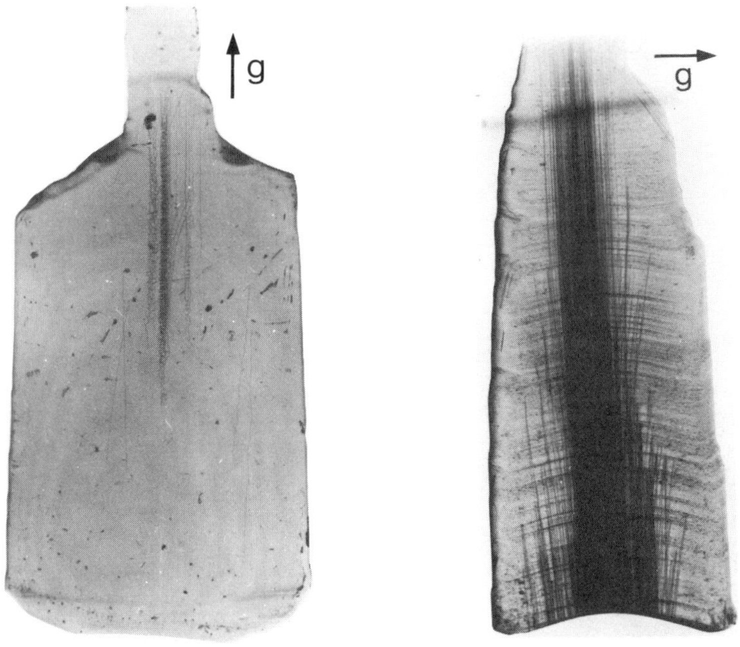

(a) (b)

Figure 2. (a) Nearly dislocation-free Czochralski benzophenone (T_m = 48°C)crystal, pulling direction [001], 48 mm long (incl. seed rod), plate 1.2 mm thick, reflection 002. Only a few dislocations, mostly originating from the seed-crystal interface, are present (*cf.* Fig.4). The black dots result from damage in the plate surface.

(b) Czochralski salol (T_m= 42 °C), imaged length 45 mm, plate thickness 1.3 mm, showing growth striations marked by tiny gas bubbles and numerous dislocations starting from them. Both CuKα radiation.

2.1.2 Propagation of dislocations during growth. A dislocation line ending on the growth face will proceed with the growing crystal. Thus its path depends on the shape and the orientation of the growth face. When crystals grow on planar (habit) faces, they consist of various growth sectors belonging to different growth faces (different growth directions). This results - under ideal conditions - in a characteristic configuration of growth dislocations which is described as follows:

(i) The dislocations start from inclusions and propagate as straight lines with directions **l** close to the growth normal **n** of the growth sector in which they lie. In general, the deviation of **l** from the growth direction **n** does not exceed angles of $\pm 20°$ (**l** and **n**: unit vectors).

(ii) The growth dislocations exhibit often sharply defined and frequently non-crystallographic preferred directions which depend on the Burgers vector **b** and on the growth direction **n**. The dependence of **l** on the growth direction becomes apparent when a dislocation penetrates a growth-sector boundary (abrupt change of growth direction **n**): Here, the dislocation line undergoes an abrupt change in its direction ("refraction" of a dislocation line by a growth-sector boundary). An example is shown in Figure 3a,b

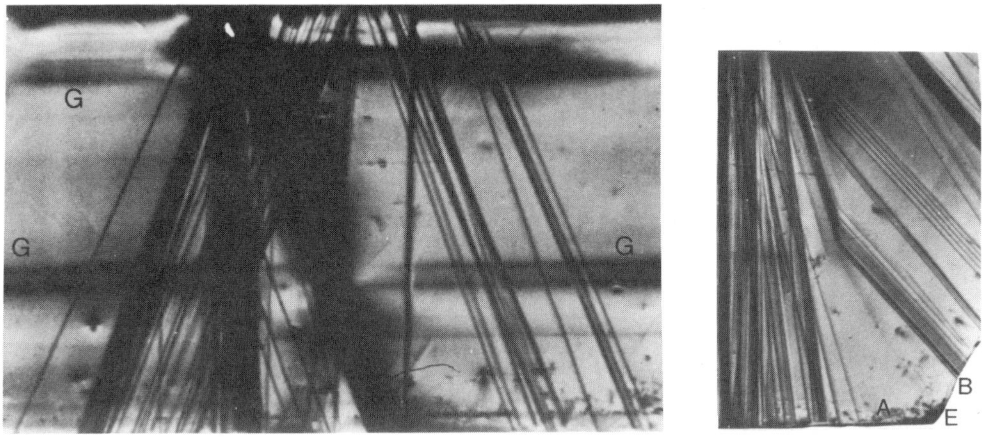

(a) (b)

Figure 3. Sections of (001) plates, (about 1.2 mm thick) cut from a benzil crystal grown from a solution in xylene by evaporation at 40°C. Reflections $20\bar{2}0$, CuKα radiation.

(a) Grown-in dislocation lines with favoured directions (three different Burgers vectors of type **b** = ⟨100⟩) in a prism-face growth sector. G: Growth bands. Section 6.0 x 8.8 mm².

(b) Change of direction of dislocation lines penetrating a growth-sector boundary (i.e. when during growth the end points of dislocation lines shift over the crystal edge E from one face A to the other face B). Section 3.8 x 6.0 mm².

The straight-lined configuration of growth dislocations is frequently perturbed by other defects generating stress (e.g. inclusions, striations), by macrosteps on the growth face and by post-growth dislocation movement (see below).

Growth dislocation lines in crystals grown on curved interfaces (e.g. Czochralski growth) also exhibit preferred directions more or less normal to the local growth front. According to the local change of the growth direction, these dislocation lines are also curved. For interfaces with convex curvature toward the melt, they diverge and grow out through the side surface of the Czochralski boule. For concave interfaces, however, they converge and accumulate along the axis of the crystal boule. This is shown in Figure 2b.

For a more detailed discussion on growth dislocations and their typical geometry the reader is referred to the literature (Klapper 1980, 1991).

2.1.3 Post-growth movement of growth dislocations. In most cases "as-grown" dislocations and dislocations moved or generated after growth ("behind" the growth front) are easily distinguished by their quite different geometrical appearance. An example showing both types of dislocations in close association in a crystal of orthorhombic lithium ammonium sulphate is presented in Figure 4. This crystal has been grown from aqueous solution at 45°C by slow evaporation. The straight grown-in dislocation (A) lines following favoured directions roughly normal to the growth face are clearly distinguished from the irregular dislocations (B) which show pinning points where the movement was stopped by unknown obstacles. It is assumed that this dislocation geometry was formed at the end of the growth experiment during inhomogeneous cooling of the crystal to room temperature.

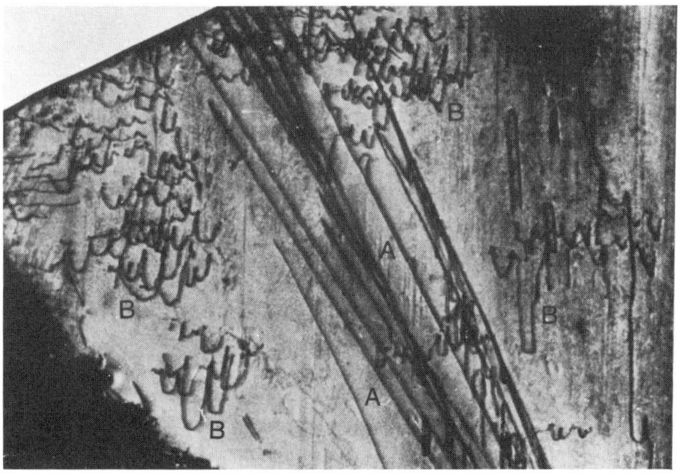

Figure 4. Section (7.2 x 10.3 mm²) of a plate (thickness about 1.1 mm) of lithium ammonium sulphate, $LiNH_4SO_4$, grown from aqueous solution by evaporation at 45°C. A: As-grown dislocation lines. B: Dislocation lines showing movement after growth. Reflection 202, MoKα.

Changes of the grown-in dislocation arrangement after growth is common in crystals grown from the melt, due to the temperature range of plasticity which practically all crystals possess at higher temperatures below the melting temperature. Even moderate stress induced by temperature gradients, occurring during growth or during the cooling period after growth, may lead to dislocation movement. Many of the growth dislocations shown in Figure 1a (benzil from supercooled melt) suffered post-growth changes in their geometry, whereas in solution-grown benzil (which is brittle at 40°C) the original grown-in dislocation arrangement is preserved (Figure 3). An example of drastic change of the original dislocation geometry is shown in Figure 5. The originally straight-lined screw dislocations have adopted complicated irregular courses with numerous pinning points where the movement of the dislocation lines was locally blocked by obstacles which could not yet be identified. In some cases the pinning points of one dislocation are aligned along a straight line, thus indicating its original course. The Burgers vectors, as determined by the **g.b** criterion (Lang 1959; Bonse et al., 1967; *cf.* Tanner, this Volume), are the shortest lattice translation [001] of the crystal. In some cases dislocations of opposite Burgers vector sense have reacted, leading to partial annihilation of dislocation segments and the formation of loops arranged along the original dislocation line direction.

Dislocations, which are not connected with the growth front and which are generated after growth will be treated in the section 3. PLASTIC DEFORMATION.

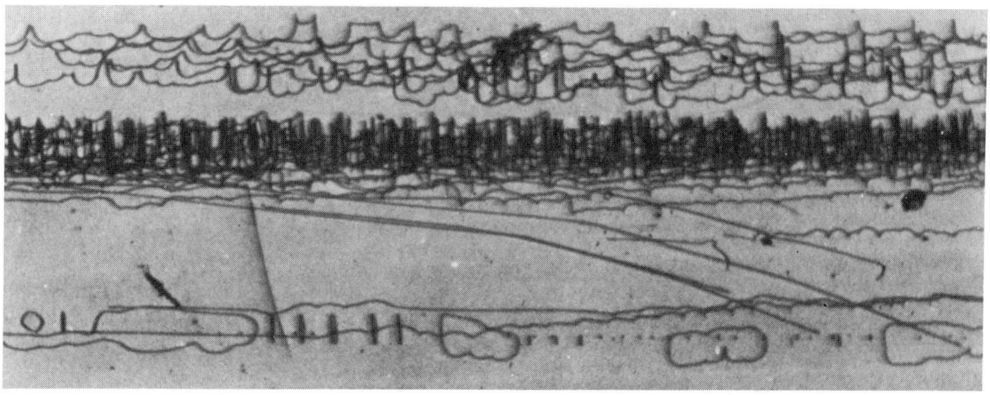

Figure 5. Enlarged section (3.8 x 10 mm^2) of the topograph Fig. 2a, showing grown-in dislocation lines (**b** = [001]) in benzophenone after post-growth movement. The growth (pulling) direction [001] is horizontal from right to left. Note that the pinning points of a dislocation at the upper margin of the figure are aligned along a line parallel to the growth direction [001]. In the lower part a row of loops has formed by partial annihilation of two dislocation lines.

2.2 Growth bands, striations and growth-sector boundaries.

Growth bands and striations originate from local variations of the impurity or dopant concentration. They are created by fluctuations of the growth conditions, such as changes of temperature, cooling rate, convection of the solution or the melt. The varying impurity or dopant concentration induces local changes of the lattice parameters which give rise to X-ray topographic contrast. These impurities may be contaminants of the solvent or of the melt, incorporated solvent components or deviations from stoichiometry.

The regions of different impurity concentration form layers coinciding with the instantaneous growth front. They appear on X-ray topographs as bands normal to the growth direction. They may occur isolated or in dense sequence (striations). Faint striations are visible in Figure 1b (besides faint pendellösung fringes arising from slight thickness variations of the specimen). Pronounced growth bands, resulting from a local change of solvent concentration in benzil, are shown in Figure 3a.

Usually, the X-ray topographic contrast vanishes in reflections with diffraction vector **g** parallel to the growth band (striations). This indicates that the displacements in the distorted regions are mainly normal to the faulted layer.

Frequently the occurrence and "intensity" of growth striations depends on the growth sector. This demonstrates the different and selective incorporation of impurities on different (i.e. symmetrically non-equivalent) growth faces. This also holds for curved and partially facetted growth fronts (e.g. in Czochralski growth). The X-ray topographic observation of striations is an excellent method for studying the (in)homogeneity of impurities or dopants and reconstructing the growth history of the crystal.

In many crystals growth-sector boundaries are fault surfaces. They appear on X-ray topographs by kinematical contrast due to lattice distortions (see Figure 1b) or by dynamical fringe contrast similar to that of stacking faults (Figure 6). A more detailed characterization of these planar defects (translation boundaries, tilt boundaries) is possible by section topography.

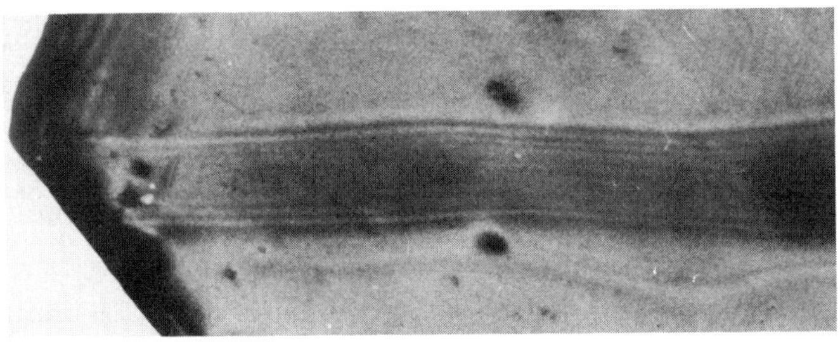

Figure 6. Section (7.2 x 2.4 mm²) of a (0001) plate of benzil, 1.3 mm thick, showing a fault boundary between a $(10\bar{1}0)$ and a $(01\bar{1}1)$ growth sector. The boundary appears by dynamical fringe contrast. Reflection $02\bar{2}0$.

3. PLASTIC DEFORMATION

Plastic deformation of crystals proceeds by the movement (glide, slip) of dislocations. The anisotropic plasticity of crystals is described by glide systems [uvw](hkl) and the critical yield stress necessary to activate them. Glide systems can be identified by the observation of dislocations and of the plane (hkl) on which they have slipped (glide plane), and by the determination of the dislocation Burgers vector [uvw]. The Burgers vector represents the direction and the amount of glide induced by one dislocation moving through the crystal.

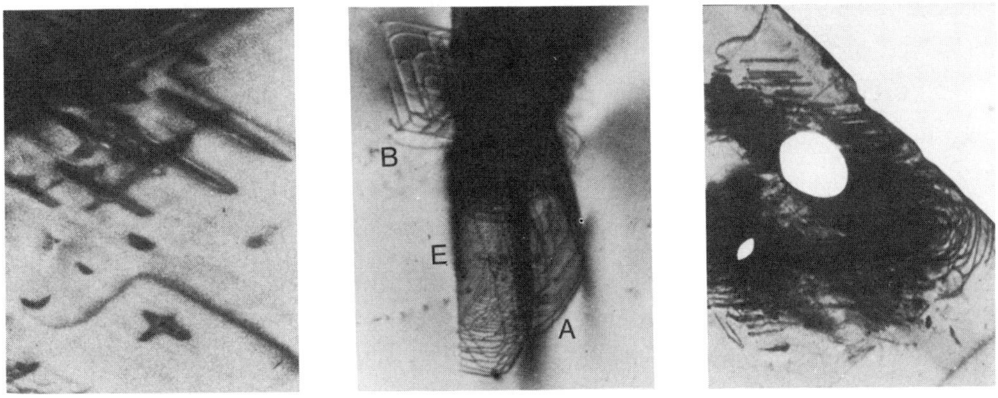

(a) (b) (c)

Figure 7. (a) Section (4.2 x 5.3 mm^2) of a (0001) plate (1.4 mm thick) of benzil grown from supercooled melt at 0.2°C below Tm = 96°C. Reflection 02$\bar{2}$0, MoKα radiation. Dislocation loops and half-loops belonging to the basal glide system ⟨100⟩(0001) were emitted from tiny incorporated particles. Note that the loops and half-loops are elongated in the slip directions ⟨100⟩. Two elongation directions are recognized, whereas half-loops elongated along the third equivalent direction are invisible according to **g.b** = 0. The contrast of dislocations is rather wide due to the relatively short wavelength of the MoKα radiation (compared with CuKα).

(b) Sequence of glide dislocation half-loops with Burgers vectors [100] (A) and [110] (B) originating from a small gas bubble in benzil. Due to a miscut of about 15° of the plate against (0001), the dislocations emerge along E through the surface. The line E is straight and sharply defined, thus indicating that the half-loops lie (within the X-ray topographic resolution) in one plane. Section (6.0 x 7.5 mm^2), reflection 20$\bar{2}$0, CuKα.

(c) Section (about 3 x 4 mm) of a (001) plate (0,8 mm thick) of sodium chlorate, $NaClO_3$. Reflection 200, MoKα. Glide dislocations with **b** = [100] generated in the stress field surrounding two inclusions originally filled with solution (now holes in the plate).

In general the plasticity of crystals is studied by deformation experiments in which stress-strain curves are recorded and slip lines on the specimen surface are analysed. The density of dislocations generated in such experiments is usually too high for X-ray topographic studies. However, glide systems of crystal can be studied without intentional deformation experiments by analysing the movement of growth dislocations after growth or, more instructively, the geometrical features of dislocations generated after growth. These dislocations are common in crystals which are plastic at room temperature or at the elevated temperatures at which they are grown (particularly in melt-grown crystals). They are connected with stress centres like inclusions, from which they are "emitted" in the shape of loops or half-loops, leading to stress relaxation by glide. Frequently a sequence of many loops or half-loops, lying in a sharply defined low-indexed plane (hkl), the glide plane, is formed. If the Burgers vector is determined, e.g. by applying the **g.b** criterion (*cf.* Tanner, this volume), the glide system [uvw] (hkl) is known. Examples are presented and explained in Figure 7. A detailed study of glide systems in benzil by X-ray topographic analysis of dislocations is given by Scheffen-Lauenroth et al. (1981).

In this connection the following observation is worth mentioning. From microhardness measurements it is known that sodium chlorate is very brittle and crackless Vickers indentations can be obtained only with small loads. X-ray topography, however, reveals the moveinent of growth dislocations, leading to similar geometries as shown for benzophenone in Figure 5, as well as the generation of dislocation loops by stress relaxation (Figure 7c) . Similar observations were made for lithium ammonium sulphate which contains many glide dislocations (Figure 4) but appears macroscopically as brittle. This controversy may be due to the higher temperatures during growth or to the different velocities of deformation which were fast in microhardness tests and slow during the cooling of the crystal (creep). In conclusion, these observations show that X-ray topography is an excellent method for studying these kinds of plastic processes.

4. REFERENCES

Authier, A., 1977, X-ray and neutron topography of solution-grown crystals, in: *1976 Crystal Growth and Materials (Current Topics in Materials Science, Vol. 2, p. 516)*, E. Kaldis and H.J. Scheel, eds., North Holland, Amsterdam.

Authier, A., 1978, Contrast of images in X-ray topography, In: *Diffraction and Imaging Techniques in Materials Science*, S. Amelinckx, R. Gevers and J. Van Landuyt, eds., North Holland, Amsterdam, p. 715.

Bhat, H.L., 1985, X-ray topographic assessment of dislocations in crystals grown from solution; in: *Progress in Crystal Growth and Characterization*, Vol. 11, p. 57, B.R. Pamplin, ed., Pergamon, Oxford.

Bonse, U.K., Hart, M. and Newkirk, J.B., 1967, X-ray diffraction topography, in: *Advances in X-Ray Analysis*, Vol. 10, p.1, J.B. Newkirk and G.R. Mallet, eds., Plenum Press, New York.

Klapper, H., 1980, Defects in non-metal crystals, in: *Characterization of Crystal Growth Defects by X-Ray Methods*, B.K. Tanner and D.K. Bowen, eds., Plenum Press, New York, p. 133.

Klapper, H., 1987, X-ray topography of twinned crystals, in: *Progress in Crystal Growth and Characterization* Vol. 14, p. 367, P. Krishna, ed., Pergamon, Oxford.

Klapper, H., 1991, X-ray topography of organic crystals, in: *Crystals: Growth, Properties and Applications*, Vol. 13, p. 109, N. Karl, ed., Springer, Berlin-Heidelberg.

Lang, A.R., 1959, The projection topograph: A new method in X-ray diffraction microradiography, *Acta Cryst.* 12: 249.

Lang, A.R., 1978, Techniques and Interpretation in X-ray topography, in: *Diffraction and Imgaing Techniques in Materials Science*, S. Amelinckx, R. Gevers and J. Van Landuyt, eds, North Holland, Amsterdam, p. 623.

Scheffen-Lauenroth, Th., Klapper, H. and Becker, R.A., 1981, Growth and perfection of organic crystals from undercooled melt. I. Benzil, *J. Crystal Growth* 55: 557.

Tanner, B.K., 1976, *X-Ray Diffraction Topography*, Pergamon Press, Oxford.

NEUTRON TOPOGRAPHY

Michel Schlenker[1] and José Baruchel[2]

[1] Laboratoire Louis Néel du CNRS, associé à l'UJF,
B.P. 166, F-38042 Grenoble, France
[2] ESRF, B.P. 220, F-38043 Grenoble, France

1. INTRODUCTION

Neutron diffraction imaging (hereafter called, following standard usage, neutron topography) is as closely related to X-ray topography as neutron diffraction is to X-ray diffraction. On the one hand the diffraction physics is nearly the same although neutrons have non-zero rest mass and, in the form under which they are used in diffraction (thermal neutrons, or nearly so), are non-relativistic, while the contrary applies for photons. The main orders of magnitude characterizing the processes are similar although the interactions are quite different. Thus, in principle, neutron topography is possible, and most of the features discussed in relation to X-ray topography in this volume are valid. On the other hand, just as neutrons, when discerningly used, are a valuable complement to X-rays for structural work, neutron topography has its own limitations, but also very special capabilities which make it worth while:

(a) neutron beams are only available in large facilities, which for most researchers means away from the "home" laboratory

(b) the available neutron fluxes are small even at a high-flux reactor and even when compared with laboratory X-ray generators (Scherm & Fåk, 1993),

(c) absorption is negligible in most materials, and

(d) magnetic scattering is a strong component.

This chapter will focus on these differences.

2. EXPERIMENTAL FEATURES AND LIMITATIONS

The effect of (a), together with the fact that nuclear reactors are, for security reasons, not widely open, is that the first attempts at neutron topography occurred late, with the work of Doi et al. (1971), Ando & Hosoya (1972), and Schlenker & Shull (1973). It is still practised at very few places in the world; one of them, instrument S20 at Institut Laue-Langevin (ILL), is open to external users.

As a result of (b), the resolution of neutron topography is poor. It was estimated to be no better than 60 μm in non-polarized work on the instrument installed at ILL Grenoble, for exposure times of hours, with roughly equal contributions from detector resolution, geometric blurring due to beam divergence, and shot noise, i.e. fluctuation in the number of diffracted neutrons reaching a pixel. The same reason leads to the technique being instrumentally simple; refinements that might lead to better resolution are discouraged by the increase in exposure time which they imply.

Typically, a neutron beam with divergence of the order of 10' is monochromated by a very good but not quite perfect crystal (mosaic spread a few minutes of arc), and the monochromatic beam illuminates the sample, which can be either a single-crystal or a grain in a polycrystal. It is advantageous, but not mandatory, to use a white beam delivered by a curved neutron guide-tube. The divergence is then limited by the guide-tube, and high-energy parts of the spectrum, which would contribute to unwanted background, as well as γ-rays, are eliminated.

After the specimen is set for a chosen Bragg reflexion with the help of a detector and counter, a neutron-sensitive photographic detector is placed across the diffracted beam, as near the sample as possible to minimize geometric blurring effects while avoiding the direct transmitted beam. Crude but comparatively fast exposures can be made with Polaroid film and an isotopically enriched ^{6}LiF (ZnS) phosphor screen. Better topographs are obtained with X-ray film associated with a gadolinium foil (if possible isotopically enriched in ^{157}Gd) acting as a (n -β) converter, or with a track-etch plastic foil with a ^{6}LiF or $^{10}B_4C$ foil or layer (n-α) converter) (Malgrange et al., 1976). Alternatively, an electronic position-sensitive neutron detector can be used for both setting and imaging (Davidson & Case, 1976; Sillou et al., 1989).

Polarized neutrons are extremely useful in the investigation of magnetic domains. The use of a polarizing monochromator, e.g. a single crystal of Cu_2MnAl placed in a magnetic field sufficient to saturate it, and a simple attachment providing a guide-field and the possibility to flip the polarization can provide this option, because the requirements are less stringent than in quantitative structural polarized neutron-diffraction work.

It is also possible to use the white beam from a curved guide-tube directly (Boeuf et al., 1975), in the same way as in synchrotron radiation X-ray topography, i.e. making a Laue diagram, each spot of which is a topograph. The technique is then instrumentally extremely simple, but background is a problem.

The ability of neutron beams to go through furnaces or cooling devices, one of the general advantages in neutron diffraction work, is of course retained in topography. It is however desirable to keep the specimen-to-film distance small (< 2 cm) .

In the observation of crystal defects, neutron topography is basically very similar to X-ray topography, in particular in the mechanism giving rise to contrast, viz. the local perturbation of wave-fields due to changes in the deviations from Bragg's condition. While individual defects can be observed [Schlenker et al. (1974), Malgrange et al. (1976)], it seems sensible to resort to neutrons only if X-rays cannot be used, i.e. if the samples have too large an absorption for X-rays.

Neutron topography also shows visually the salient dynamical interference effect, viz. Pendellösung, in the form of fringes (Kikuta et al, 1971; Malgrange et al.,1976; Tomimitsu and Zeyen, 1978)].

The quality of the neutron source, in terms both of intensity and of background level, is a key factor for neutron topography, and so is beam-time availability. For any substantial topographic work, it is desirable to use a dedicated instrument, which can however be simple, and be placed downstream of existing instruments using reasonably transparent monochromators. All the work on neutron topography proper was performed on steady-state sources, i.e. reactors. Pulsed neutron beams were used, in a related but essentially different approach, in mapping the strain distribution, e.g. around welds, in industrial equipment (Withers et al., 1994).

3. SPECIAL CAPABILITIES

3.1. Application to investigations of heavy crystals

As indicated in (c) in §1, the absorption of neutrons, due to nuclear capture reactions, is very small in most materials, and the absorption cross-section is not related to atomic number as it is for X-rays. The elements that stand out as reasonably strong neutron absorbers, say with $1/\mu$ smaller than 1 mm at $\lambda = 1.8$ Å, with μ the linear absorption coefficient for the natural mixture of isotopes, are (International Tables for Crystallography, vol. C, 1992) B, Rh, Cd, Sm, Eu, Gd, Dy, Ir, Hg and Pu.

As a result, it is possible to investigate the defect distribution in samples which are too absorbing for X-rays, because they are too thick and/or contain elements with too high an atomic number, i.e. heavy crystals. There is no problem with thin crystals of high-density materials, where neutron topography is quite convenient for observing the gross defect structure (Tomimitsu and Doi, 1974, Baruchel et al., 1978, Tomimitsu et al., 1983, Kvardakov et al., 1992), or the spatial modulation of distortion due to vibration, for example in quartz (Michalec et al., 1975) and resonant magnetoelastic effects (Kvardakov et al., 1991a).

The observation of very thick (cm-range) crystals by the standard, wide-beam, technique of neutron topography implies a superposition of the contributions of sizeable portions of the crystal. It is therefore convenient to restrict observation to a virtual slice, exactly as in Lang's method of X-ray section topography in low-absorption cases (Schlenker et al., 1975; Davidson et al., 1976). Figure 1 shows an example of such an observation, where the virtual slice as seen with neutrons is compared with a white-beam X-ray topograph of the actually cut slice.

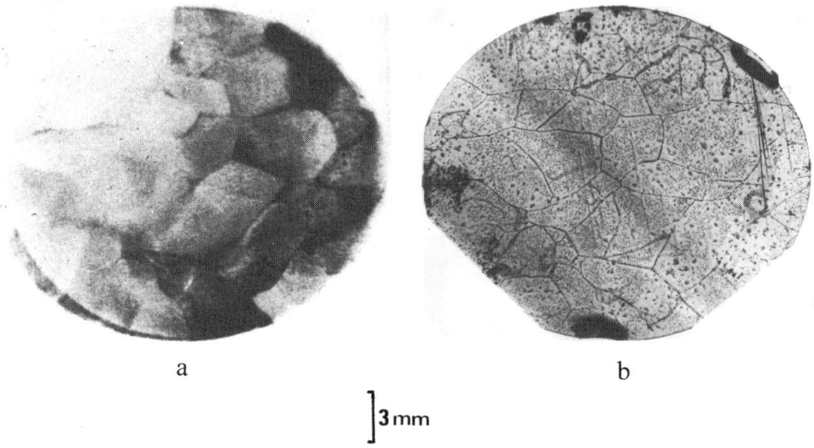

Figure 1. a) Neutron section topograph of an as-grown Fe-Si Bridgman-grown single crystal, showing the subgrains in a virtual slice. b) X-ray white beam topograph of the actually cut slice (Schlenker et al., 1975).

This method is useful in the process of preparation of monochromator crystals (Hustache, 1979). It has been applied in investigations of large crystals of copper-based alloys (Tomimitsu et al., 1983), superalloys, and α-$LiIO_3$ (Bouillot et al., 1982), one of the materials where spectacular increases in the diffracted intensity occur, both in X-ray and in neutron diffraction, under the effect of an electric field..

Another interesting potentiality is the investigation of crystals where X-rays induce a reaction, for example polymerization (Dudley et al., 1990).

3.2. Investigation of magnetic phase transitions and magnetic domains

The strong contribution of electronic magnetic moments in materials to neutron diffraction (item (d) in §1) makes magnetic neutron scattering a unique tool in determining magnetic structures on the unit-cell level. When a single-crystal specimen contains regions with different orientations of the magnetic structures, i.e. magnetic domains, or coexisting phases with different magnetic structures, they will be imaged because of the local variations in structure amplitude, hence in diffracted intensity, they entail. This contrast mechanism is almost foreign to X-ray topography, where it can be used to image twins and ferroelectric domains in only a few situations (Klapper, 1987).

3.2.1. Phase coexistence. Neutron topography thus is a valuable tool in the investigation of the coexistence of different magnetic phases, for example heli- and ferromagnetic, at a first-order phase transition, which can be reached either by temperature changes or by applying a magnetic field (Baruchel, 1989). One example is that of MnP, discussed in the chapter on magnetic and phase transition applications of topography (Baruchel et al., this volume). Another interesting example is the Morin transition (antiferromagnetic - weak magnetic) in α-Fe_2O_3 (hematite), where the transition can be stretched out over tens of K in temperature (Kvardakov et al., 1991b), an effect attributed to impurity distribution (Baruchel et al., 1987). Neutron topography has the advantage that it is usually possible to characterize very directly which region is in which phase. Thus, a helimagnetic phase will produce satellite reflections, and a topograph set on one of them will show only the parts of the crystal that are helimagnetic. Nevertheless, since there is always a discontinuous change in lattice parameters at the first-order transition, the coexisting phases can be distinguished with X-rays too. In the X-ray case, however, the requirements on crystal perfection are higher, since the image of the interphase wall competes with that of defects.

3.2.2. "Conventional" domains. Ferromagnetic domains can be imaged by neutron topography using polarized neutrons (Schlenker & Shull, 1973). In the simplest case where the magnetization in the domains would be along the incident beam polarization, antiparallel (180° domains) and perpendicular to the scattering vector, the structure amplitudes would be proportional to ($F_N \pm F_M$) in the domains, with F_N and F_M the nuclear and magnetic contributions respectively, one sign for one set of domains and the other for the oppositely magnetized one. Thus the contrast between domains, for given incident beam polarization, will be the flipping ratio $\mathcal{R}$ as discussed in the chapter on the dynamical theory of neutron diffraction in this volume. The contrast will be reversed when the polarization is flipped. A naive expectation would be that, using unpolarized neutrons, there should be no (area) contrast between 180° domains, and that 180° domain walls would not be visible, because, just as in X-ray topography, they are not associated with a change in lattice distortion except for relaxation effects near the surfaces. While the first conclusion is true, the second is not: 180° domain walls are seen very well as lines of enhanced diffracted intensity on unpolarized neutron topographs (Schlenker et al.,1978; Nakatani et al., 1992). Observing ferromagnetic domains with neutrons is however of restricted value because so many other and more efficient techniques are available. The competition is not really with synchrotron radiation topography, because this is an indirect technique which reveals ferromagnetic domains only via magnetostriction effects. It is rather with simple, direct and "home laboratory" techniques such as magneto-optics in or around the visible spectrum. However, except for transparent materials, these do not allow observations in the bulk.

An alternative form of imaging, where neutrons are just transmitted by the sample, should be mentioned in this connection. If the sample is placed between identical perfect crystals set for non-dispersive diffraction, or, equivalently, inside a channel-cut type monochromator crystal, or an interferometer with the reference beam blocked off, 180° ferromagnetic domain walls are visible because they behave as a phase step for the neutrons (Schlenker et al., 1980). Under the name of neutron radiography with refraction contrast, this technique, essentially a form of Schlieren imaging, was developed by Podurets et al. (1991), who were able to image internal domain walls in samples 10 mm thick.

Neutron topography is of outstanding value for the observation of antiferromagnetic domains of various kinds, because it is the only direct method, and often simply the only one available, to visualize them. The reason is that an antiferromagnetic magnetic structure does not couple to fields with macroscopic wavelengths. Hence optical methods in or around the visible range are not effective, nor are methods based on inhomogeneous magnetic fields from the domains - there is none. The effect on neutron scattering can take several forms. Figure 2 shows a rather clear example, in the case of NiO which had been well documented previously through the use of *indirect* techniques based on lattice distortion associated with exchange striction and magnetostriction.

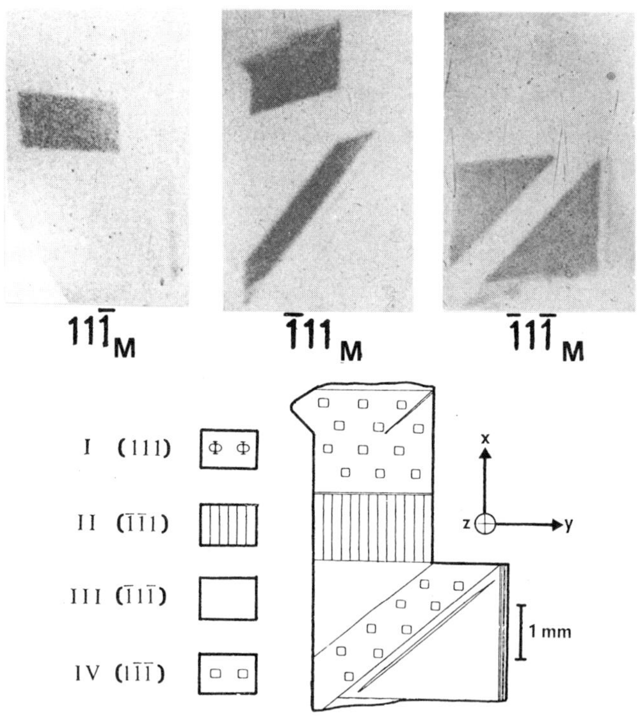

Figure 2. **k** (propagation vector) domains in NiO. The sample was set in turn for each for the 111-type magnetic (superstructure) reflections. Each topograph thus shows the area with the propagation vector along the diffraction vector. The schematic diagram of the domain structure is shown (Baruchel et al., 1977).

Both the **k**-domains, where **k** is the propagation vector characterizing the magnetic moment arrangement, and the S (spin-direction) domains, could be directly imaged and characterized (Baruchel et al., 1977, 1981). **k**-domains correspond to different orientations of the {111} planes which contain aligned magnetic moments, and which are stacked with antiparallel orientations: thus neighboring such planes are different for a probe sensitive to the magnetic moment, and new, purely magnetic or superstructure, reflections appear in the antiferromagnetically ordered phase, i.e. below the Néel temperature. When the crystal is set for such a reflection, say $\frac{\bar{1}}{2}\frac{1}{2}\frac{1}{2}$ if referred to the crystallographic unit cell, or $\bar{1}11$ in the magnetic cell with parameter doubled, the only part of the crystal which will diffract is that where the propagation vector is $\frac{\bar{1}}{2}\frac{1}{2}\frac{1}{2}$, hence only one of the four **k**-domains.

As shown in Figure 2, this provides straightforward characterization, as well as imaging, of the domains. Similary, S domains appear directly due to the difference in the magnetic scattering amplitude associated with the fact that the effective component of the magnetic moment, that perpendicular to the scattering vector, changes.

3.2.3. "Exotic" domains. The pioneering work by Ando & Hosoya (1972, 1978) and Davidson et al. (1974) showed spin-density wave domains in antiferromagnetic chromium through the satellite reflections they produce. The same principle, using polarized neutrons with polarization close to the propagation vector of the magnetic structure, made it possible to visualize chirality domains (right/left-handed helix) in helimagnets (Baruchel et al., 1990) because, for given incident polarization of the beam, one set of chirality domains only contributes to one of the satellite reflections. Fig. 3 shows such chirality domains in terbium, where the helimagnetic phase is sandwiched in temperature between the ferromagnetic, low temperature phase, and the paramagnetic one above. Conspicuously different shapes are associated with the domains depending on whether they are reached on warming from the ferro- or on cooling from the paramagnetic phase.

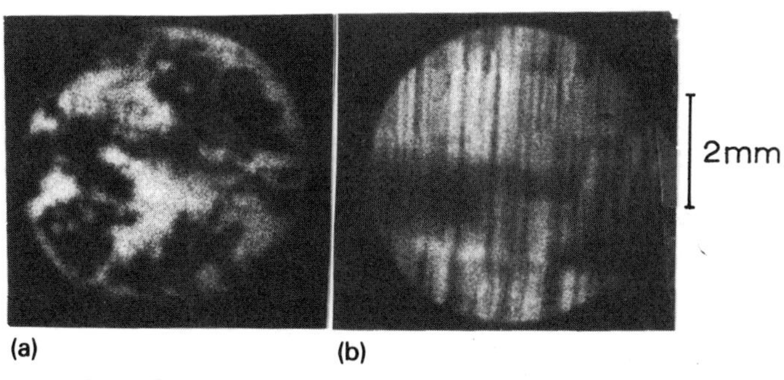

Figure 3. Polarized neutron topographs showing chirality (right-hand / left-hand spin rotation) domains in helimagnetic terbium: a) as reached from the paramagnetic phase; b) as reached from the ferromagnetic phase (Baruchel et al., 1981)

Following a pioneering investigation, based on a one-dimensional scan, by Alperin et al. (1962), 180° antiferromagnetic domains could be observed for the first time using polarized neutron topography (Baruchel et al., 1990) in MnF_2, a material with the tetragonal rutile structure. Here the magnetic moments are all along the *c* axis, and there is no doubling of the unit cell in the antiferromagnetic phase. The difference between the 180°, or time-reversed, domains, is that the magnetic moment at the apices is up in one, down in the other, with the moment at the center of the unit cell being in all cases opposite. Because the direction along which the F ions are positioned around the apex and center sites is different by 90°, these domains are physically different for a probe that can sense both the up or down direction of the magnetic moment and the chemical environment. Mixed reflections, i.e. those involving a nuclear and a magnetic contribution, and polarized neutrons, are thus expected, and indeed found, to distinguish the magnetic domains. In fact one such reflection, 210, also has its magnetic and nuclear contributions to the structure amplitude almost equal, resulting in excellent area contrast, around 20 K. Figure 4 shows such domains, using this situation.

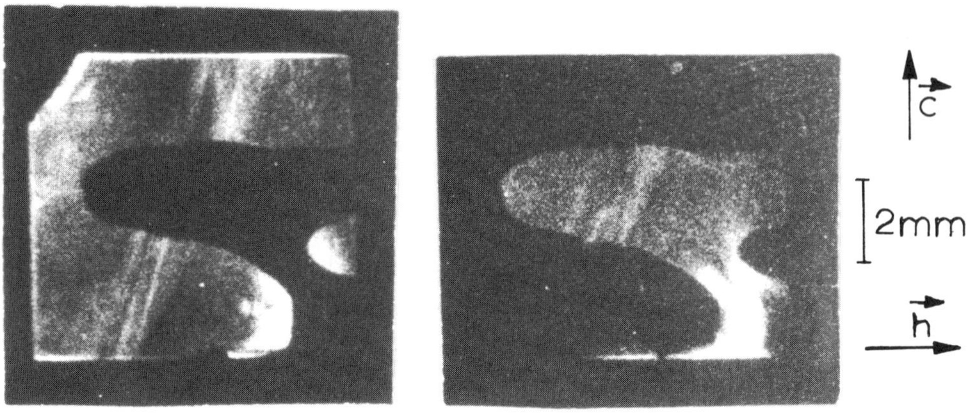

a **b**

Figure 4. Polarized neutron topographs of 180°, or time-reversed, domains in antiferromagnetic MnF_2. The polarization of the incident beam is switched between a and b.

The possibility of observing several kinds of antiferromagnetic domains for the first time opened up new questions, not completely answered yet, about their behavior. In particular, MnF_2 shows fascinating memory effects in the shape and position of the 180° domain walls when the sample is cycled across the Néel temperature, with or without an applied field. Some correlation was apparent with the impurity contents, in particular oxygen (El Kadiri et al., 1986), but no satisfactory model has yet been proposed.

In the context of dynamical diffraction, the central subject of this book, it is worth noting that, unlike the topographic observation of defects, the investigation of magnetic domains by neutron topography does not require very high crystal perfection for the contrast mechanism itself; good crystals are needed primarily because they allow antiferromagnetic domains to be big and simple.

Several subjects appear to be in want of a close investigation of antiferromagnetic domains, hence of neutron topography. One is the class of materials in which antiferromagnetism couples with electric order, in particular in magnetoelectrics (Siratori, 1994). Another is the set of intricate memory effects in substituted antiferromagnets, which have been studied by Jaccarino and co-workers (Belanger et al., 1982) and Shapira et al. (1984).

References

Alperin, H.A., Brown, P.J., Nathans, R., and Pickart, S.J., 1962. Polarized neutron study of antiferromagnetic domains in MnF_2. *Phys. Rev. Lett.*, **8**: 237-239

Ando, M., and Hosoya, S., 1972, Q-switch and polarization domains in antiferromagnetic chromium observed with neutron diffraction topography. *Phys. Rev. Lett.*, **29**: 281-285.

Ando, M., and Hosoya, S., 1978, Size and behavior of antiferromagnetic domains in Cr directly observed with X-ray and neutron topography. *J. Appl. Phys.*, **49**: 6045-6051

Baruchel, J., Schlenker, M., and Roth, W.L., 1977, Observation of antiferromagnetic domains in nickel oxide by neutron diffraction topography, *J. Appl. Phys.*, **48**: 5-8

Baruchel, J., Schlenker, M., Zarka, A., and Petroff, J.F., 1978, Neutron diffraction topographic investigation of growth defects in natural lead carbonate single crystals. *J. Cryst. Growth*, **44**: 356-362

Baruchel, J., Schlenker, M., Kurosawa, K. and Saito, S., 1981, Antiferromagnetic S-domains in NiO: 1. Neutron magnetic topographic investigation,. *Phil. Mag.B*, **43**: 853 -860; 2. Effect of crystallographic twinning on the visibility of domains by X-ray topography, *Phil. Mag.B*, **43**, 861-868

Baruchel, J., Palmer, S.B. and Schlenker, M., 1981, Observation of chirality domains in terbium by plarized neutron diffraction topography. J. Physique, **42**: 1279-1283

Baruchel J., Clark G., Tanner B.K. and Watts B.E., 1987, Neutron topographic investigation of the Morin transition in flux grown crystals of α-Fe_2O_3, *J. Magn. Magn. Mat.* **68**: 374-378

Baruchel, J., 1989, The contribution of neutron and synchrotron radiation topography to the investigation of first-order magnetic phase transitions. *Phase Transitions*, **14**: 21-29.

Baruchel, J., Schlenker, M., and Palmer, S.B., 1990, Neutron diffraction topographic investigations of "exotic" magnetic domains. *Nondestr. Test. Evalu.*, **5**: 349-367

Belanger, D.P., King, A.R. and Jaccarino, V., 1982, Random-field effects on critical behavior of diluted Ising antiferromagnets. Phys. Rev. Lett., **48**: 1050-1053

Boeuf, A., Lagomarsino, S., Rustichelli, F., Baruchel, J., and Schlenker, M., 1975. White beam neutron topography. *Phys. Stat. Sol. a*, **31**: K91-K93 + plates

Bouillot J., Baruchel J., Remoissenet M., Joffrin J and Lajzerowicz J., 1982, Electric field related extinction reduction in diffraction experiments on alpha-LiIO3, *J. Physique*, **43**: 1259-1266

Davidson, J.B., and Case, A.L., 1976. Applications of the fly's eye neutron camera: diffraction tomography and phase transition studies. *Proc. Conf. on Neutron Scattering, ORNL, USERDA CONF* 760601-P2: 1124-1135

Davidson, J.B., Werner, S.A., and Arrott, A.S., 1974. Neutron microscopy of spin density wave domains in chromium. *AIP Conference Proc.*, C.D. Graham, J.J. Rhyne, ed., **18**: 396-400

Doi, K., Minakawa, N., Motohashi, H., and Masaki, N., 1971. A trial of neutron diffraction topography. *J. Appl. Cryst.*, **4**: 528-530

Dudley, M., Baruchel, J., and Sherwood, J.N., 1990, Neutron topography as a tool for studying reactive organic crystals: a feasibility study. *J. Appl. Cryst.*, **23**: 186-198

El Kadiri M., Baruchel J., Rodriguez F., Moreno M. and Henry J.Y., 1986, Impurity related memeory effect of the antiferromagnetic domains in MnF2, *J. Magn. Magn. Mat.* **54-57**: 853-854

Hustache, R., 1979. Selection de blocs monocristallins dans un cristal de beryllium par topographie neutronique. *Nucl. Instr. and Meth.* **163**: 151-156

International Tables for Crystallography, vol. C, 1992, Kluwer Academic Publishers, Dordrecht

Kikuta, S., Kohra, K., Minakawa, N. and Doi, K., 1971, An observation of neutron Pendellösung fringes in a wedge-shaped silicon single crystal, *J. Phys. Soc. Jap.*, **31**:954-955

Klapper, H., 1987, X-ray topography of twinned crystals. *Prog. Crystal Growth*, 14: 367-401

Kvardakov, V.V., and Somenkov, V.A., 1991a, Neutron diffraction study of nonlinear magnetoacoustic effects in perfect crystals of $FeBO_3$ and α-Fe_2O_3. *J. Moscow Phys. Soc.*, **1**:33-57

Kvardakov, V.V., Sandonis, J., Podurets, K.M., Shilstein, S.Sh. znf Baruchel, J., 1991b, Study of the Morin transition in nearly perfect crystals of hematite by diffraction and topography, *Physica B*, **168**: 242-250

Kvardakov, V.V., Somenkov, V.A., Shil'shtein, S.Sh, 1992, Study of defects in cuprate single crystals by the neutron topography and selective etching methods. *Superconductivity*, **5**: 623-629

Malgrange, C., Petroff, J.F., Sauvage, M., Zarka, A., and Englander, M., 1976. Individual dislocation images and Pendellösung fringes in neutron topographs. *Phil. Mag.*, **33**: 743-751

Michalec, R., Mikula, P., Sedláková, L., Chalupa, B., Zelenka, J., Petržilka, V., and Hrdlička, Z., 1975, Effects of thickness-shear vibrations on neutron diffraction by quartz single crystals. J. Appl. Cryst., **8**: 345-351

Nakatani, S., Tomimitsu, H., Takahashi, T., and Kikuta, S., 1992, Phase shift of neutrons in magnetic domains observed by interferometry; *Jpn. J. Appl. Phys.*, **31**: L1137-L1139

Podurets, K.M., Sokol'skii, D.V., Chistyakov, R.R. and Shil'shtein, S.Sh., 1991, Reconstruction of the bulk domain structure of silicon iron single crystals from neutron refraction images of internal domain walls. *Sov. Phys. Solid State*, **33**: 1668-1672

Scherm, R. and Fåk, B., 1993, Neutrons. in *Neutron and Synchrotron Radiation for Condensed Matter Studies, vol. 1: Theory, Instruments and Methods. HERCULES (Higher European Research Course for Users of Large Experimental Systems), Grenoble.* J. Baruchel, J.L. Hodeau, M.S. Lehmann, J.R. Regnard, C. Schlenker, ed., Les Editions de Physique, Les Ulis and Springer-Verlag, Heidelberg.

Schlenker, M., and Shull, C.G., 1973, Polarized neutron techniques for the observation of ferromagnetic domains, *J. Appl. Phys.*, **44**: 4181-4184

Schlenker, M., Baruchel, J., Pétroff, J.F. and Yelon, W.B., 1974, Observation of subgrain boundaries and dislocations by neutron diffraction topography, *Appl. Phys. Lett.*, **25**: 382-384

Schlenker, M., Baruchel, J., Perrier de la Bathie, R., and Wilson, S.A., 1975. Neutron diffraction section topography: observing crystal slices before cutting them. *J. Appl. Phys.*, **46**: 2845-2848

Schlenker, M., Linares-Galvez, J., and Baruchel, J., 1978, A spin-related contrast effect: visibility of 180° ferromagnetic domain walls in unpolarized neutron diffraction topography, *Phil. Mag. B*, **37**: 1-11

Schlenker, M., Bauspiess, W., Graeff, W., Bonse, U. and Rauch, H., 1980, Imaging of ferromagnetic domains by neutron interferometry, *J. Magn. Magn. Mat.*, **15-18**: 1507-1509

Shapira, Y., Oliveira, N.F., Jr. and Foner, S., 1984. Effects of random fields on the phase transitions and phase diagram of $Mn_{0.75}Zn_{0.25}F_2$. *Phys. Rev. B*, **30**: 6639-66489

Sillou, D., Baruchel, J., Kuroda, K., Michalowicz, A., Guigay, J.P., and Schlenker, M., 1989, First experiments on flipping ratio mapping with the "multi-PM" position-sensitive detector. *Physica B*, **156-157**: 581-583

Siratori, K., 1994, Magneto-electric effect and solid state physics, *Ferroelectrics*, **161**: 29-41

Tomimitsu, H., and Doi, K., 1974, A neutron diffraction topographic observation of strain field in a hot-pressed germanium crystal. *J. Appl. Cryst.*, **7**: 59-64

Tomimitsu, H., and Zeyen, C., 1978, Neutron diffraction topographic observation of twinned silicon crystal. *Jap. J. Appl. Phys.*, **3**: 591-592

Tomimitsu, H., Doi, K., and Kamada, K., 1983, Neutron diffraction topographic observations of substructures in Cu-based alloys. *Physica*, **120B**: 96-102

Withers, P.J., Harris, I.B., Edwards, L., Wang, D.Q., Johnson, M.W., Priesmeyer, H., Larsen, J. and Rustichelli, F., 1994. ENGIN - a neutron spectrometer for internal stress measurement in engineering structures. in *EEC Conference on Lightweight Structures*, p.75-80, Noordwijk, ESA WPP 070

APPLICATION OF DIFFRACTION TOPOGRAPHY TO THE STUDY OF MAGNETIC DOMAINS AND PHASE TRANSITIONS

José Baruchel [(1)] and Michel Schlenker [(2)]
(1) European Synchrotron Radiation Facility
BP 220, 38043 Grenoble, France
(2) Lab. Louis Néel, CNRS-UJF
166X, 38042 Grenoble, France

1. DIFFRACTION TOPOGRAPHY AND MAGNETISM

X-ray and neutron diffraction topography are valuable tools for the investigation of macroscopic 'magnetic defects', i.e. disruptions in the three dimensional periodicity of the magnetic structure (Roth 1970). These extended defects are interesting both in themselves and because they play a key role in the macroscopic magnetic properties, just as e.g. plastic properties are related to crystal defects. A crystal structure is characterized by the position **r** and type of the atoms, which appear in the diffraction formulaes through Z for X-rays, or the scattering length, b_{coh} when averaged over the isotopes, for neutrons. To define a magnetic structure, the position **r** and magnitude $|\mathbf{m}|$ of the magnetic moments, but also their direction, are needed. The additional directional parameter leads to purely magnetic defects, with no crystallographic counterpart, such as domain walls, or interfaces between coexisting magnetic phases (Figure 1). In contrast to all crystal defects with non-zero dimensionality, two dimensional discontinuities within the magnetic structure can be thermodynamically stable. They can result in crystal defects through the change in lattice distortion, related e.g. to magnetostriction, between neighbouring domains or phases.

The application of topographic techniques to the observation of magnetic domains started when Polcarova and Lang (1962) observed ferrromagnetic domain structures in silicon-iron by Lang's method, confirming its high strain sensitivity and its ability to explore phenomena not accessible by other techniques (observation of domains within the bulk of non optically transparent crystals). Ando and Hosoya (1972), using neutron topography, made the first observation of spin-density-wave domains in antiferromagnetic chromium, and found a disagreement of several orders of magnitude for the domain sizes between experiment ($\approx 10^{-3}$cm^3) and the calculation ($\approx 10^{-16}$cm^3) based on the then accepted model.

Neutron and X-ray topography, especially with synchrotron radiation, have proved to be complementary tools for the investigation of magnetic domains of all kinds (ferro-, ferri- and antiferromagnetic) and of magnetic phase coexistence in single crystals (see Baruchel (1993) and references therein). Thanks to the magnetic interaction of neutrons with the magnetic structure, neutron topography can show and characterize *directly* magnetic inhomogeneities. In X-ray topography, the main contrast mechanism is the variation in distortion across a wall or phase boundary, whose connection with the departure from Bragg's law, which characterizes a defect from the diffraction point of view (Tanner, this volume) was treated by Polcarova and Gemperlova (1969) and Kleman and Schlenker (1972). Hence X-ray topography, using Lorentz (charge) scattering, is an *indirect* method in the case of magnetic investigations. Its resolution and speed are however enormously superior to neutron topography. A direct observation of magnetic domains was recently shown by Kawata and Mori (1995) to be feasible with X-rays by taking advantage of the weak magnetic interaction of X-rays.

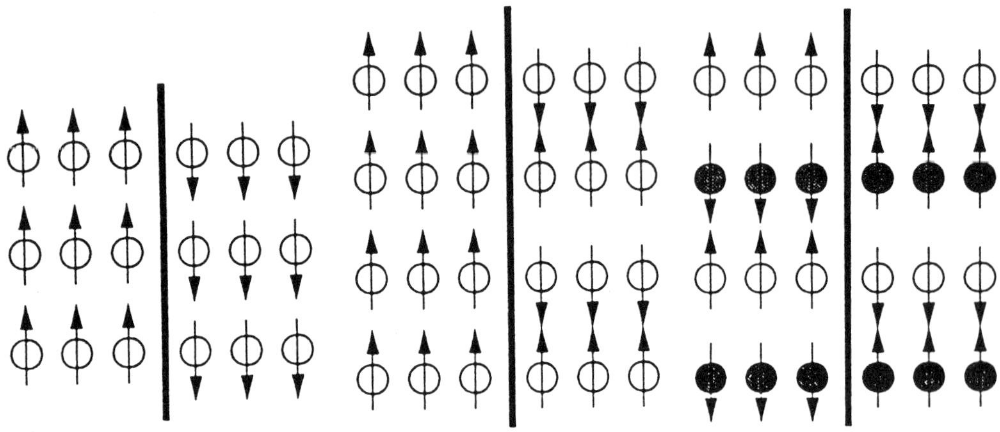

a b c

Figure 1. Simplified view of several magnetic defects: a) 180° ferromagnetic wall b) 180° antiferromagnetic wall c) antiferro-ferromagnetic interface during a phase coexistence.

The main drawback in the application of these techniques to magnetic crystals is the restriction to single crystals with high crystal quality, where the images associated with the magnetic boundaries are not blurred by those related to the defects. When nearly perfect magnetic crystals are available, unique information can be obtained, as will be shown in the case of MnP, which we will take as the example throughout this chapter.

2. MAGNETIC PROPERTIES OF MnP

The magnetic properties of the orthorhombic metallic compound MnP attracted considerable attention. Figure 2 shows schematically the three magnetic arrangements we will be concerned with in the present paper: ferromagnetic, helimagnetic and fan. MnP is helimagnetic for $T<T_S$ = 47K, and ferromagnetic, with easy magnetization axis along **c** (convention a>b>c) in the temperature range $T_S<T<T_C$ = 291K. Figure 3a shows that rather small fields (<0.3 Tesla) applied along **c** induce, for $T<T_s$, a transition to the ferromagnetic phase (Huber and Ridgley, 1964). The **a** axis is both the hard one in the ferromagnetic phase and the axis of the helix in the helimagnetic one. When a field is applied along the intermediate **b** axis, MnP exhibits a complex magnetic phase diagram (Obara et al. 1980), partly shown on Figure 3b. This phase diagram displays a triple point T where the helimagnetic, ferromagnetic and fan phases are simultaneously present within the crystal. The fan phase magnetization is nearly perpendicular to that of the ferromagnetic phase. All the transitions are first order, and hence lead to phase coexistence.

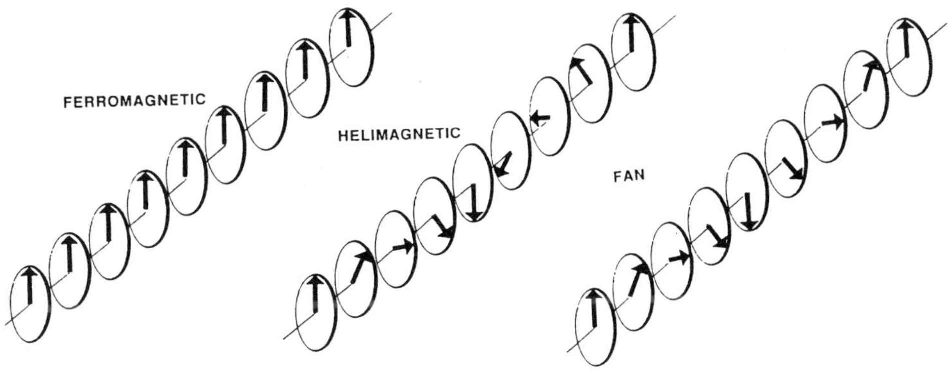

Figure 2. Schematic representation of the ferromagnetic, helimagnetic and fan arrangements

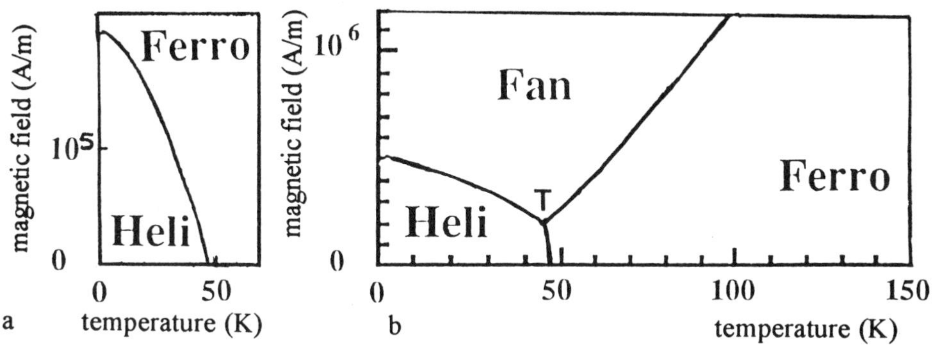

Figure 3. Phase diagram of MnP when the magnetic field is applied along a) the easy axis **c** ; b) the intermediate axis **b**: a triple point T is observed. Note that the magnetic field scales are different for a) and b).

3. CRYSTAL QUALITY OF THE MnP CRYSTALS

The samples, grown at Tohoku University (Sendai, Japan) and kindly given by the late Prof. Y. Ishikawa, were nearly perfect single crystal plates of MnP, with main surfaces parallel to (001), (010) and (100). Their thickness lies in the 0.1 - 0.4 mm range.

A test of the perfection of the arrangements of both the crystallographic and the magnetic lattices is the observation of Pendellösung oscillations by using polarized neutrons and measuring the flipping ratio R (i.e. the ratio of the integrated reflectivities ρ_+ and ρ_- corresponding to incident neutron polarization parallel or antiparallel to the magnetization, saturated perpendicularly to the scattering vector) as a function of the wavelength (Baruchel et al. 1982). In this situation two different structure factors F_+ and F_- can be introduced, and consequently two extinction distances Λ_+ and Λ_- . The oscillations of ρ_+ and ρ_- thus occur with different periods, and their ratio R also displays oscillations. Figure 4 shows the experimental result obtained at ILL (Grenoble) for the (100) MnP crystal, together with the best fits obtained by using Kato's Statistical Dynamical Theory (SDT), or Kulda's Random

Elastic Deformation (RED) model. The oscillation amplitude is smaller than expected for the perfect crystal, but the parameters obtained in both fits show the high quality of the sample (Kulda et al. 1991). The static Debye Waller factor E, which can vary between 0 and, in the perfect crystal limit, 1, is 0.97 in our case.

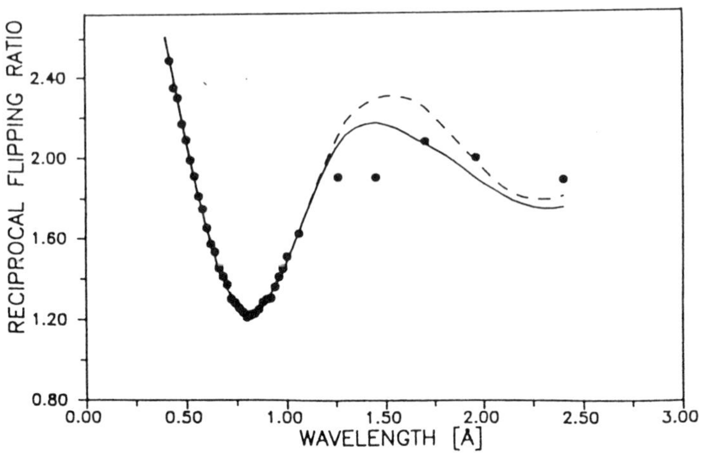

Figure 4. Observed values of the reciprocal of the flipping ratio R^{-1} and the calculated curves fitted according to the RED (full line) and the SDT (dashed line) models for a (100) MnP crystal.

4. CHIRALITY DOMAINS

A chirality domain is a region of a helimagnetic crystal where the sense of the helix is constant (right- or left-handed) (Figure 5). The helimagnetic arrangement gives rise, in neutron diffraction, to magnetic satellites around a nuclear peak. When diffraction is performed on planes perpendicular to the helix axis, and the incoming neutrons are polarized along this axis, only one type of domains contributes, for a given polarization and satellite peak, to diffraction, and is thus imaged on the topograph (Schlenker and Baruchel, this volume)

Chirality domains were first observed in Tb (Baruchel, Palmer and Schlenker, 1981). Figure 6 shows that in MnP they have the shape of stripes with walls parallel to **c**, i.e. to the magnetization in the ferromagnetic phase (Patterson et al. 1985). This feature, which is only observable by polarized neutron topography, stems from the fact that the helimagnetic phase is reached only from the ferromagnetic one, and that the first order ferro-helimagnetic transition starts and propagates in the domain walls present in the ferromagnetic phase.

The size of these stripe-shaped domains is related to the magnetic history of the sample: eliminating the ferromagnetic domain walls reduces the number of nucleation centers, and leads to large chirality domains. Indeed cooling an (010) plate through the transition either with no applied field or under a field of 0.05T parallel to the easy axis produces an average domain volume an order of magnitude bigger in the applied field case. Further thermal cycles, with no applied field, lead to "big" (or "small") domains if the highest temperature reached corresponds to the ferromagnetic (or to the paramagnetic) phase. This suggests that the ferromagnetic domain walls retain the chirality of the spiral domains, which are restored on cooling back to the helimagnetic phase (Baruchel, Palmer and Patterson 1988).

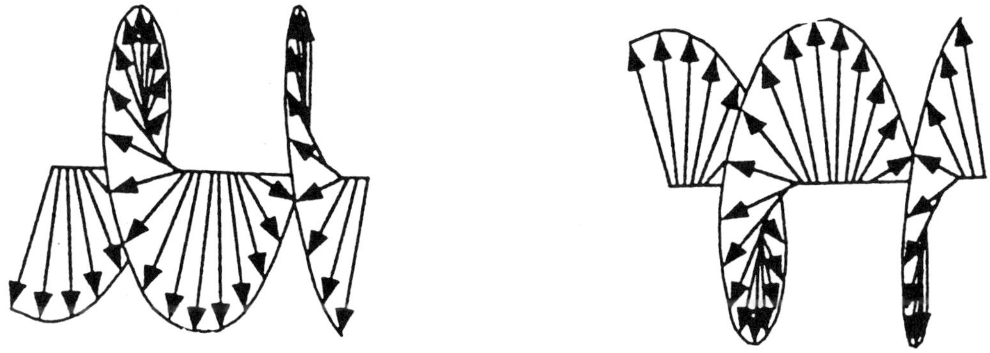

Figure 5. Schematic representation of chirality domains (right- or left-handed) in a helimagnet.

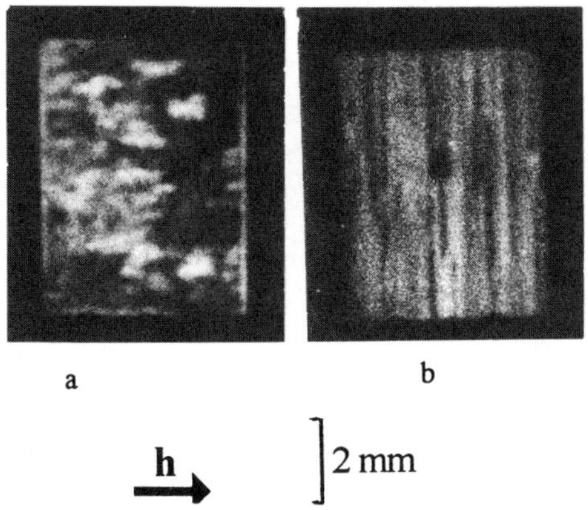

Figure 6. Polarized neutron topographs of a (001) MnP sample a) nuclear peak b) magnetic satellite peak: the stripe shaped chirality domains are visible. White means more illuminated.

A puzzling feature observed on the topographs is the memory effect of the domain patterns observed on successive coolings: the locations of the walls are roughly reproducible, and a given region of the sample turns always left, or always right (Baruchel, Schlenker and Palmer 1990).

5. MAGNETIC PHASE COEXISTENCE

The coexistence of two magnetic phases in a given single crystal sample leads to a rich variety of situations which were only scarcely investigated experimentally, mainly by optical methods on transparent samples. Diffraction topography offers the possibility of extending this kind of investigation to non transparent materials, and leads to valuable new information on these often complex transitions. In the case of purely structural transitions, the orientation of the interfaces is mainly governed by elastic compatibility, the initial and transformed phases fitting together along a plane leading to minimum elastic strain (Khatchaturyan, 1967; Salje 1990). In a magnetic transition, a magnetostatic term arises in addition whenever at least one of the phases displays a macroscopic magnetization. While magnetostatics is a familiar ingredient in ferromagnetism, the situation here is unusual in that the geometry, including the size, of the ferromagnetic region is an adjustable internal variable. The observation of the coexistence pattern allows the investigation of the competition between the different energy terms, and a test of the theoretical treatments developed to account for the "intermediate state" (Bar'yakhtar et al., 1986). Understanding the morphology of the phase transition appears important as a clue to the physics of chirality domains, and in particular their puzzling memory effects.

5.1 'Real-time' experimental technique

The experiments were performed using a either a 'white' synchrotron radiation beam (at LURE, Orsay or at ESRF, Grenoble), or a monochromatic beam from a steady, or vibrating, 220 reflection of a silicon monochromator (at ESRF). The use of Sofretec cameras fitted with scintillators allowed the real-time observation of the phase coexistence. The sample was cooled in a helium closed-circuit refrigerator. A remote controlled film-holder was set on the diffracted beam path, between the crystal and the video camera, to record the topographs without shutting off the synchrotron beam. This proved necessary in white beam because the presence or absence of the beam modifies substantially the temperature of the sample. Kodak Industrex M (at LURE) or SR (at ESRF) film was mainly used.

5.2 Helimagnetic-ferromagnetic coexistence

The helimagnetic-ferromagnetic transition is associated with a variation of distortions between the phases: the lattice parameter variations at the heli-ferromagnetic transition temperature of MnP are $\frac{\Delta a}{a} = 1.7\ 10^{-5}$, $\frac{\Delta b}{b} = -5.8\ 10^{-5}$, $\frac{\Delta c}{c} = 5.0\ 10^{-5}$ (Okamoto et al., 1968; Ishizaki et al., 1970). Hence X-ray topography is a convenient way of observing the phase coexistence and its evolution either when heat is provided or removed, or under changes in the magnetic field. The interfaces were found to display various shapes (plane, zig-zag, stripes) depending on the thickness and orientation of the crystal, the temperature and the applied magnetic field. This corresponds to a competition between the various energy terms involved in the transition. In the field driven case, the magnetostatic energy is dominant, and the observed pattern minimizes this energy term. The same transition occurs very differently in the thermally driven case with no magnetic field applied; now other energy terms (elastic, interface) influence the coexistence, which is nearly reproduced through several thermal cycles (Baruchel, Patterson and Palmer, 1988).

Figure 7 shows, for a (010) plate shaped crystal 0.4 mm thick crystal, the heli-ferromagnetic transition observed by recording neutron topographs on a) the nuclear 2 0 0 reflection, where both phases are present on the image and b) the $2+\tau$ 0 0 magnetic satellite, which only displays the helimagnetic region. In the former case, the ferromagnetic region is the more illuminated one, because the nuclear and magnetic contributions add, whereas only the nuclear contribution is present for the helimagnetic region. As the exposure times of neutron topographs are long, this experiments implies the retention of the same phase coexistence pattern for about two days.

Figure 8 shows white beam synchrotron radiation topographs, recorded in a few seconds at LURE. The interface varies when the transition is a) thermally driven; b) thermally driven under a magnetic field ($1.2\ 10^4$ A/m); and c) field driven ($2.4\ 10^4$ A/m, T<47 K). In case a) the magnetostatic energy is very much reduced with respect to the appliedfield case, because domains are present in the ferromagnetic region. The images of the interfaces are broad, indicating that they are inclined with respect to the surface. One of them is in addition clearly pinned by a defect. Figure 8b shows that bright lines, or

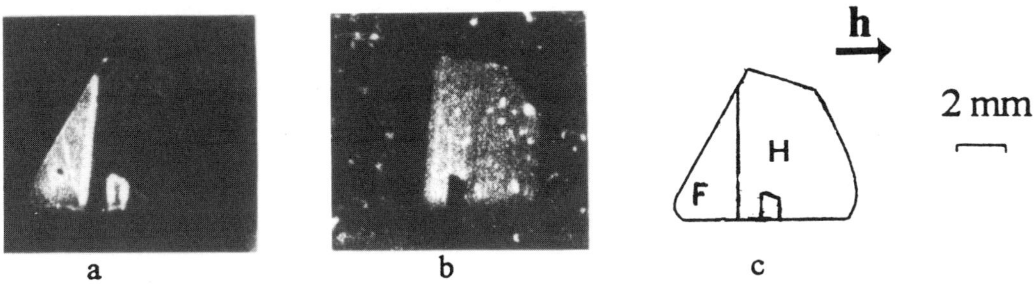

Figure 7. Neutron topographs (white means more illuminated) showing the heli-ferromagnetic phase coexistence in a 0.4 mm thick MnP a) nuclear reflection b) magnetic reflection c) scheme of the phase coexistence.

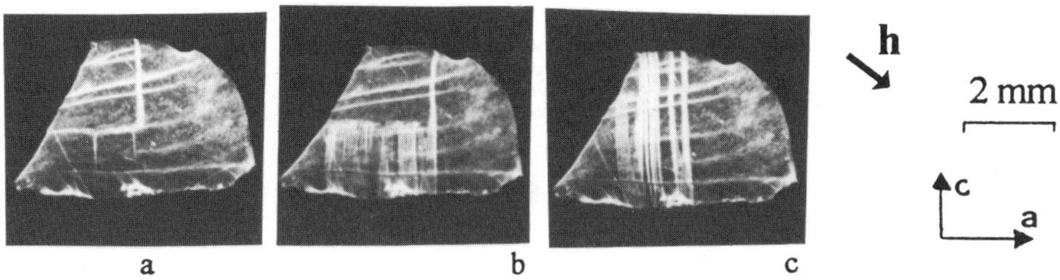

Figure 8. Synchrotron radiation topographs performed in a) the thermally driven (T≈ 47K) b) thermally driven with a small applied magnetic field and c) field driven (T< 47 K) cases. Note the printing convention, unusual for X-ray topography: here white means more illuminated, for a better comparison with Figure 6.

"needles", parallel to **c** appear on this pinned boundary observed on Figure 7a when a small field is applied during the transition. The observed interface results from an increased magnetostatic energy term. Under a higher field (Figure 8c), and at a lower temperature, this magnetostatic term becomes dominant and the needles occupy the whole sample.

An interpretation of the observed interfaces was given by Sandonis et al. (1993) assuming that the interface adopts a zig-zag shape around the plane of minimal deformation, as indicated on Figure 9, and calculating the size h of the needles that minimize the energy. This model leads to a dependence of the reduced dimension h/t (where t is the thickness of the sample) upon the volume of the ferromagnetic phase and a dimensionless parameter, characteristic of the material, which corresponds to the ratio of the magnetostatic energy over the sum of the other relevant energy terms present in the interface. It accounts for the two limiting cases observed: a stripe shaped alternation for thick crystals (the model predicting for this situation a needle size h larger than the crystal size), and a flat interface for thin crystals in zero magnetic field (a case corresponding, within the model frame, to a dimension h approaching zero).

More details can be obtained by taking advantage of the possibilities of third generation synchrotron sources like the ESRF: indeed monochromatic 'real time' experiments are feasible (Medrano and Barrett, unpublished). Figure 10 topographs show a detail of the field driven transition, using either a) a small ($\Delta\lambda-\Delta\theta$) window: only the ferromagnetic region is in Bragg position on the topograph, or b) a larger ($\Delta\lambda-\Delta\theta$) window, obtained by vibrating (≈100 Hz) the monochromator. A monochromatic beam and real time section topograph corresponding to the case a) was performed. It shows the ferromagnetic bands within the crystal bulk, mainly located in the neighbourhood of the surfaces.

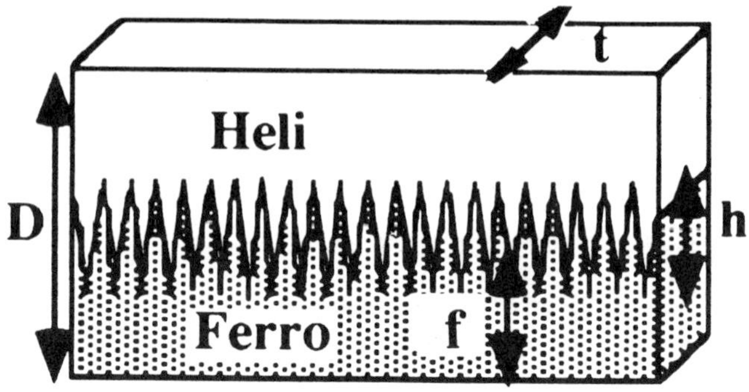

Figure 9. Model used for the calculation of the interface: it assumes that the interface adopts a zig-zag shape around the plane of minimal deformation (from Sandonis et al. 1993)

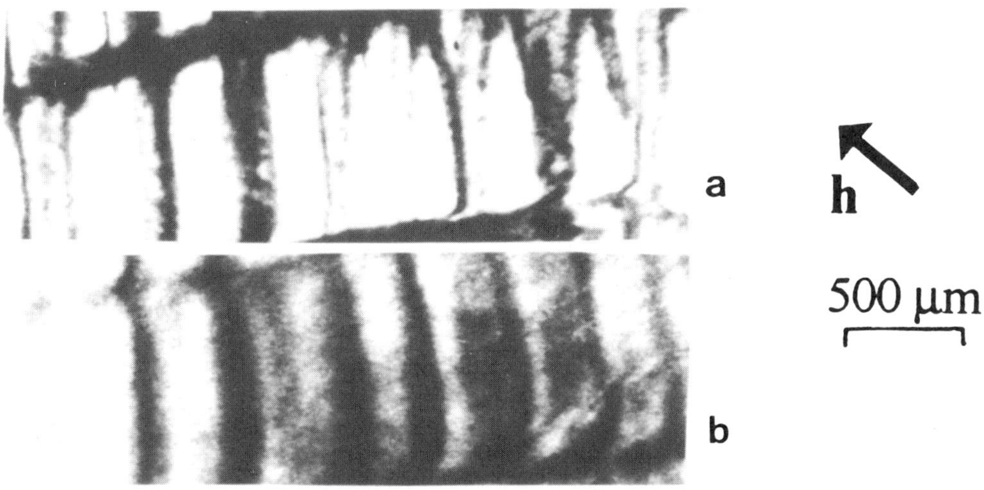

Figure 10. Monochromatic beam topographs ($\lambda \cong 0.3$ Å) recorded during the heli-ferromagnetic phase transition in a 0.4 mm thick (010) MnP, performed by using the a) 220 reflection from a steady silicon monochromator b) same, but vibrating monochromator.

5.3 Phase nucleation at a triple point.

The investigation of the MnP phase transitions was pursued by studying the phase coexistence in the neighbourhood of the triple point (coexistence of the fan, heli- and ferromagnetic phases). The aim of this study was the observation of the nucleation of the fan phase in the presence of the helimagnetic and ferromagnetic ones.

The triple point actually corresponds, as for the heli-ferromagnetic coexistence, to coexistence over a *range* of applied magnetic fields. Since at least one of the phases exhibits a magnetization, the applied magnetic field produces a demagnetizing field, opposite to the applied one, within the crystal. Therefore the coexistence lines transform into coexistence bands in the (H_{appl} *vs* temperature) phase diagram, and the triple point changes into an extended coexistence region. Figure 11 shows the phase diagram obtained topographically for the (001) sample by detecting, on a video display, the first emergence of the new phase (lower line), and the disappearance of the old one (upper line).

This 'real time' observation indicates that the nucleation of the fan phase occurs at the interface which separates the two phases already present within the crystal. Figure 12a shows the heli-ferromagnetic coexistence, at a field just below the triple point. The interface appears as a double line in this white beam image. A slight increase of the field leads to the nucleation of the fan phase (Figure 12b), and to its further propagation, mainly along the heli-ferromagnetic interface. This is a 'wetting'-like process (Landau and Lifschitz 1967).

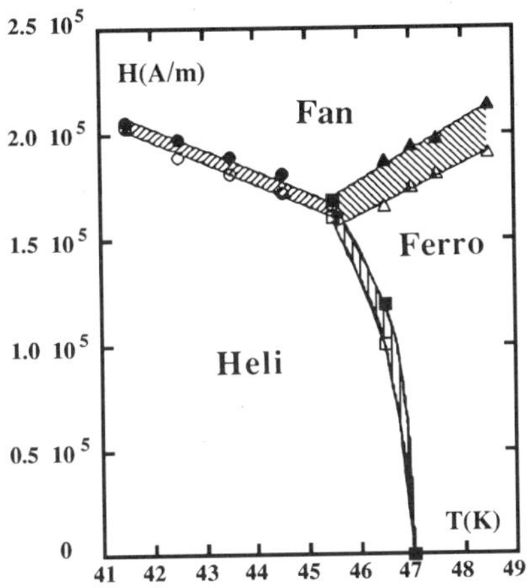

Figure 11. Magnetic phase diagram observed for the (001) crystal. The measurements indicate the applied field phase coexistence range, as observed on the video monitor.

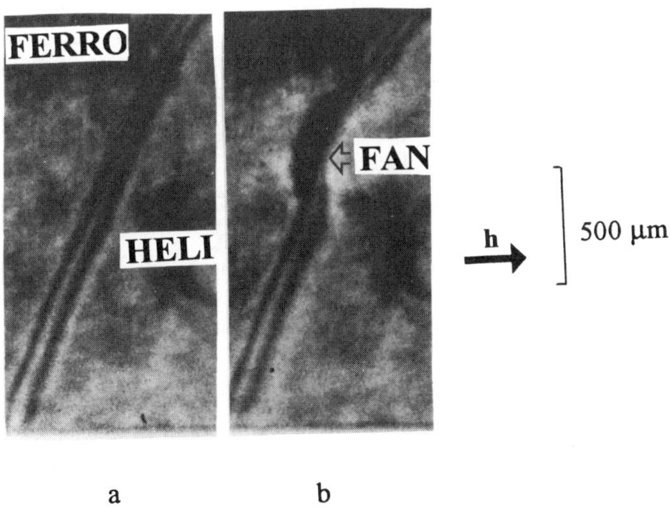

Figure 12. Nucleation of the fan phase between the crystal heli and ferromagnetic ones a) the interface shows up as a double line b) the growing fan phase is the dark lens shaped structure.

6. SHORT CONCLUSION

What is the future of neutron and synchrotron topography for this type of investigation? Neutron topography, although it is slow and has poor resolution, remains an invaluable tool to image directly (antiferro)magnetic domains and phase coexistence. The instrument S20, at ILL, was the only one in the world mainly devoted to this type of investigation. It appears important to keep this 'special' instrument operational. On the other hand, all types of 'real time' topographic investigations, and magnetic phase coexistence observations in particular, will greatly benefit from the advent of third generation synchrotron facilities, through the reduction in exposures times (down to the 10^{-2} s range at the ESRF), the high energy of the radiation, and the possibility of performing in real time both white beam and monochromatic beam projection and section topographs (see Baruchel, this volume). This opens up new fields of investigation, which will surely lead to novel results in the very near future.

REFERENCES

Ando M. and Hosoya S., "Q-switch and polarization domains in antiferromagnetic chromium observed with neutron diffraction topography" *Phys.Rev.Lett.* **29**, 281 (1972)

Baruchel J., Palmer S.B. and Schlenker M., "Observation of chirality domains in terbium by polarized neutron diffraction topography" *J. Physique* **42**, 1279 (1981)

Baruchel J., Guigay J.P., Mazuré C., Schlenker M., Schweizer J., "Observation of Pendellösung effect in polarized neutron scattering from a magnetic crystal" *J. Physique* **43**, C7-101 (1982)

Baruchel J., Patterson C. and Palmer S.B., "The heli-ferromagnetic phase coexistence in MnP : a synchrotron radiation and neutron diffraction topography investigation" *Phil. Mag.* **B57**, 505 (1988)

Baruchel J., Palmer S.B. and Patterson C., "New features about chirality domains: influence of the ferro-helimagnetic transition" *J. Physique* **49**, C8-1893 (1988)

Baruchel J., Schlenker M. and Palmer S.B., "Neutron diffraction topographic investigations of "exotic" magnetic domains" *Nondestr.Test. Eval.* **5**, 349 (1990)

Baruchel J.,"X-ray and neutron topographical studies of magnetic materials" *Physica B* **192**, 79 (1993)

Bar'yakhtar V.G., Bogdanov A.N. and Yablonskii D.A., "Effect of the magnetoelastic interaction of an intermediate state of magnetically ordered crystals" *Sov. Phys. Sol. St.* **28**, 1876 (1986)

Huber E.E. and Ridgley D.H., "Magnetic properties of a single crystal of manganese phosphide" *Phys. Rev.* **135**, A1033 (1964)

Ishizaki A., Komatsubara T. and Hirahara E., "Magnetostriction in manganese phosphide single crystal" *Prog. Theor. Phys.* suppl.**46**, 256 (1970)

Kawata H. and Mori K., "X-ray magnetic Bragg scattering topography from Fe_3O_4" *Rev. Sci. Instr.* **66**, 1407 (1995)

Khachaturyan A.G., "Some questions concerning the theory of phase transformations in solids" *Sov. Phys. Sol. St.* **8**, 2163 (1967)

Kleman M. and Schlenker M., "The use of dislocation theory in magnetoelasticity" *J. Appl. Phys.* **43**, 3184 (1972)

Kulda J., Baruchel J., Guigay J.P. and Schlenker M., "Extinction effects in polarized neutron diffraction from magnetic crystals. I- Highly perfect MnP and YIG samples" *Acta Cryst* **A47**, 770 (1991)

Landau L. and Lifschitz E., '*Physique Statistique*,' Mir (Moscow) 1967

Obara H., Endoh Y., Ishikawa Y.and Komatsubara T., "Magnetic phase transition of MnP under magnetic field" *J.Phys.Soc.Japan* **49**, 928 (1980)

Okamoto T., Kamigaichi T., Iwata N. and Tatsumoto E., "Anisotropic linear thermal expansions of MnP single crystal" *J.Phys.Soc.Japan*, **25**, 1730 (1968)

Patterson C., Palmer S.B., Baruchel J. and Ishikawa Y., "Observation of domains in the helical phase of MnP" *Sol. St. Comm.*, **55**, 81 (1985)

Polcarova M. and Lang A.R., "X-ray topographic studies of magnetic domain configuration and movements" *Appl.Phys.Lett.*, **1**, 13 (1962)

Polcarova M. and Gemperlova J., "Distortion of an Fe-Si single crystal and x-ray topographic contrast due to a 90° ferromagnetic domain wall" *Phys. Stat. Sol.*, **32**, 769 (1969)

Roth. W.L., in '*The Chemistry of Extended Defects In Non Metallic Solids*', North Holland, Amsterdam (1970)

Salje E.K.H. '*Phase transitions in ferroelastic and co-elastic crystals*', Cambridge University Press (1990)

Sandonis J., Baruchel J., Gomez Sal J.C., Palmer S.B. and Zontone F., "Observation by synchrotron radiation topography of the phase coexistence at the triple point in MnP" *J. Phys. D: Appl. Phys.*, **26**, A115 (1993)

NEW POSSIBILITIES OF DIFFRACTION TOPOGRAPHY AT THIRD GENERATION SYNCHROTRON RADIATION FACILITIES

José Baruchel
European Synchrotron Radiation Facility
BP 220, 38043 Grenoble, France

1. INTRODUCTION

Third generation synchrotron radiation (SR) facilities like the ESRF provide new experimental possibilities for diffraction topography. Some of them are straightforward; the high energy photons (50-120 keV) allow the investigation of heavy and/or bulky samples in transmission, and the photon flux is such that real-time experiments at the 0.01 s time scale are feasible. However the availability of high energy photons and the brilliance of the source have additional consequences. The high energy photons are associated with a small intrinsic diffraction width, and consequently with the possibility of studying defects inducing very weak distortion fields. On the other hand, the high brilliance corresponds to a small source size with dimensions in the 0.1 mm range for the ESRF ID19 wiggler 'topography' beamline, devoted to imaging and high resolution diffraction. The ID19 experimental station is relatively far away from the source, at 145 m, to compensate for the small opening angle of the radiation (Baruchel, Draperi and Zontone 1993). The "geometrical" resolution of topographs $d \times S / D$ (S being the source size, D the source-to-sample and d the sample-to-detector distances) is thus greatly improved compared with the topographic stations at older SR machines. Practically, this enhanced geometrical resolution allows observation of details in the topographs with an acceptable resolution (less than 10 μm) at distances as great as one meter from the sample, and consequently the use, as a new parameter, of the crystal to detector distance to characterize a given defect.

The small angular size of the ESRF sources, and in particular their transverse coherence, leads to novel 'phase imaging' possibilities which have also implications for the diffraction topographic work (Cloetens, Barrett, Baruchel, Guigay and Schlenker 1996). These features show that not only the interaction of the incident beam with the crystal, but also the propagation of the incoming and diffracted beams outside the crystal must be considered when analysing the contrast observed on the topographs.

We describe in the present paper some of these new possibilities for diffraction topography , as well as their associated drawbacks, related to third generation SR machines.

X-ray and Neutron Dynamical Diffraction: Theory and Applications
Edited by Authier *et al.*, Plenum Press, New York, 1996

They are illustrated by experiments mainly performed at the ESRF BM5 beamline, and, very recently, at the ID19 'Topography' beamline.

2. INVESTIGATION OF HEAVY OR BULKY CRYSTALS

One particular feature illustrated by the first white beam diffraction topographic experiments performed at the ESRF is the high energy corresponding to some of the recorded Bragg spots ($\approx$ 50-120 keV). This allows the high resolution investigation of heavy or bulky samples as illustrated either by the observation of magnetic domains in a 1.4 mm thick Fe3%Si crystal ($\lambda \approx 0.15$ Å, E $\approx$ 83 keV, $\mu t \approx 0.6$), or by a section topograph of 1cm thick Si, which displays the usual Kato's Pendellösung fringes ($\lambda \approx 0.21$ Å, E $\approx$ 59 keV, $\mu t \approx 0.7$) (Baruchel et al. 1994).

Figure 1 shows a white beam 021 topograph of a 0.5 mm thick (001) HgI_2 crystal, performed by using a photon energy slightly smaller than the one corresponding to the Hg absorption edge. Dislocations are revealed in the form of lines mainly oriented along $[110]$ and $[1\bar{1}0]$ (Gastaldi et al. 1996). A 'liquid filter' solution of KI + HgI_2 was set in the white beam to absorb photons with energies such that their μt is high for the HgI_2 crystal, in order to reduce the sample heat load (further details in section 3.2).

The X-ray topographic observation of similar features (Kato's fringes, domains, defects,...), in transmission, was achieved in the past on samples 10 times thinner. The thick or heavy samples were, up to now, the exclusive domain of neutron diffraction topography, a technique which implies low resolution ($\approx$ 60 μm) and long exposure times (hours).

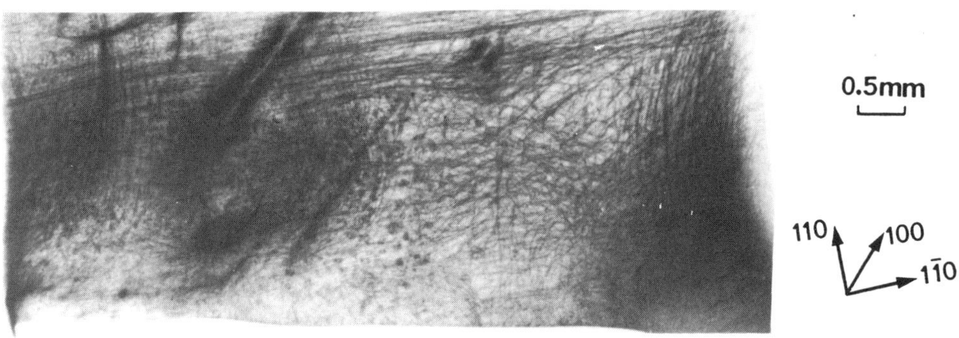

Figure 1. White beam 021 topograph of a 0.5 mm thick HgI_2 crystal; $\lambda \approx 0.21$ Å, E $\approx$ 59 keV, $\mu t \approx 1.95$ (courtesy J. Gastaldi)

3. REAL TIME EXPERIMENTS

3.1. Short exposure times, bulky sample environment

A 'real time' experiment is such that the recording of the information 1) requires a time shorter than the typical evolution time of the physical process and 2) can be repeated in such a way that this evolution can be followed. The availability of SR, through the reduction of the exposure times from several hours to a few seconds, allowed investigation, in 'real time', of a range of physical processes. These include defect movements, phase transitions, crystal growth, and when using stroboscopy, wall movements, resonator vibrations, etc .

The experiments performed at the ESRF (Baruchel et al. 1994, Barrett et al. 1995) show that very short exposure times, down by about a factor 10^2-10^3 below the standard exposure time at other SR topographic setups, can be achieved. Two types of CCD camera, equipped with scintillators, are available at ID19: 1) an intensified camera, which can record 25 images/second, with a field of about 20×20 mm^2, a resolution of 40 μm, and a dynamic range of ≈ 200; 2) a Peltier cooled camera, which can record 3 images/second, with a field of ≈ 10×10 mm^2, a resolution of 10 μm and a dynamic range of about 10^4.

Very often the investigated dynamical phenomena result from the variation of an applied external parameter (temperature, field, stress,...) which implies a bulky sample environment (furnace, cryostat, electromagnet, traction device,...), and consequently an enhanced crystal to detector distance. It is practically important to retain a good resolution under these conditions. Figure 2 shows that the width of the direct image of a screw dislocation, measured at BM5, is not very enhanced when the crystal-to-detector distance is increased up to 1 meter . This corresponds to the small angular size of the source seen from one point of the sample (10^{-5} radians at BM5, and 10^{-6} radians in the ID19 case).

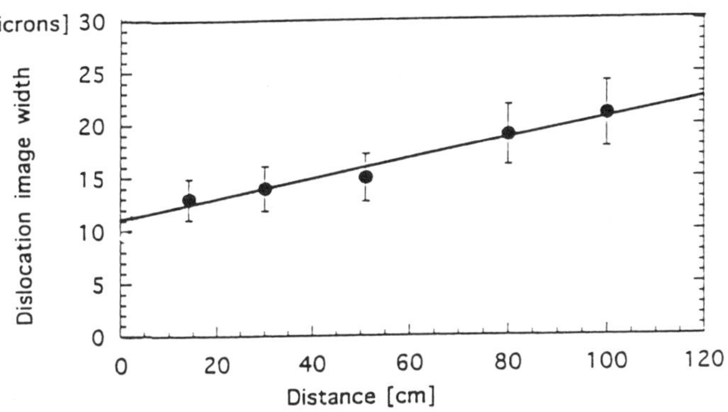

Figure 2. Evolution versus the sample-to-film distance of the image width of a screw dislocation in germanium (220 reflection, λ ≈ 0.19 Å, E ≈ 65 keV, μt ≈ 0.24)

3.2. Heat Load in white beam diffraction topography

White beam diffraction topography (which produces several diffraction topographs simultaneously and exhibits reduced exposure times with respect to its monochromatic counterpart) is usually foreseen when planning 'real time' experiments. This is even more necessary for crystal growth investigations, where the orientation of the sample is not known in advance. When considering the flux available from an insertion device at the ESRF, it might be thought that exposure times could be reduced to the few microseconds level. This is unfortunately not the case. Indeed the inhomogeneous heat load associated with the absorbed power can produce an unacceptable distortion of the sample and the topograph becomes meaningless. This is often the case because of the limited size of the beam in the vertical direction (14 mm as a maximum for ID19), and because the high energy photons are concentrated in the center of this beam. The non uniform distribution of the absorbed power density along the vertical direction can create a gradient of temperature in the sample, and consequently of distortion which is observed on the topographs.

Figure 3a indicates a simple model to describe heat load effects. The bump which is generated does not affect too much, for symmetry reasons, the Bragg planes in the middle part of the illuminated area but distorts the outer areas and consequently produces a large amount of additional intensity on the topograph. This model also suggests a criterion for fixing the maximum absorbed power which does not affect the image, based on the physical idea that the image at the z position is a 'good' one if the angular variation $\delta\theta$ at this position is less than the intrinsic diffraction width ω_D of the considered reflection ($\delta\theta$ / ω_D < 1). Zontone and Mancini (1996) treated the various limiting cases as a function of the absorbed power, the linear thermal expansion coefficient, and the thermal conductivity of the sample. The steady case is usually reached for 'non absorbing' crystals with rather high thermal conductivity, like silicon, whereas the transient state has often to be considered for thermal insulators. Figure 3 shows the same topograph of a quartz crystal recorded either under different beam conditions (b-c) or on different sensitivity films (b-d): the heat load effect is clearly (Figure 3b) or faintly (Figure 3d) visible, or not at all observed (Figure 3c).

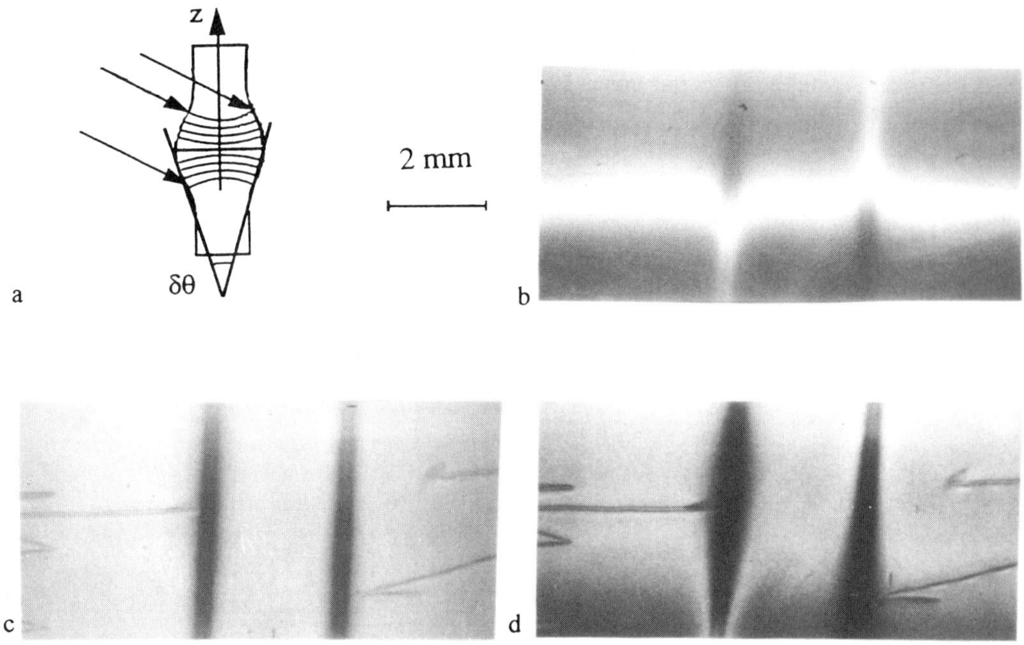

Figure 3. Heat load effect in quartz, $3\bar{1}\bar{2}1$ reflection, λ = 0.41 Å, E = 30 keV a) schematic model of the distorsion. b) non-chopped beam (100% transmission), c) chopped beam (2% transmission), same actual exposure time as for (b) d) non-chopped beam, recorded on a 100 times faster film than for (b) (exposure time divided by 100) (courtesy J. Härtwig).

When working with a white beam, tailoring the beam is often necessary. In the case of a wiggler source this can be achieved by changing the gap. Filters are nearly always necessary to eliminate the unused soft part of the spectrum, and to absorb the photons corresponding to the absorption edges of the heavy elements present in the sample before they reach the crystal. The best filter for a given material is the material itself. As this is not always feasible, a 'liquid filter' containing the elements present in the investigated crystal appears to be a good solution, as already discussed in fig. 1. However, when working with low energy photons, a 'chopper' can be used to achieve the best compromise between heat-load and exposure-time (see Figure 3) and avoid the filter-related-hardening of the spectrum.

3.3 Examples

3.3.1. Magnetic domains in magnetite using multiple diffraction. The 71° and 109° magnetic domain walls evolution under a magnetic field was followed using a video camera in a (110) platelet shaped crystal of Fe_3O_4 grown at the Saitama University (Japan). Spectacular anomalous contrast was observed on the very weak reflection 711, enhanced through simultaneous reflection ('Umweganregung') of the 400 and 311 reflections: the domains then showed up as **area contrast** (Figure 4), although their difference in distortion, due to magnetostriction, is small. ($\lambda_{111} \approx 8.10^{-5}$) (Schlenker, Barrett, Miyamoto, unpublished).

3.3.2. Inconmensurate phase of quartz. Real time investigations can also be carried out by using white beam **section** topographs, as in the case of the inconmensurate phase of quartz at around 848K, where Dolino, Bacheimer and Medrano (unpublished) followed the position of very weak satellites, and related the phase transition temperatures to the local crystalline quality. They showed in addition that the shape and size of the domains within a given phase depend on the crystal quality, and exhibit a stong relationship with the growth band pattern. Let us note that this kind of work involves the use of strongly absorbing and precise slits, which are not trivial to manufacture and set. Heat load is much weaker for this technique.

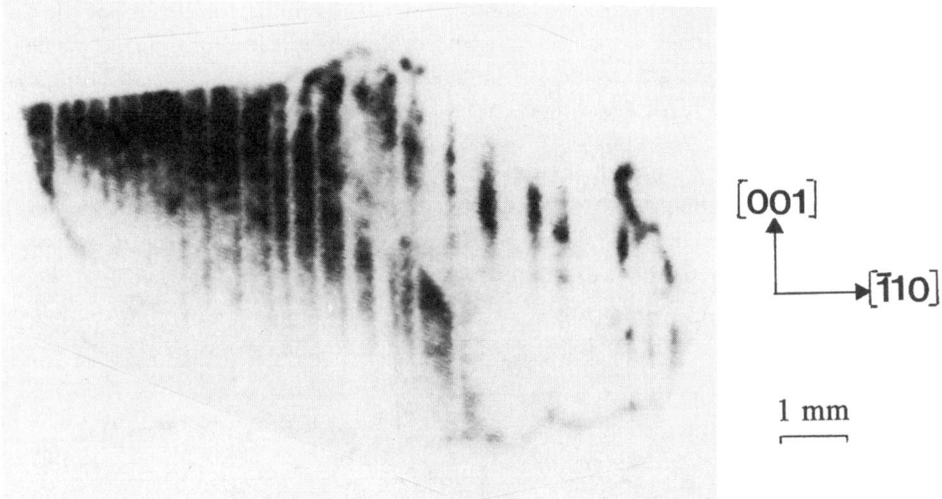

Figure 4. Domains in magnetite: area contrast is observed on the very weak reflection, 711, enhanced through simultaneous reflection ('Umweganregung') of the 400 and 311 reflections (λ = 0.23 Å Å, E = 54 keV).

4. USING THE CRYSTAL TO DETECTOR DISTANCE AS A PARAMETER

In most of the published contrast-mechanism models the source is assumed to give an incoherent illumination of the sample and, in general, the propagation of the diffracted beam from the crystal to the detector is neglected. We will treat in section 6 the effects associated with the spatial transverse coherence of the incident beam. The present section is devoted to the use of the information provided by varying the crystal to detector distance *d*. We already mentioned that a rather large *d* (1 m, as compared to 1-2 cm in Lang's method, and about 10 cm on the previous SR topographic setups) involves no subtantial geometrical degradation of the image when working at the ID19 ESRF beamline. This allows the observation of what we are actually interested in, the broadening of the image resulting from the distortions of the crystal itself.

This idea is not new. Indeed Tanner Midgley and Safa (1977) have demonstrated in pioneering work that using a SR source where the geometrical criterion is not resolution limiting then the direct image width of an edge dislocation may be observed to change with variations of the sample-to-film distance. This was explained as being due to the change of the beam divergence induced by the defect through orientation contrast. They concluded that the image width of an edge dislocation is sensitive to the sign of the Burgers vector **b**. Unfortunately problems with stability prevented Tanner and coworkers from studying these effects in more detail, and the reported observations remained unique until the ESRF machine was ready.

In recent experiments, we turned our attention towards the direct image of edge (or with a strong edge component) dislocations, and looked for contrast changes in shape and width as a function of the crystal orientation and sample-to-film distance. Figures 5a and b show the images of a dislocation composed of three segments running along [111], [110] and [111] of Burgers vector **b** 1/2[110], obtained by exchanging the entrance and exit surfaces for the X-ray beam. The contrast of segments A and C (Figure 5c) changes from simple to double, and vice versa. This corresponds to orientation contrast, because the edge component of **b** changes sign with respect to the dislocation line when going from A to C. Figure 5d shows the evolution of the width of the segments A and C as a function of the sample-to-film distance ('focusing/defocusing' behaviour). The evolution of the direct image width of a dislocation as a function of the sample-to-film distance can allow a complete determination of the Burgers vector, i.e. in sign and modulus, a topic treated by Zontone, Mancini, Barrett, Baruchel, Härtwig and Epelboin (1996), and in a problem in this volume.

Section topographs also benefit from observation as a function of the crystal-to-detector distance. This applies to surface layers very weakly misoriented (≈1 arc second) with respect to the bulk, as shown when investigating implantation associated polarization-reversed-layers in $LiNbO_3$ (Rejmankova, Baruchel and Moretti 1996).

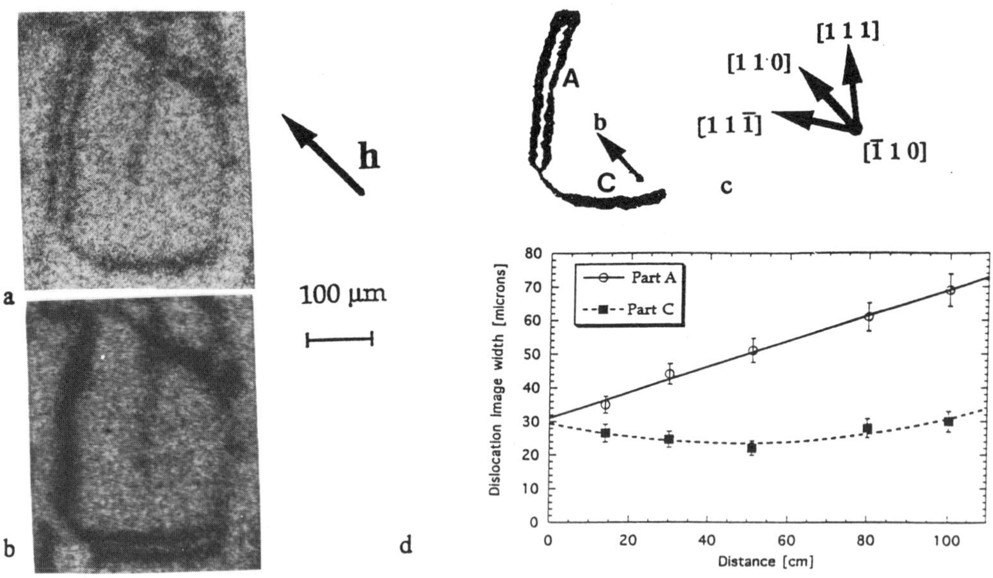

Figure 5. a) b) a contrast change from single to double is observed for the dislocation segments A and C (on c) when exchanging entrance and exit surfaces; crystal-to-film distance d≈50 cm; Ge crystal, 220 reflection, (λ=0.19Å, 65 keV, μt=0.24) d) Dislocation width as a function of d for the A and C segments.

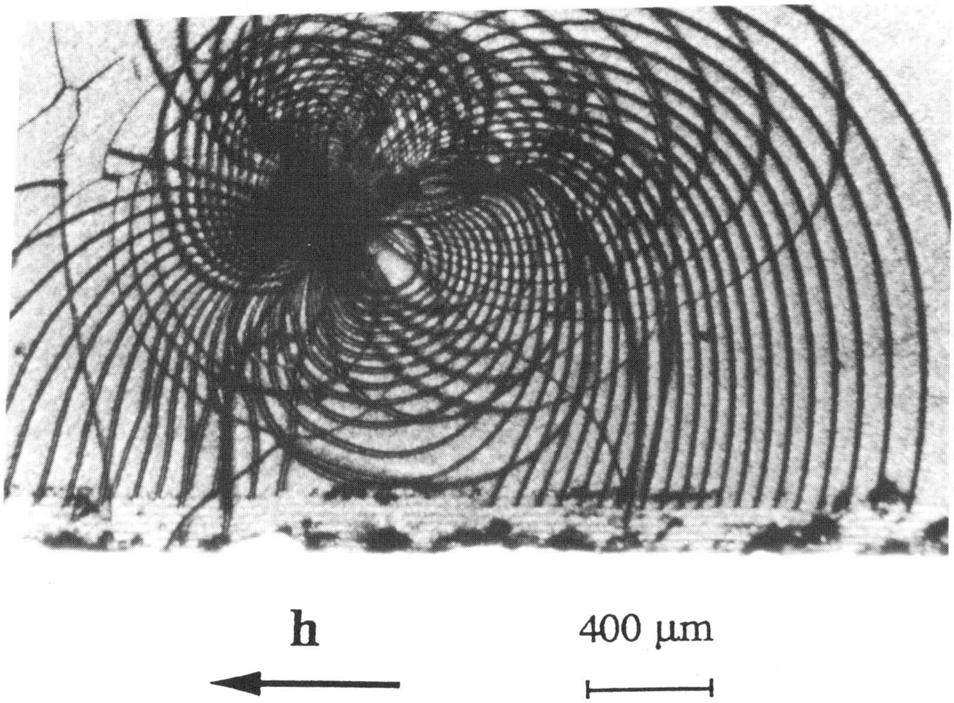

Figure 6. Dislocation network in 6H-SiC, 111 reflection, λ=0.21Å, 59 keV (courtesy E. Prieur)

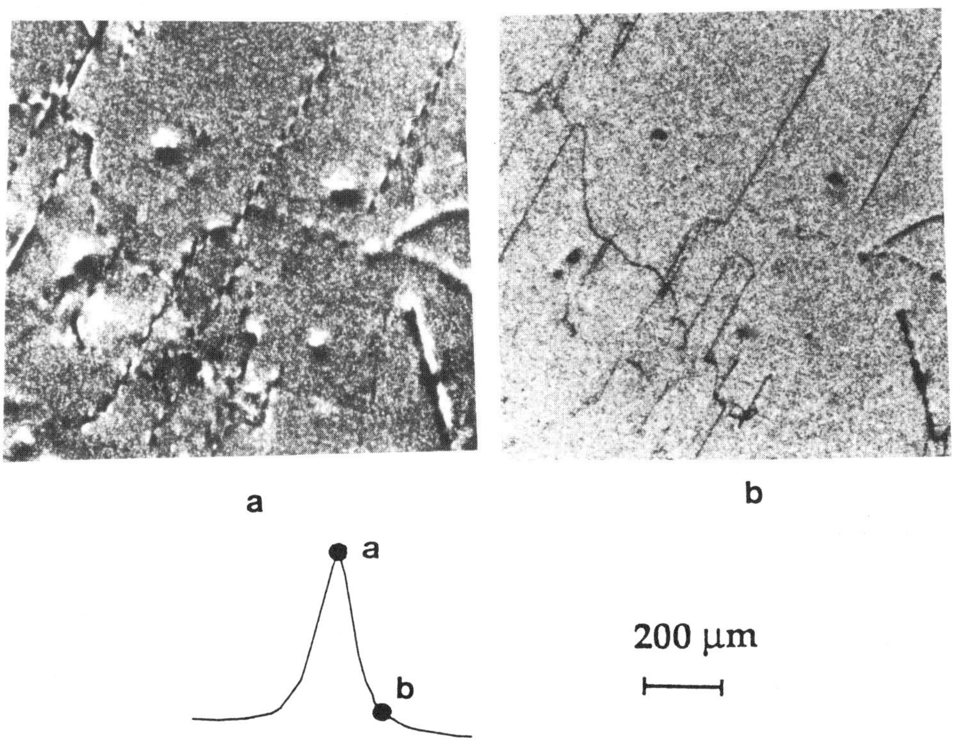

Figure 7. a) peak and b) 'weak beam' topographs of a Ge crystal; 111 reflection (λ≈0.35 Å, E≈35.5 keV) .

5. HIGH RESOLUTION DIFFRACTION TOPOGRAPHIC EXPERIMENTS

The expression 'high resolution diffraction topographic experiments' is found in the literature to describe at least two distinct kinds of experiments: either those where a very high spatial resolution is achieved, or the ones where a very high sensitivity to weak distortions and/or gradients of distortion is required (plane wave or 'ultra'-plane wave topography). Both types of experiment benefit from the third generation SR possibilities.

5.1. High energy diffraction topography

Figure 6 illustrates one of the effects associated with high energy photons in white beam topography. The 222 topograph of a 6H-SiC crystal, produced with 67 keV photons, exhibits double contrast dislocation images. Such a feature, reported in the past (Authier 1967) is very frequently observed at the ESRF. It is related to the fact that the intrinsic reflection width ω_D decreases when decreasing the photon wavelength. Consequently, the regions which are at the origin of the direct image correspond to very weak deformations (of the order of ω_D) with respect to the perfect crystal. Therefore, the use of high energies is a simple way to study these very weak distortions, when varying in addition the photon energy diffracted by the investigated Bragg spots.

5.2. High flux and 'weak beam' diffraction topography

In double-crystal topography the enhanced flux means that images can be formed either from very misoriented parts of the sample ('weak beam') or from weaker scattering processes, retaining in both cases short exposure times. 'Weak beam' topography leads to narrower defect images, hence raising the density at which individual defects can be resolved. When working at the ESRF the use of a Si crystal as a monochromator can provide, when investigating a Ge crystal, a "nearly plane wave" with a small spectral window. This is shown (Figure 7) on topographs recorded on a) the peak and b) the low angle wing of the rocking curve (exposure time in the second range). An integrated image can be obtained by either vibrating the monochromator or rotating the crystal during the exposure.

Another interesting possibility related to the high flux and weaker scattering processes, which is not connected with high resolution, is the use of resonant magnetic scattering to perform images of magnetic domains using a contrast mode arising directly from the magnetization or the magnetic arrangement (Kawata and Mori, 1995).

6. COHERENCE PROPERTIES OF THE BEAM

At the ESRF phase objects can lead to spurious contrast on X-ray diffraction images. The small angular size of the X-ray source leads to a high spatial coherence of the photon beam. Indeed the transverse coherence length $\lambda x D/2S$, (with S the source size and D the source-to-sample distance) is, in the ID19 case and for $\lambda \approx 1$Å, about 0.1 mm. This can produce phase images of objects present in the beam (beryllium windows, dust, ...) which are nearly non-absorbing at the photon energies used. Their images are related to the variations of phase associated with the variations in thickness. Such phase images can disturb the observation and interpretation of the topographic ones, as shown on Figure 8a where the horizontal striations correspond to small variations in thickness ($\approx$ 5 μm) of the 500μm thick beryllium window. It is compulsory, to avoid these effects, to use clean slits for section topography, and carefully polished filters and windows. However if this appears not to be possible, a random phase plate, which provides an effective way of tailoring the beam divergence, can be added to obtain better topographic results, as shown on Figure 8b (Cloetens et al. 1996).

The usual requirement in diffraction topography is that the surfaces of the investigated crystal (and/or monochromator) should be strain free; no specific requirement is made in what concerns their flatness. This is not any more true when dealing with high quality crystals and using a 'coherent' x-ray beam. Figure 9 shows the image of a surface scratch, which is mainly a phase contrast image, the direct image component being very weak.

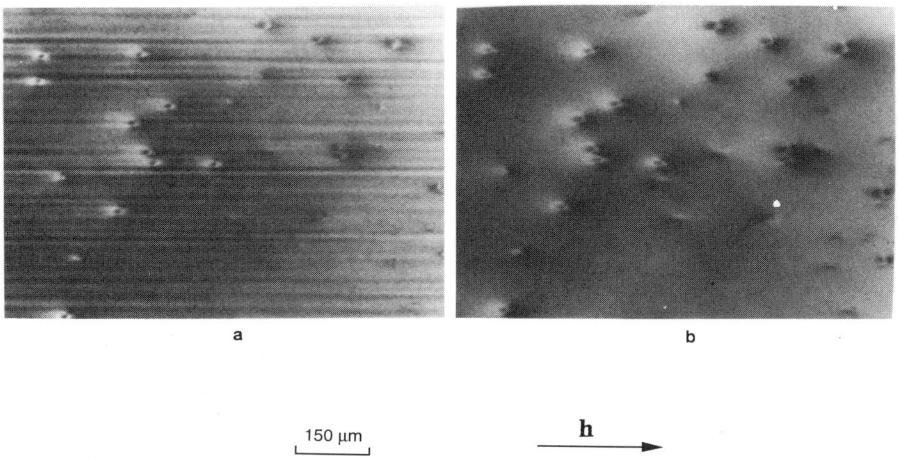

Figure 8. Monochromatic beam 022 topographs (λ=0.37Å, E=33.5 keV) of a Simox sample (540 μm Si substrate + 0.4 μm SiO_2 amorphous layer + 10 μm epitaxic Si capping layer) a) no random phase screen (RPS) and b) with a 2 mm thick ash wood RPS. Dislocations in the top layer are visible in (a) as well as corrugations on a beryllium window. The use of the RPS in (b) blurs completely the phase image of the beryllium window and affects only slightly the image of the dislocations (from Cloetens et al. 1996)

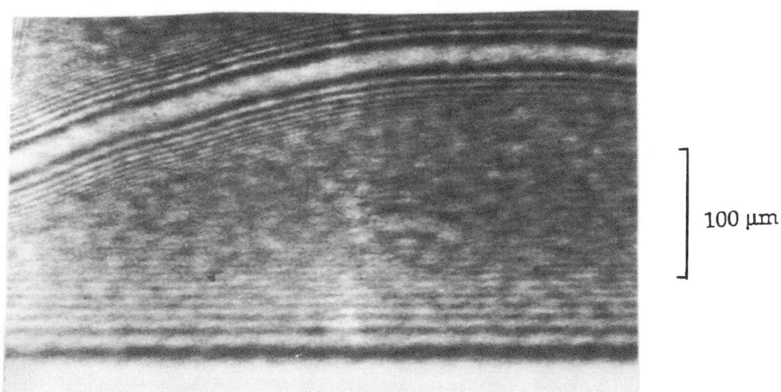

Figure 9. Phase image of a surface scratch on a (111) silicon crystal; Bragg symmetrical setting; sample-to-film distance 92 cm (λ=0.53Å, E=21 keV). The fringes associated with the edge of the crystal are also visible (courtesy P. Cloetens).

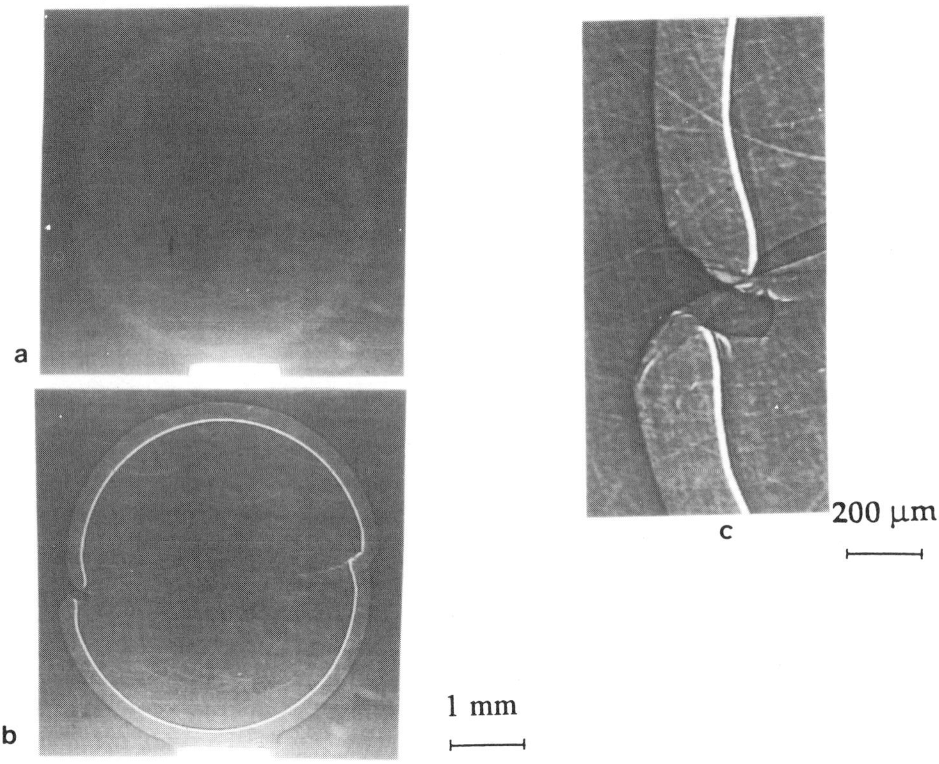

Figure 10. a) Absorption (sample-to-film distance 3 cm) and b) c) phase contrast 'propagation' (sample-to-film distance 92 cm) images of two halves of a polystyrene sphere (λ=0.53Å, E=23 keV) (courtesy P. Cloetens)

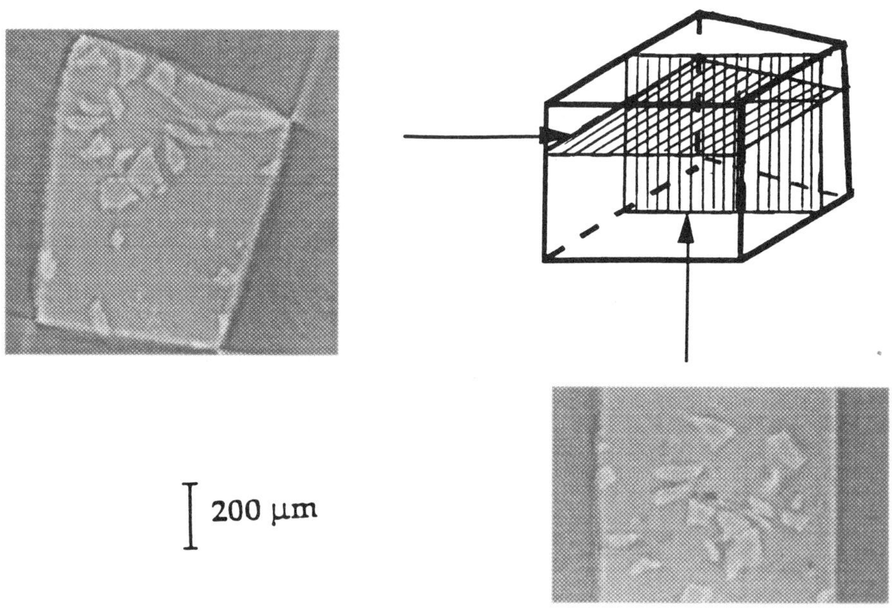

Figure 11. Phase-contrast-based tomographic reconstruction slices, obtained from 600 projection images, of an aluminium sample containing SiC particles (λ=0.53Å, E=23 keV) (courtesy M. Pateyron, J.Y.Buffière)

Phase images can be used to investigate materials of interest in a very simple way. The techniques used so far to study this phase variation are based on a double crystal or a Bonse-Hart type interferometer device. The current method only involves, apart from the beam from the ESRF source, a perfect crystal monochromator in symmetrical Bragg setting, and a detector (CCD camera or film) placed at a distance d from the specimen such that $d \geq a^2/ 2\lambda$, *a* being the size of the detail to be observed (*d* is about 0.4 m when a $\approx$ 5 μm). This 'propagation' imaging technique corresponds to Fresnel diffraction. Its principle is identical to the defocusing mode of electron microscopy and to in-line Gabor holography in classical optics.

This new technique gives the possibility of imaging objects involving negligible absorption of hard X-rays but appreciable variations in optical path length due to thickness or compositional variations. The first experiments performed at the ESRF on organic (bones, plants) or inorganic (test materials with inclusions, holes, cracks) samples appear very promising (Snigirev et al., 1995, Cloetens et al. 1996). As an example, Figure 10 shows the absorption (d=3 cm, maximum contrast 6%) and phase contrast 'propagation' (d=92 cm) images of two halves of a polystyrene sphere. The 'edge' contrast observed on figures 10b and 10c derives from the expected Fresnel fringe structure. A problem is devoted, in this volume, to the qualitative explanation of the observed contrast.

This phase imaging recording reveals novel possibilities when extended to the three dimensional image reconstruction (tomography). This involves the use of a camera with good spatial resolution as detector, a setup for specimen rotation, and the implementation for the X-ray case of the mathematical algorithms produced for the visible light and electron cases. We are presently still limited to the use of the absorption-based reconstruction programs, which nevertheless can provide valuable results in many cases, in particular when the 'edge' contrast is the dominant one. Figure 11 shows an example of such a reconstruction; the sample is an aluminium matrix containing SiC particles with sizes in the 100 μm range (Buffière, Cloetens, Pateyron, Peix, unpublished).

7. SOME CONCLUSIONS

The experiments described illustrate the new possibilities for diffraction topography at third generation SR machines. They include exposure times down to the 10^{-2} second range, high resolution connected with the small source size, and availability of short wavelengths ($\leq$ 0.1 Å). The high flux allows 'real time' topographical investigations, even in a monochromatic beam. It is on the other hand easy to switch from the "low divergent wave" condition to the "integrated image" one.

The good spatial resolution retained even when setting the film far from the sample is important in that it allows the use of the sample-to-detector distance as an additionnal parameter to characterize the crystal, and it makes the setting easier in bulky sample environments.

The large spatial transverse coherence length of the photons ($\approx$100 μm) reaching the sample opens the promising field of phase contrast images, corresponding to phase jumps in the crystal. Its extension to tomography provides a new tool for materials science.

REFERENCES

Authier A., "Contrast of dislocation images in x-ray transmission topography", *Adv. X-ray Analysis*, **27**, 9 (1967)

Barrett R., Baruchel J., Härtwig J. and Zontone F., "The present status of the ESRF diffraction topography beamline: new experimental results", *J. Phys.D: Appl. Phys.*, **28**, A250 (1995)

Baruchel J., Draperi A. and Zontone F., "Present status of the topography and high-resolution diffraction beamline at the ESRF", *J. Phys. D: Appl. Phys.*, **26**, A9 (1993)

Baruchel J., Epelboin Y., Gastaldi J., Härtwig J., Kulda J., Rejmankova P., Schlenker M. and Zontone F., "First topographic results at the European Synchrotron Radiation Facility", *Phys. Stat. Sol. (a)*, **141**, 59 (1994)

Cloetens P., Barrett R., Baruchel J., Guigay J.P. and Schlenker M., "Phase objects in synchrotron radiation hard x-ray imaging", *J. Phys.D: Appl.Phys.*, **29**, 133 (1996)

Gastaldi J., Rossberg A., Smolski I. and Le Lay G., "New possibilities for the characterization of defects in α-HgI_2 crystals by synchrotron x-ray topography", *Nucl. Instr. Meth.*, in press

Kawata H. and Mori Y. "X-ray magnetic Bragg scattering topography from Fe_3O_4", *Rev. Sci. Instr.* **66**, 1407 (1995)

Rejmankova P., Baruchel J. and Moretti P., "Investigation of hydrogen implanted $LiNbO_3$ under a DC electric field by synchrotron radiation topography", *Physica B*, in press Snigirev A., Snigireva I., Kohn V.G., Kuznetsov S.M. and Schelokov I., "On the possibilities of x-ray phase contrast microimaging by coherent high energy synchrotron radiation", *Rev . Sci. Instr.*, **66**, 5486 (1995)

Zontone F., Mancini L., "Treatment of the heat load associated contrast in synchrotron radiation topography", *Il Nuovo Cimento*, in press

Zontone F., Mancini L., Baruchel J., Härtwig J. and Epelboin Y., "New features of dislocation images in third-generation synchrotron radiation topographs", *J. Synch. Rad.*, in press

X-RAY STANDING WAVES

J. R. Patel[1,2]

[1]Stanford Synchrotron Radiation Laboratory
[1]Stanford Linear Accelerator Center, Stanford, CA 94309-0210, USA
[2]Lawrence Berkeley National Laboratory, Berkeley, CA 94720,USA

1. INTRODUCTION

One of the more fascinating aspects of dynamical Bragg diffraction of x-rays is the excitation of standing wave fields inside the crystal. Direct experimental manifestations of standing wave effects were evident in the discovery of anomalous transmission in the case of the Laue geometry by Borrmann (1941) and in studies of fluorescent scattering in the Bragg geometry initiated by Batterman (1964). Experiments in the Bragg geometry (i.e., the diffracted beam exits the crystal through the same face that the incident beam enters) had the peculiar advantage that only a single standing wave–type eigenfunction is excited in the crystal and this field could be manipulated experimentally to *probe* atom sites in the diffracting crystal. It was later pointed out by Golovchenko, Brown, and Batterman (1974) that studies directly within the Bragg *band gap** could lead to comprehensive atom-location information. Subsequent work has demonstrated that both surface and near-surface impurities can be localized to approximately one hundredth of a lattice constant (Andersen, Golovchenko and Mair, 1976).

*We refer to the angular region where no itinerant solutions to Maxwell's equations exist within the crystal as the Bragg band gap. This is the region normally associated with the high reflectivity characterized by the Darwin-Prins reflection curve.

X-ray and Neutron Dynamical Diffraction: Theory and Applications
Edited by Authier *et al.*, Plenum Press, New York, 1996

2. THEORY

For a quantitative understanding of standing wave effects, we need to know the amplitude of the standing wave field and its phase. The amplitude of the standing wave field in a crystal under conditions of Bragg diffraction can easily be determined by considering the interaction of the incident and diffracted beams. Consider a plane wave $\mathbf{E}_0 \exp(-2\pi i\mathbf{K}_0 \cdot \mathbf{r})$ incident on a crystal face. In Bragg diffraction we also have a diffracted beam $\mathbf{E}_H \exp(-2\pi i\mathbf{K}_H \cdot \mathbf{r})$. For plane waves we can sum the incident and diffracted beams to give the total amplitude of the electric field in the crystal.

$$\mathbf{E} = \mathbf{E}_0 \exp(-2\pi i\mathbf{K}_0 \cdot \mathbf{r}) + \mathbf{E}_H \exp(-2\pi i\mathbf{K}_H \cdot \mathbf{r}) \tag{1}$$

where $\mathbf{E}_0$ and $\mathbf{E}_H$ are the incident and diffracted beam amplitudes and $\mathbf{K}_0$ and $\mathbf{K}_H$ = 1/l are the incident and diffracted wave vectors. To obtain the intensity of the standing wave field $\mathbf{E}^2$ we multiply (1) by its complex conjugate, remembering that for Bragg diffraction $\mathbf{K}_H - \mathbf{K}_0 = \mathbf{H}$ where $\mathbf{H}$ is the relevant reciprocal lattice vector $\mathbf{H}$ = 1/d, d being the spacing of Bragg planes. We obtain

$$\mathbf{E}^2 = \mathbf{E}_0{}^2 + \mathbf{E}_\mathrm{H}{}^2 + 2\mathbf{E}_H\mathbf{E}_0 \cos(\phi - 2\pi\mathbf{H} \cdot \mathbf{r}) \tag{2}$$

dividing through by $E_0{}^2$

$$(\mathbf{E}/\mathbf{E}_0)^2 = 1 + (\mathbf{E}_H/\mathbf{E}_0)^2 + 2(\mathbf{E}_H/\mathbf{E}_0)\cos(\phi - 2\pi\mathbf{H} \cdot \mathbf{r}) \tag{3}$$

where ϕ is the phase of the standing wave field. Equation 3 states that there are planes of constant intensity when $\mathbf{H} \cdot \mathbf{r}$ is constant, r being any position vector projected on the reciprocal lattice vector $\mathbf{H}$. Since $\mathbf{H}$ = 1/d, the periodicity of the standing wave field has the spacing d. One can readily show that the normalized values of the standing wave field Equation 3 ranges from 0 to 4 depending on whether the node or anti-nodes of the standing wave field coincide with the atomic planes. For nodes coincident with the atomic planes, the atoms see an electric field intensity = 0. Whereas at the atomic planes $(\mathbf{E}/\mathbf{E}_0)^2 = 4$ or four times the average intensity far from the Bragg reflection.

The probability of fluorescence radiation from a crystal during Bragg reflection is proportional to the standing wave field intensity at the relevant reflecting planes. In an actual measurement, the fluorescence yield is normalized to the yield far from the Bragg reflection. Following Hertel et al. (1985) and Vlieg et al. (1991), we can write the fluorescence yield as

$$\mathbf{Y}(\theta) = c\int \mathbf{I}(\theta, \mathbf{r})\rho_u(\mathbf{r})d\mathbf{r} \tag{4}$$

where c is a normalization constant far from the Bragg reflection and $\rho_u(\mathbf{r})$ the electron density distribution in one unit cell of the fluorescing atom. With the assumption that the surface has N atoms per unit cell at positions $\mathbf{r}_j$ and there is no disorder, the atomic distribution (or position) is given by the usual δ function representation

$$\rho_u(\mathbf{r}) = \sum_{j=1}^{N} \delta(\mathbf{r} - \mathbf{r}_j) \tag{5}$$

Substituting Equations 3 and 5 into Equation 4 and normalizing the yield to the total number of atoms we get letting $R(\theta) = (\mathbf{E}/\mathbf{E}_0)^2$

$$(Y(\theta)/NCI_0) = 1 + R(\theta) + 2\sqrt{R(\theta)}\frac{1}{N}\sum_j cos(\phi - 2\pi\mathbf{H}\cdot\mathbf{r}_j) \tag{6}$$

We want this to be identical to the shape of fluorescence yield curve

$$1 + R(\theta) + 2\sqrt{R(\theta)}F\cos(\phi - 2\pi P) \tag{7}$$

where F, the coherent fraction, and P, the position of the atoms, are the two adjustable parameters in fitting the fluorescence yield curve. The condition that Equation 7 is identical to 6 is

$$F\cos(\phi - 2\pi P) = \frac{1}{N}\sum_j \cos(\phi - 2\pi\mathbf{H}\cdot\mathbf{r}_j) \tag{8}$$

writing the cos as the sum of two exponentials, it follows that

$$F\exp(-2\pi iP) = \frac{1}{N}\sum_j \exp(-2\pi i\mathbf{H}\cdot\mathbf{r}_j) = \mathbf{F}_{SW} \tag{9}$$

This quantity we may call the x-ray standing wave structure factor $\mathbf{F}_{SW}$.

To specify the phase of the standing wave field during Bragg reflection, we consider the Darwin reflectivity. We have seen in the earlier chapter by Authier on the dynamical theory of x-rays from perfect crystals that the Darwin reflectivity assuming no absorption is given by

$$(E_H/E_0)^2 = \left|\eta \pm (\eta^2 - 1)^{1/2}\right|^2 \tag{10}$$

where η is a dimensionless parameter proportional to the deviation from the exact Bragg condition. In its simplest form, assuming symmetric reflection $\eta = [\Delta\theta\sin 2\theta + \Gamma\mathbf{F}_0]/\Gamma\mathbf{F}_H$, where $\Delta\theta$ = deviation from the exact Bragg position, $\Gamma = r_e \lambda^2 / \pi V$, r_e is the electron radius, λ the wavelength and V the unit cell volume, $\mathbf{F}_0$ and $\mathbf{F}_H$ are the zeroth and **H** component of the structure factor. In the region $-1 \langle \eta \langle 1$ we can write equation 10 as

$$(E_H / E_0)^2 = | \eta \pm i(1-\eta^2)^{\frac{1}{2}}|^2 \qquad (11)$$

where $\tan^{-1}\left[(1-\eta^2)^{\frac{1}{2}} / \eta\right]$ is the η or $\Delta\theta$ dependence of the phase ϕ of the standing wave field. With the phase of standing wave field ϕ obtained from the reflectivity measurement, we can calculate the intensity of the standing wave field at any point on the reflectivity curve from (3).

3. EXPERIMENTAL

Standing wave experiments are usually performed at synchrotrons. However, the experimental arrangement, with some sophisticated modification, is essentially the same as proposed by Batterman (1964). In Fig. 1 we show a schematic of his experimental setup. This is the familiar parallel double crystal arrangement with the one added feature that a fluorescence detector is placed in front of the crystal to record the fluorescence yield, which is proportional to the standing wave field, as the crystal is rocked through its Bragg position. Since the reflectivity occurs over a very small angular range, drift of the reflection curve during the long fluorescence counting times is a severe problem. To counter this, feedback schemes analog or digital have usually been employed. This consists of rocking the crystal at a fixed frequency, say 1 Hz, and applying a correction signal after each sweep to correct for inevitable drift. The fluorescence data are collected automatically in a suitable multi-channel detector and the reflectivity recorded simultaneously.

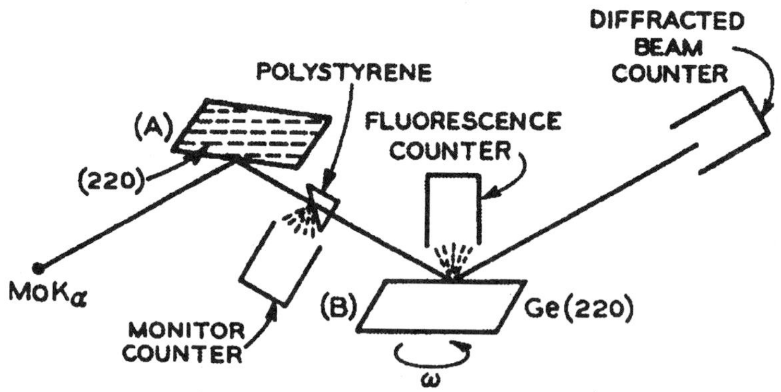

Figure 1. Parallel arrangement of double-crystal spectrometer with asymmetrically cut first crystal.

At this stage it is useful to ask what the standing wave experiment actually measures. The simplest way to express the standing wave experiment is that it measures the amplitude and phase of the relevant Fourier component of the charge density. In the usual diffraction experiment, we measure only elastically scattered intensities and the phase information is lost. In the standing wave experiment we record, in addition to the elastically scattered reflected beam, the inelastic fluorescence response during reflection. Resort to an inelastic channel preserves the phase information necessary for atom location.

Let us illustrate this by an example. Consider the (111) planes of the diamond structure such as silicon. The double nature of these planes is well known and their spacing is d/4 of the (111) spacing. If we consider the origin to be at the middle of the double plane, the atomic planes for silicon are roughly at ± 0.4Å from the origin. We can, therefore, express the electron charge density in the z direction as a Fourier series

$$\rho(\mathbf{z}) = \sum_{H} \mathbf{A}_H \exp(2\pi i \mathbf{H} \cdot \mathbf{z}) \tag{12}$$

where $\mathbf{A}_H$ is the **H**th Fourier coefficient and **H** the reciprocal lattice vector. We express the Fourier coefficients in the usual way

$$\mathbf{A}_H = \int_{-d/2}^{d/2} \rho(\mathbf{z}) \exp(-2\pi i \mathbf{H} \cdot \mathbf{z}) dz \tag{13}$$

Following the usual procedure, one can represent the electron density $\rho(\mathbf{z})$ by delta functions

$$\rho(\mathrm{z}) = a_1 \delta(\mathrm{z} + 0.4) + a_2 \delta(\mathrm{z} - 0.4) \tag{14}$$

where $a_1 = a_2 = 1/2$ is the atomic fraction on each plane. Since we are considering only two planes, we can replace the integral in Equation 13 by a sum and write for the (111) Fourier coefficients

$$A_{111} = a_1 \delta(z + 0.4) \exp(2\pi i 0.4 H) + a_2 \delta(z - 0.4) \exp(-2\pi i 0.4 H) \tag{15}$$

Expressing the exponentials in Equation 15 in cosine and sine terms and remembering that $H_{111} = 2\pi / d \approx 2$, we obtain

$$A_{111} = \cos[2\pi(\Delta d / d)] = \cos(0.8) = 0.7 \tag{16}$$

This is the amplitude of the Fourier component of the charge density Δd = 0.4 and d = 3.14; the phase is 0 or 1 modulo 2π since the sine terms drop out for charge planes located symmetrically from the origin. A_{111} is the quantity F, called the coherent fraction in Equation 7.

The experimental evidence for just such a situation is shown in Figure 2 (Patel et al., 1986). A monolayer of germanium was deposited on a clean silicon surface by molecular beam epitaxy (MBE) and diffused into the silicon crystal at 900° C for 0.5 hr. The measurement shows that phase *P* or POS = 1 in agreement with the result of the Fourier analysis, whereas the amplitude or the coherent fraction F = 0.67, which is close to the amplitude A_{111} term in the analysis 0.7. If we correct for the thermal vibration amplitude, the agreement is exact within F = ± .01.

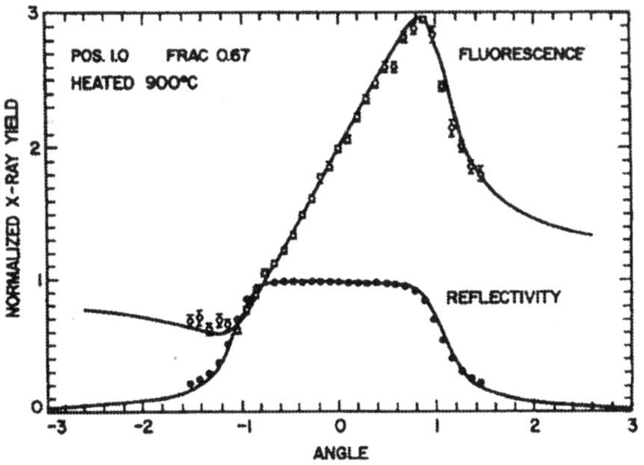

Figure 2. Reflectivity and fluorescence yield for Ge diffused into Si at 900°C for 0.5 h. Solid curves are fits by dynamical theory.

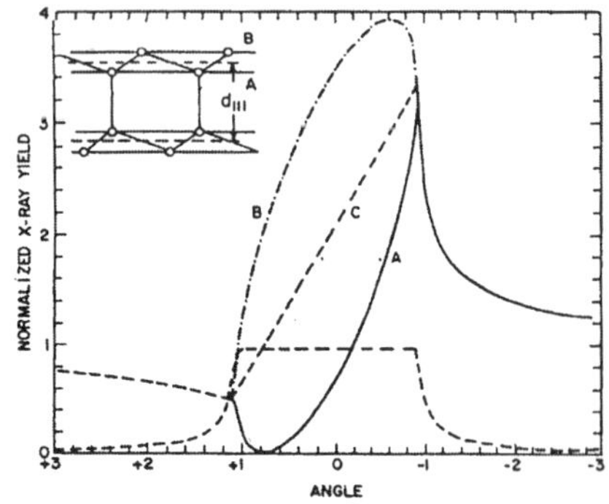

Figure 3. Theoretical fluorescence yield curves and Darwin reflectivity (no absorption for GaAs (111). The yield curves A & B correspond to planes shown in the inset.

Before confronting the actual standing wave experimental results, we need to familiarize ourselves with the nature of the fluorescence yield for the diamond structure and obtain a feel for the accuracy with which atoms can be located in crystals or at surfaces or interfaces. In Figure 3 (Patel and Golovchenko, 1983) we show the ideal fluorescence response from the double (111) planes of the diamond structure assuming no absorption (Darwin case). The ideal Darwin reflectivity is shown as a function of the dimensionless angular parameter η. If the atoms on planes A and B in the inset Figure 3 were, for example, Ga and As atoms, respectively, then the fluorescent responses for A and B are indicated by the solid curve A and the dot-dashed curve B. The Ga fluorescence A shows a dip, while the As atoms B show a peak in fluorescence response. This can be readily understood by considering the standing wave position as we sweep through the Bragg reflection. Dynamical theory predicts that on the low angle side of the Darwin reflectivity, nodes of the standing wave field coincide with the dashed line (inset) between planes A and B. As reflection proceeds and we move to higher angles, the nodes move down into the crystal and pass through plane A which results in zero field intensity at A and, hence a dip in the fluorescence intensity. Somewhat later the antinodes pass through plane B which results in a maximum in the fluorescence yield. If the atoms on planes A and B are the same as is the case for Ge diffused in Si Figure 2, the response will be the average of A and B shown by curve C, which is identical in form to the experimental yield in Figure 2. The position values P are normalized to the d_{111} spacing shown in the inset Figure 3. Hence, $P = 1$ corresponds to d_{111}. For plane B, $P = 1.125$ since B is displaced $d_{111}/8$ from the midposition. Similarly, $P = 0.875$ corresponds to plane A. One can readily see that the fluorescence intensity differences between atomic planes A and B (d/4) apart can be as high as a factor of 100. Under average experimental conditions, the position vector P can be determined with an accuracy $P = \pm.01$ or .01 of the d spacing. In what follows we will observe many examples of similar fluorescence responses under actual experimental conditions.

4. ATOM LOCATION AT SURFACES

While the standing wave technique is applicable to atom location studies at surfaces, interfaces and in bulk crystals, we shall confine ourselves to *contributions that the standing wave method has made to the study of surfaces*. It must be emphasized at the outset that because the fluorescence yield in a pure crystal comes from many thousands of atomic layers the signal from the surface is almost impossible to isolate. Even such other signal channels as electron yield or Auger electrons involve several atomic layers, and the surface signal is masked by the background. Thus, native reconstructed surfaces cannot directly be studied by the standing wave method. In the studies to be described, we will deal exclusively with *adsorbates species at surfaces different from the underlying atoms*. With the proper choice of adsorbate systems, unambiguous information on surface atom location can be obtained. Amongst the many adsorbates we have investigated, arsenic on Si (111) has yielded such an unambiguous adsorbate site and the agreement with theoretical calculations the closest that we have obtained.

4.1. Arsenic on Si (111)

Except for some early experiments in UHV with a rotating anode source, the bulk of our surface science experiments were performed at a specially dedicated UHV x-ray standing wave beam line X-15A at NSLS Brookhaven. The UHV chamber was capable of operation in the high 10^{-11} torr range. Samples could be cleaned and characterized by Auger and LEED. Known amounts of adsorbates could be deposited in a separate deposition chamber and the sample transferred to the x-ray chamber where a specially adapted stage could orient the sample accurately for Bragg reflection from planes parallel to the surface or other crystallographic planes with normals inclined to the surface. In all these operations, specimens remained in a UHV environment.

The incident x-ray beam was prepared in a separate double crystal monochromator chamber operating in a helium atmosphere. The first monochromator crystal was asymmetrically cut and served a dual purpose of narrowing the incident angular width on the specimen and spreading out the heat load from the white beam. With this arrangement, we obtained even for the most sensitive 400 reflections from silicon narrow, well-defined flat-top reflection curves.

An important consideration in applying x-ray standing wave methods to problems of atom location in surface physics concerns the inherent lack of surface specificity of fluorescent or other secondary signals used to detect impurities. One must typically be assured by alternative surface techniques that all impurities being studied are at the crystal boundary rather than in the bulk. In the case studied here, such concerns are completely alleviated by recognizing the following general symmetry argument. Atomic locations occupied by impurities in the bulk crystal may generally be assumed on average to satisfy the symmetry group of the crystal. Thus, for example, atomic positions of impurities *in* a centrosymmetric crystal should yield average impurity positions with this same symmetry. For atoms *on* the surface, this symmetry is broken and average measured impurity positions need not satisfy the above constraints. More interestingly, if experimental measurements show that such unsymmetrical results are obtained, one is generally assured that the atoms are in a nonsymmetrical environment such as the crystal surface. The results we present below show exactly this behavior and no further demonstration that all the impurity atoms are at the surface is necessary.

The reflectivity and As fluorescence yield for a crystal prepared as described above is shown in Figure 4 (Patel et al., 1987). The experimental points, together with the error bar, are shown in the figure. The solid line represents the theoretical fit to the data calculated from x-ray dynamical theory. Both the position P and coherent fraction F are shown with F being defined as the fraction of atoms at the value of P indicated. For an unrelaxed bulk terminated upper half of the silicon (111) double plane, $P = 1.125$ and $F = 1$. With the origin midway between the (111) double planes $P = 1$ corresponds to the d spacing of the (111) planes. The experimentally determined value for the As position in Figure 1 is $P = 1.18$ and $F = 0.98$. From the P values indicated, the As atoms lie about 5% of the (111) d spacing above the position for perfectly terminated bulk silicon. The absence of a signal associated with As atoms at additional sites near plane (b) inset Figure 4, which is consistent with the symmetry of the bulk, is a manifestation of the general argument previously given that here we are only observing signals from atoms at the surface. The reason for the strong As maximum has been discussed earlier with reference to Figure 3.

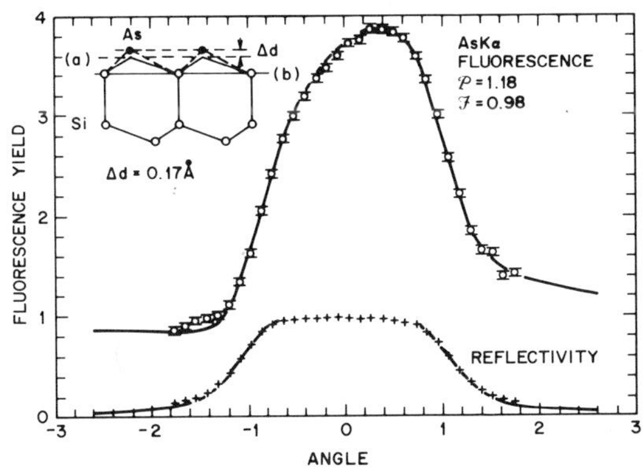

Figure 4. Reflectivity and fluorescence yield for As on silicon (111) deposited according to procedure outlined in the text. As coverage 0.93 ML, shown on a reduced angular scale. Full width at half-maximum of the reflectivity curve ≈ 3 s of arc.

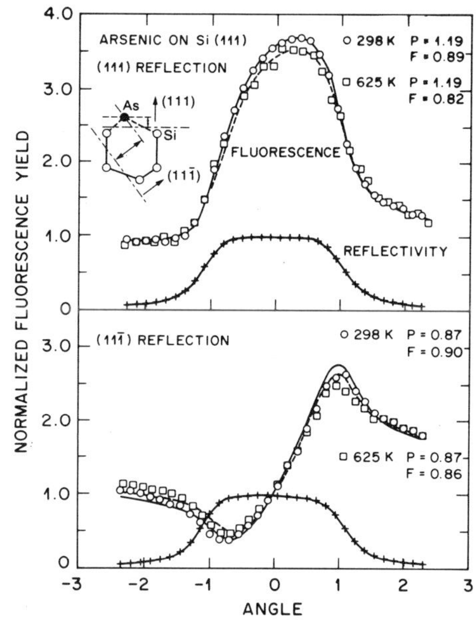

Figure 5. Reflectivity and fluorescence yield for As on Si(111). Surface registration established from P values for (111) and (11$\bar{1}$) reflections.

4.1.a. Solution to the Surface Registration Problem. The measurement in the previous section establishes the position of the As atoms normal to the (111) planes. No information on the lateral position in the As plane can be inferred from this measurement alone. If, however, the reciprocal lattice vector of the reflection does not lie perpendicular to the surface, the impurity can be located with respect to a plane that intercepts the surface at an angle. Such a measurement, together with the distance measured normal to the surface, constitutes a solution to the registration problem (Golovchenko et al., 1982) since by triangulation the impurity position along the surface in the direction of the reciprocal-lattice projection is completely determined.

Measuring the 11$\bar{1}$ reflection for As on Si (111), we obtain the result shown in the lower half of Figure 5. Ignoring for the moment the temperature effects, note now that the As fluorescence signal shows a dip whereas for the 111 reflection Figure 4 we observe a peak. The reason for the two very different fluorescence signals for the same As position can be understood readily from the [110] projection of the (111) planes Figure 6. For (111) the As atoms lie in the top half whereas for the (11$\bar{1}$) planes the As atoms lie in the bottom half of the respective (11$\bar{1}$) plane. We expect, therefore, from previous arguments, that the 11$\bar{1}$ reflection will show a dip in the fluorescence response. From the two measurements in Figures 4 and 5, the vertical and lateral position is determined unequivocally. The As atom moves upward from the extrapolated bulk Si position with no lateral shift.

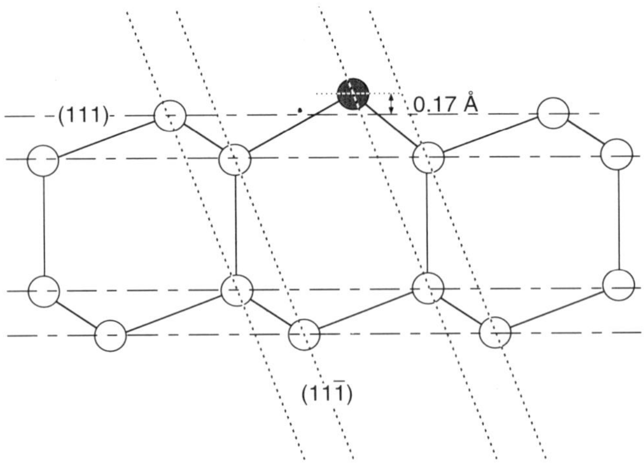

Figure 6. ($\bar{1}$10) projection of the (111) and (11$\bar{1}$) planes. Surface registration of As established from P_{111} and $P_{(11\bar{1})}$

We compare now our findings in Figure 4 with recent first-principles pseudopotential total-energy minimization calculations of the As terminated surface. Such sophisticated calculations are highly accurate in cases where the unit cell of the surface structure is simple, such as the (1 x 1) case for As on silicon (111). Assuming that this is the case, as borne out by LEED observations, Uhrberg et al. (1987) have calculated the As position on an As terminated silicon surface. His value of 0.19 Å for Δd (see inset in Figure 4) can be compared to our experimental x-ray standing wave result of Δd = 0.17 ±0.03 Å. All displacements are indicated in Å from the bulk terminated silicon positions. In a more recent calculation, Hybertsen and Louie find a value of 0.16 Å for Δd. Our experimental value of 0.17 Å lies between the theoretical estimates. Both theoretical results are well within experimental uncertainty and agree with each other within the expected precision of these calculations for surface geometry.

4.2. Gallium on Silicon (111)

In contrast to the relatively straightforward As result, the situation for Ga on Si(111) is much more complex. In a classic paper on trivalent metals on Si(111) Lander and Morrison (1964) point out: "The structural chemistry of these systems is the most complex so far encountered in low-energy electron diffraction." At low coverages up to 1/3 ML, Ga on Si(111) exhibits a $\sqrt{3}x\sqrt{3}$ reconstruction as demonstrated by the LEED results. An explanation for this result in line with intuitive expectations, based on gallium's trivalency, would place the gallium atoms above three silicon atoms, themselves the last plane of a (111) silicon double layer. Such an arrangement of gallium atoms completely terminates all silicon surface dangling bonds with a coverage of 1/3 ML. While alternative structures may be envisioned, even this simple adatom model has nontrivial alternatives for the gallium atom positions because there are two types of threefold triangular sites above the surface. The first, the so-called hollow or H_3 site, has no atom directly below in the lower part of the (111) silicon surface double layer. The other site, the T_4, is also triangular but contains a silicon atom, directly below in the bottom of the double layer.

The T_4 model is very nonintuitive. If lattice relaxations are not allowed, then an adatom in a T_4 site is prevented, by the presence of the second layer atom, from being close enough to the first layer atoms to form strong bonds. However, for the $Si(111)-(\sqrt{3}x\sqrt{3})$ Al surface, first-principles total energy calculations (Northrup, 1984) predicted that, in fact, the T_4 site was lower in energy than the H_3 site by 0.3 eV per adatom. Although there is now strong evidence, based on transmission electron diffraction and tunneling microscopy, that Si adatoms on the Si(111) (7×7) surface occupy the T_4 sites, no such conclusive experimental determination of the bonding site has been made for the Group III adatoms on Si(111). Thus, the generality of the T_4 site for adatom bonding on Si(111) is an open question.

After deposition of 0.1 ML of Ga on a clean Si(111) surface in UHV, the specimen was transferred in UHV to the x-ray chamber. The position and structural perfection of the gallium layer were measured using x-ray standing waves (Martinez et al., 1992). The sample was annealed at 840 K until the gallium overlayer developed an extremely high room temperature coherent fraction F_{111} (normal to the surface) of 0.98 ± 0.01. The position of the layer, P_{111}, was 0.58 ± 0.01 which corresponds to a mean gallium atom height of 1.46 ± 0.03 Å above the upper half of the bulk extrapolated Si(111) double layer, in excellent agreement with earlier standing wave results (Zegenhagen et al., 1988). These same measurements were carried out for the $(11\bar{1})$ plane which is inclined at 70.53 deg to the (111) surface. $F_{11\bar{1}}$ and $P_{11\bar{1}}$ were found to be 0.99 ± 0.01 and 0.34 ± 0.01, respectively. The standing wave raw data, fits, and atomic position are shown in Figure 7. The measurement using both atomic planes allows a unique quantitative assignment of the gallium atoms to the T_4 site.

Total energy calculations for group III elements on Si(111) $\sqrt{3}x\sqrt{3}$ have been described (Zegenhagen et al., 1989). The local-density-functional approach employing a momentum space pseudopotential formalism was used. Centrosymmetric supercells containing eight layers of silicon and two of Ga were employed. Equilibrium geometries for the two possible surface structures (T_4 and H_3) were obtained using the Hellman-Feynman theorem to find the minimum energy configuration for both. The T_4 structure was found to be lower in energy than H_3 by 0.38 eV per adatom. Substantial surface relaxations are predicted for both models and play the major role in determining the lower energy of the T_4 site. Measured relative to the bulk extrapolated silicon surface plane, the vertical height of the Ga in the T_4 site is 1.34 Å or 0.16 Å lower than the x-ray

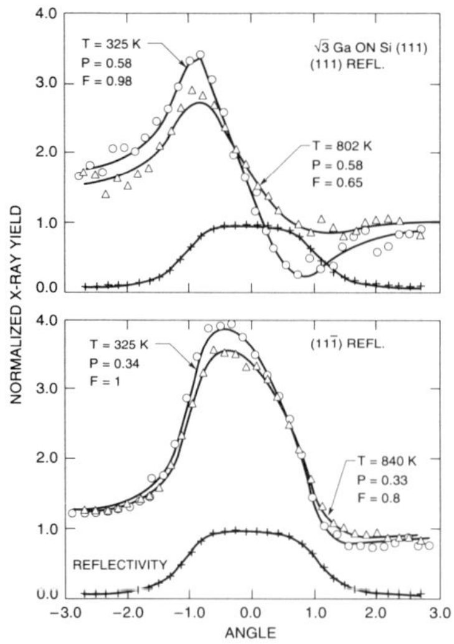

Figure 7. Reflectivity and fluorescence yield for Ga on Si(111) $\sqrt{3} \times \sqrt{3}\,R30$. (a) (111) reflection. (b) $(11\bar{1})$ reflection.

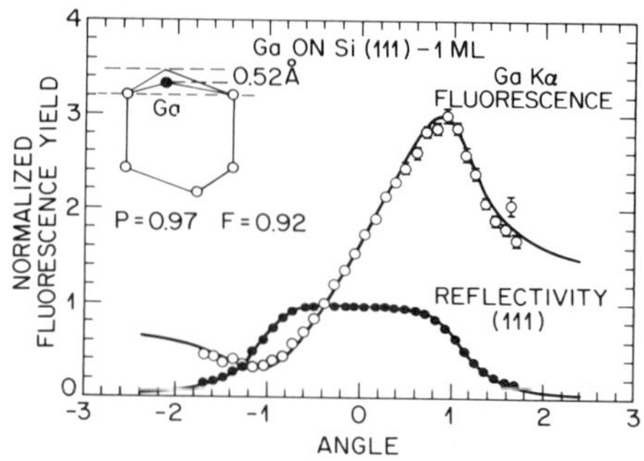

Figure 8. Reflectivity and fluorescence yield for Ga on Si(111) deposition temperature $T \approx$ 580°C. The $(\bar{1}10)$ projection of the (111) planes showing the Ga position is indicated in the inset.

determination. The calculated H_3 position is even lower and corresponds to a height of 1.17 or 0.34 Å below our measured value. It is significant that for As on Si(111) where the surface is (1×1) or unreconstructed, the agreement between theory and experiment is exact. For Ga on Si(111) $\sqrt{3} \times \sqrt{3}$ the reconstruction is strong with heavily distorted bonds between Si and Ga on the T_4 site. The agreement with theory is quite unsatisfactory, and further theoretical understanding and development appears necessary.

For higher coverages there is a dramatic shift of the Ga sites from the T_4 to substitutional sites on the topmost silicon (111) double plane (Zegenhagen et al., 1988). The reflectivity and fluorescence yield for a Si(111) crystal with approximately 1 ML Ga is shown in Figure 8. The position value P = 0.96 indicates that Ga atoms occupy a position slightly below the middle of the extrapolated double (d111) surface plane. All the Ga atoms are in highly coherent sites as shown by the high coherent fraction F = 0.92. The coherent fraction up to 1 ML depends on deposition conditions. Above 1 ML the coherent fraction decreases with increasing coverage.

The deduced average Ga position is 0.10 Å below the extrapolated last maximum in the (111) Fourier component of the silicon charge density at the surface which lies between the closely spaced (111) double plane (see dashed line inset Figure 4). It is also 0.50 ± 0.02 Å below the top half of the extrapolated silicon (111) double plane as indicated in the inset of Figure 8. The strong inward relaxation in the threefold substitutional site is understood in terms of the rehybridization of the Ga atoms from sp^3 to sp^2 enhanced by a small outward relaxation of the subsurface Si atoms. Thus the outer double layer of Si and Ga atoms relaxes towards a planar sp^2 bonded network.

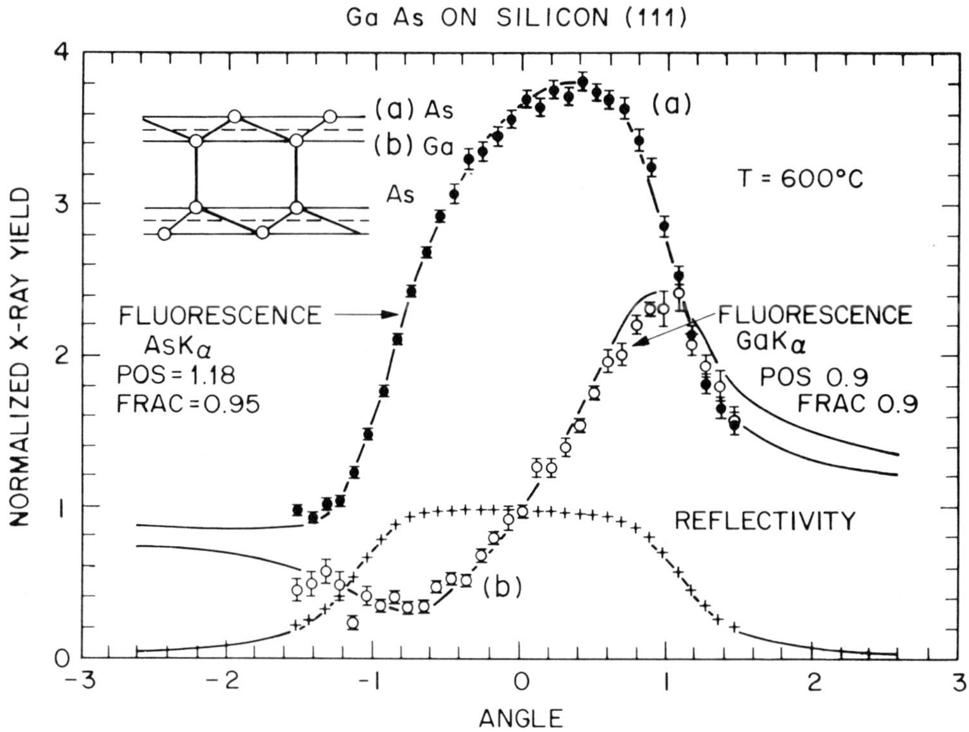

Figure 9. Reflectivity and fluorescence yield for 0.5 ML of GaAs epitaxially grown on a Si(111) surface. Mo *Ka* radiation 111 reflection.

4.3. GaAs on Silicon (111)

When a beam of Ga and As atoms is incident on clean Si(111) one might well ask which atom occupies the top (111) layer. X-ray standing wave measurements (Patel et al., 1987) can provide a definitive answer to this question as demonstrated in Figure 9 (Patel et al., 1987). For a perfect termination of the bulk silicon lattice [plane (a) in inset], P = 1.125 and F = 1.0. For As, the experimental result is P = 1.18 and F = 0.95. Thus, on average, the As atoms occupy a position ≈ 5% higher relative to the Si(111) *d* spacing than the perfectly terminated silicon lattice. The coherent fraction 0.95 indicates the overwhelming tendency for As atoms to lie in the *upper half* of the (111) double layers.

The position and coherent fraction for the Ga atoms are also indicated in Figure 9. The position of the lower half of the (111) plane in the perfectly terminated silicon bulk lattice is P = 0.875 and F = 1.0 for Ga; the experimental result is P = 0.90 and F = 0.90. Thus, on average, the Ga position is ≈ 3% higher than that for a perfectly terminated silicon position relative to the Si(111) *d*-spacing and the Ga atoms occupy mainly the *bottom half* of the (111) double plane. From the F value observed we conclude that at most 5%–10% of Ga atoms can lie on the upper half of the (111) plane (a) inset Figure 9.

REFERENCES

Andersen, S.K., Golovchenko, J.A., and Mair, G., 1976, New applications of x-ray standing wave fields to solid state physics, *Phys. Rev. Lett.* 37:1141.

Batterman, B.W., 1964, Effect of dynamical diffraction in x-ray fluorescence scattering, *Phys. Rev.* 133:759.

Bormann, G., 1941, Uber extinktionsdiagramme von quarz, *Physikal. Zeit.* 42:157.

Golovchenko, J.A., Brown, W.L., and Batterman, B.W., 1974, Observation of internal x-ray wave fields with an application to impurity lattice location, *Phys. Rev.* B10:4239.

Golovchenko, J.A., Patel, J.R., Kaplan, D.R., Cowan, P.L., and Bedzyk, M.J., 1982, Solution to the surface registration problem using x-ray standing waves, *Phys. Rev. Lett.* 49:560.

Hertel, N., Materlik, G., and Zegenhagen, J., 1985, X-ray standing wave analysis of bismuth implanted in silicon (110), *Z. Phys.* B58:1999.

Lander, J.J., and Morrison, J., 1964, Surface reactions of silicon with aluminum and with iridium, *Surf. Sci.* 2:553.

Martinez, R.E., Fontes, E., Golovchenko, J.A., and Patel, J.R., 1992, Giant vibrations of impurity atoms on a crystal surface, *Phys. Rev. Lett.* 69:1061.

Northrup, J.E., 1984, Si (111) √3×√3 al: An adatom-induced reconstruction, *Phys. Rev. Lett.*, 53:683.

Patel, J.R., and Golovchenko, J.A., 1983, X-ray-standing-wave atom location in heteropolar crystals and the problem of extinction, *Phys. Rev. Lett.* 50:1858.

Patel, J.R., Freeland, P.E., Golovchenko, J.A., Kortan, A.R., Chadi, D.J., Quian, Guo-Xin, 1986, Normal displacements on a reconstructed silicon (111) surface: an x-ray-standing wave study, *Phys. Rev. Lett.* 57:3077.

Patel, J.R., Golovchenko, J.A., Freeland, P.E., and Gossmann, H-J, 1987, Arsenic atom location on passivated silicon (111) surfaces, *Phys. Rev.* B36:7715.

Patel, J.R., Freeland, P.E., Hybertsen, M.S., Jacobson, D.C., and Golovchenko, J.A., 1987, Location of atoms in the first monolayer of gaas on si, *Phys. Rev. Lett.* 59:2180.

Uhrberg, R.I.G., Bringans, R.D., Olmstead, M.A., Bachrach, R.Z., and Northrup, J.E., 1987, Electronic structure, atomic structure and the passivated nature of the arsenic-terminated si(111) surface, *Phys. Rev.*, B35:3945.

Vlieg, E., Fontes, E., and Patel, J.R., 1991, Structure analysis of si (111)–(√3×√3)R30°/ag using x-ray standing waves, *Phys. Rev.* B43:7185.

Zegenhagen, J., Hybertsen, M.S., Freeland, P.E., and Patel, J.R., 1988, Monolayer growth and structure of ga on si (111), *Phys. Rev.* B38:7885.

Zegenhagen, J., Patel, J.R., Freeland, P.E., Chen, D.M., Golovchenko, J.A., Bedrossian, P., and Northrup, J.E., 1989, X-ray standing wave and tunneling microscope location of gallium atoms on a silicon surface, *Phys. Rev.* B39:1298.

X-RAY STANDING WAVE STUDIES OF BULK CRYSTALS, THIN FILMS AND INTERFACES

S. Lagomarsino

Istituto Elettronica Stato Solido - CNR
V. Cineto Romano, 42 - 00156 Roma (Italy)

1. INTRODUCTION

The basic principles of the x-ray standing wave technique and its application to surface studies have been treated by J. R. Patel in this same volume. In this chapter applications to bulk crystals, thin films and interfaces will be described. Some interesting perspectives coming from non-standard geometries such as back-reflection, grazing incidence and multiple diffraction will be also given.

2. BULK CRYSTALS

The first demonstration of the possibility to detect the e.m. field modulation due to the standing wave pattern has been given by Batterman (1964) on a Ge bulk crystal. The same author (Batterman, 1969) applied this technique to localization of As atoms uniformly distributed in silicon crystals. Later on Golovchenko et al. (1974) made similar measurements with a slight but important difference, consisting in performing the experiment on a Si crystal where the As atoms were contained within a distance from the surface that was smaller than the extinction length for the probing X-rays. One of the main problems in applying the x-ray standing wave technique to bulk crystals is in fact the extinction effect: when the crystal is in diffraction, the penetration of x-rays is strongly dependent on the incidence angle, due to the depletion of the incident intensity in favor of the diffracted one, and has a minimum at the diffraction peak, as discussed in the chapter by Authier (this volume). The total absorption coefficient of the e.m. field μ_{tot} can be expressed therefore by the sum of two terms: $\mu_{tot} = \mu_{ph} + \mu_{ext}$, where μ_{ph} is the normal photoelectric absorption and μ_{ext} is related to the extinction effect. Moreover, the attenuation by the crystal matrix of the secondary radiation excited by the standing wave must also be taken into account. The fluorescence yield at any depth t in the crystal is therefore given by:

$$Y(\theta,t)= C\exp\left[-(\mu_{tot} + \mu_F / sin\beta)t\right]\left[1+R(\vartheta)+2\sqrt{R(\theta)}F\cos(\phi - 2\pi P)\right] \qquad (1)$$

where θ is the deviation from Bragg angle, C is a proportionality constant, μ_F is the absorption coefficient of the fluorescence radiation, β is the angle between the crystal surface and the center of the detector for fluorescence, R the reflectivity, F and P respectively the coherent fraction and the coherent position of the atoms, and Φ the phase of the complex quantity E_H/E_0, $E_0$and E_H being respectively the diffracted and the incident wave fields (see the chapter by Patel, this volume). In order to obtain the total fluorescent intensity, $Y(\theta,t)$ must be integrated in the whole thickness T where the scattering centers are present. The exponential term, the only where the thickness t is present, becomes:

$$-\frac{1}{\mu_{tot} + \mu_F / sin\beta}\left\{ \left[\exp-(\mu_{tot} + \mu_F / sin\beta)T\right]-1 \right\} \qquad (2)$$

If the exponent is much less than 1 then the fluorescence intensity is simply proportional to the thickness T. On the other side, if $\mu_F/\sin\beta$ is much larger than μ_{tot} then the extinction effect will be negligible because the depth from which the fluorescence originates will be much smaller than the extinction length. In all the other cases, the extinction effect will contribute with a characteristic modulation which is superposed on the standing wave modulation due to the nodes and antinodes movement. It is clear that the same arguments hold for photoelectrons and for any other secondary effects excited by the standing wave field.

A good example capable to illustrate many features of the XSW method applied to bulk crystals is that of garnet crystals (Lagomarsino et al., 1984). Garnets have a quite complex structure with a unit cell containing eight formula unit of the type $A_3B_2C_3O_{12}$ where A, B and C indicate sites with different coordination with oxygen, respectively dodecahedral (A), octahedral (B) and tetrahedral (C). In Gadolinium Gallium Garnets (GGG) the Ga occupy octahedral and tetrahedral sites, while Gd the dodecahedral sites only. A distinct difference between Ga and Gd fluorescence modulation has been demonstrated with a (444) reflection (Lagomarsino et al. 1984). In fact in this case the atoms in dodecahedral sites and in tetrahedral sites coincide with the diffracting planes, while the atoms in octahedral sites are displaced by d/2 with respect to the (444) diffracting planes, where d is the interplanar distance. The fluorescence yield for Ga atoms is therefore the sum of two contributions out of phase between them, weighted following the concentration of Ga in the two sites, i.e. 3:2. In this first application of the XSW method to garnets (Lagomarsino et al., 1984), the measurements were strongly affected by primary extinction effect. In order to overcome this problem, in a subsequent experiment the term $\mu_F/\sin\beta$ in eq. 1 has been made larger by looking at the crystal surface with the detector mounted at a grazing outgoing angle (small angle β). See also Patel and Golovchenko (1983). Figure 1 shows the Ga fluorescence yield (closed points: experimental; full line: theoretical) together with the reflected intensity with the detector placed in front of the sample (see inset).

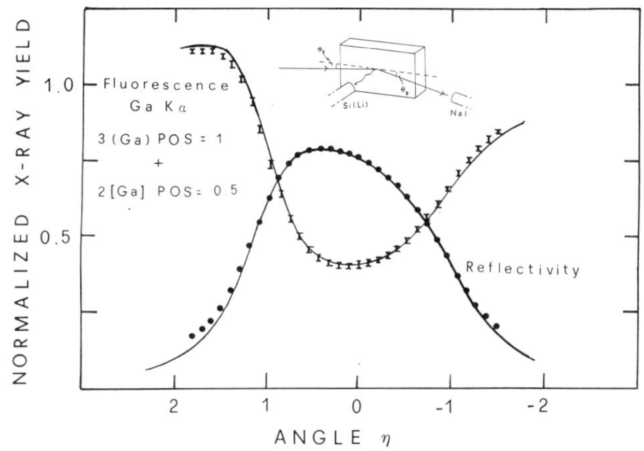

Figure 1. Reflectivity and fluorescence intensity modulation of the Ga Kα radiation for the (444) Mo Kα reflection with the Si(Li) detector placed in front of the sample(see inset)

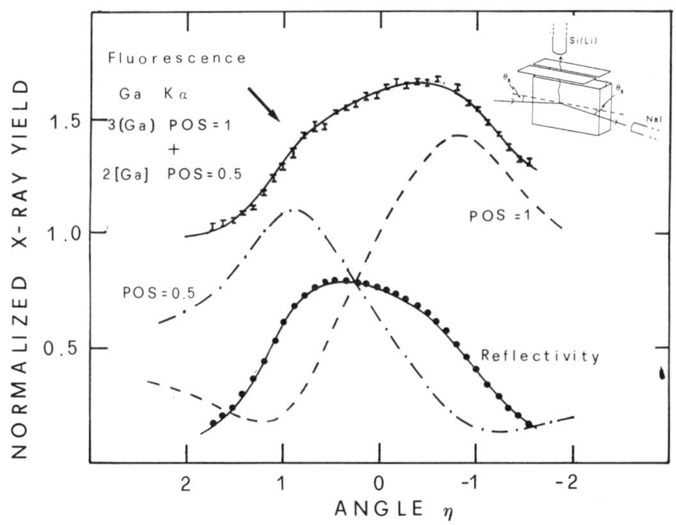

Figure 2. Reflectivity and fluorescence modulation of Ga Kα radiation with the Si(Li) detector placed to catch only the fluorescence coming out of the sample at a glancing angle (see inset). The theoretical individual curves for octahedral and tetrahedral sites are shown.

Figure 2 shows instead the same Ga fluorescence yield, but with the solid state detector positioned so that only the fluorescence radiation which comes out of the sample at a very grazing angle β with respect to the crystal surface can reach the detector (see inset). In this way the depth from which the fluorescence can come is strongly limited, and the extinction effect has less influence on the fluorescence yield, as can be easily seen comparing figure 1 with figure 2. In this last the theoretical curves for octahedral and tetrahedral sites are shown separately. It is interesting anyway to note that theory can account in a very accurate way for the extinction effect.

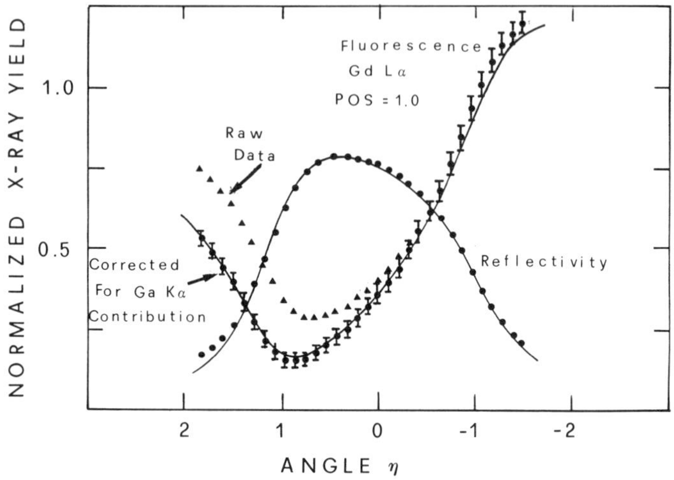

Figure 3. Reflectivity and fluorescence intensity modulation of the Gd Lα radiation for the (444) Mo Kα reflection for the same geometry as figure 1. The results obtained with (closed circles) and without (triangles) the correction for secondary interelement fluorescence effect are shown.

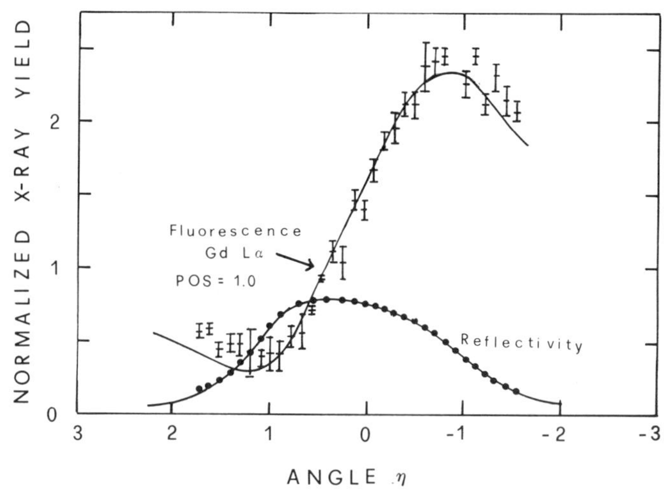

Figure 4. Reflectivity and fluorescence intensity modulation of the Gd Lα radiation for the (444) Mo Kα reflection for the same geometry as figure 2. The correction for secondary interelement fluorescence effect has been carried out.

To analyze the Gd fluorescence we have to take into account another effect that can be very important in some cases: the secondary interelement fluorescence effect, i.e. the excitation of an element (in this case Gd Lα) by the fluorescence radiation of another element (in this case Ga Kα). If the two elements occupy different sites the angular modulation can be strongly affected. This effect can be taken into account in the simulation. Figure 3 shows the Gd fluorescence yield as seen with the solid state detector placed in front of the sample, with and without the correction for the interelement fluorescence effect. As can be seen the agreement, with the correction, is excellent. Figure 4 reports the same Gd fluorescence but with the other geometry, allowing to overcome the problem due to the primary extinction.

Other methods have been proposed to bypass the primary extinction problem: Bedzyk, Materlik and Kovalchuk (1984) choosed to detect, instead of fluorescence, the photoelectrons excited by the standing wave field, which have a strong absorption coefficient. This method has been extensively applied to studies of perturbed crystalline surface (Kovalchuk et al., 1986) using gas-proportional photoelectron detectors that have a very large acceptance angle but poor energy resolution. An interesting method has been developed by Mukhamedzanov et al., (1991). They measured in time coincidence both the photoelectrons and the fluorescence radiation, combining in this way the capability of fluorescence to be atomic specific and the high absorption coefficient of photoelecrons, thus probing only a small depth of the crystal. They applied this method to study impurity in garnet crystals.

The interest in bulk crystal studies relies on the fact that through the x-ray standing wave analysis both the modulus *and* the phase of the structure factor can be directly determined. Authier (1986) has shown, for example, that the asymptotic limit of the position of the nodes far from the reflection region is given by the phase of the structure factor. In non-centrosymmetric crystals also the position of the nodes at the centre of the reflection region is strongly absorption dependent, giving the possibility to precisely measure the imaginary part of the structure factor (see also Bedzyk and Materlik, 1985).

Another field of interest is related to ion implantation: in many cases it is very important to know the exact location of the implanted ions and the deformation profile of the substrate. An XSW analysis can yield information about location and local deformation (Hertel et al., 1985), or can implement the information generally obtained with x-ray diffraction, enabling to distinguish between different distortion profiles that give the same kinematical diffraction curve (Bocchi et al., 1994).

3. THIN FILMS AND BURIED INTERFACES STUDIES

One of the most interesting properties of the XSW is their ability to give information about the structure of buried interfaces. One of the first applications in this field dealt with silicide films epitaxially grown on silicon substrates (Vlieg et al., 1986; Lagomarsino et al.,

1989). In this case the main problem was the determination of the coordination of the metal atoms with Si. The analysis of the XSW signal for a thin film is more complex than for an interface composed of one monolayer, because one have to take into account the mismatch between substrate and film. The XSW pattern has the periodicity of the lattice planes of the substrate, and in a first approximation we can assume that the presence of the film doesn't disturb this pattern (as we will see in the following this assumption is valid only in certain conditions). The positions of the overlayer atoms with respect to the standing waves nodes therefore will be different from layer to layer. Assuming that the fluorescence is taken as probe, the fluorescence yield Y(θ) is still given by equation (1), but the contribution of the different layers must be summed up, each with its own position with respect to the standing wave field. Assuming a sharp interface and a constant lattice parameter of the overlayer, the parameters P and F can be expressed as:

$$P = \frac{Z}{ds} + \frac{N-1}{2}\,\delta d/d_s \qquad \text{and} \qquad F = fc\,\frac{sin(\pi N\delta d/d\ )}{Nsin(\pi\delta d/d\ )} \tag{3}$$

d_s is the interplanar distance of the relevant substrate planes, $\delta d/d_s$ the lattice mismatch, N the number of planes in the overlayer and f_c a parameter related to the static and dynamic disorder. Z, the most interesting parameter related to the coordination of metal atoms with silicon, is the distance between the last silicon and the first metal layers. The extinction effect will not be taken into consideration because the thickness of the epitaxial films is in general much less than the extinction depth, and the fluorescence intensity will be simply proportional to the film thickness. The Co and Ni silicides both have a fluorite structure and have been demonstrated to grow coherently on (111) Si substrates with two possible orientations: the same orientation as the substrate (case A) or rotated by 180° with respect to the surface normal (case B). $CoSi_2$ grows generally with B orientation, while Ni silicide can grow with A or B orientation depending on the growth conditions. Three different models have been proposed for the coordinations of the metal atoms at the interface: fivefold, sevenfold and eightfold. Figure 5 shows the scheme of the three models for Co silicide, and figure 6 shows an example of experimental result (Lagomarsino et al., 1989) on a film with N = 16 ± 1. In this specific case the lattice mismatch has been independently measured with X-ray diffraction. The closed points are the experimental data and the full line refers to a best fit with P = 0.77 ± 0.1 and F = 0.35 ± 0.05. Considering the values of N and $\delta d/d_s$, the experiment gives a value of the Z parameter (i.e. the distance between the Si diffracting planes and the Co atoms) of Z = 2.79 Å, to be compared with the theoretical values for fivefold and eightfold coordination (Z = 2.75 Å), and for sevenfold coordination (Z = 3.52 Å). In figure 6 the expected curve in the case of a sevenfold coordination is shown with the dashed line. The result from XSW clearly excludes the possibility of a sevenfold coordination, but unfortunately is not able to distinguish between the fivefold and the eightfold ones which have the same distance Z. Similar results have been found by other groups on the same system, giving us confidence about the reliability of this method.

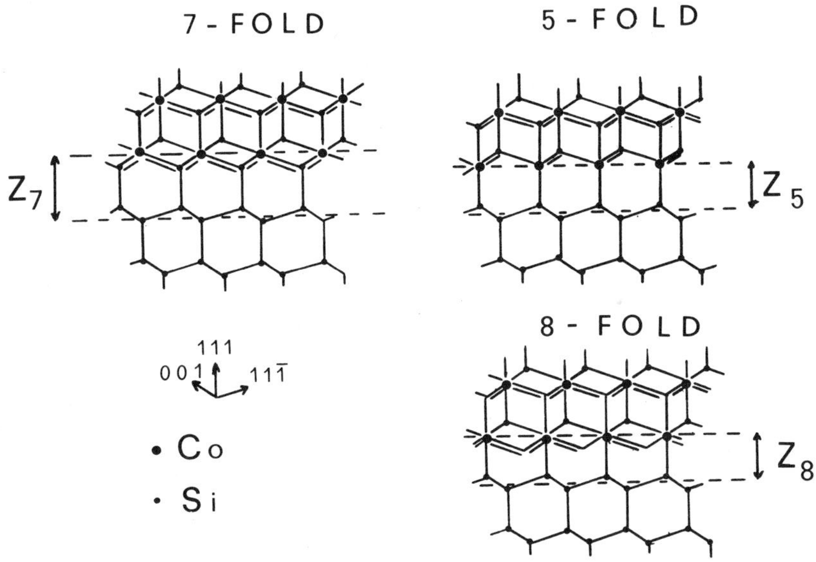

Figure 5. Schematic model of the type B $CoSi_2$/Si(111) interface: the sevenfold, fivefold and eightfold coordination of the metal atoms are shown at the interface.

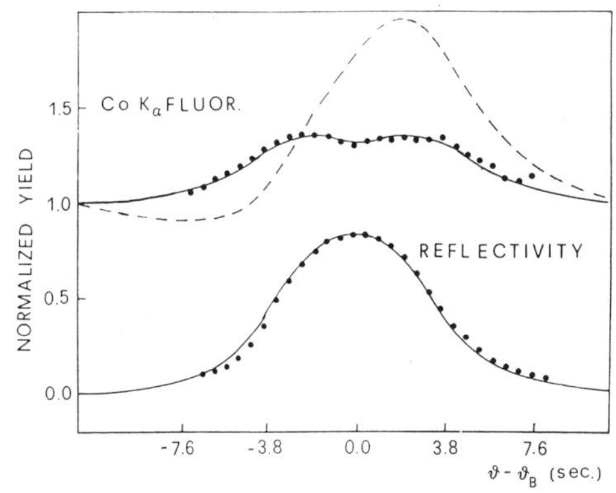

Figure 6. XSW experimental results (closed points) compared with simulation curves for F = 0.35 and P = 0.77 (full line) relative to a situation close to the fivefold or eightfold coordination and for F = 0.35 and P = 0.99 (dashed line) corresponding to the sevenfold ccordination.

In the above analysis it has been assumed that the standing wave pattern generated in the substrate is not influenced by the presence of the thin film. In fact Authier et al. (1989) have shown that this is true only for very thin films (of the order of ten unit cells). When the thickness is larger, the nodes are hooked to the substrate undeformed planes only at the exact Bragg angle. For the other angles the standing wave spacing is different from that of the substrate, but at the same time is never equal to that of the epilayer deformed planes. If the strain profile is known, the exact position of the nodes can be deduced, and the atomic positions can be measured as in the usual XSW analysis.

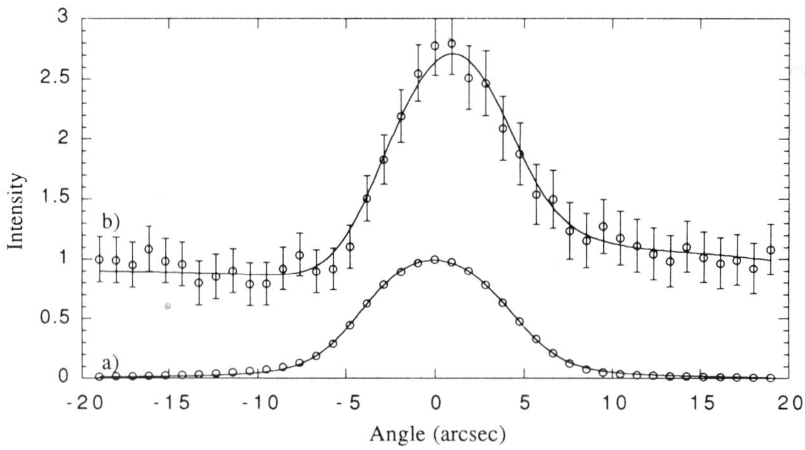

Figure 7. Experimental results (open circles) and simulation curves for In Lα fluorescence in a GaAs/InAs/GaAs interface recorded in correspondence with the (400) reflection of the GaAs substrate. The best fit results in values of P = 1.17 ± 0.02 and F = 0.58 ± 0.07.

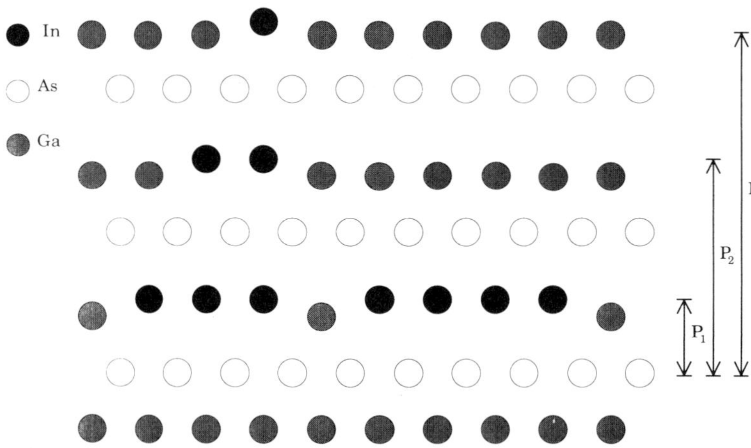

Figure 8. Schematic model of the GaAs/InAs/GaAs interface: from the XSW and XRD results a model with 75% of the total In atoms in the first position, 20% in the second one and 5% in the third one has been proposed.

Another example of the applications of the XSW technique to the analysis of interface structure is concerned with segregation studies on III-V heterointerfaces. Segregation phenomena are particularly important in semiconductor interfaces, because they can alter the composition profile of the heterostructure and as a consequence also its electronic properties. Giannini et al. (1993) have carried out an analysis on this subject combining XSW and high resolution x-ray diffraction experiments on a buried interface composed of 1.2 ML InAs sandwiched between a GaAs substrate and a GaAs cap layer 300 Å thick. Figure 7 shows the experimental result together with the best fit for the In fluorescence yield as a function of the incidence angle. The P and F values obtained by the best fit are P = 1.17 and F = 0.58. The low value of F can be explained in terms of multiple sites occupancy with respect to the standing wave periodicity.

In fact, if atoms can occupy different positions P_i each with probability f_i, the resulting F and P values in equation 1) are given by (f_c is the parameter related to the static and dynamic disorder):

$$\tan(2\pi P) = \frac{\sum_i f_i sin(2\pi P_i)}{\sum_i f_i \cos(2\pi P_i)} \tag{4}$$

$$F = f_c \left[\left(\sum_i f_i sin(2\pi P_i) \right)^2 + \left(\sum_i f_i \cos(2\pi P_i) \right)^2 \right]^{1/2} \tag{5}$$

In our specific case, considering the coverage measured with X-ray diffraction (1.2 ML), the experimental result can be interpreted by assuming that 75% of the total amount of In atoms are incorporated at the first position (P_1 = 1.15), 20% at the second (P_2 = 3.45) and 5% at the third one (P_3 = 5.75). A schematic model of the interface is shown in figure 8. A limited segregation therefore occurs in this case where a special procedure was adopted to limit segregation. It is interesting that detailed information about this important aspect of heterointerfaces can be obtained with the combined use of XSW and high resolution diffraction.

4. NON-STANDARD GEOMETRIES

Most of the XSW studies have been carried out on perfect crystals, because the lattice imperfections can alter the coherence of the wave fields inside the crystal. A method that makes use of reflections with $\theta_B = 90°$ has been proposed by Woodruff et al. (1987) to bypass this limitation. In fact at $\theta_B = 90°$ the rocking curve is quite unsensitive to crystal imperfections because its natural width is in the range of minutes of arcs. It is clear that equations valid also in these extreme cases must be used. Instead of an angle scan, an energy scan is used to move the standing wave nodes. In this way imperfect crystals such as metal crystals can be used also as substrates (Woodruff et al., 1987). Other advantages are that different Bragg planes can be studied using a tunable monochromator in synchrotron radiation sources and that no very accurate adjustments are necessary to obtain reliable data (Hashizume, 1992).

Another interesting geometry is the grazing incident conditions: in this way planes perpendicular to the surface can be put in diffraction, and their standing wave pattern can be used to get direct information about location of atoms in the plane (Cowan et al., 1986).

Finally, XSW measurements have been performed in multiple diffraction geometry: the first experiment dealt with simultaneous excitation of the 333 and 115 planes in a bulk Ge crystal (Greiser, 1987). Later on Kazimirov et al. (1993) performed experiments under the conditions of three beam 111/220 diffraction, detecting photoelectrons in order to overcome the problem of extinction. This geometry allows direct two-dimensional localization of atom positions in the crystal lattice.

REFERENCES

Authier, A., 1986, Angular dependence of the absorption-induced nodal plane shifts of x-ray stationary waves, *Acta Cryst.*, A42:414

Authier, A. , Gronkowski J. and Malgrange, C., 1989, Standing waves from a single heterostructure on GaAs- a computer experiment, *Acta Cryst.*, A45:432

Batterman, B.W., 1964, Effect of dynamical diffraction in x-ray fluorescence scattering, *Phys. Rev.* 133A:759

Batterman, B.W., 1969, Detection of foreign atom sites by their x-ray fluorescence scattering, *Phys. Rev. Lett.*, 14:703

Bedzyk, M.J., Materlik G. and Kovalchuk, M.V., 1984, X-ray standing wave modulated electron emission near absorption edges in centrosymmetric and noncentrosymmetric crystals, *Phys. Rev.* B30:2453

Bedzyk, M.J. and Materlik G. , 1985, *Phys. Rev.* B32:6456

Bocchi, C., Franzosi, P., Imamov, R.M., Lomov, A.A.., Maslov, A.V., Mukhamedzhanov, E.Kh. and Yakovchick, Yu.V., 1994, Investigation of lattice distortions in InP crystals implanted with Fe+ ions by means of high-resolution x-ray diffraction and x-ray standing-wave methods, *J. Appl. Phys.*, 76:7239

Cowan, P.L., Brennan, S., Jack, T., Bedzyk, M.J. and Materlik, G., 1986, *Phys. Rev. Lett.*, 57:2399

Giannini, C., Tapfer, L., Lagomarsino, S., Boulliard, J.C., Taccoen, A., Capelle, B., Ilg, M., Brandt, O. and Ploog, K.H., 1993, X-ray standing wave and high resolution x-ray diffraction study of the GaAs/InAs/GaAs(100) heterointerface, *Phys. Rev.* B48:11496

Golovchenko, J.A., Batterman, B.W., Brown, W.L., 1974, Observation of internal x-ray wave field during Bragg diffraction with an application to impurity lattice location, *Phys. Rev.* B10:4239

Greiser, N., and Materlik, G. ,1986, *Z. Phys.*, B66:83

Hashizume, H., Sugiyama, M., Niwa, T., Sakata, O. and Cowan, P.L ,1992, Backreflection x-ray standing waves and crystal truncation rods as structure probe for epilayer-substrate systems, *Rev. Sci. Instrum.* 63:1142

Hertel, N., Materlik, G. and Zegenhagen, J., 1985, X-ray standing wave analysis of Bismuth implanted in Si(110), *Z. Phys. B* 58:199

Kazimirov, A.Yu., Kovalchuk, M.V., Kohn, V.G., Kharitonov, I.Yu., Samoilova, L.V., Ishikawa, T., Kikuta, S. and Hirano, K., 1993, Multiple diffraction in x-ray standing wave method: photoemission measurements, *Phys. Stat. Sol.*, 135:507

Kovalchuk, M.V. and Kohn, V.G., 1986, X-ray standing wave-a new method of studying the structure of crystals, *Sov. Phys. Usp.*, 29:426

Lagomarsino, S., Scarinci, F. and Tucciarone, A., 1984, X-ray standing waves in garnet crystals, *Phys. Rev.* B29:4859

Lagomarsino, S., Nikolaenko, A., Scarinci, F., D'Angelo, S., Derrien J. and Veuillen, J.Y, 1989, Structural study of the $CoSi_2$/Si(111) buried interface, *Surf. Science*, 211/212:692

Mukhamedzhanov, E.Kh., Maslov, A.V., Imamov, R.M., Bzhaumikhov, A.A., Fedorov, E.A. and. Kobzareva, S.A , 1991, Localization of impurity atoms in garnet crystals by the double-channel x-ray standing waves method, *J. Appl. Cryst.*,

Patel, J.R. and Golovchenko, J.A., 1983, X-ray standing wave atom location in heteropolar crystals and the problem of extinction, *Phys. Rev Lett.*, 23:1858

Vlieg, E., Fischer, A.E.M.J., Van der Veen, J.F., Dev, B.N. and Materlik, G., 1986, Geometric structure of the $NiSi_2$-Si(111) interface: an x-ray standing wave analysis, *Surf. Science*, 178:36

Woodruff, D.P., Seymour, D.L., McConville, C.F., Riley, C.E., Crapper M.D. and Prince, N.P., 1987, *Phys. Rev. Lett.*, 195:237

X RAY STANDING WAVES: THERMAL VIBRATION AMPLITUDES AT SURFACES

J. R. Patel[1,2] and E. Fontes[3]

[1]Lawrence Berkeley National Laboratory,
Berkeley, CA 94720
[2]Stanford Synchrotron Radiation Laboratory,
[2]Stanford Linear Accelerator Center
Stanford, CA 94309-0210
[3]CHESS, Cornell University,
Ithaca, NY 14853

INTRODUCTION

In the previous chapter, we have seen that the standing wave method measures the amplitude and the phase of the relevant Fourier component of the charge density in the crystal. The phase information, lost in the usual x-ray diffraction intensity measurements, is preserved by recording an inelastic channel (i.e., fluorescence) as the standing wave moves through the atomic planes during Bragg reflection. Also in the previous chapter, we have used the phase information to accurately locate atoms at surfaces. In the present chapter, we will utilize the Fourier amplitude exclusively and show how these amplitudes can be used to measure directly the thermal vibration amplitudes. While knowledge of thermal vibration amplitudes is in itself useful for a quantitative evaluation of bonding of atoms at surfaces, its variation with temperature has provided information on defect mediated vibrational motion and the onset of surface mobility. Indications that the atoms are not situated at a rigid single fixed position has already been observed for As and Ge adsorbates on (111) silicon surfaces. In spite of the fact that the atoms in these two cases lie in a single plane, the coherent fraction *F*, which measures the amplitude of the Fourier coefficient, is never 1.0 but reaches a maximum of 0.98. This 2% deviation from unity corresponds to an average thermal vibration amplitude $(\langle u^2 \rangle)^{1/2} = 0.12$ Å at room temperature. This value may be compared to the corresponding vibration value of Si atoms in the bulk of 0.08 Å. Thus, we can already deduce from the results in the last chapter

X-ray and Neutron Dynamical Diffraction: Theory and Applications
Edited by Authier *et al.*, Plenum Press, New York, 1996

that vibration amplitudes at surfaces are 50% higher than the bulk values. In what follows, we relate quantitatively the F value observations to the thermal displacements $\langle u^2 \rangle$.

THEORY

We have shown that one may write the standing wave structure factor F_{sw} in terms of F and P equation (9) in the previous chapter. A more rigorous treatment of the fluorescence yield, again for atoms all in a single plane, should include a Debye Waller factor. This is analogous to the situation for the x ray diffraction structure factor $\overline{F} = \Sigma f_j e^{-M_j} exp(-2\pi H \bullet r_j)$, where f is the atomic structure factor and $exp(-M_j)$ the Debye Waller factor. Using this, we write for equation (9) in the previous chapter

$$F\,exp(-2\pi iP) = \frac{1}{N}\sum_j exp\left(-2\pi iH \bullet r_j\right)exp(-M) \tag{1}$$

where $M = \left\langle 2\pi H \bullet u^2 \right\rangle / 2.$

For atoms located at a single plane equation (1) simplifies to

$$F = exp - (2\pi H \bullet u)^2 / 2 \tag{2}$$

For the sake of comparison in Figure 1, we show the difference between $F = cos[2\pi(\Delta d / d)]$ (previous chapter, equation [15]) versus a Gaussian displacement distribution function for the

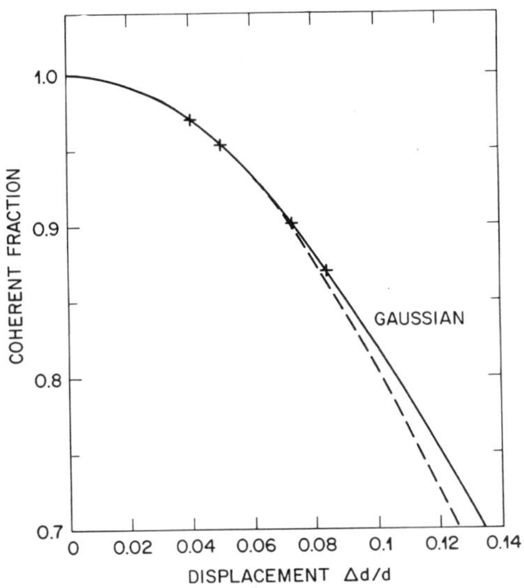

Figure 1. Coherent fraction versus $\Delta d/d_{111}$ for Silicon showing the difference between $F = cos[2\pi(\Delta d / d)]$ dashed, and a Gaussian displacement about the mean position, Equation (2), solid.

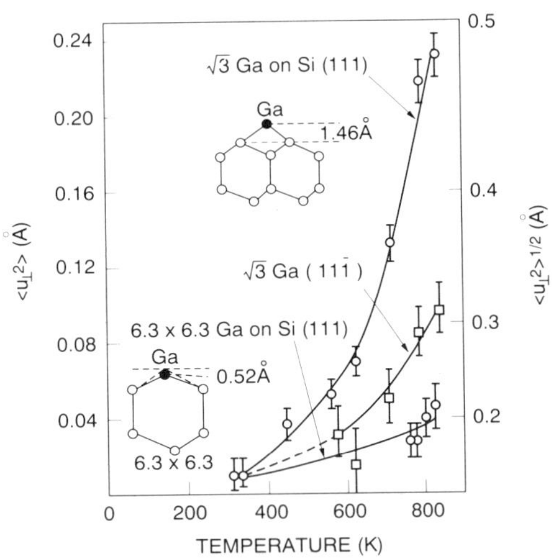

Figure 2. Temperature dependence of displacement $\langle u^2 \rangle$ of Ga atoms on a Si (111) $(\sqrt{3}x\sqrt{3})$ R30 and (6.3 x 6.3) surfaces. The insets show Ga location for the $(\sqrt{3}x\sqrt{3})$ and (6.3 x 6.3) reconstruction.

thermal vibration case. For coherent fractions F close to unity, the two functions are indistinguishable. Differences become apparent when $\Delta d = 0.08\, d$. Thus, if we measure F as a function of temperature, the mean squared vibration amplitude $\langle u^2 \rangle = (\Delta d)^2$ can be determined directly. In this chapter, we use the parameter F to directly measure displacement for vibrating atoms at adsorbates on silicon.

GALLIUM ON SILICON (111)

The position of gallium on a silicon (111)$\sqrt{3}$x$\sqrt{3}$ surface has been unambiguously determined as described in the previous chapter (Zegenhagen et al., 1987). Here we will consider only the temperature dependence of the standing wave results. It is apparent from Figure 7(a) in the previous chapter, that while the average position P = 0.56 remains the same for the two temperatures shown, 325° K and 802° K, the coherent fraction F decreases drastically from F = 0.98 at 325° K to F = 0.65 at 802° K. These results reflect amplitudes normal to the plane. For the $11\bar{1}$ reflection 70.35° from the surface normal the lateral vibration components parallel to the surface dominate. This is evident in Figure 7(b) in the previous chapter where the change in F is much smaller. The high value of F = 0.99 for $11\bar{1}$ at low temperature assures us that atoms are not highly disordered in the lateral direction by the large thermal vibrations normal to the surface.

In Figure 2, we show $\langle u^2 \rangle$ as a function of temperature for the [111] and $[11\bar{1}]$ direction (Martinez et al., 1992). The most striking feature of the data is the extremely large 0.47 ± 0.02 Å vibration amplitude normal to the surface obtained at 830° K. Equally important is the strong deviation from a linear temperature dependence predicted by Debye model. It is remarkable that in spite of these very large vibrational amplitudes, there is no shift, to within 0.03 Å accuracy, in the mean gallium atom position at elevated temperatures.

By comparison, the vibrational amplitude along $[11\bar{1}]$ is seen to reach a maximum of 0.33, ± 0.02 Å. Again, no shift in mean atom position is observed to the highest temperature of the measurement.

The large magnitude of the above vibrational amplitudes can perhaps be appreciated best by comparison with that for the incommensurate (6.3 x 6.3) reconstruction obtained by increasing the gallium coverage to about 1 ML. Gallium atoms in this layer are believed to bond with, and become nearly coplanar with, a single layer of Si(111) atoms. Only the perpendicular component can be determined from x-ray standing waves in this incommensurate surface; the results are shown in Figure 1. Incidently, perpendicular amplitudes for surface substitutional arsenic atoms are very similar to the Ga 6.3 x 6.3 data.

Returning to the $(\sqrt{3}$x$\sqrt{3})R30$ data, we can draw a number of preliminary conclusions. It is not surprising that there is no shift in the mean position parallel to the surface at elevated temperature, due to the high symmetry of the T_4 binding site in this plane. It is surprising, however, that there is no shift perpendicular to the surface. This suggests that cubic terms in the binding potential anharmonicity normal to the surface must be quite small. A possible explanation may involve significant lateral relaxation of the neighboring silicon atoms as the gallium atom moves towards the surface.

Doak (1989) has measured the spectrum of phonons on the gallium $(\sqrt{3}x\sqrt{3})R30$ surface using low-energy atomic helium atom scattering. Surface phonon dispersion curves measured along the $[11\bar{2}]$ azimuth display a flat optic mode at approximately 11.5 meV, and a acoustic mode with bends and flattens abruptly near the Brillouin-zone center, reaching a plateau of approximately 10 meV at the zone boundary. Assuming that the weight of these low-lying vibrational eigenstates is concentrated primarily on the gallium atoms, their flat dispersion is likely the result of (anisotropic) Einstein oscillator motion of nearly independent gallium adatoms. The large separation of gallium adatoms (6.65 Å) tends to support this argument. With this assumption, the resonant frequency Ω of the vibrations normal to the surface is

$$\Omega = [K_b T/M(u_\perp^2)]^{1/2} \tag{3}$$

where M is the mass of gallium. Using our room-temperature value of $(u_\perp^2)$, we obtain good agreement with the zone-boundary frequency obtained by Doak (1989). If one chooses to model the extra large vibration amplitudes obtained at larger temperatures as resulting from a "softening" of a harmonic gallium binding potential, one obtains the results plotted in Figure 3. The softening is dramatically shown as a reduction in the normal resonant mode frequency by nearly a factor of 2.6 at 830° K. We believe either large shifts or broadening of the helium inelastic scattering line (relative to its room-temperature values) will be observed when this system is studied at elevated temperatures.

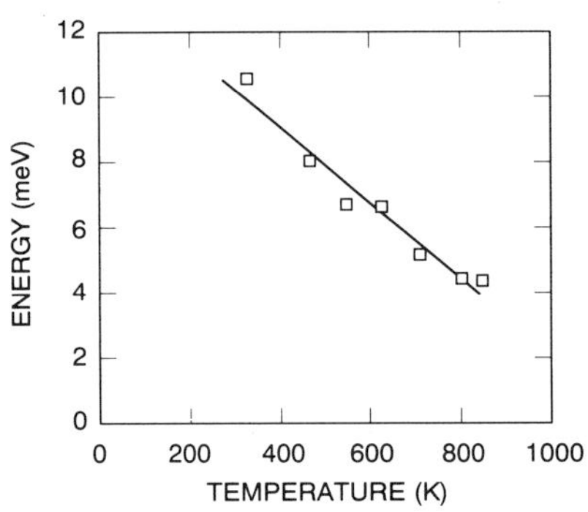

Figure 3. Temperature dependence of the resonant frequency for Ga ($\sqrt{3}x\sqrt{3}$) on Si(111) calculated using equation (3).

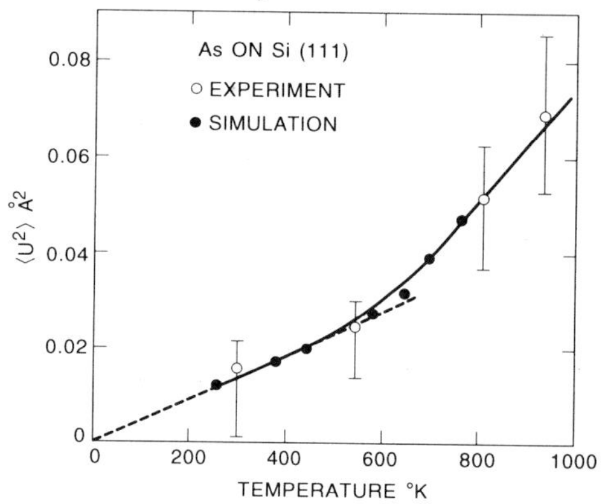

Figure 4. Temperature dependence of <u²> for Arsenic on Si(111) (1 x 1).

ARSENIC ON SILICON (111)

For arsenic atoms on silicon (111), the reconstruction is not as drastic as in the case of Ga. We also see from Figure 5 in the previous chapter that changes in the coherent fraction both for (111) and $(11\bar{1})$ are not as large as the case for $\sqrt{3}$ Ga over the same temperature range. In Figure 4, we show a plot of <u²> for As on Si (111) versus temperature (Fontes et

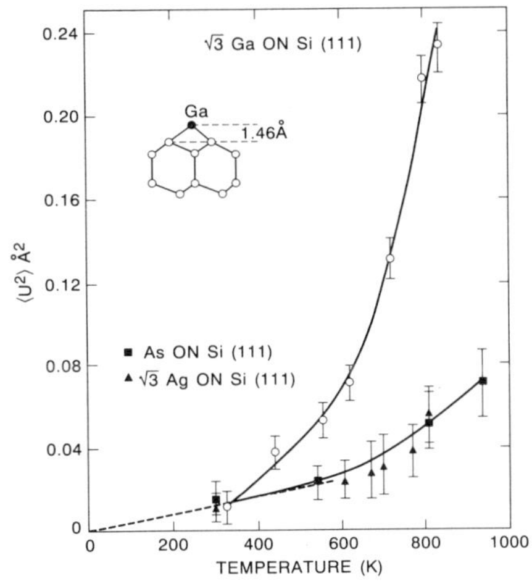

Figure 5. Comparison of $\langle u^2 \rangle$ versus temperature for various adsorbates indicated on Silicon(111).

al.). While the experimental data are meager, and the error bars quite large, we observe a difference in behavior between the low and high temperature range, around 600° K. The behavior of As on Si is normal, or Debye-like below 600° K. We see that neglecting zeropoint vibration, the $<u^2>$ data can be extrapolated through 0° K. This is expected from Debye Theory, which strictly speaking should not apply for an anisotropic case (i.e., surface atoms). Nevertheless, it is remarkable that this behavior is noted for a number of adsorbates on silicon, as will be shown later.

To understand the reason for the change in behavior of $<u^2>$ around 600° K, we have resorted to a molecular dynamics simulation of the As covered surface. Without getting into details of the molecular dynamics model, we have assumed a three body Stillinger-Weber (1985) potential which predicts well the melting behavior of Si. With this potential, and the As atom position from our standing wave results, the As-Si bond strength was adjusted to match the $<u^2>$ observed at room temperature. After which the temperature was allowed to vary and $<u^2>$ determined for a series of temperatures. In the first trials, while the experimental lower temperature data were well replicated, the points at temperature above 600° K were well below the experimental values in Figure 4. It was decided then to introduce defects in the surface above 600° K. As atoms were removed from silicon lattice sites in the model to create vacancies at the surface. The simulated $<u^2>$ were larger under these conditions and appeared consistent with the experimental data as shown by the solid points in Figure 4. It appears then that at temperatures in the 600° K range, the mobility of surface atoms is significant. Strong evidence that this situation is real comes from molecular beam experiments on silicon where good crystalline growth is observed at temperatures as low as 300° K.

We have extended thermal vibration measurements to other atoms on Si. In Figure 5, we summarize some of this data for adsorbates on silicon (111). The difference between the Ga $\sqrt{3}$ data and all of the other adsorbates is striking. While most of the adsorbates indicate normal Debye-like behavior, namely extrapolation of the low temperature data through 0° K, the behavior of Ga$\sqrt{3}$ is anomalous in that it exhibits no such low temperature behavior. At high temperatures, the amplitude is 0.47 Å, as compared to a value of 0.27 Å for the other adsorbates. This anomalous behavior of Ga$\sqrt{3}$ on Si has not been explained. Molecular dynamics, in this case, has had little success and exhibits instability of the T_4 site which is not true in nature. Shown in Figure 5 are the As and Ag data, where high coherent fractions have been observed. We also have data for Cu, Pb, and Ga (6.3 x 6.3) where the initial coherent fractions are not as high due to the corrugation at the initial impurity atom plane distribution. While such curves are displaced slightly upwards from the lower curve of Figure 5, if plotted directly, they lie almost exactly on a master curve, shown in Figure 4, if the data are normalized to the observed room temperature vibration amplitude $\langle u^2 \rangle = 0.12$ Å. It is curious that there are no large changes of $\langle u^2 \rangle$ with the atomic mass *M*. However, one can show from elementary arguments for a simple harmonic oscillator that the mass term drops out. Setting the thermal energy k_BT equal to the kinetic energy of the atom $Mv^2/2$, $\langle u^2 \rangle$ can be expressed as $\langle u^2 \rangle = k_BT/k_S$, where k_B is Boltzmann's constant, T the temperature, and k_S the bond stiffness. This conclusion, we are informed, is quite general and not confined to the harmonic approximation. Since $\langle u^2 \rangle$ for the vastly disparate masses is so similar, we can only conclude that k_S, the bond stiffness, is not very different for the various impurities. This has not as yet been theoretically justified.

ARSENIC ON SILICON(001)

In measuring the thermal vibration amplitudes of Si atoms at the Si(001) surface, one is confronted by the fact that, in addition to the vibration amplitude of individual atoms, there is superimposed a high-frequency (of order 10^{10} Hz) switching motion of the reconstructed dimer pair at the surface. This switching motion occurs because the energetics of the buckled dimer can be described by asymmetric double-well potential where one of the dimer atoms is either in an up or a down state. In addition to the dimer motion, the average dimer separation is not fixed but seems to change with temperature.

One means for simplifying the Si thermal vibration measurement is to modify the surface with an impurity. For adsorbates on the Si(001) surface, though, where dimerization is the dominant mechanism for reducing the number of dangling bonds, $\langle u^2 \rangle$ can easily be determined only if the impurity produces stable dimers (preferably symmetric) at the surface. Arsenic dimers on the Si(001) surface are similar to Si or Ge dimers, but have the added virtue of being planar (not buckled) at room temperature. In this study, we report direct thermal-vibration-amplitude measurements of As on Si(001).

Figure 6 shows the room temperature measured (004) and (022) reflectivities as open squares and the corresponding As Kα normalized fluorescence yields as open circles (Franklin et al., 1994). The solid lines are calculated from the dynamical diffraction theory and fitted to the data.

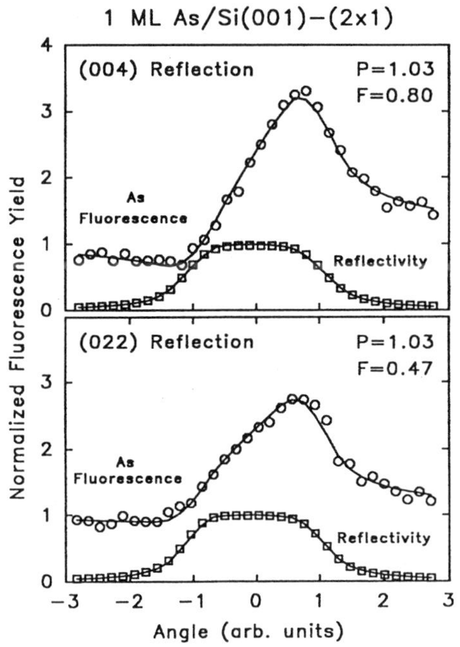

Figure 6. Reflectivity and fluorescence yield for Arsenic on Si(001). The two reflections, 004 and 022, establish Arsenic registry with the Si substrate.

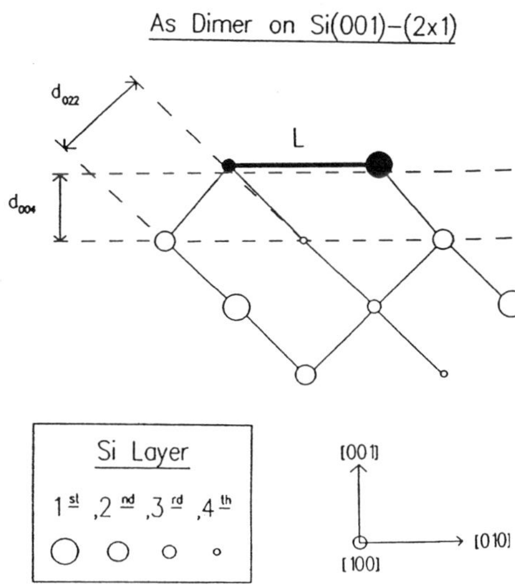

Figure 7. [$\bar{1}$00] projections of the (001) and (110) planes illustrating the As dimer registration from the P values in Figure 6.

From Figure 7, we can see the relevant geometry where the Si atoms (open circles) are presented in ideal bulk-like positions, and the As atoms (solid circles) are shown in a projected view. Thus, our measured coherent position of $P_{004} = 1.03$ (Figure 6) indicates that the position of the center of mass of the As atoms, relative to the (004) planes, is 3% of a d spacing above these bulk-extrapolated planes ($d_{004} = 1.36$ Å).

Using our four experimentally determined values ($P_{004} = 1.03 \pm 0.01$, $F_{004} = 0.80 \pm 0.01$, $P_{022} = 1.03 \pm 0.01$, and $F_{022} = 0.47 \pm 0.02$), and symmetry considerations of the dimer, we were able to uniquely determine by triangulation the As atom positions and dimer bond length $L = 2.58 \pm 0.04$ Å for symmetric As dimers and these dimers sit $Z = 1.40 \pm 0.01$ Å above the top (004) bulk-extrapolated plane.

The dimer length determined from the standing wave result is in excellent agreement with other experimental determinations while the theory predicts a length 0.06 Å shorter.

Assuming a Gaussian distribution for the thermal motion of the symmetric As dimer atoms, we have obtained $\langle u_{004}^2 \rangle$ directly from the normalized F_{004} values. A plot of the (004) (solid circles) and (022) (open circles) components of the mean-square vibration amplitude of the As atoms as a function of temperature is shown in Figure 8. The solid line through the points in a linear least-squares fit to the $\langle u_{004}^2 \rangle$ versus T data, for $T < 650°$ K. For $T > 650°$ K, a second-order polynomial term was added, though only to improve the goodness of the fit. The dashed line is a theoretical vibration-amplitude result from Mazur and Pollmann (1990) for the average (004) component (i.e., normal component) of the Si dimers on Si(001). The

Alerhand, Joannopulos (1989), and Mele theoretical results (not shown) have a slope slightly less than predicted by Mazur and Pollmann (1990).

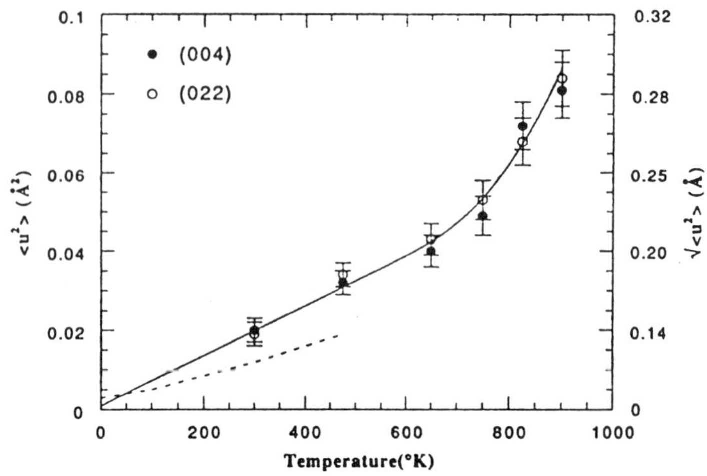

Figure 8. Mean squared amplitude $\langle u^2 \rangle$ as a function of temperature. The dashed line is the theoretical result of Mazur and Pollman, (1990).

It is apparent from Figure 8 that the vibrational amplitude shows a different behavior above roughly 650° K. From 300° K to 650° K, $<u^2_{004}>$ versus T is linear with a slope of $(6.3 \pm 0.3) \times 10^{-5}$ Å^2/K^2, which extrapolates to about zero at $T = 0$ K, neglecting the small zero-point oscillations of order 0.003 Å^2 from Mazur and Pollmann (1990). This follows the well known harmonic behavior predicted by Debye, and we refer to the behavior in this regime as Debye-like. Theoretical results predict a linear behavior in this regime with a slope of 3.6 x 10^{-5} Å^2. Note that our results are for As adatoms and not Si adatoms. Above 650° K, $<u^2_{004}>$ varies much more rapidly with temperature. We emphasize that the high-temperature behavior is not due to the onset of anharmonicity. If anharmonic effects were significant, we should observe a change in the coherent-position (P_{004}) values. Within the experimental uncertainty, no such change was observed, although the coherent fraction (F_{004}) varied by a factor of 2 in the same temperature range. In addition, since the relative coherent position did not change as a function of temperature, the thermal coefficient of expansion of the surface structure in the [001] direction is not significantly different from that of bulk silicon.

The change of slope for the (001) surface occurs in the same temperature range as the (111) surface indicating no significant anisotropy for the onset of mobility of surface atoms. Though molecular dynamics simulations have not been attempted for the dimer terminated Si(001) case, defects (i.e., vacancies) are responsible for change in slope around 600° K.

We now turn to the question of the anisotropy of the normal and in-plane measurements of the thermal vibration amplitudes on the Si(001) surface. For the Si(001)(2x1) surface, a convenient off-normal reciprocal-lattice vector is the (022), which is inclined at 45° to the (001) surface. The projection of the normal component of the thermal vibration amplitude along [022] is substantial and, if the anisotropy is small, it will be difficult to separate the normal and in-plane components of <u^2>. We find that the anisotropy (i.e., the [022] component in Figure 8) is indeed rather small since the (004) and (022) values follow each other closely. Our results are in agreement with the theoretically calculated anisotropy for the Si(001)–(2x1) surface, which is about 6% at room temperature, increasing to 10% at 450° K.[2] This is in contrast to the Si(111) surface were we have observed a marked anisotropy for Ga and As absorbed species as a function of temperature.

GERMANIUM ON SILICON(001)

Since the earliest low-energy electron diffraction (LEED) studies the detailed nature of the dimer adatom arrangement on the technologically important Si(001) surface has been studied intensively by all manner of surface probes and techniques. These studies have in turn stimulated sophisticated theoretical efforts using ab initio total energy and molecular dynamics approaches to locate the surface dimer positions and subsurface distortions. In spite of these efforts, ambiguities persist regarding the symmetric or asymmetric nature of surface dimers, particularly at room temperature.

On the experimental side, two questions need clarification: (a) what is the magnitude of the dimer asymmetry, if any, and (b) are the dimers at room temperature asymmetric or symmetric? In this section, we describe a direct quantitative x-ray standing wave measurement of the surface dimer displacement and asymmetry on the well-known Si(001)/Ge (2x1) reconstructed surface at room temperature and, to our knowledge, we present the first characterization of the dimers at high temperatures. With suitable preparation, a monolayer of Ge atoms deposited by molecular beam epitaxy at high temperature onto a clean UHV Si(001) surface substitutes for the topmost Si adatoms, forming Ge dimers. Although a Ge-substituted dimer surface is not a replica of either the bare Si(001) or Ge(001) surfaces, theoretical calculations suggest that the adatom stability of both Si and Ge(001) surfaces is similar. Substituting Ge for the Si dimer atoms at the (001) surface does not perturb the observed (2x1) reconstruction as seen by LEED.

Figure 9 shows the x-ray reflectivity and Ge fluorescence yields for the (004) and (022) reflections for samples with less than 1 monolayer (ML) coverage at room temperature (Fontes et al., 1993). The solid lines are fits to standard dynamical diffraction theory which yields F and P values. $P_{004} = 0.98$ implies that the center of mass of the Ge dimers is 2% below the lattice spacing of the bulk extrapolated (004) lattice planes. To obtain lateral registry, we have examined the (022) reflection with planes inclined 45° to the (001) surface [Figure 9(b)]. The inset in Figure 9(b) shows the intersection of the (004) and (022) P values slightly below the bulk positions of the two planes. Thus, the dimer atoms are almost equally displaced above and below the bulk (001) plane position. The subsurface relaxation normal to the surface is a small $0.02d_{004} \approx 0.03$ Å.

Although the four experimentally derived parameters (P_{004}, P_{022} and F_{004}, F_{022}) are not sufficient to determine the six Cartesian coordinates of the two Ge atoms, if the two P values

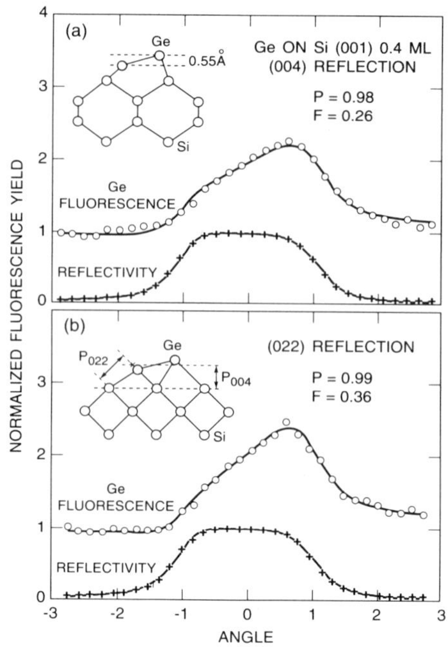

Figure 9. (a) reflectivity and fluorescence yield for Ge on Si(001) 0.4 ML; (b) 022 reflectivity showing dimer-pair registration.

are used to constrain the position of the dimer center of mass, the two F values can be used to calculate the vector displacement between the two atoms. A two-parameter model for a planar dimer is shown in Figure 10(a); the dimer is parametrized by the bond length L between the Ge atoms and the angle θ between that bond and the surface. Using $F_{004} = 0.26 \pm 0.01$ and $F_{022} = 0.36 \pm 0.01$. a χ^2 minimization yields a best fit for $\theta = (12.1 \pm 0.2)°$ and $L = 2.60 \pm 0.04$ Å. These parameters determine the dimer height displacement $\Delta z = 0.55 \pm 0.02$ Å. Of course, a symmetric dimer would have $\Delta z = 0$.

This analysis does account for degeneracy in dimer domain and tilt orientations, and assumes that all Ge atoms are in perfectly formed dimers. Ge atoms in random sites or disordered clusters would reduce the measured coherent fraction. There is no evidence yet for cluster growth at submonolayer coverages. Our observations of no change in F for three different submonolayers coverages (Fontes et al., 1993) strongly support the conclusion that if Ge islands are present, little or no disorder is introduced; otherwise, we would expect a decrease in F with coverage. On the contrary, we find consistent evidence of highly ordered layer-by-layer Ge growth for coverages up to 3 ML. F increase steadily up to 3 ML, where $F = 0.46$ is 77% larger than the 1 ML value. For 2 ML, (3 ML) only 1/2 (1/3) of the atoms are in the dimer configuration and one (two) Ge atom layers are in well-ordered lattice sites, thereby leading to an increase in F. In addition, the relationship between the coherent fractions for different reflections, $F_{022} = 0.36$ and $F_{004} = 0.26$, is completely explained by the geometric model of perfectly formed Ge dimers. If surface disorder were appreciable, there would be no reason to expect consistency between these two measurements, or between different sample preparations.

In the above analysis, both dimer atoms have the same isotropic vibration amplitude of $(<u^2>)^{1/2} = 0.12$ Å, a value found in other experiments and matching theoretical calculations (Alerhand et al., 1989). As demonstrated in previous sections, we can measure $<u^2>$ from a systematic study of F versus temperature. Remarkably, at 500° C, we find $F_{004} = 0.31$, a 20% *increase* over the room temperature coherent fraction. This result is both surprising and puzzling since previous experiment and theory indicate that vibration amplitudes should increase with increasing temperature, thereby decreasing F. In terms of the static model above, for F_{004} to increase, the displacement Δz must decrease significantly. Using $<u^2>=0.042$ Å^2 at 500° C, estimated from both theory (Alerhand et al., 1989) and experiment (Martinez et al., 1992) (Fontes et al., unpublished) Δz = 0.44 Å at 500° C versus 0.55 Å at room temperature. Using only this simple static model, this decrease in Δz may indicate a phase transition at elevated temperatures from an asymmetric to symmetric (2x1) surface.

We expect, though, that the essential physics of the dimer is more closely described by a dynamic oscillator model. Qualitatively, the energetics of the angular position of the dimer can be given by a simple symmetric double-well potential [Figure 10(a)].

$$E(\theta) = \frac{E_1}{\theta_0^4}(\theta^4 - 2\theta_0^2\theta^2), \tag{4}$$

where E_1 is the dimer flipping barrier and θ_0 is the angle of minimum energy. Since we have no direct information otherwise, we use for this model the L and θ_0 equilibrium values measured at room temperature. A more accurate model requires full temperature dependence of E_1, L, and θ_0. The standing wave measurement is both a time and ensemble average,

$$F_{004} = e^{-1/2q^2<u^2>} \int_{-\pi}^{\pi} d\theta P(\theta) \cos[\pi L \sin\theta / 1.36], \tag{5}$$

where the probability amplitude for finding a dimer at angle θ is given by

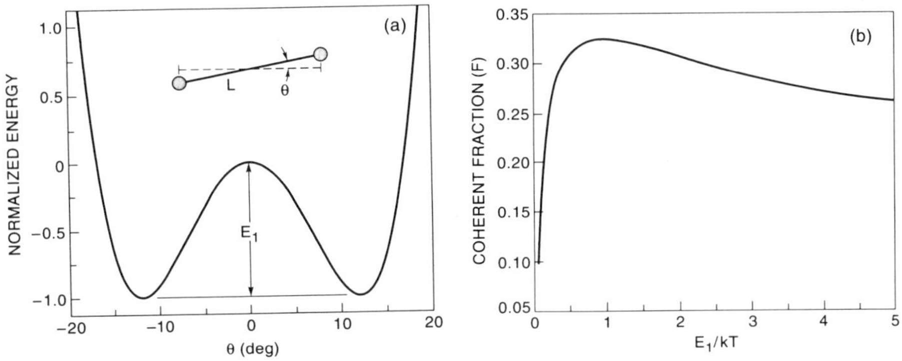

Figure 10. (a) dimer energy versus θ showing double well potential; (b) coherent fraction F versus temperature showing increase in F with T at low temperature.

$$P(\theta) = e^{-E(\theta)/kT} / \int_{-\pi}^{\pi} e^{-E(\theta')/kT} d\theta' \quad (6)$$

The behavior of F_{004} as a function of E_1/kT, Figure 10(b) shows rather unexpected behavior. At the lowest temperatures, the dimers are frozen at θ_0, so F approaches asymptotically [to the right of Figure 10(b)] the value given by the static model. As the temperature increases, the probability of finding a dimer with $\theta < \theta_0$ increases whereby both the average Δz and the dimer asymmetry decreases and F increases. At the highest temperatures, larger angular (θ) excursions and large thermal vibrations $<u^2>$, lead to a net decrease in F. This general behavior is independent of the actual details of the potential; the potentials given by Stillinger (1992) and Kockanski and Griffith (1991) produce similar curves.

Thus, it is possible that F could either increase or decrease with increasing temperature, depending on the activation of the dimer oscillator, which in turn depends on the energy barrier for dimer flipping. The STM results argue that dimers are actively flipping at room temperature. A recent calculation by Dabrowski and Scheffler (1992) shows an energy barrier of 0.1 eV. From our observation that F increases by 20% upon increasing temperature from 30 to 500° C, we conclude that $[E_1/\mathrm{kT}] >> 1$ at room temperature.

While none of the theoretical calculations addresses directly the heteroepitiaxial case of Ge on Si, a survey of those findings raises interesting quantitative questions.

In general, the theoretical calculations deviate most from the experimental findings concerning the lattice expansion disagreeing with $P \approx 1$ for two reflections. The predications range from $\Delta z = 0.27$ to 0.69 Å for Si, compared to our measurement of $\Delta z = 0.55$ Å. Interestingly, the Needels, Payne, and Joannopoulos (1987) calculation for Ge dimers on Ge(001) gives $\Delta z = 0.57$ Å, in remarkable (fortuitous) agreement with our measurement. Although the error range for the theoretical calculations is not specified, it would appear that Ge on Si(001) most closely mimics Ge dimers on the intrinsic Ge surface. The experimental Ge dimer bond length of 2.6 Å is about 6–7% larger than the bulk Ge-Ge bond, and is considerably larger than theoretical calculations, which tend to be close to the bulk value for Si of 2.35 Å.

ACKNOWLEDGEMENTS

Any project at synchrotron light sources is never an individual effort. I have been fortunate to have as my collaborators outstanding individuals whose names follow. B.W. Batterman for introducing me to the dynamical theory of x rays and the early standing wave results. J.A. Golovchenko for involving me in the art and science of standing waves for surface studies. Brian Kincaid who designed and built X-15A at NSLS. J. Zegenhagen for his help in designing and building a dedicated UHV standing wave facility X15A at NSLS, and for the early UHV studies on surfaces. E. Fontes, my coauthor, for initiating the surface vibration studies. G. Franklin for vibration studies on (001) surfaces. Finally, it is a pleasure to acknowledge P.E.

Freeland, who worked closely alongside me during my long association with the Bell Telephone Laboratories; and NSLS Brookhaven for its cooperative staff and excellent beams.

REFERENCES

Alerhand, O.L., Joannopulos, J.D. and Mele,E.J., 1989, Thermal amplitudes of surface atoms on Si(111) 2x1 and Si(001) 2x1, *Phys. Rev*. B39:12622.

Dabrowski, J., Scheffler, M., 1992, Self-consistent study of the electronic and structural properties of the clean Si(001)(2x1) surface, *Appl. Surf. Sci.* 56–58:15.

Doak, R.B., 1989, Structure and dynamics of strong chemisorption on Si(111) as measured with helium atom scattering, *J. Vac. Sci. Technol.* B7:1252.

Fontes, E., Gilmer, G.H., Golovchenko, J.A., and Patel, J.R., unpublished work.

Fontes, E., Patel, J.R., and Comin, F., 1993, Direct measurement of the asymmetric dimer buckling of Ge on Si(001), *Phys. Rev. Lett.* 70:2790.

Franklin, G.E., Fontes, E., Qian, Y., Bedzyk, M., Golovchenko, J.A., and Patel, J.R., 1994, Thermal vibration amplitudes and structure of As on Si(001), *Phys. Rev.* B:7483.

Kochanski G.P. and Griffith, J.E., 1991, A ginzberg-landau model of dimers on the Si(001) surface, *Surf. Sci. Lett.* 249:L293.

Martinez, R.E., Fontes, E., Golovchenko, J.A., and Patel, J.R., 1992, Giant vibrations of impurity atoms on a crystal surface, *Phys. Rev. Lett.* 69:1061.

Mazur, A. and Pollmann, J., 1990, Anisotropy of the mean-square displacements at the Si(001)-(2x1) surface, *Surf. Sci.* 225:72.

Needels, M., Payne, M.C., and Joannopulos, J.D., 1987, Ab initio molecular dynamics on the Ge(100) surface, *Phys. Rev. Lett.* 58:1765.

Stillinger, F.H. and Weber, T.A., 1985, Computer simulation of local order in condensed phases of silicon, *Phys. Rev.* B31:5262.

Stillinger, F.H., 1992, Microscopic model for dimers buckling on Si(001), *Phys. Rev.* B46:9590.

Zegenhagen, J., Patel, J.R., Freeland, P.E., Chen, D.M., Golovchenko, J.A., Bedrossian, P., and Northrup, J.E., 1987, X-ray standing-wave and tunneling microscope location of gallium atoms on a silicon surface, *Phys. Rev.* B39:1298.

LONG-PERIOD X-RAY STANDING WAVES

S. Lagomarsino

Istituto Elettronica Stato Solido - CNR
V. Cineto Romano, 42 - 00156 Roma (Italy)

1. INTRODUCTION

Long-period X-ray standing waves can be generated essentially in two ways: through diffraction in a medium with a large periodicity or through total reflection by surfaces. In the first case a strict analogy exists with the usual XSW technique, described in the contribution of Patel (this volume). In correspondence to Bragg diffraction peaks of the periodic structure, such as for example a synthetic multilayer made of inorganic or organic molecules or a crystalline superlattice, an interference between the incident and the diffracted beams will take place, giving rise to a stationary wave having the same period of the structure. Going through the diffraction peak the nodes and antinodes move inward causing a modulation of the e.m. intensity in the structure itself.

In the case of total reflection the situation is quite different. It is worthwhile to recall that for X-rays the index of refraction n is less than 1 and is given by: $n = 1 - \delta - i\beta$, where $\delta = N_e r_e \lambda^2/2\pi$ and $\beta = \mu\lambda/4\pi$ (N_e = electron concentration, r_e= classical electron radius, λ = wavelength, μ = absorption coefficient). This implies that x-ray beams incident on a surface are totally reflected if the incidence angle α is smaller than a critical angle α_c given by: $\alpha_c = (2\delta)^{1/2}$. If this happens a strong interference takes place between the incident and the totally reflected beams, giving rise to a stationary wave ***above*** the reflecting surface, as pointed out by Bedzyk et al (1989). This wave has equintensity planes parallel to the surface and a periodicity D given by: $D = \lambda/(2 \sin\alpha)$. Therefore the periodicity is not constant but depends on the incident angle. At $\alpha = 0$ the first antinode is at infinity. As the incidence angle increases, the first antinode moves towards the surface with the other antinodes following with the periodicity determined by α, like a compressing bellows. At the critical angle the first antinode coincides with the reflecting surface, while for larger angles the reflected beam starts decreasing and consequently also the amplitude of the s.w. field above the surface dies off.

Material studies have been carried out using both methods (often in conjunction) to create long-period standing waves. In what follows some examples are described.

2. SYNTHETIC MULTILAYERS

Artificial multilayers can be obtained by periodically alternating deposition of different materials. A condition to have strong enough reflectivity is that the two materials have a quite different atomic number Z. The period can typically range between few tens to few hundreds Å. The first demonstration of the applicability of the XSW technique to such periodic structures has been given by Barbee and Warburton (1984) on a platinum-carbon multilayer (ML) with a period of 31 Å. If the interfaces are sharp and the period consant, the ML can be considered as a crystalline structure in one dimension. As in the usual XSW technique, in correspondence to Bragg diffraction peaks the interference between incident and diffracted beams gives rise to a stationary wave having the same period as the periodic ML. Going through the diffraction peak the nodes and antinodes move inward displacing the maximum of intensity from one material to the other, passing through the interfaces. Several interesting applications of this technique have been reported in literature. For example, Kortright and Fischer-Colbrie (1987) reported standing wave enhanced scattering in W-C periodic structures, and Heald and Tranquada (1989) standing-wave assisted EXAFS (Extended X-ray Absorption Fine Structure) studies in a Ni-Ti multilayer. Both studies pointed out that selective information on the materials composing the ML can be extracted with the use of the standing wave field. Density evaluation of deposited films on multilayers has been carried out by Zheludeva et al (1992), while interface roughness and interdiffusion has been studied in Ni/C multilayers by Kawamura and Takenaka (1994) and by Chernov et al (1995). Instead Bedzyk et al. (1986) used the standing wave field generated in a Platinum-Carbon ML to study electrochemical interfaces. A layer of iodide was adsorbed on top of the platinum-terminated multilayer, followed by electrodeposition of Cu. The XSW analysis showed that the copper atoms were directly adsorbed on Pt, and therefore that the iodide layer was displayed by the subsequent electrodeposition of copper.

3. LANGMUIR-BLODGETT FILMS

Langmuir-Blodgett (LB) films are formed by organic molecules characterised by a long hydrophobic chain. The molecules can be deposited on solid substrates forming ordered structures (for a review see Roberts, 1985), with periods of the order of tens of Å. LB films can be studied with long-period standing waves using one or a combination of the following methods:

a) Standing waves generated in the LB film itself.

b) Standing waves generated by an underlying multilayer having a periodicity similar to the LB film period

c) Standing wave generated in total reflection

Aim of these studies was the determination of the structural parameters of the organic layers. In all cases a fluorescent marker was used, therefore the LB films contained a metallic ions whose fluorescence could be measured as a function of the incidence angle.

A combination of methods a) and c) have been used by Zheludeva et al. (1990) to study with a standard X-ray generator an LB film composed of about 100 monolayers of lead stearate covered by one monolayer of barium behenate. Bedzyk et al. (1988) studied a Zinc Arachidate bilayer deposited over a W/Si multilayer covered with one monolayer of Cd Arachidate, recording Cd and Zn fluorescence in correspondence to the Bragg diffraction peak of the multilayer (fixed periodicity of 25 Å), and in the total reflection region. Measurements at different temperatures indicated clearly an inward collapse of the Zinc atom layer due to some temperature-induced change of the structure. In this example the mean position and width of the zinc atom layer was determined with a precision of ± 0.3 Å.

4. SUPERLATTICES

In multilayers the structure of the individual components can be amorphous, and an ordered structure crystalline-like exists only in one direction. In superlattices the components have themselves a crystalline nature, and a superperiod is introduced by the periodic deposition of different materials. The importance of superlattices is mainly due to their novel optoelectronic properties. The most common materials used for superlattices are the III-V compounds and, more recently, the group IV elements (mainly Si and Ge). Their properties are strongly influenced by the structure of the interfaces that need a characterization at the atomic level. To this purpose an extension of the X-ray standing wave method has been very recently applied to SiGe superlattices (Lagomarsino et al., 1995; Castrucci et al., 1996). Many structural information can be extracted by X-ray diffraction: in general an intense peak is obtained from the substrate and satellite peaks from the superlattice structure. The angular distance between the substrate and the zero-th order satellite peaks is essentially related to the average lattice mismatch. The distance between the satellite peaks of different orders yields the superlattice periodicity. Interference fringes in between the main peaks are related with the total thickness of the structure. In every case only the modulus of the structure factor can be obtained. XSW measurements can give in addition information about the phase of the structure factor.

X-ray Standing Waves are generated in correspondence to the substrate peak, with the same periodicity of the substrate lattice. Due to the lattice mismatch these S.W. go rapidly out of phase with respect to the superlattice periodicity, and are therefore unuseful. However S.W. are generated also in correspondence to the superlattice satellite peak.

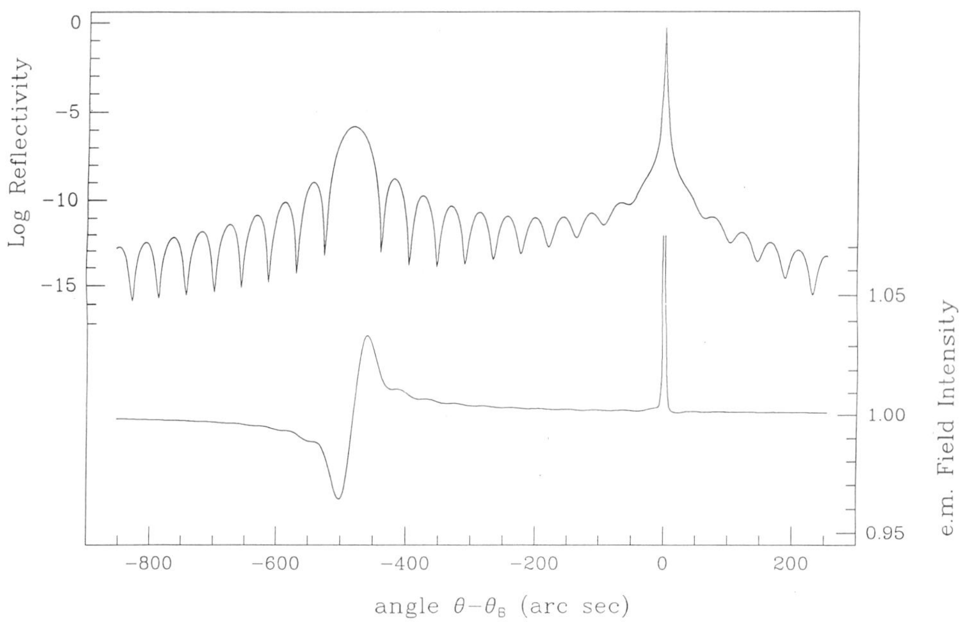

Figure 1. Diffracted intensity and e.m. field for a pseudomorphic superlattice (Ge_2Si_{18})x75 on Si(100). Incident beam energy: 15 KeV.The region close to the (400) substrate peak is shown

Figure 1 shows the calculated diffracted intensity (in log scale) and the e.m. field intensity in the angular range comprising between the substrate and the first satellite peaks for an ideal superlattice 75 x $(Si_{18}Ge_2)$ grown epitaxially on a Si(100) substrate.The modulation of the e.m. field at the angular position of the satellite peak is obviously much weaker than that corresponding to the substrate peak, due to the reduced reflectivity. The calculation has been carried out considering the unit cell of the superlattice as composed by 18 Si layers and 2 Ge layers, calculating its structure factor and computing the reflectivity and the e.m. field applying the recursive formalism used by Halliwell et al. (1984). The zero-th order satellite peak becomes in this way the 20-th order diffraction peak of the superlattice structure.

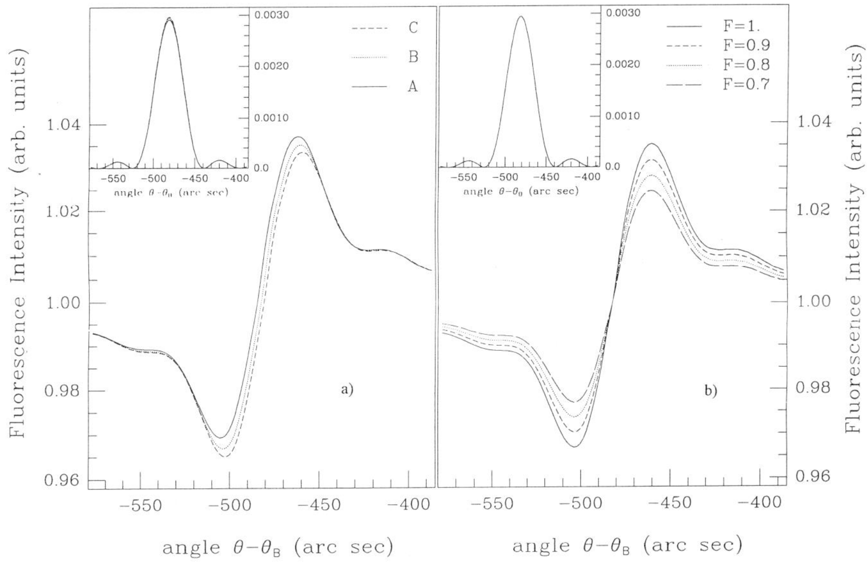

Figure 2. a): diffracted intensity (in the inset) and Ge fluor. modulation for the same kind of structure of fig. 1) for three different values of the distances between Si and Ge atoms (see text). b): diffracted intensity (in the inset) and Ge fluor. modulation for four different values of the coherent fraction (see text).

Two kinds of specific information can be obtained by an XSW analysis in superlattices, in analogy with the standard technique, i.e. the position of the atoms with respect to the superlattice structure and the crystalline order. Figure 2a) shows the simulated Ge fluorescence as a function of angle in the angular region of the satellite peak for three different values of the perpendicular distances between the Si and Ge layers at the interface. Because in the structure there are two monolayers of Ge, we have to consider three distances: $d_{Si\text{-}Ge}$, $d_{Ge\text{-}Ge}$ and $d_{Ge\text{-}Si}$ Due to the elastic deformation not necessarily $d_{Si\text{-}Ge}$ and $d_{Ge\text{-}Si}$ are equal. The values for the three cases reported in figure 2a) are:

Case A:	$d_{Si\text{-}Ge}$ = 1.4581 Å	$d_{Ge\text{-}Ge}$ = 1.4581 Å	$d_{Ge\text{-}Si}$ = 1.3577 Å
Case B:	$d_{Si\text{-}Ge}$ = 1.4298 Å	$d_{Ge\text{-}Ge}$ = 1.4581 Å	$d_{Ge\text{-}Si}$ = 1.3861 Å
Case C:	$d_{Si\text{-}Ge}$ = 1.4079 Å	$d_{Ge\text{-}Ge}$ = 1.4581 Å	$d_{Ge\text{-}Si}$ = 1.4079 Å

In all these cases the Si-Si layer distance is: $d_{Si\text{-}Si}$ = 1.3577. Distinct fluorescence curves are obtained for the three different cases, but the same reflectivity curves, as shown in the insert of figure 2a). In figure 2b) curves related to four different values of the coherent fraction F are shown. Again distinct fluorescence curves are obtained, while the reflectivity curve doesn't change. These examples show that useful information about the interface structure and the overall superlattice quality can be obtained by means of this method.

Figure 3 shows experimental data compared with the computed ones. The superlattice, nominally 75 x ($Si_{18}Ge_2$), has been grown at Daimler-Benz by Molecular Beam Epitaxy (MBE), and the experiment has been carried out at the D5 beam line at ESRF.

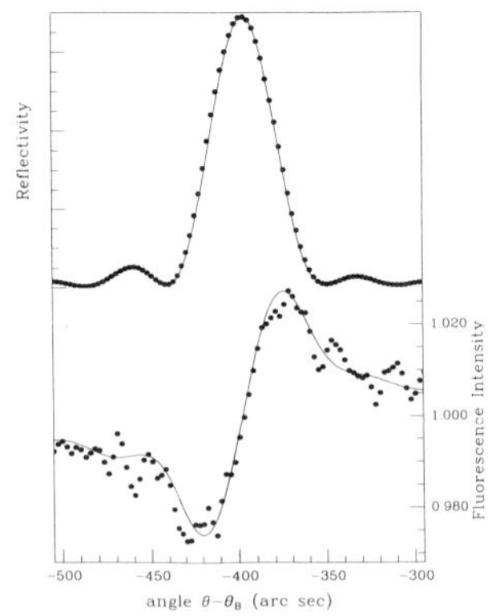

Figure 3. Experimental (closed points) and simulation curves (full line) of reflectivity and Ge fluorescence as a function of incidence angle.

A good agreement between experimental data and simulation curves has been obtained considering 72 superlattice periods, a linearly varying composition of the second Ge layer (from 100% to 63.7% from the first to the 15-th period), a pseudomorphic growth with values of interplanar distances at the interface corresponding to case B above described, and a quite good overall order ($F_c = 0.85$).

This example shows that the XSW technique can be effectively applied to superlattice structures, obtaining useful information about the intimate structure at the interface. Even if in this specific case the standing wave periodicity is quite short, dealing with the 20-th order superlattice peak, the same procedure can be applied to any order, with different values of periodicities.

5. GIREX

When a thin light film (thickness t ≈ 100 nm) is deposited onto a heavier substrate, a strong resonance in the e.m. field intensity inside the film can take place in the glancing angle incidence region (Wang et al., 1992 ; de Boer, 1991). This technique, that we will call GIREX (Glancing Angle Resonance Enhanced X-rays), can be used to study thin films and to produce intense sub-micrometer X-ray beams. The principle is the following: as pointed out before, the critical angle is proportional to the square root of the electron density of the reflecting materials. Therefore lighter materials have smaller critical angles. In the case of a light thin film on a heavier substrate, for some incidence angles the x-ray beam can penetrate in the thin film, but is totally reflected at the substrate interface. As a consequence a standing wave is formed in the thin film, with a periodicity which depends on the incidence angle. When the S.W. periodicity is equal to an integer fraction of the thin film thickness, then a resonance takes place due to the constructive interference between the beam transmitted in the thin film and the beam reflected at the interface film-vacuum. The resonance can enhance the e.m. field intensity inside the film up to tens or hundreds times the incoming intensity. Figure 4 shows the calculated e.m. field, for the first three resonances, inside a C film 985 Å thick deposited onto a Ni-covered glass. The field enhancement can be used for two main applications: structural studies of thin films and production of submicrometer x-ray beams.

5.1 Thin film studies

The existence of nodes and antinodes in the structure of the e.m. field inside the film allows accurate studies, and in particular the determination of the film thickness and the localization of foreign atoms. As an example we will describe a study on amorphous C films deposited onto a Ni-covered glass, where a Ti marker has been placed at different depths. (Di Fonzo et al, 1996). The measurements have been performed using a standard laboratory x-ray source. Three C layers about 1000 Å thick were deposited by sputtering. One of these samples had a Ti layer 5 Å thick in the center, while the others had 2 Å Ti at respectively 1/3 (≈328 Å) and 2/3 (≈ 656 Å), starting from the surface, of the total film thickness.

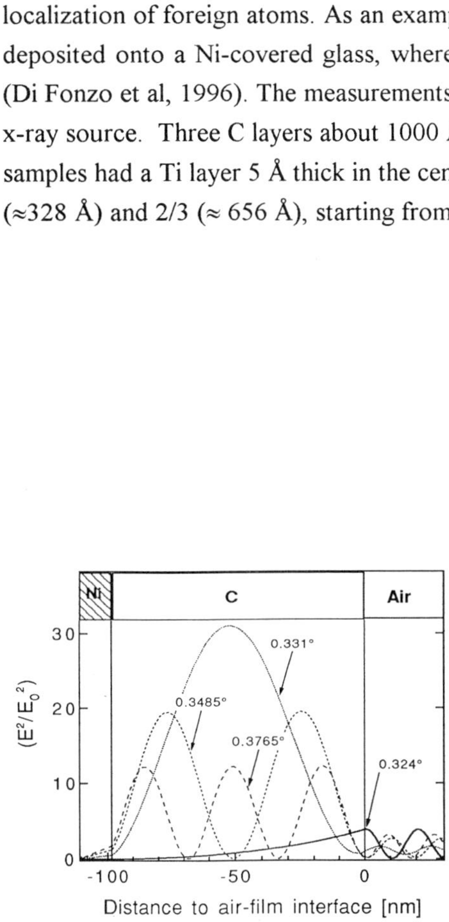

Figure4. e.m.. field intensity vs. depth in C film 985 Å thick on a Ni-covered glass for incidence angles corresponding to the first three resonances (0.331°, 0.3485°, 0.3765° respectively) The curve at 0.324° corresponds to the critical angle for C Incident photon energy: 5.4 KeV.

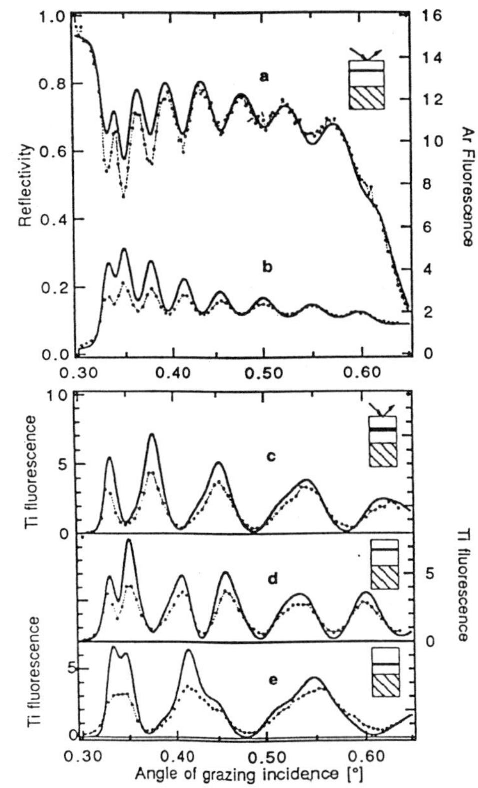

Figure 5. Exp. data (closed points) and simulation curves for (a): reflectivity; (b) Ar fluorescence; (c) Ti fluor. for 5 Å Ti at the center of the C film; (d) Ti fluor. for 2 Å Ti at ≈328 Å from the surface; (e) Ti fluor. for 2 Å Ti at ≈ 656 Å from the surface as a function of incidence angle.

Moreover, Ar contamination was present in the films due to the deposition procedures. The reflected intensity was measured with a Na(I) scintillation detector, while Ti and Ar fluorescence was measured with a Si(Li) detector. Figure 5a) reports the angular dependence of the reflectivity, which is almost identical for the three samples.The simulation curve has been calculated taking into account all reflection and transmission processes of the monochromatic incident planar waves at the different interfaces (N-1 for a N layers system). The formalism developed by Parratt (1954) has been used for the calculation. Figure 5b) reports the angular dependence of the Ar fluorescence. Also in this case the same results are obtained for the three samples. The simulation curve (full line) has been calculated considering an homogeneous distribution of the Ar atoms in the film. It is worthy to note that at each minimum in the reflectivity corresponds a maximum in the Ar fluorescence. This proves the enhancement of the e.m. field at resonance. The agreement between theoretical and experimental features is good, mainly for the angular positions of the resonance peaks. In figures 5c) - 5e) the Ti fluorescence for the three different positions of the Ti marker in the C film is shown (full line: simulation curve; closed points: experimental data). The differences in the spectra allow us to determine the positions of the marker layer. For example, fig. 5c refers to the sample with the Ti layer in the center of the C film, where the even resonances have nodes and the odd resonances antinodes (see figure 4). Accordingly Ti fluorescence is excited only at odd resonances. The other two samples have the Ti marker at symmetric positions with respect to the center of the film, i.e. at 1/3 and 2/3 of the total thickness. At these positions the 3rd and 6th orders of the standing wave patterns have nodes, thus the fluorescence intensity has a minimum for these orders. It is interesting to note that a difference exists between the two samples, due to the movement of the standing wave pattern as a function of the incidence angle. One important point is the accuracy in the determination of the total film thickness and of the marker position: from the angular position of the reflectivity minima in figure 5a) a total thickness value of 985 Å ± 10 Å has been derived. The marker position has no influence on the reflectivity minima positions, but determines instead the structure of the Ti fluorescence features. From the best fit (see figures 5c-5e) the Ti position can be derived within 10 Å. Therefore, even with a standard X-ray source, thickness and marker position can be evaluated with about 1% accuracy.

5.2 Submicrometer X-ray beam production

The resonance effect can be used not only to study thin film structures, but also to produce submicrometer beams. In fact in resonance the X-ray beam entering the structure is reflected back and forth at the interfaces, and is therefore trapped in the thin film which acts as a waveguide (see figure 6). The beam exiting from the waveguide should then have one dimension equal to the thin film thickness, i.e. of the order of 100 nm. Such a beam has been actually obtained and characterized (Lagomarsino et al., 1996, Feng et al, 1996). The structure is similar to that used for the marker experiment, but an important difference relies on the presence of a very thin metallic cap layer on top of the C film: this improves significantly the reflectivity for the beam trapped in the waveguide but at the same time is thin enough to allow the incoming beam to pass through without too much attenuation. It is clear that a beam can be seen at the waveguide exit only when the resonance takes place. Figure 7a) shows the sample reflectivity (only the region correspondig to the first two resonances) and figure 7b) the intensity coming out from the waveguide. The figures refer to the first experiment, carried out at the microfocus beamline (BL1) of ESRF, where the efficiency was of the order of 10^{-4}. Subsequent experiments, whose analysis is not yet completed, yielded a significantly improved efficiency and a monochromatized flux in the range of 10^8 ph/s has been obtained in a beam of 130 nm x 0.5 mm. The dimensions of the beam have been measured by means of transmission through a lithographic test pattern. It is also interesting to study the structure of the beams exiting from the waveguide: in fact two beams come out at the waveguide end, due to the reflection at the two interfaces. Careful analysis of the exiting beams indicates that the two beams are coherent. Several applications in the field of microimaging, microdiffraction and microspectroscopy are envisaged.

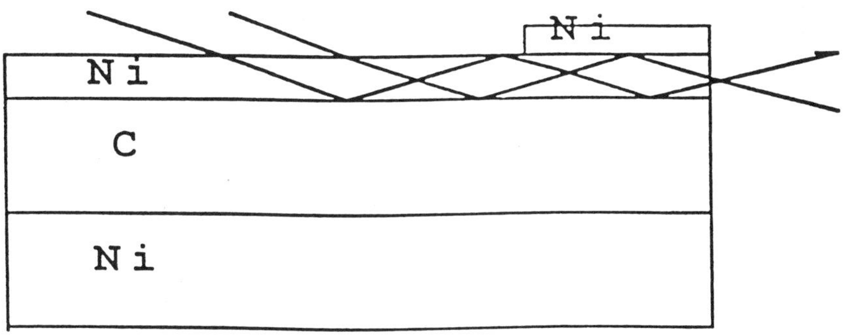

Figure 6. Scheme of the waveguiding effect

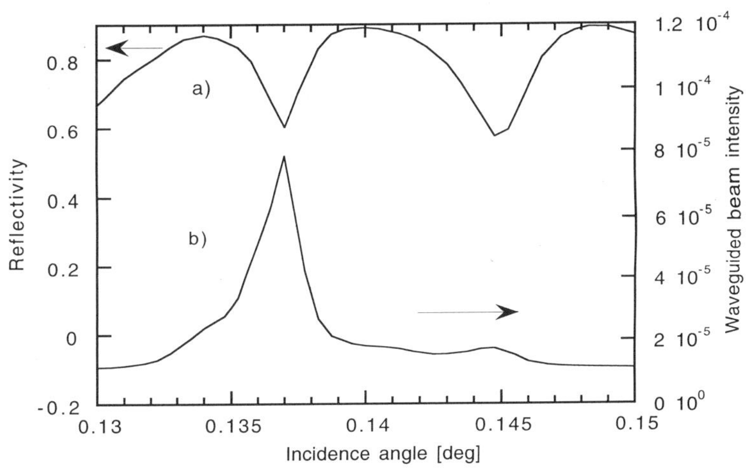

Figure 7. Reflectivity (a) and intensity exiting from the waveguide (b) as a function of the sample angular position.

REFERENCES

Barbee, T.W. and Warburton, W. K., 1984, X-ray evanescent and standing-wave fluorescence studies using a layered synthetic microstructure, *Mat. Lett.* 3:17

Bedzyk, M.J., Bilderback, D., White, J., Abruna, H.D. and Bommarito, M.G.,1986, Probing electrochemical interfaces with x-ray standing waves, *J. Phys. Chem.*, 90:1926

Bedzyk, M.J., Bilderback, D.H., Bommarito, G.M., Caffrey, M. and Schildkraut, J.S., 1988, X-ray Standing Waves: a molecular yardstick for biological membranes, *Science* 241:1788

Bedzyk, M. J., Bommarito, G.M., Schildkraut, J.S., 1989, X-ray standing waves at a reflecting mirror surface, *Phys. Rev. Lett.*, 62:1376

Castrucci, P., Lagomarsino, S., Calicchia, P., and Cedola, A., 1996, X-ray Standing Wave study of Si/Ge superlattices, *Appl. Surf. Science*, in press

Chernov, V.A., Chkhalo, N.I., Dolbnya, I.P. and Zolotarev, K.V., 1995, The application of the x-ray standing wave method to study Ni/C layered structures obtained by laser-assisted deposition, *Nuclear Instr. and Methods in Phys. Res.* A 359:175

De Boer, D.K.G., 1991, Glancing-incidence x-ray fluorescence of layered materials, *Phys. Rev* B44:498

Di Fonzo, S., Jark, W., Lagomarsino, S., Cedola, A., Muller, B.R and Pelka, J.B., 1996, Electromagnetic field resonance in thin amorphous films: a tool for non-destructive localization of thin marker layers by use of a standard X-ray tube, *Thin Solid Films*, in press

Feng, Y.P., Sinha, S.K., Fullerton, E.E., Grubel, G., Abernathy, D., Siddons, D.P. and Hastings, J.B., 1995, X-ray Fraunhofer diffraction patterns from a thin-film waveguide, *Appl. Phys. Lett.*, 67:3647

Halliwell, M.A.G., Lyons, M.H. and Hill, M.J., 1984, The interpretation of X-ray rocking curves from III-V semiconductor device structures, *J. Crystal Growth* 68:523

Heald, S.M. and Tranquada, J. M., 1989, Standing-wave-assisted extended x-ray absorption fine-structure study of a Ni-Ti multilayer, *J. Appl. Phys.*, 65:290

Kawamura, T. and Takenaka, H., 1994, Interface roughness characterization using x-ray standing waves, *J. Appl. Phys.* 75:3806

Kortright, J.B. and Fischer-Colbrie, A., 1987, Standing wave enhanced scattering in multilayer structures, *J. Appl. Phys.* 61:1130

Lagomarsino, S., Castrucci, P., Calicchia, P., Cedola, A. and Kazimirov, A., 1995, X-ray standing wave and high resolution diffraction study of SiGe superlattices, in *MRS Vol. 379- Strained Layer Epitaxy-Materials, Processing and device applications*, MRS publisher

Lagomarsino, S., Jark, W., Di Fonzo, S., Cedola, A., Mueller, B., Engstrom, P. and Riekel, C., 1996, Submicrometer x-ray beam production by a thin film waveguide, *J. Appl. Phys.*, 78

Parratt, L.G., 1954, Surface studies of solids by total reflection of x-rays, *Phys. Rev.* 95, 359

Roberts, G.G., 1985, An applied science perspective of Langmuir-Blodgett films, *Adv. in Phys.* 34:475

Wang, J., Bedzyk, M.J. and Caffrey, M., 1992, Resonance-Enhanced X-rays in thin films: a structure probe for membranes and surface layers, *Science* 258:775

Zheludeva, S.I., Lagomarsino, S., Novikova, N.N., Kovalchuk, M.V. and Scarinci, F., 1990, X-ray Standing Waves in Bragg diffraction and in total reflection regions using Langmuir-Blodgett multilayers, *Thin Solid Films*, 193:395

Zheludeva, S.I. Kovalchuk, M.V., Novikova, N.N., Bashelhanov, I.V., Salaschenko, N.N., Akhsakhalyan, A.D. and Platonov, Y.Y., 1992, Thickness and density determination of ultrathin solid films comprising multilayer x-ray mirrors by x-ray reflection and fluorescence study, *Rev. Sci. Instr.* 63:1130

THEORETICAL DESCRIPTION OF MULTIPLE CRYSTAL ARRANGEMENTS

Václav Holý, and Petr Mikulík

Dept. of Solid State Physics, Masaryk University,
Kotlářská 2, 611 37 Brno,
Czech Republic

1. INTRODUCTION

In a multiple crystal arrangement (MCA), the X-ray beam irradiating the sample is conditioned by crystal monochromators and collimators, and the beam emitted by the sample is analyzed by a crystal analyzer. In this paper, the function of a MCA will be described. The diffraction process in the crystals will be calculated by means of the conventional dynamical diffraction theory (see the Chapter by Authier in this Volume), so that the highly asymmetric diffraction will not be taken into account. Geometrical optics will be used for the description of the propagation of the X-rays between the crystals.

The result of a MCA experiment is given by the convolution of the reflectivity of the sample with the device function of the MCA. Several works have dealt with the device function of MCA. The basis of the theory of double-axis and triple-axis diffractometers has been published already in early sixties in the pioneering paper by Bubáková et. al. (1961). Monolithic monochromators frequently used in MCA have been described thoroughly by Kikuta and Kohra (1970), Kikuta (1971) and Matshushita et al. (1971). Bartels (1983) analyzed the function of the Bartels monochromator used in most of the triple-axis diffractometers. The device function of a triple-axis diffractometer has been studied by Brügemann et al. (1992) and Neumann et al. (1994). The role of the slits in a MCA has been analyzed by van der Sluis (1994).

2. DYNAMICAL DIFFRACTION FROM ONE PERFECT CRYSTAL

2.1. Direction of the diffracted radiation

If a plane, monochromatic wave irradiates a perfect crystal, the diffracted wave is plane and monochromatic as well. The direction of the wave vector $\mathbf{K}_h$ of the diffracted

X-ray and Neutron Dynamical Diffraction: Theory and Applications
Edited by Authier *et al.*, Plenum Press, New York, 1996

wave is connected with the wave vector $\mathbf{K}_0$ of the incoming radiation by the boundary condition

$$\mathbf{K}_{h\|} = \mathbf{K}_{0\|} + \mathbf{h}_{\|} \tag{1}$$

following from the continuity condition of the wave vectors at the vacuum–crystal surface. The subscript $_{\|}$ denotes the component parallel to the surface, $\mathbf{h}$ is the diffraction vector. Eq. (1) is valid for the Bragg case diffraction, or for the Laue case and a plan-parallel crystal. In the following, we restrict ourselves to the Bragg case diffraction on a semiinfinite crystal, since the Laue case has a limited importance in MCAs.

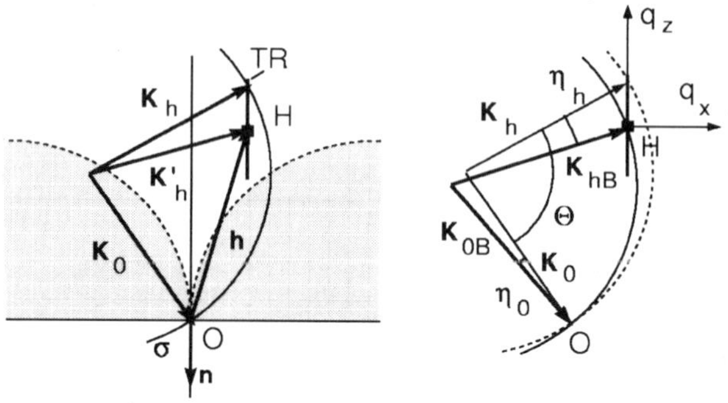

Figure 1. The Ewald construction of the diffraction from a perfect crystal. TR means the truncation rod, the reciprocal lattice points in the shaded area represent the Laue case of diffraction. $\mathbf{K}_{0,h}$ are the wave vectors of the primary and diffracted beams, $\mathbf{K}'_h = \mathbf{K}_0 + \mathbf{h}$.

Since the scattering is elastic,

$$|\mathbf{K}_h| = |\mathbf{K}_0| = 2\pi/\lambda \tag{2}$$

must hold (λ is the wavelength in vacuum). From (1) and (2) the well-known Ewald construction follows that yields the direction of $\mathbf{K}_h$. As shown in Figure 1, the ending point of $\mathbf{K}_h$ lies at the intersection of the Ewald sphere with the truncation rod going through the reciprocal lattice point H ($\overrightarrow{OH} = \mathbf{h}$) perpendicular to the crystal surface.

If we denote $\mathbf{K'}_h = \mathbf{K}_0 + \mathbf{h}$, the vertical component $(\mathbf{K}_h - \mathbf{K'}_h)_\perp$ is proportional to the deviation of the crystal position from the diffraction maximum, i.e. it determines the diffracted intensity.

Simple geometrical considerations following from Figure 1 provide the expression

$$\eta_h = -\frac{\gamma_0}{\gamma_h}\eta_0, \quad \Theta = 2\theta_B + \eta_0 \left(1 - \frac{\gamma_0}{\gamma_h}\right) \tag{3}$$

for the angle Θ between the primary $\mathbf{K}_0$ and diffracted $\mathbf{K}_h$ waves and for the angular deviation η_h of the diffracted wave from its direction at the diffraction maximum. These angles depend on the angular deviation η_0 of the primary beam from its direction at the maximum and on the direction cosines $\gamma_{0,h}$ of the vectors $\mathbf{K}_{0B}$ and $\mathbf{K}_{hB}$ with respect to the internal surface normal. The vectors $\mathbf{K}_{0,hB}$ are the vectors $\mathbf{K}_{0,h}$ at the diffraction maximum (see Figure 1), $2\theta_B$ is the angle between them following from the Bragg equation

$$\sin\theta_B = h/(2K) \tag{4}$$

not corrected by X-ray refraction.

Eq. (3) has been derived assuming $\eta_0 \ll \theta_B$. and it is *not affected* by X-ray refraction, influencing only the dependence of the diffracted intensity on η_0. It follows only from (1,2) so that it is valid both within the kinematical and dynamical theories.

Eq. (3) has interesting consequences. For instance, in the symmetrical Laue case $\gamma_0 = \gamma_h$ holds and, therefore, $\eta_h = -\eta_0$ and $\Theta = 2\theta_B = \text{const}$. Since $\gamma_h = -\gamma_0$ in the symmetrical Bragg case, we obtain $\eta_h = \eta_0$ and $\Theta = 2(\theta_B + \eta_0)$.

2.2. Crystal reflectivity

We introduce the crystal reflectivity $\mathcal{R}(\eta_0)$ as a ratio of the energy *fluxes* of the diffracted and primary beams. In the Bragg case on a semiinfinite crystal, the conventional dynamical diffraction theory yields

$$\mathcal{R}(\eta_0) = \frac{I_h S_h}{I_0 S_0} = \left|\frac{\chi_h}{\chi_{\bar{h}}}\right|^2 \left|y \pm \sqrt{y^2 - 1}\right|^2, \tag{5}$$

where $\chi_{0,h,\bar{h}}$ are the Fourier coefficients of the crystal polarizability,

$$y = \frac{2\eta_0 \sin(2\theta_B) + \chi_0(1 - b)}{2C\sqrt{|b|}\sqrt{\chi_h \chi_{\bar{h}}}}$$

is the effective deviation from the diffraction maximum, $b = \gamma_h/\gamma_0$ is the asymmetry factor ($b < 0$ in the Bragg case), and C is the linear polarization factor. The sign in Eq. (5) must be chosen so that $\mathcal{R} < 1$. The ratio of the cross-sections of the beams is $S_h/S_0 = |b|$. The term $\mathrm{Re}(\chi_0)(1 - b)$ in the expression for y represents the shift of the diffraction maximum due to X-ray refraction. This shift depends on the diffraction asymmetry, for instance, $b = 1$ in the Laue case and the maximum is not shifted.

The dependence $\mathcal{R}(\eta_0)$ is called input (intrinsic) reflection curve of the crystal. Its width (FWHM) depends on the asymmetry factor as follows

$$(\Delta\eta_0)_{in} = \frac{2C}{\sin(2\theta_B)}\sqrt{|b|}\mathrm{Re}(\sqrt{\chi_h \chi_{\bar{h}}}) \equiv (\Delta\eta)_{in}^{sym}\sqrt{|b|}, \tag{6}$$

where $(\Delta\eta)_{in}^{sym}$ is the FWHM in the symmetrical case. Since the deviation η_h is proportional to η_0, from (3) we can calculate the dependence $\mathcal{R}'(\eta_h)$ of the reflectivity on the deviation of the diffracted wave (output intrinsic reflection curve). Its FWHM is

$$(\Delta\eta_h)_{in} = \Delta\eta_0/|b| \equiv (\Delta\eta)_{in}^{sym}/\sqrt{|b|}. \tag{7}$$

For instance, if the angle of $\mathbf{K}_0$ with the surface is smaller than that of $\mathbf{K}_h$, $|b| > 1$ and $(\Delta\eta_h)_{in} < (\Delta\eta_0)_{in}$.

2.3. A single crystal as a collimator

According to Figure 2, the crystal is irradiated by a divergent wave emitted from a point focus of an X-ray tube. The spectral density of the wave is given by the function $S(\lambda - \lambda_0)$ and we assume that the spectral range (spectral line) is sufficiently narrow ($|\lambda - \lambda_0| \ll \lambda_0$). Each ray falling on the crystal is characterized by its deviation η_0 of its direction from the diffraction maximum and the deviation $\delta\lambda = \lambda - \lambda_0$ of its wavelength from the maximum λ_0 of the spectral line. If the energy flux of each primary ray is unity, the energy flux of the diffracted wave $\mathcal{J}_h$ is given by the reflectivity $\mathcal{R}$ (5), its wavelength is same and its direction is determined by the deviation η_h (3).

The connection between η_0, η_h and λ follows from (3) and from the Bragg law (4). It can be expressed in a simple geometrical way, if we assume $|\eta_{0,h}| \ll \theta_B$. The points in the $(\eta_0, \delta\lambda)$ plane corresponding to the rays diffracting from the crystal create a stripe

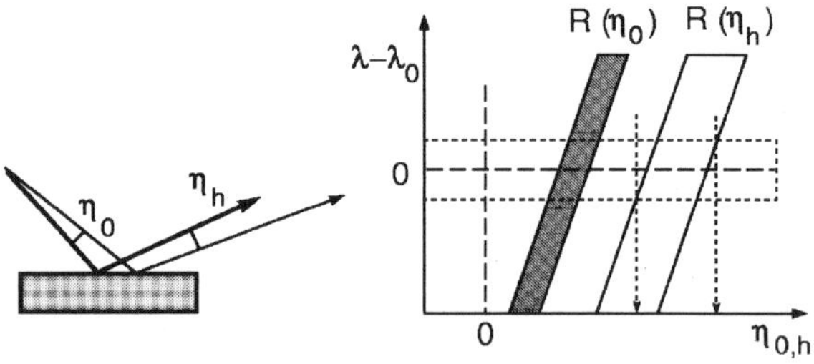

Figure 2. Single crystal as a collimator (left) and its DuMond graph (right).

of the width $(\Delta\eta_0)_{in}$ making the angle $\arctan(\lambda_0/\tan\theta_B)$ (so called DuMond graph). Similarly, the diffracted rays correspond to the points in the $(\eta_h, \delta\lambda)$ plane lying in a stripe of the width $(\Delta\eta_h)_{in}$. If the spectral width of the primary radiation is denoted by a horizontal stripe (see Figure 2), from the graph the divergence of the rays emitted by the crystal can simply be deduced, it is denoted by vertical dashed arrows. Therefore, the beam emitted by the crystal has the same spectral density as the primary beam, its divergence, however, is drastically reduced. From Figure 2, this divergence can be estimated to

$$\Delta\eta_h \approx (\Delta\eta_h)_{in} + \frac{\Delta\lambda}{\lambda_0}\tan\theta_B \tag{8}$$

where $\Delta\lambda$ is the FWHM of the spectral line.

If the detector behind the crystal collects all the rays, the signal measured by the detector is

$$\Phi = \int_{\eta_0^{min}}^{\eta_0^{max}} d\eta_0 \int d(\delta\lambda)\, S(\delta\lambda)\mathcal{R}(\eta_0 - \frac{\delta\lambda}{\lambda_0}\tan\theta_B) \tag{9}$$

Here we have assumed that the reflectivity $\mathcal{R}$ depends on λ only by means of the dependence $\theta_B(\lambda)$, i.e. we have neglected the slow dependences of $\gamma_{0,h}$, $\chi_{0,\pm h}$ and C on λ. We have also neglected the influence of the vertical divergence (perpendicular to the plane of the picture), the role of this divergence will be treated separately in Section 4. The limits $\eta_0^{min,max}$ of the horizontal divergence are given by the slits in the MCA.

3. MANY CRYSTAL ARRANGEMENTS

3.1. Two separate crystals

In principle, there are two basic arrangements of a pair of the crystals according to Figure 3. Their corresponding DuMond graphs are in Figure 3 as well. The signal of the detector placed behind the second crystal will be (at the same assumptions as in the previous section)

$$\Phi = \int_{\eta_{01}^{min}}^{\eta_{01}^{max}} d\eta_{01} \int d(\delta\lambda)\, S(\delta\lambda)\mathcal{R}_1(\eta_{01})\mathcal{R}_2(\mp\frac{\eta_{01}}{b_1} \pm \alpha_2 \pm \delta_1 - \delta_2) \tag{10}$$

where η_{01} is the angular deviation of the ray falling on the first crystal. α_2 is the rotation angle of the second crystal as shown in Figure 3. The position $\alpha_2 = 0$ corresponds to the diffraction maximum for $\eta_{01} = 0$, and $\delta\lambda = 0$. The upper (lower) signs correspond to the arrangement (a) or (b), respectively. $\delta_{1,2}$ mean the changes in the Bragg angles

$$\delta_j = \frac{\delta\lambda}{\lambda_0}\tan\theta_{Bj},\ j = 1, 2 \tag{11}$$

where θ_{Bj} is the Bragg angle of the j-th diffraction for $\lambda = \lambda_0$.

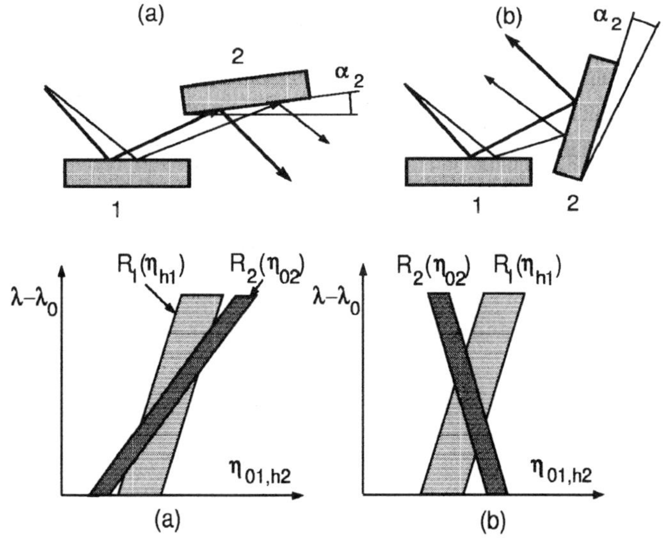

Figure 3. Two arrangements of the pair of crystals, double crystal diffractometer (a) and spectrometer (b) and their DuMond graphs. The stripes correspond to the rays diffracted by the first crystal ($\mathcal{R}_1(\eta_{h1})$) and to the rays falling on the second crystal ($\mathcal{R}_2(\eta_{02})$).

As a special case, let us assume that the Bragg angles $\theta_{B1,2}$ of both crystals are same. Then, for arrangement (a), the term in (10) containing $\delta\lambda$ vanishes and the detector signal *does not depend* on the spectral line $S(\delta\lambda)$. In the DuMond graph, the stripes $\mathcal{R}_1(\eta_{h1})$ and $\mathcal{R}_2(\eta_{02})$ are parallel. This arrangement is called *nondispersive* and it is often used for high–resolution X-ray diffractometry. In arrangement (b) this term never vanishes (*dispersive setting*) so that this arrangement is always sensitive to $S(\delta\lambda)$, therefore it is used for X-ray spectrometry. The advantage of the dispersive setting is that the spectral width of the radiation emitted from the second crystal is reduced. This is used in the Bartels monochromator described in Section 3.3.

The purpose of most double crystal arrangements is to measure the intrinsic reflection curve of the second crystal (sample). From (10) it follows that the measured signal Φ equals the convolution

$$\Phi(\alpha_2) = \int_{\eta_{02}^{min}}^{\eta_{02}^{max}} \mathrm{d}\eta_{02} G(\eta_{02}) \mathcal{R}_2(\eta_{02} + \alpha_2) \tag{12}$$

of the intrinsic reflection curve $\mathcal{R}_2$ of the measured sample with the device function

$$G(\eta_{02}) = |b_1| \int \mathrm{d}(\delta\lambda)\, S(\delta\lambda) \mathcal{R}_1((\delta_1 - \delta_2 - \eta_{02})b_1) \tag{13}$$

where we have assumed the arrangement (a) in Figure 3. If the Bragg angles of both the crystals are same, the device function is proportional to the intrinsic reflection curve of the first crystal. The measured signal equals the convolution of the reflection curves $\mathcal{R}_{1,2}$ and the spectral width of the primary radiation plays no role. Thus, the arrangement (a) can be used for measuring the reflectivity of the sample only if the materials of the sample and the monochromator have similar Bragg angles.

3.2. Two diffractions in one crystal block

Two successive diffractions can be performed in the same crystal block in the arrangements according to Figure 4.

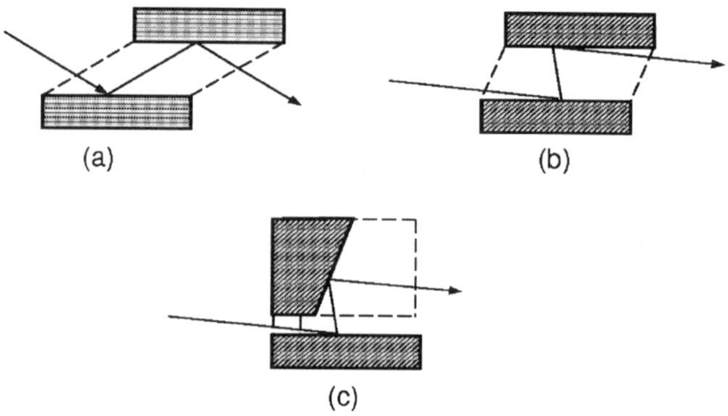

Figure 4. Monolithic twofold collimators.

The diffracting parts of the block must be cut so that both diffractions have the maximum for the same ray. Eq. (5) provides the condition for the diffraction maximum

$$\eta_0^{max} = -\frac{\mathrm{Re}(\chi_0)}{2\sin(2\theta_B)}(1-b) \tag{14}$$

From (10) we obtain the necessary misorientation α^{max} of the diffracting net planes for the second diffraction

$$\alpha^{max} = -\frac{\mathrm{Re}(\chi_0)}{2\sin(2\theta_B)}\left(\frac{1}{b^{(1)}} - b^{(2)}\right) \tag{15}$$

In arrangement (a) in Figure 4, both the diffractions are symmetrical, i.e. $b^{(1)} = b^{(2)} = -1$, and $\alpha^{max} = 0$. In arrangement (b) the asymmetries of the diffractions are opposite ($b^{(2)} = 1/b^{(1)}$) and $\alpha^{max} = 0$ again. In arrangement (c) $b^{(1)} < -1, b^{(2)} < -1$ holds and $\alpha^{max} \neq 0$. Therefore, in this case, the second part of the crystal block must be slightly rotated. This rotation can be achieved, most conveniently, by an elastic deformation of the "neck" connecting both parts of the block.

In a double-axis arrangement, the first crystal can be replaced by a monolithic block. The measured signal is given by Eq. (12) again, the device function being

$$G(\eta_{02}) = \int \mathrm{d}(\delta\lambda) S(\delta\lambda) \mathcal{R}_1^{(1)}(b_1^{(1)} b_1^{(2)}(\eta_{02} - \delta_1 + \delta_2)) \mathcal{R}_1^{(2)}(b_1^{(2)}(\delta_1 - \delta_2 - \eta_{02})) \tag{16}$$

where $b_1^{(1,2)}$ are the asymmetry factors of the successive diffractions in the monolithic block. Comparing Eq. (13) with (16) we find the substantial advantage of the monolithic collimator comparing with the single crystal. In the arrangements (a) and (b), the width of the device function is the same as in the case of one diffraction, however, the tails of $G(\eta_{02})$ are substantially reduced. The arrangement (c) provides an extremely narrow device function. In the case of arrangement (a), $\mathcal{R}_1^{(1)} = \mathcal{R}_1^{(2)}$ holds and the device function is a convolution of the spectral density with the square of the reflectivity $\mathcal{R}_1^{(1)}$. Again, if the Bragg angles of the monolithic collimator and the sample are same, the spectral distribution does not affect the measured signal.

3.3. Triple-axis diffractometer

Let us deal with two arrangements of a triple-axis diffractometer shown in Figure 5. The first and the second crystals in the arrangement (a) are usually the collimators and the third crystal is the sample. Since the first two crystals are in the dispersive setting, they can reduce not only the divergence, but also the spectral width of the radiation.

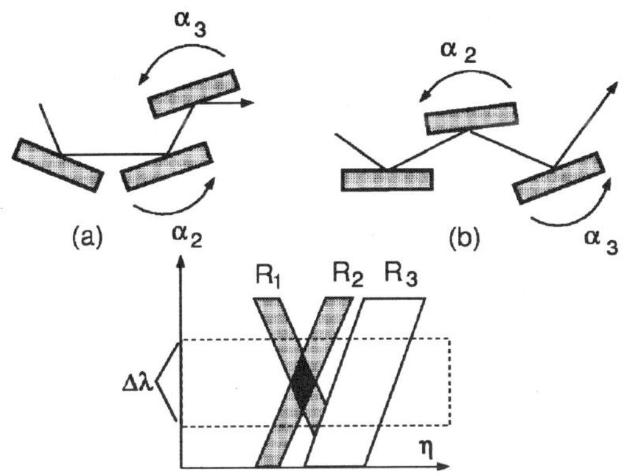

Figure 5. The arrangements of a triple-axis diffractometer and the DuMond graph of the arrangement (a).

In the arrangement (b), the first crystal is a collimator, the second one is the sample and the third crystal acts as a directionally sensitive analyzing element (analyzer). This arrangement makes it possible to measure not only the intensity of the beam diffracted by the sample, but also its direction. This is important especially if diffuse X-ray scattering takes place in the sample.

The measured signal in both arrangements is (the upper signs refer to (a), the lower signs to the arrangement (b))

$$\Phi(\alpha_2, \alpha_3) = \int_{\eta_{01}^{min}}^{\eta_{01}^{max}} d\eta_{01} \int d(\delta\lambda)\, S(\delta\lambda)\mathcal{R}_1(\eta_{01})$$

$$\times\, \mathcal{R}_2(\pm\frac{\eta_{01}}{b_1} \mp (\alpha_2 + \delta_1) - \delta_2)\mathcal{R}_3(\mp\frac{\eta_{01}}{b_1 b_2} \mp \alpha_2(1 - \frac{1}{b_2}) \pm \alpha_3 + \delta_2 - \delta_3 + \frac{\delta_2 \pm \delta_1}{b_2}) \quad (17)$$

The first two crystals in (a) are used as a monochromator (the Bartels monochromator) and, usually, they are replaced by the monolithic blocks shown in Figure 5a. Then, the measured signal is a convolution of the reflectivity of the sample

$$\Phi(\alpha_3) = \int_{\eta_{03}^{min}}^{\eta_{03}^{max}} d\eta_{03} \mathcal{R}_3(\eta_{03} + \alpha_3) G(\eta_{03})$$

with the device function of the Bartels monochromator

$$G(\eta_{03}) = \int d(\delta\lambda)\, S(\delta\lambda)\mathcal{R}_1^2(-\eta_{03} - \delta_1 - \delta_3)\mathcal{R}_2^2(\eta_{03} - \delta_1 + \delta_3) \quad (18)$$

where we have assumed $b_1 = b_2 = -1$, $\alpha_2 = 0$ and $\theta_{B1} = \theta_{B2}$. From (18) the substantial advantage of the Bartels monochromator follows. Even if the Bragg angles of the monochromator (θ_1) and sample (θ_3) are different, the device function is not broadened by the spectral width. This is demonstrated in Figure 5, where its DuMond graph is plotted. The beam emitted by the monochromator corresponds to the intersection of the stripes $\mathcal{R}_1$ and $\mathcal{R}_2$ (black area) that may be substantially smaller in λ direction than the spectral width $\Delta\lambda$ of the primary radiation denoted by the horizontal dotted lines. Therefore, in the contrast to a single collimator (Section 3.1), the same Bartels monochromator can be used for measuring the reflectivities of various samples.

In Figure 6 the device functions of a single–crystal collimator and the Bartels monochromator are compared. If the collimator and the sample have the same Bragg angles (the full lines), the arrangement is nondispersive and the device function equals the intrinsic (output) reflection curve of the collimator. If the Bragg angles are different, the spectral profile $S(\delta\lambda)$ broadens the device function. In the case of a single collimator (the left panel), the width of the device function is a sum of the intrinsic width and the contribution of the dispersion after (8). The Bartels monochromator transmits only a part of the spectral line and its device function (the right graph) is nearly not changed by the dispersion.

In the arrangement (b) of Figure 5, the signal can be expressed by the device function as well

$$\Phi(\alpha_2, \alpha_3) = \int_{\eta_{02}^{min}}^{\eta_{02}^{max}} \mathrm{d}\eta_{02} \mathcal{R}_2(\eta_{02}) G(\eta_{02} - \alpha_2, \eta_{02} - b_2(\alpha_2 - \alpha_3)) \tag{19}$$

and the device function is

$$G(\alpha, \beta) = |b_1| \int \mathrm{d}(\delta\lambda)\, S(\delta\lambda) \mathcal{R}_1(b_1(-\alpha + \delta_1 - \delta_2)) \mathcal{R}_3(-\frac{1}{b_2}\beta - \delta_1 + \delta_2) \tag{20}$$

(we have assumed $\theta_1 = \theta_3$).

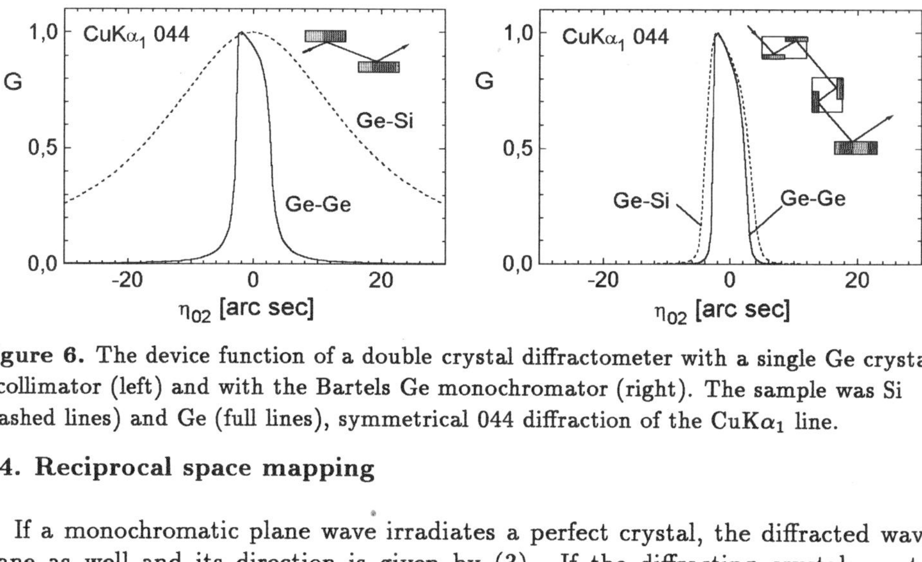

Figure 6. The device function of a double crystal diffractometer with a single Ge crystal as a collimator (left) and with the Bartels Ge monochromator (right). The sample was Si (dashed lines) and Ge (full lines), symmetrical 044 diffraction of the CuKα_1 line.

3.4. Reciprocal space mapping

If a monochromatic plane wave irradiates a perfect crystal, the diffracted wave is plane as well and its direction is given by (3). If the diffracting crystals contains randomly placed defects, the diffracted wave is divergent due to diffuse X-ray scattering. The intensity of the scattered wave can be expressed being the function of the wave vector transfer

$$\mathbf{Q} = \mathbf{K}_h - \mathbf{K}_0 = \mathbf{h} + \mathbf{q}$$

In the arrangement according to Figure 5b, the distribution $\Phi(\mathbf{q})$ can be measured. The rotation angles α_2, α_3 of the sample and the analyzing crystal, respectively, are connected with the wave vector transfer by the formula

$$\mathbf{q} = (q_x, q_z) = K(\alpha_2(\gamma_0 - \gamma_h) + \alpha_3\gamma_h, \alpha_2(\delta_0 - \delta_h) + \alpha_3\delta_h) \tag{21}$$

where the coordinate axes $q_{x,z}$ are defined in Figure 1 and $\delta_{0,h}$ are the sines of the angles of $\mathbf{K}_{0,h}$ with the internal surface normal of the sample. Especially, in the symmetrical Bragg case we obtain

$$\mathbf{q} = K((2\alpha_2 - \alpha_3)\sin\theta_B, \alpha_3\cos\theta_B) \tag{22}$$

Rotating the sample and keeping constant the position of the analyzer, we measure a linear scan (so called ω-scan), whose trajectory in reciprocal plane can be approximated by a straight line perpendicular to $\mathbf{h}$. The trajectory of the $\omega/2\theta$-scan (rotation of the sample and of the analyzer by the double angular velocity) is a straight line crossing the origin of reciprocal space. Rotating both the sample and the analyzer independently, we can scan an area of reciprocal plane (*reciprocal space map*). Inverting Eq. (21) we obtain

$$\alpha_2 = \frac{q_x\delta_h - q_z\gamma_h}{K\sin(2\theta_B)}, \quad \alpha_3 = \frac{q_x(\delta_h - \delta_0) + q_z(\gamma_0 - \gamma_h)}{K\sin(2\theta_B)} \tag{23}$$

Putting these relations into Eq. (20) we obtain the device function of the triple-axis diffractometer expressed in $q_{x,z}$ coordinates. This function determines the resolution of the diffractometer in reciprocal plane.

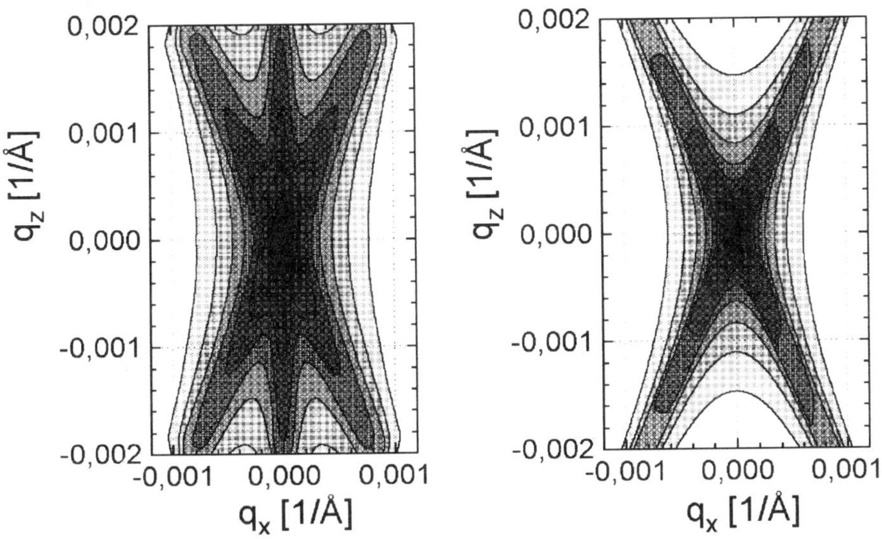

Figure 7. The device function of a triple axis diffractometer (left) and the detector signal (right). The diffractometer contains three Ge crystals, symmetrical diffraction 022 Cu$K\alpha_1$. The equi-intensity contours are logarithmic, their step is the half of the decade.

Figure 7 shows an example of the device function calculated for the case of a successive diffraction from three Ge single crystals. The arrangement is nondispersive, so that the device function is a simple product of the reflectivities of the first and the third crystals. The device function has the characteristic shape in reciprocal plane, its maximum is stretched along the lines $q_x = \pm q_z\tan\theta_B$. These lobes are caused by the tails of the intrinsic reflection curve of the collimator (line $q_x = q_z\tan\theta_B$) and of the analyzing crystal (line $q_x = -q_z\tan\theta_B$). Therefore, these lobes are called collimator and analyzer streaks, respectively. Performing the convolution of the device function with the reflectivity of the ideal sample after Eq. (19), an additional streak occurs (Figure 7, the right panel) perpendicular to the surface. This streak originates in the truncation rod of the sample.

4. CONCLUDING REMARKS

In the previous sections, we assumed that the primary beam was linearly polarized. A completely unpolarized beam (as is the case of a beam from an X-ray tube) can be expressed as an incoherent superposition of the S– and P–polarized components. Thus, for instance, Eq. (12) reads

$$\Phi(\alpha_2) = \int_{\eta_{02}^{min}}^{\eta_{02}^{max}} \mathrm{d}\eta_{02} \left[G^S(\eta_{02})\mathcal{R}_2^S(\eta_{02}+\alpha_2) + G^P(\eta_{02})\mathcal{R}_2^P(\eta_{02}+\alpha_2)\right]$$

Usually, the measured detector signal is normalized to the signal measured without the sample

$$\Phi(\alpha_2) = \frac{\int \mathrm{d}\eta_{02} \left[G^S(\eta_{02})\mathcal{R}_2^S(\eta_{02}+\alpha_2) + G^P(\eta_{02})\mathcal{R}_2^P(\eta_{02}+\alpha_2)\right]}{\int \mathrm{d}\eta_{02} \left[G^S(\eta_{02}) + G^P(\eta_{02})\right]} \tag{24}$$

Similar expressions can be found for a triple-axis diffractometer.

The vertical divergence of the primary wave in the direction perpendicular to the scattering plane (i.e. the plane of the pictures) slightly modifies the formulas. Let us consider two rays, one ray lies in the scattering plane and the other makes a small angle ψ with it. The angular deviation of the latter ray from the diffraction position differs from that of the former ray by $-\psi^2 \tan(\theta_B)/2$. Thus, all the above formulas for the detector signal can be modified aby adding the integral $\int \mathrm{d}\psi$ over the vertical divergence and the factor $-\psi^2 \tan(\theta_B)/2$ into the arguments of all the reflectivities $\mathcal{R}$.

In conclusion, we have found the formulas for the device functions of various multiple crystal arrangements. In all the arrangements, the detector signal can be expressed as a convolution of the intrinsic reflection curve of the sample with the appropriate device function.

REFERENCES

Bartels, W.J., 1983, Characterization of thin layers on perfect crystals with a multipurpose high-resolution X-ray diffractometer, *J. Vac. Sci. Technol.* B1:338.

Brügemann, L., Bloch, R., Press, W., and Tolan, M., 1992, Resolution investigations of X-ray three-crystal diffractometers, *Acta Cryst.* A48:688.

Bubáková, R., Drahokoupil, J., and Fingerland, A., 1961, A contribution to the theory of the triple crystal diffractometer, *Czech. J. Phys.* B11:205.

DuMond, J.W.M, 1937, Theory of the use of more than two successive X-ray crystal reflections to obtain increased resolving power, *Phys. Rev.* 52:872.

Kikuta, S., and Kohra, K., 1970, X-ray crystal collimators using successive asymmetric diffractions and their applications to measurements of diffraction curves, I. General considerations, *J. Phys. Soc. Japan* 29:1322.

Kikuta, S., 1971, X-ray crystal collimators using successive asymmetric diffractions and their applications to measurements of diffraction curves, II. Type I collimator, *J. Phys. Soc. Japan* 30:222.

Matsushita, T., Kikuta, S., and Kohra, K., 1971, X-ray crystal collimators using successive asymmetric diffractions and their applications to measurements of diffraction curves, III. Type II collimator, *J. Phys. Soc. Japan* 30:1136.

Neumann, H.-B., Rütt, U., Bouchard, R., Schneider. J.R., and Nagasawa, H., 1994, The resolution function of a triple-crystal diffractometer for high-energy synchrotron radiation in nondispersive Laue geometry, *J. Appl. Cryst.* 27:1030.

van der Sluis, P., 1994, Slits and high-resolution X-ray diffraction, *J. Appl. Cryst.* 27:1015.

RECIPROCAL SPACE MAPPING

Paul F Fewster

Philips Research Laboratories
Cross Oak Lane
Redhill RH1 5HA
U.K.

1. INTRODUCTION

Crystals that are highly perfect will diffract over a very small angular range and therefore require very high angular resolution x-ray techniques to extract detailed structural information. The aim of this paper is to discuss the use of reciprocal space mapping or diffraction space mapping as a route to extracting very detailed structural information from high quality crystals. In fact much information is obscured by standard diffraction profile analysis methods unless significant assumptions are made to interpret the data. Reciprocal space mapping has not been confined to highly perfect crystals but has yielded detailed understanding of polycrystalline samples, (Fewster and Andrew, 1996). The conventional "powder diffraction" methods yield little detailed information because of the swamping effects of the ill-defined diffraction probe. The well defined probe of the high resolution diffractometer to be described is important for analysing ill-defined samples. Only when the analysist is sure of the assumptions about the structure should a less well defined probe be used. The aim of this paper though is to show how an enormous quantity of structural information from multiple crystal diffractometric methods can be extracted using reciprocal space mapping. The relationship between the resultant diffraction pattern and the real structure will be discussed and how with the aid of additional tools it is possible to come closer to a truer picture of the structure. This paper will be confined to high angular resolution techniques. Low resolution diffraction space mapping has its advantages for mapping very large areas of reciprocal space and for very imperfect samples (Fewster and Andrew, 1996, 1993a).

The advantages of reciprocal space mapping in high resolution mode will be compared with the more conventional double and triple crystal profiles and how misinterpretations can be made by simple profiles. Reciprocal space mapping is a general term and this has been extended to image the shape of the three-dimensional diffraction pattern at very high angular resolution. The procedures for collecting this data will be discussed so that the most appropriate experimental arrangement can be designed.

2. THE NATURE OF RECIPROCAL SPACE

All structures are three-dimensional and therefore have a three-dimensional reciprocal lattice. The transformation is very simple in that the reciprocals of any length (inter-planar spacing, multilayer periodicity, etc.) can be used to create a space of reciprocals that turns out to be quite a convenient representation of the structure for interpreting diffraction data. So now a real space plane of atoms will create a line in reciprocal space and an infinite three-dimensional object will create a single spot in reciprocal space. Hence the "average" and more general characteristics of the molecular structure are contained close to the origins of reciprocal space, whereas the details are contained in the more remote regions. If now a weight is associated with every reciprocal lattice point, that is related to the scattering power of the plane associated with it in real space, then it is possible to reconstruct the molecular structure. This molecular structure information is contained in the position of this reciprocal lattice point and its associated weight. However the shape of the reciprocal lattice point contains information on the size of the diffracting volume, i.e. the number of atomic plane repeats contributing to the weight of this point. This shape obviously contains information on the layer dimensions in a multilayer structure, etc. Also if there are defects that disrupt the atomic plane giving rise to a given reciprocal lattice point then there will be other length scales, distance between defect or distortions, etc., that will also influence the shape of the reciprocal lattice point. Reciprocal space can therefore be considered in terms of a distribution of scattering that has the characteristics of weight, position and shape. This paper will consider how it is possible to measure this complex reciprocal lattice by using X-ray diffraction and begin to interpret the resulting pattern.

The length scales within reciprocal space are often tailored to suit the physical property to which the investigator is interested. In general the crystallographer will use the straightforward one over length scale whereas in other branches of physics the user prefers the 2π over length scale. Also it is not uncommon for investigators to use a reciprocal space length scale of X-ray wavelength over the real space length. These various forms necessarily alter the definition of the radius of the diffracting sphere and will be discussed in the following section along with the various transformations.

3. THE NATURE OF DIFFRACTION SPACE

The concept of the reciprocal lattice presents a very tangible approach to understanding the nature of diffraction, because the response of the diffracted X-rays to simple angular movements of the diffractometer have an almost inverse relationship with the real structure length scales. For example a long lattice periodicity will produce a closely spaced array of diffraction peaks whereas short lattice periodicity will produce a widely spaced array of diffraction peaks. This is basic from Bragg's equation:

$$2\theta = 2\sin^{-1}\left(\frac{\lambda}{2d}\right)$$

where λ is the X-ray wavelength, d the interplanar spacing (periodicity) and 2θ is the scattering angle. This equation is not linear except over small angular ranges and therefore for many high resolution small-angular-range experiments The length scales can be transformed with a simple linear expression from diffraction space to reciprocal space (if the experiment is carried out with equi-spaced angular movements).

Firstly consider Figure 1, where a regular array of reciprocal lattice points (arranged as a two dimensional net in the diffraction plane for simplicity) is illustrated with an incident wave vector of magnitude $1/\lambda$ pointing towards the origin of reciprocal space. This incident

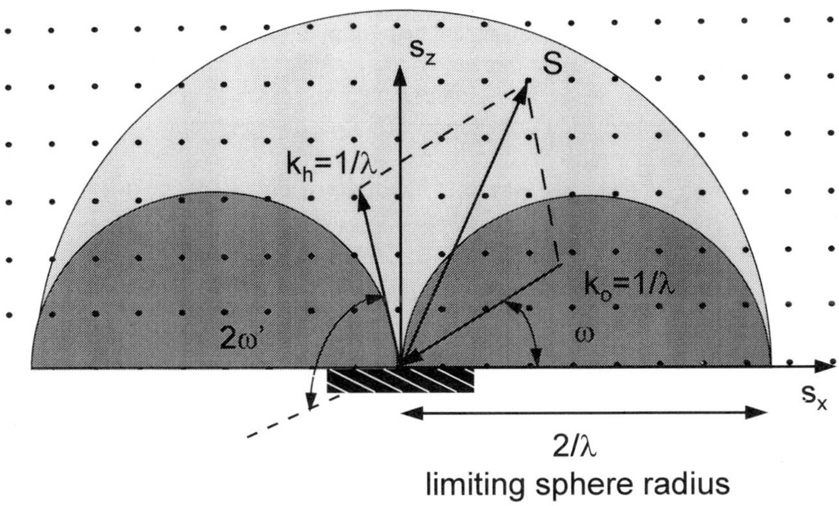

Figure 1. The region of reciprocal space that is accessible to X-rays of wavelength λ in the Bragg case. The incident beam vector $\mathbf{k}_0$ makes an angle of ω with respect to the sample surface and the scattered beam $\mathbf{k}_h$ makes an angle $2\omega'$ with respect to the incident beam direction. The region being investigated is at the position S in reciprocal space.

wave vector $\mathbf{k}_0$ is inclined to the crystal surface by the angle ω, the normal to the crystal surface is given by $\mathbf{s}_z$ and the direction parallel to the surface is given by $\mathbf{s}_x$. Any feature in reciprocal space at a position **S** from the origin, provided that it is contained within the limiting sphere, can be investigated by adjusting the incident beam direction and looking for the scattered beam in the direction of $\mathbf{k}_h$. The direction of this scattered beam is at an angle of $2\omega'$ to the incident beam direction. The scattering angle $2\omega'$ corresponds to any arbitrary position of the detector and 2θ is used purely to represent twice the Bragg angle that is specific to the case when the Bragg condition is satisfied. Also in this discussion the scattering angle, i.e. the angle between the scattered beam and the incident beam, will be used and not the Bragg angle θ since the latter is not in general a diffractometric angle. Since only Bragg case scattering is considered, i.e. the beam enters and leaves the sample by the top surface the following condition must be satisfied:

$$0<\omega<2\omega'$$

This gives rise to the further inaccessible regions in reciprocal space marked in Figure 1 by two darkened hemispheres. Suppose now that our position of interest **S** is being investigated and is found with angular coordinates (ω, $2\omega'$) then its position in reciprocal space coordinates is given by:

$$(s_z,s_x)=\frac{1}{\lambda}([\sin(2\omega'-\omega)+\sin\omega],[\cos\omega-\cos(2\omega'-\omega)])$$

As in Figure 1 and in the above expression the reciprocal lattice to have a dimension of purely one over length and therefore to satisfy the Bragg equation the Ewald sphere has a radius of $1/\lambda$. To transform this to "q-space" then the above expression is multiplied by 2π since $\mathbf{q} = 2\pi\mathbf{s}$ and **S** should be substituted for **Q** giving the reciprocal lattice vector which now has the dimension of 2π over length. For those who prefer to work with an Ewald sphere of unitary radius then the above expression should be multiplied by λ, which now means that the dimensions of our reciprocal lattice depend on the wavelength of the incident beam. The reciprocal lattice vector now has a dimensionless magnitude of wavelength over length. For consistency the following discussion will keep to an Ewald sphere of radius $1/\lambda$ and reciprocal lattice vectors of one over length.

If a small region of reciprocal space is investigated then the angular ranges ($\Delta 2\omega' - \Delta\omega, \Delta\omega$) can be transform to reciprocal space coordinates by

$$(\Delta s_z, \Delta s_x) = \frac{1}{\lambda}(\cos[2\omega' - \omega](\Delta[2\omega'] - \Delta[\omega]) + \cos\omega\,\Delta\omega), \sin[2\omega' - \omega](\Delta 2\omega' - \Delta\omega) - \sin\omega\,\Delta\omega)$$

similarly this can be converted to "q-space", etc. This expression can sometimes be convenient since from inspection of the reciprocal space map it is possible obtain length scales associated with reciprocal space features by inserting the relevant angular separations. However it is important to be aware that diffraction effects considerably confuse the picture in detail and so a simple transform may not result in the correct real structure length scales, (Fewster, 1993).

It is therefore clear to see that the dimensions of the reciprocal lattice of a structure that is accessible is dependent on choosing a suitable wavelength, this defines the size of the limiting sphere, and by choosing a diffraction space probe that is comparable or smaller than the features of interest.

4. THE INTERACTION VOLUME OF THE DIFFRACTOMETER PROBE

The diffractometer probe defines the resolution and the ultimate length scale (large) that can be investigated. The smallest length scale is determined by the size of the Ewald sphere whose radius is given by the reciprocal of the X-ray wavelength, this gives a value of $\lambda/2$. Hence from the above equation it is clear that for studying very small regions of reciprocal space the angular spread of the incident beam $\Delta\omega$ and that accepted by the detector $\Delta 2\omega'$ has to be small. The history of high resolution diffraction is briefly reviewed to indicate how this high angular resolution has been achieved and the consequential evolution of the diffractometer probe for the various instruments.

The double-crystal diffractometer (two-crystal two-reflection diffractometer) was the real first high resolution diffractometer, Compton (1917), built to understand the reflectivity of crystals. The concept is simple and transparent and an understanding of reciprocal space is unnecessary, which probably explains its wide appeal. Figure 2 shows the (+m, -n) geometry and the paths of two X-ray waves of different wavelengths. +m represents the incident wave being scattered in the $+2\omega'_1$ direction from the diffraction plane order m of the collimating crystal and similarly for -n for the sample crystal except that the wave is scattered in the opposite direction. The first crystal, being "perfect" will diffract any one wavelength provided that it is incident within the intrinsic diffracting profile (defined by dynamical theory) and this is true for any wavelength. Provided that all the wavelengths are not too dissimilar, i.e. close to the CuKα lines for example then the shapes of all these diffraction profiles are very similar. This first crystal therefore collimates each wavelength. If the second crystal is set to diffract at the same Bragg angle then both wavelengths will diffract simultaneously and the wavelength dispersion has be reduced to that accepted by the intrinsic diffraction width of the crystal reflections. However to investigate a different diffraction plane or a different sample that diffracts with a different Bragg angle the first crystal has to be changed to match that of the sample. This was a serious draw-back although for routine analysis this was not a major problem. Fewster (1985) partially overcame the difficulties of the double-crystal diffractometer by devising a very rapid method of realignment: also in that paper an automatic method for computer control was presented that is still applicable for any high resolution instrument.

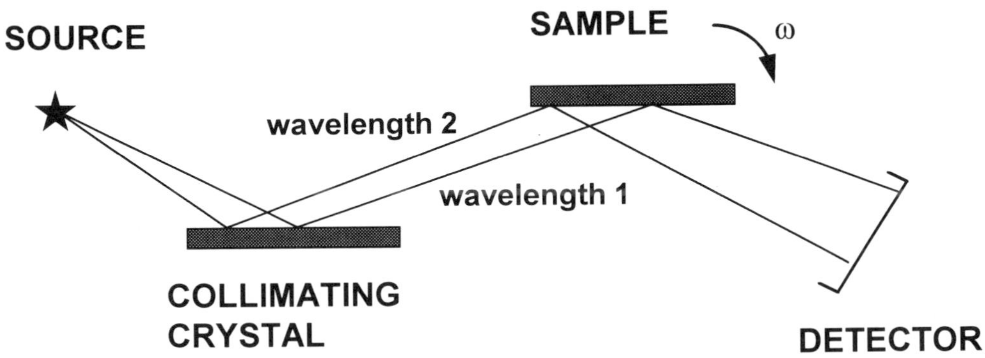

Figure 2. The geometry of the double-crystal diffractometer showing how different wavelengths are diffracted simultaneously for identical scattering angles from the collimating and sample crystals.

In the early 1980's Bartels (1983) made use of the ideas of DuMond (1937) and others to construct a diffractometer that removed the inconvenience of having to change crystals to investigate different materials and different diffraction planes. In place of the collimating crystal Bartels used two channel-cut monochromators (two-crystal four-reflection monochromator) the first was set as a non-dispersive double crystal diffractometer (+m, -m), whilst the second was set in the opposite configuration (-m, +m). The combination of these two channel cut crystals with the -m to -m component between the first and second crystals creates a wavelength sensitive combination, i.e. scanning the second channel-cut crystal against the first will produce a broad wavelength spectrum. However if the second crystal is kept stationary then it will select a very narrow wavelength band whose width is governed by the intrinsic diffraction width of the channel-cut crystals. This is therefore a very effective monochromator that also has a very narrow angular divergence in the diffraction plane. It is important to be aware however that as with any diffractometer it is not possible to state that any instrument has a simple "probe" width that convolutes with the intrinsic diffraction profile of the sample. This arises from the fact that all monochromators have a spectral wavelength bandpass, otherwise there will be no intensity, hence not only is there horizontal divergence that can be controlled by the appropriate choice of crystals but also a contribution to the angular divergence of the beam from this bandpass. Unfortunately because the relationship between the wavelength spread and consequential broadening involves a tangent function the diffraction width increases to quite large values for high scattering angles, Figure 3. Although the reciprocal space resolution does not worsen at this rate since for a constant separation of two features in reciprocal space their angular separation in diffraction space will increase approximately as the inverse of the cosine of the half scattering angle. The wavelength dispersion increases as the tangent of the half scattering angle. We can only reduce this variation in the angular divergence by choosing monochromators with smaller intrinsic diffraction widths, however the proportional change in the instrument function can be worsened although the absolute range in this width is reduced.

However the double-crystal diffractometer and the two-crystal four-reflection monochromator discussed above are unsuitable for interpreting complex diffraction from all but the most perfect of samples that are not curved. After depositing layers all samples will be bent to an extent and no samples are really perfect. Nor are we able to map out reciprocal space with these instruments.

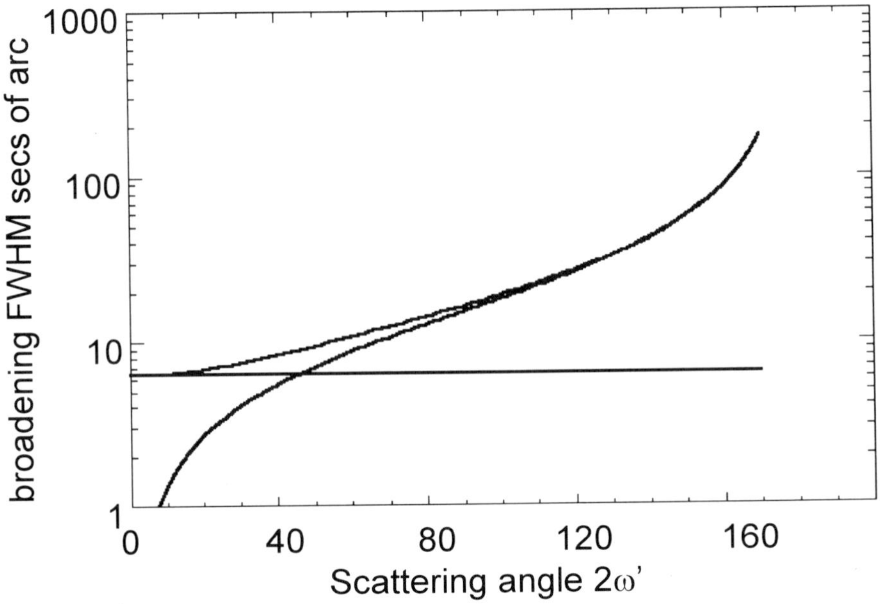

Figure 3. The variation in the full width at half maximum of the total divergence from the two crystal four reflection monochromator as a function of the scattering angle from the sample. The horizontal divergence is a constant and is represented by the straight line whereas the wavelength contribution dominates the overall broadening at high 2ω'

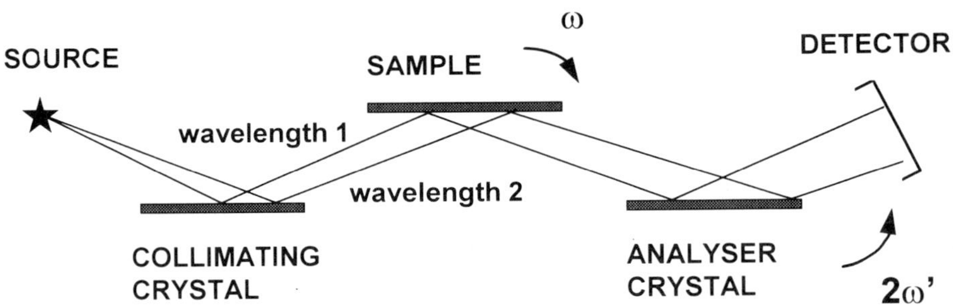

Figure 4. The geometry of the triple crystal (three-crystal three-reflection) diffractometer showing the different paths for different wavelength radiations.

The conventional triple-crystal diffractometer (three-reflection three-crystal diffractometer), Figure 4, overcomes the problem of having a bent sample and will produce a significantly improved diffraction profile. The reason for the improvement with this geometry arises from the analyzer acting as a detector with a very narrow angular acceptance, this will be discussed further in the next section. Unfortunately this diffractometer has the same draw back of a rapidly varying instrument function away from the position when all three crystals have the same scattering angle. The alignment is no longer as trivial as the double-crystal diffractometer and therefore has only been used by a few research groups around the world. Iida and Kohra (1979) have shown how such an instrument could be used to map the intensity in diffraction space but the most significant problem here is that very large instrumental artifacts, (collimating crystal and analyzer crystal streaks) are introduced.

Hence we have the elegant concept of the triple-crystal diffractometer for diffraction space mapping yet it is not versatile enough to accommodate new materials or reflections and also it is not really suited to obtaining reciprocal space maps.

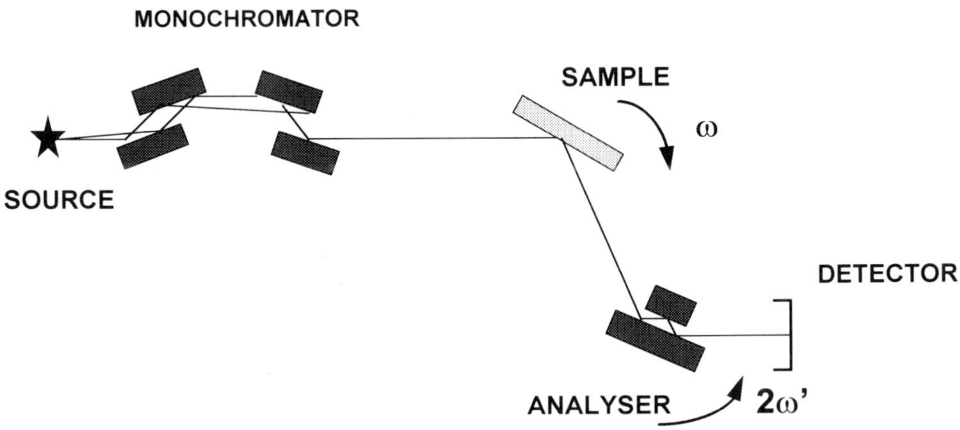

Figure 5. The geometry of the high resolution four crystal eight reflection diffractometer that is used to create a near "δ-function" instrument probe..

The High Resolution Multiple-Crystal Multiple-Reflection Diffractometer, HRMCMRD Figure 5, was built to overcome all these problems with the intention of creating a near "δ-function" in diffraction space, (Fewster, 1989). So now we have an instrument that can accommodate any material and reflection, with the two-crystal four-reflection monochromator, and create diffraction space maps that have the minimum of artifacts. The results discussed in this paper have all been obtained with the monochromator with two 220 Ge channel-cut crystals and a 220 Ge channel cut analyzer crystal, since it offers a good compromise between the instrument function and the intensity. The instrumental artifacts of the three-crystal three-reflection diffractometer arise from the shape of the intrinsic diffraction profile, Figure 6. The introduction of a two and three-fold reflection in these channel cut crystals is to effectively multiply these profiles together and reduce the tails in the diffraction profile, yet the peak intensity remains little changed because the peak reflectivity is very close to unity, (Bonse and Hart, 1965). For an instrument that has a dynamic range of intensity $\sim 10^7$ three multiple reflections is sufficient yet two multiple reflections is insufficient. In practice there is a limit to the quality of the crystals of whatever material, since at these levels >4 orders of magnitude below the peak intensity the influence of crystal imperfections, point defects and other micro-defects as well as thermal diffuse scattering begin to be apparent. Therefore there is little advantage in increasing the number of multiple reflections and what is more the peak intensity will be reduced. Hence a good compromise is the three reflection analyzer. The monochromator effectively undergoes four multiple reflections and therefore the contribution of the tails to the diffraction is negligible. Hence with this combination as in Figure 5 the result is a well defined instrument function, with a minimum of artifacts and an absolute intensity of $\sim 10^6$ photons per second passing from the monochromator and through the analyzer at $\omega=0$, $2\omega'=0$ position (CuKα_1 2kW sealed source). With a good detector with low levels of electronic noise it is possible to obtain 7 orders of intensity by counting for 10 seconds at each data point.

All the discussion so far has been concerned with defining the instrument function in the plane of diffraction, but the axial divergence must also be considered, i.e. out of the plane of diffraction. This is important to understand for accurately aligning the sample on the diffractometer, for interpreting the diffraction space maps and ultimately to be able to carry out three-dimensional diffraction space mapping that is discussed later in this chapter. The contribution of the axial divergence to the diffraction profile broadening in the diffraction plane is very small because of the nature of the monochromator, (Fewster, 1989); the size of the beam gives rise to an apparent vertical divergence but our interest is the vertical divergence for a "pure" monochromatic beam from a point source. An exaggerated instrument function for the double-crystal diffractometer and the HRMCMRD is illustrated in Figure 7. The various components of divergence and wavelength dispersion that contribute to this shape are included. Effectively this instrument function gives the region of the Ewald sphere that is accessible or will pick up intensity at any one time. The Ewald sphere as it is drawn is for a single wavelength (it has a radius of $1/\lambda$) but this wavelength dispersion has only be included in the instrument function. An axially divergent wave incident on crystal planes that are tilted out of the diffractometer plane (i.e. the diffraction plane in the aligned condition) can generate a scattered wave that will be detected at a diffractometer angle that is a projection of the true angle. Therefore it is important to align the sample so that the normal to the diffracting plane lies at the centre of the axial divergence that is in the plane of the diffractometer. If it is not then the signal will not occur at the correct ω, $2\omega'$ settings, (Fewster, 1985). The axial divergence also contains much of the intensity in the probing beam since the other two directions are confined to ~10"arc. each. The detector will only receive those photons that are travelling within the horizontal and axial divergence and have a wavelength bandpass accepted by the monochromator. The analyzer then only accepts those scattered x-rays that are within its allowed horizontal and axial divergence. The latter is usually unconstrained and matches that of the monochromator. Reducing the axial divergence can also overcome ambiguities in the interpretation and will be discussed latter.

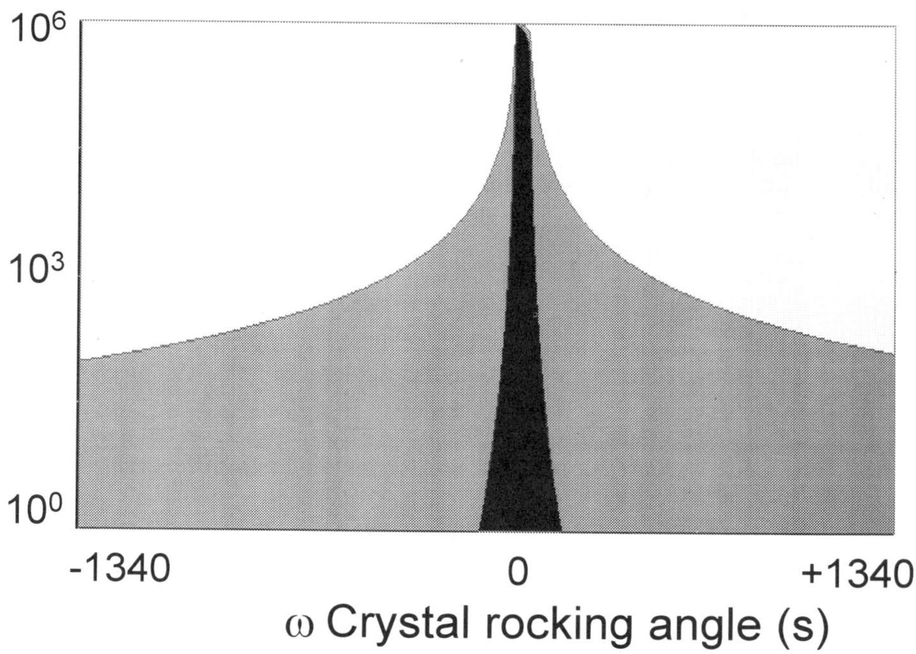

Figure 6. The calculated profile for a single reflection from a Ge crystal (grey curve) and that from a 3x reflected channel cut crystal (black curve). Note the enormous reduction in the tails of the reflection curve, $CuK\alpha_1$ 220 reflection.

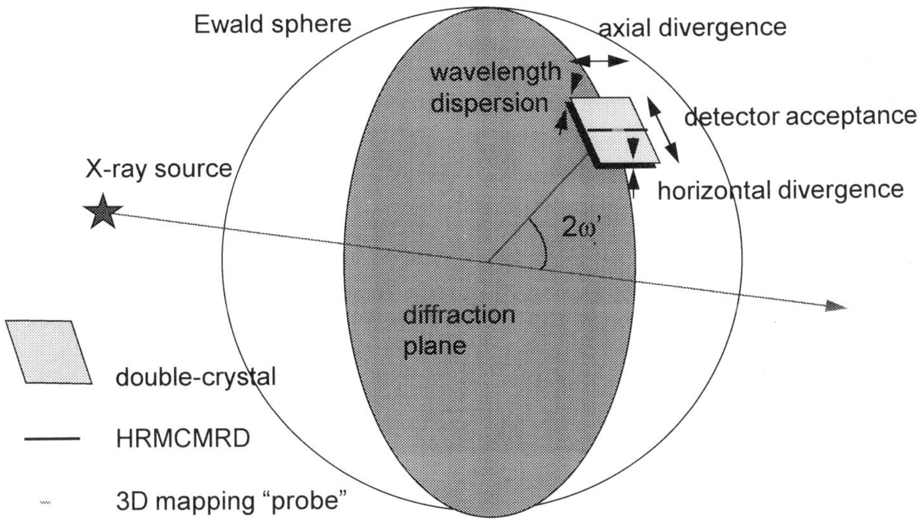

Figure 7. The three dimensional diffraction space probe for several data collection methods, showing the various parameters that influence its shape.

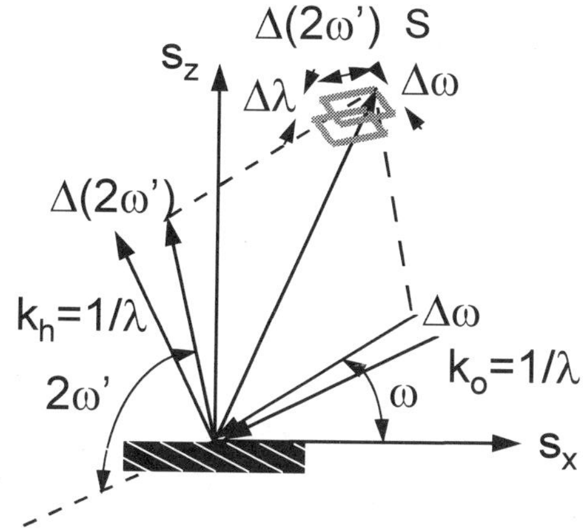

Figure 8. The origin of the instrument function created by a divergent incident beam, a finite acceptance into the detector and a spread in wavelengths.

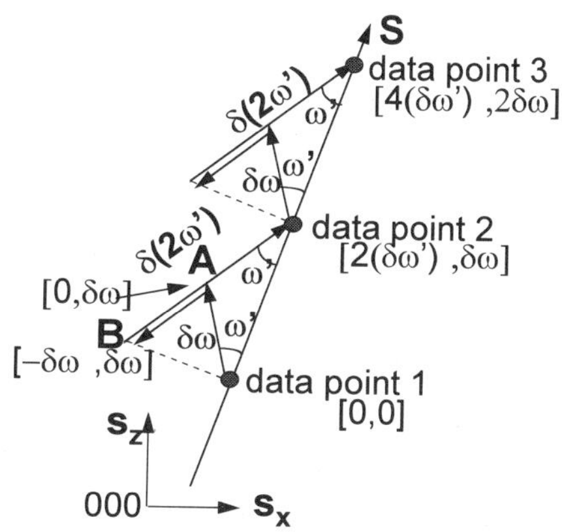

Figure 9. The angular movements required to bring about a scan through reciprocal space.

5. HIGH RESOLUTION DIFFRACTION SPACE METHODS

From Figure 1 we can see that we can define a very small region in reciprocal space by restricting the horizontal divergence of the incident beam and limit the angular acceptance of the scattered radiation entering the detector. There is also a spread in wavelengths giving a spread in $\mathbf{k}_0$ and $\mathbf{k}_h$ and hence a smearing of the instrument function, Figure 8. Remembering that this instrument function does vary over the accessible region of diffraction space we then choose the most appropriate data collection procedure. Collecting data along the radial direction is convenient since in general the relationship between the diffractometer stepping angles is very simple, i.e. $\Delta\omega = \Delta 2\omega'$. Collecting data along an arbitrary direction may result in a rather zig-zag motion of the probe depending on the smallest increment of the motors involved in the movement.

The angular movements to achieve a general scan through diffraction space will be discussed with reference to Figure 9. Suppose that the diffractometer angles are set at the position corresponding to data point 1, Figure 8, then to move to data point position 2 the incident angle is altered through an angle $\delta\omega$ to arrive at point A, but in so doing the diffracted beam direction has been increased by $\delta\omega$. Therefore the diffracted beam direction now appears at an angle further from our data point 2 (at point B). Hence the effective motion in ω is perpendicular to the vector $\mathbf{S}$. If the detector assembly is driven through an angle $\delta(2\omega')$ the probe will arrive at the data point 2 position. Suppose that this angle is chosen to be equal to twice $\delta\omega$ (i.e. $\delta(2\omega') = 2\,\delta\omega$) then from the geometry of Figure 9 the data points will be radial from the origin of reciprocal space 000. This scan is often referred to as the $\omega/2\omega'$ or $2\omega'/\omega$ scan. When this ratio is not met then it is possible to undertake any arbitrary scan direction. It becomes clearer therefore to think in simple terms that by rotating in ω from the data point 1 to point B and rotate our detector through an angle twice this increment to collect the second data point. After completing the scan and returning to data position 1, implementing an offset in ω and repeating this scan procedure a map of the intensity distribution in two dimensions can be measured. The step size and offset should not exceed the width of the instrument function otherwise information can be lost unless assumptions are made about the variation between these points. Also it is wasteful of time to collect data with step sizes well below the dimensions of the instrument function.

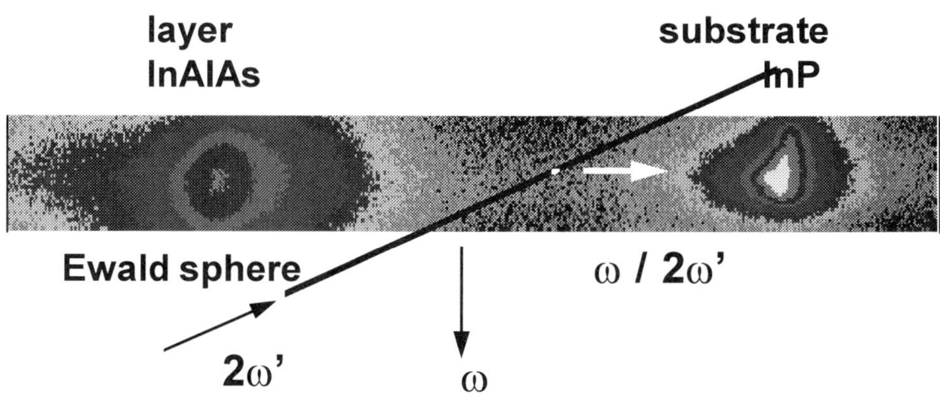

Figure 10. A map of the distribution of scattered intensity in diffraction space indicating the different scans and probe sizes for the double crystal diffractometer and the HRMCMRD. AlInAs layer on an InP substrate 004 reflection and CuKα_1 radiation.

A diffraction space map close to the 004 reflection from a single layer of AlInAs on an InP substrate is shown in Figure 10. This Figure gives an indication of the probe (instrument function size) and how it differs from the double crystal diffractometer "open detector" probe. It is very clear from this example that the intensity collected from an open detector system collects scattered photons from a large region and it clearly becomes difficult to associate the intensity measured with a feature in reciprocal space. However with the small probe of the HRMCMRD the situation becomes much clearer in terms of the conventional profile shape,

Figure 11, and the additional information contained in the diffraction space map, Figure 10. The diffraction profile of Figure 11 from the HRMCMRD can easily be matched to a calculated shape since this represents the intensity profile along a radial direction from the origin of reciprocal space: consequently the angular separation and shapes can be associated with changes in the deviation parameter $\alpha_H(\omega)$ in dynamical theory. Whereas the profile from the open-ended detector experiment includes the above information and the distribution of tilts, sample curvature and contributions to the diffuse scattering from imperfections. However the diffraction space map includes all this information directly in an "un-convoluted" form.

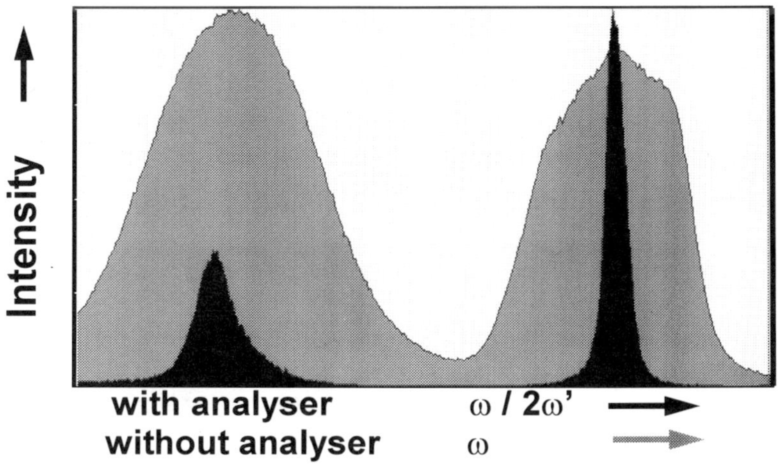

Figure 11. The resulting profiles for the two different scan types of Figure 10. The scan with an analyser separates the strain from the finite lateral correlation length, mosaic spread and sample curvature effects.

It is clear also from this map, Figure 10, that the dimension and shape of the probe is very important and this is more obvious when we look at the influence of the instrument function of the three-crystal three-reflection diffractometer on a perfect sample, Figure 4, compared with the five-crystal eight-reflection diffractometer (HRMCMRD), Figure 5, shown in Figure 12. In the former case we have a much larger probe or region of uncertainty from where the scattered intensity is emanating.

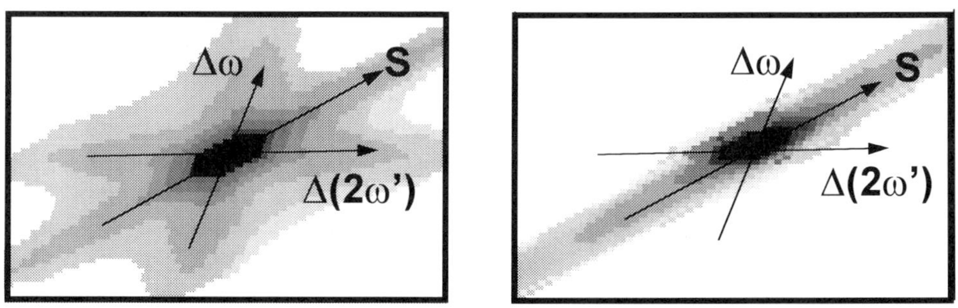

Figure 12. The influence of the instrument function on a perfect crystal sample; left - the broadening and streaking caused by a three crystal three reflection diffractometer, and right - the negligible broadening caused by the instrument function of the HRMCMRD.

It is most important to be aware that the intensity measured at any one point though cannot be simply considered in two dimensions, since our sample is a three dimensional object the diffraction pattern must also be three-dimensional. With this type of measurement the diffraction space probe is diagrammatically illustrated in Figure 7 where we can see that it is rather like a rod with dimensions of ~10"arc x ~10"arc x 0.4°. It is the larger dimension that captures most of the intensity and effectively integrates and keeps the measured intensity high. For a relatively well aligned sample, in terms of tilt angle to the diffraction plane this is not too serious and the diffraction space map produces a very good representation of the intensity distribution. However for an imperfect sample, where for example the epitaxy is poor, significant variations in intensity along this rod will be averaged and inadequate projections of the scattered intensity can occur. This can result in the projected positions of the reciprocal lattice features appearing at the wrong angular positions. To overcome this three-dimensional reciprocal space mapping has been developed and is performed by reducing the axial divergence and creating a series of diffraction space maps that section through the reciprocal lattice features, (Fewster & Andrew, 1995a, 1996). Between each diffraction space map the sample is tilted through an angle $\Delta\chi$ so that a different region of the scattering from the sample is detected within the bounds of the probe axial dimension of $2\delta\psi$. In general the step sizes should be such that $\Delta\chi \leq 2\delta\psi$, so that the data collection time is optimised without missing reciprocal lattice features.

6. DATA COLLECTION STRATEGY

The reason for mapping the intensity in two- or three-dimensions may well be obvious from the above arguments, but in this section the concentration will be on a fairly general example of where the advantages over more traditional methods is very significant. Because the reciprocal space mapping method effectively deconvolutes the various contributions of the structure (strain and tilts, etc.), this has proved to be a very powerful method for measuring these parameters unambiguously. It is also used extensively for separating all the various strain components in an epitaxial layer for example. The diffuse scattering is also separated and therefore can be studied without the interference of strong scattering.

The example considered will describe how structural information on a partially relaxed epitaxial layer can be extracted. Firstly the appropriate experiments should be considered for extracting the information of interest. Clearly there are many unknowns (lattice parameters a,b,c for the layer, relative layer to substrate tilts in two perpendicular directions and the distribution of these two tilts) and also there are lateral variations in defect density, mosaic spread, mosaic block size and distortions emanating from the interface. Figure 13 gives the various structural parameter to diffraction pattern relationships for an arbitrary reflection hkl. Although there are many correlated components by judicious choice of reflections the various components can be separated. Suppose that a reflection is chosen from diffracting planes that are approximately parallel to the surface then the in-plane lattice parameter component becomes negligible and the shape is only dependent on the macroscopic tilt. There will also be tilting in the orthogonal direction and therefore to obtain the general macroscopic tilt two reflections are required. To obtain the two in-plane lattice parameters two reflections from planes not parallel to the surface, whose projections onto the surface are approximately orthogonal, should be measured. These reflections can also be used to separate the microscopic tilting from the lateral correlation lengths, etc. From these arguments the most appropriate reflections can be chosen.

In general it is better to use radially scanned maps for a general reflection where the step $\Delta\omega = 0.5\,(2\Delta\omega')$, because this fixes the direction to the reciprocal lattice origin and is likely to be more precise mechanically. However scans along the surface normal can save time by reducing the mapped area provided that the relaxation and tilts in the sample are known to be small. The choice of step size relates entirely to the instrument function profile and should be smaller than the full width at half maximum for the various directions. The count time and scan ranges depend on the information sought and the material perfection. The more imperfect the weaker the scattering, etc. Remember though that the eye is a very good integrator and when the intensity has been plotted, the map will reveal very weak features that are invisible in single scans.

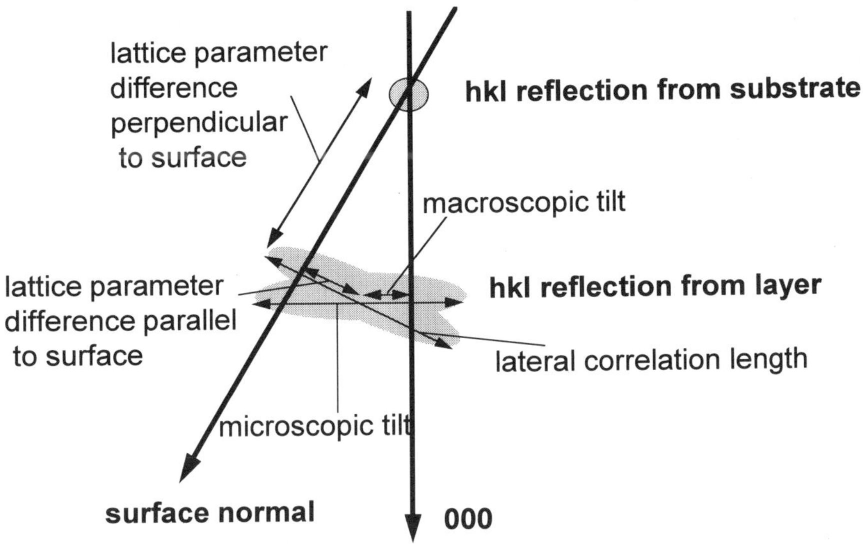

Figure 13. A schematic of the various parameters contributing to the shape of the reciprocal space features for a general reflection from a partially relaxed epitaxial layer on a substrate.

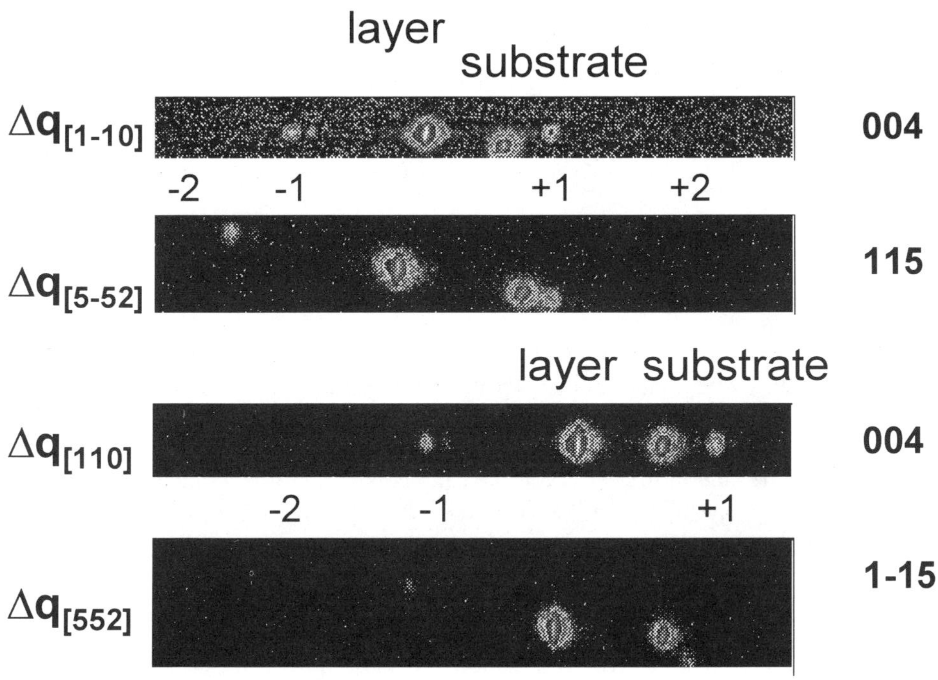

Figure 14. This illustrates a typical data set from a partially relaxed structure with tilts, etc. These are all radial scans from a sample of a 6μm InGaAs layer separating the GaAs substrate from a superlattice structure on top, $CuK\alpha_1$ radiation.

7. DATA INTERPRETATION

The advantages of collecting data with a well defined instrument function is that it simplifies the comparison with the simulated profiles. If the material is "perfect" within the "region of uncertainty" of the instrument function then the comparison with the dynamical model should be very nearly exact. As the structure deviates from perfection then the conventional equations of dynamical theory (Takagi, 1962, 1969; Taupin, 1964) become inadequate on their own because more information concerning the structure needs to be included. We have to consider aspects of wave-field coherence averaging as in the statistical theory of Kato (1980) as well as the additional distortions created by the imperfections.

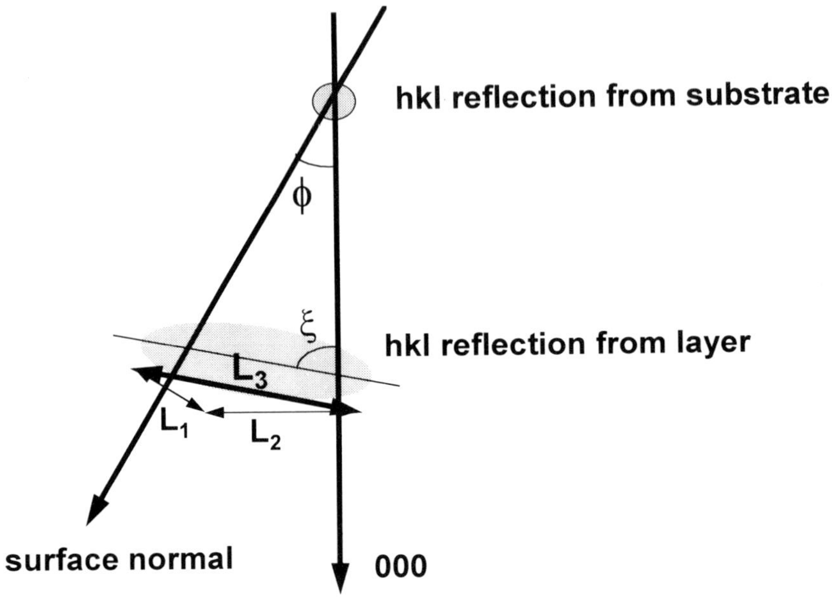

Figure 15. A schematic of the distribution of scattering in diffraction space and the length, L_3, and angles required to separate the microscopic tilting and lateral correlation lengths for a general reflection.

The procedure for analysing a partially relaxed structure will be discussed and how it is possible to extract a significant amount of information. A typical data set of a complex partially relaxed structure is given in Figure 14, from which the macroscopic tilts can be extracted from the 004 reflections, since the surface normal is along [001]. These maps also give the lattice parameter c along [001] by comparison with that of the substrate. The 115 reflections contain a component of the in-plane lattice parameter as well as this tilt and since the latter is known the former can be obtained, Fewster and Andrew (1993b). So at this stage the macroscopic tilt (by resolving the two orthogonal contributions), the apparent lattice parameters normal and in the surface plane can be extracted in a straightforward manner: whereas the microscopic tilt and lateral correlation lengths are less easily distinguished. This last point will be considered more carefully. Ideally the reciprocal space map should be simulated but this is beyond the scope of this article and therefore a more pragmatic approach will taken.

Microscopic tilts for regions of the sample that have a common lattice parameter will lie along the perpendicular to the radial direction, whereas any in-plane correlation length will be perpendicular to the surface normal, Figure 13. Therefore by measuring a reciprocal space map of the intensity scattered from crystal planes that are a long way from the surface normal then these various components can be separated. It must be remembered however that mosaic blocks that contribute to the correlation lengths, because of their finite lateral size, will also contribute to this diffraction broadening because of their relative orientations. The overall shape then becomes quite a complex convolution in two dimensions. By considering the broadening due to microscopic tilting and finite dimensions it is possible to untangle these two effects.

From the diagram in Figure 13 it is clear that there are two contributions to the profile width namely microscopic tilts and lateral correlation lengths. Because of the smearing effect and convolution described above the shape observed will be approximately elliptical with an axis that is intermediate between these two extremes. From simple trigonometry, Figure 15, the ratio of these two contribution is given by:

$$\frac{lateral\text{-}correlation\text{-}length}{microscopic\text{-}tilt}=\frac{L_1}{L_2}=-\frac{\cos\xi}{\cos(\phi+\xi)}$$

and the actual measured broadening, L_3, in terms of L_1 is given by:

$$\frac{L_3}{L_1}=\frac{\sin\phi}{\cos\xi}$$

and hence from these relationships the individual contributions of L_1 and L_2 are obtained on the ω scale. The angle ξ is the angle between the line through the axis of the ellipse of the layer profile and the radial direction towards the origin, Figure 15. The derivation of the lateral correlation length is simply $(\Delta s_x)^{-1}$ where:

$$\Delta\omega_{L_1}=L_1(\cos\phi+\sin\phi\tan\theta)$$

and

$$\Delta(2\omega')_{L_1}=\frac{L_1\sin\phi}{\cos\theta}$$

whereas the microscopic tilt is simply:

$$\Delta\omega_{L_2}=L_2$$

It is therefore obvious that the map of the intensity from diffraction planes that are significantly tilted with respect to the surface plane it is possible to obtain the most accurate estimates of these two parameters, i.e. large ϕ.

Therefore with a combination of reciprocal space maps all the parameters described in Figure 13 can be determined. However care has to be exercised with changes in lattice parameter between mosaic blocks and the influence of an extended axial divergence because of the influence of projecting the diffraction on to the plane of the diffractometer. The former may only be of concern if there is a correlated and progressive change in lattice parameter with mosaic block tilting, Fewster (1991a). To really extract these individual contributions we will need to map the intensity in three dimensions as described by Fewster and Andrew (1995a) and then carry out the analysis on each individual section. A similar procedure would have to taken if there are significant tilts out of the diffraction plane that can be checked by limiting the axial divergence and comparison. The three-dimensional reciprocal space map will always be the most general case and represents the ideal procedure for each analysis, however the two-dimensional map or in some cases the "pseudo" one-dimensional profile from a double crystal diffractometer may be sufficient and this depends on the sample problem to be resolved.

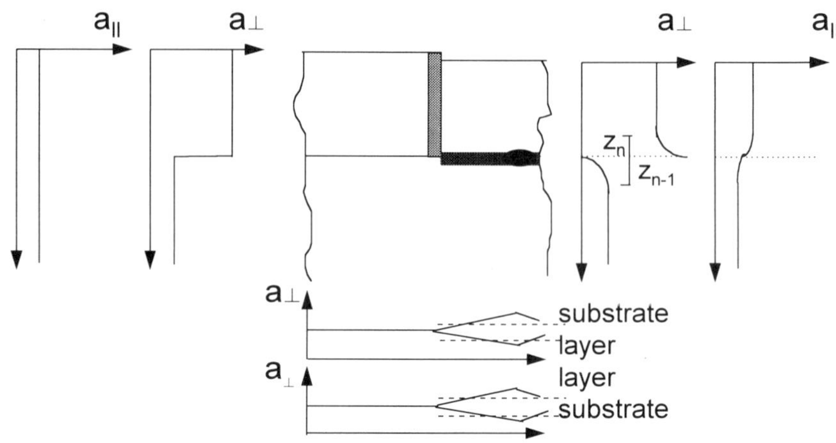

Figure 16. The variation in lattice strain experienced by regions remote and close to a defect. The calculation is based on a two-column approximation, i.e. a perfect and distorted region.

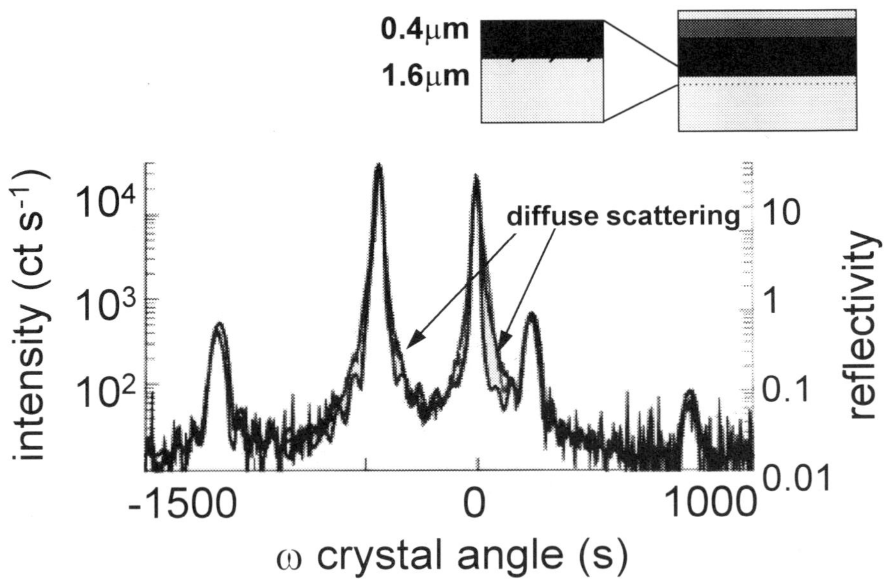

Figure 17. The comparison of the calculated intensity that gives a near perfect fit to the model including the strain variations of Figure 16 and taking into account the wave-field disruption with the standard dynamical model. The latter gives poor agreement in the peak height and cannot model the diffuse scattering required to estimate the interface disruption.

There is also additional information in the diffraction space map and that is the weak diffuse scattering that emanates from a range of defect types, each contributing in their own way. If the dominating defects are expected to be at the interface, i.e. misfit dislocations arising from relaxation in the above analysis then there will be diffracting plane distortion and disruption of the x-ray wave-fields, Fewster (1992). The distortions will take the form illustrated in Figure 16, whereas the coherence of the wave-field above and within the distorted region will tend to average and appear incoherent with that from the remainder of the structure. This incoherency is then included by adding the intensity of the incoherently related regions so that the phase information is lost. By selecting a reflection from a set of planes parallel to the surface, mapping the scattered intensity and by projecting this intensity onto the radial direction the influence of the finite correlation lengths parallel to the surface and that of the microscopic and macroscopic tilts will be eliminated. Therefore the resulting profile is only sensitive to strain and composition as a function of depth, correlation lengths normal to the surface and distortions. Hence the strain at the interface can be reconstructed by modelling this diffraction profile, Figure 17.

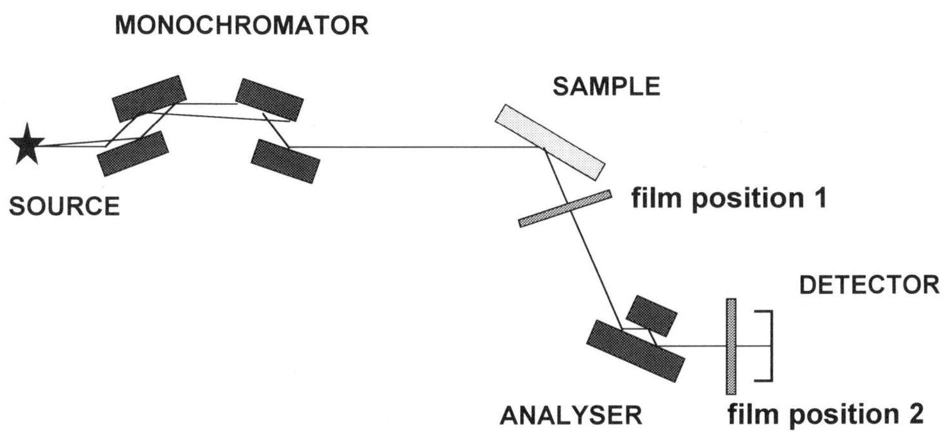

Figure 18. The film positions are shown for carrying out multiple crystal topographs. Position 1 can be used for well defined defects, i.e. easily interpretable, whereas position 2 is very useful for associating diffraction space map features with the contributing defects or real space features. The position 2 is used for diffuse scattering topography.

8. PROCEDURES TO AID INTERPRETATION

In this section some additional developments that increase the understanding of our reciprocal space maps will be discussed. Clearly if there are regions of our sample that have significant tilting then it may be important to know the lateral dimensions of these regions or their location on the sample. This can be achieved by combining these reciprocal space maps with topography, (Fewster, 1991a, 1991b). There are two modes for this, either by placing a film in the diffracted beam from the sample or after the analyzer, Figure 18. In the former case the topographs results in an image of the distribution of intensity across the sample, such that at any position on it corresponds to the integrated intensity along the Ewald sphere, whereas in the latter case the integration volume is highly specific in reciprocal space, Figure 10. For understanding the microscopic tilts in the example above then placing the film after the sample will probably be sufficient. However if knowledge of the origin of a region of scattering in our reciprocal space map is of interest then placing the film after the analyzer would be most appropriate. For example for understanding the origin of the diffuse scattering then this is perfectly possible, Fewster and Andrew (1993c).

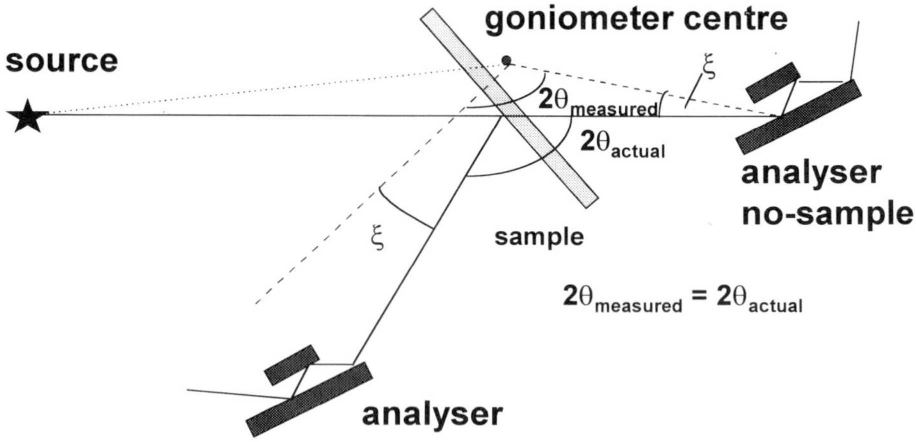

Figure 19. The geometry of the method to obtain precision lattice parameters on an absolute wavelength scale. The position of the sample centre and the incident beam direction with respect to the goniometer centre does not influence the precision.

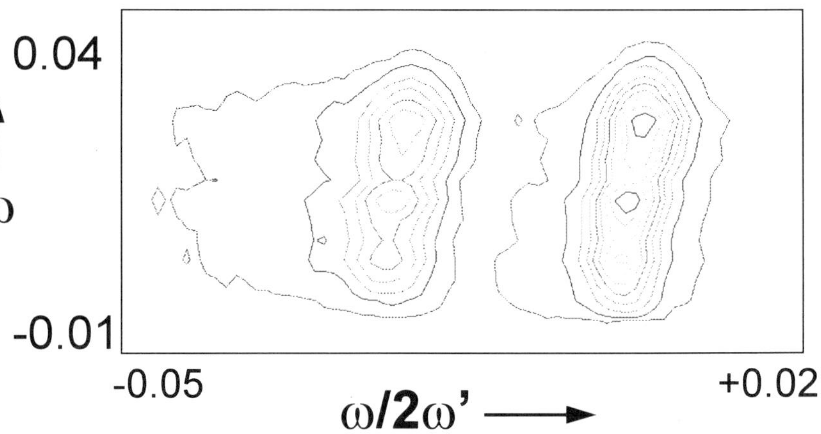

Figure 20a. The distribution of scattering close to the 004 reflection for a mosaic AlGaAs layer on a GaAs substrate showing three distinct contributing grains, $CuK\alpha_1$ radiation.

Another very useful method that can be combined with reciprocal space mapping is the application of precision lattice parameter measurement, Fewster and Andrew (1995b). The principle of the method is illustrated in Figure 19, and relies on the angular selectivity of the diffractometer and not on the spatial position of the beams. From Figure 19 and simple trigonometry it is clear that the precision is purely dependent on the angular position of the diffracted peak and the direct beam, the diffractometer gearwheel and wavelength selection and not on whether the sample is positioned in the goniometer centre or whether the incident beam intersects the goniometer centre. The accuracy is comparable to the method of Bond (1960), although since this new method requires a single measurement on one position on the sample reciprocal space maps can be placed on an absolute scale. Thus reliance on tabulated values of lattice parameters for internal calibration is unnecessary.

It is very important to be aware of the three dimensional nature of the diffraction pattern when interpreting any profile or map. The instrument function described above integrates the intensity over a finite three dimensional volume and associates it with a specific position in reciprocal space. Therefore to associate a lattice parameter, for example, with a feature in a diffraction space map we should determine the full coordinates of the position of interest or ensure that it lies in the plane of the diffractometer as described by Fewster and Andrew (1995b). Consider for example the two-dimensional map of Figure 20(a) where the scattering from mosaic blocks of a GaAs crystal are projected onto the diffractometer plane. The three-dimensional map resolves the "true" view of reciprocal space and from this the individual tilts and lattice parameters can be determined, Figure 20(b).

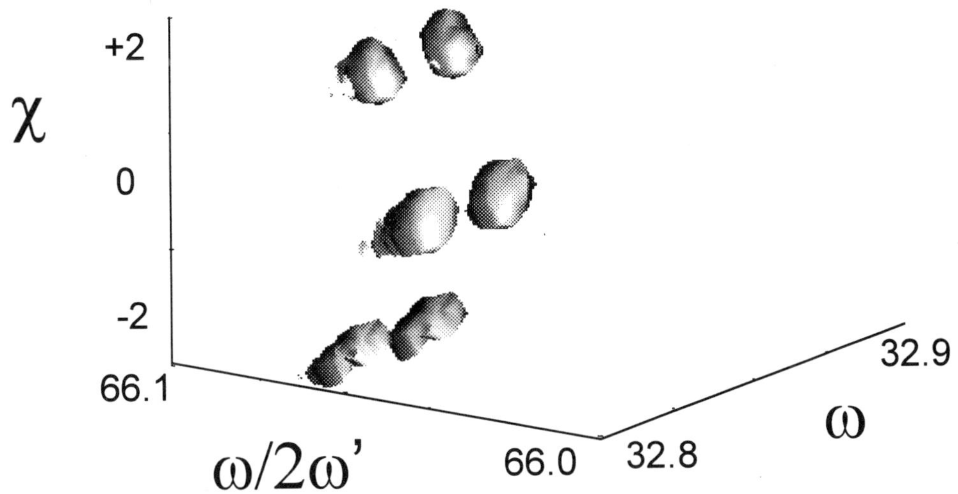

Figure 20b. The true picture in reciprocal space where the individual mosaic blocks are separated by two tilt components. From this type of three-dimensional maps it is possible to detedrmine the real quality of the epitaxy.

9. CONCLUSION

It is important to remember that X-ray diffraction is a full three-dimensional process and that the instrument function has a three-dimensional nature. The closest to obtaining the distribution of scattering from a structure is by three-dimensional reciprocal space mapping, however with care very valuable information is contained in the two-dimensional reciprocal space map. This information allows the full evaluation of partially relaxed epitaxial layer structures and diffuse scattering studies, etc. However the use of single profiles has to be treated with considerable caution since it requires an almost complete understanding of the sample, since generally for these experiments the instrument function has a large acceptance volume. Reciprocal space mapping methods therefore greatly aid interpretation of the structural properties of materials, since the instrument function is so well defined, and when used in combination with topography and precision lattice parameter measurements more complex structural problems can be addressed.

10. REFERENCES

Bartels, W.J., 1983, Characterization of thin layers on perfect crystals with a multipurpose high resolution X-ray diffractometer, *J. Vac. Sci. Technol. B*, 1:338

Bond, W.L., 1960, Precision lattice constant determination, *Acta Cryst.* 13:814

Bonse, U. and Hart, M., 1965, Tailless X-ray single crystal reflection curves obtained by multiple reflection, *Appl. Phys. Lett.* 7:238

Compton, A.H., 1917, The reflection coefficient of monochromatic X-rays from rock salt and calcite, *Phys. Rev.* 10:95

DuMond, J.W.M., 1937, Theory and use of more than two succssive X-ray crystal reflections to obtain increased resolving power, *Phys. Rev.* 52:872

Fewster, P.F., 1985, Alignment of double-crystal diffractometers, *J. Apl. Cryst.* 18:334

Fewster, P.F., 1989, A high-resolution multiple-crystal multiple-reflection diffractometer, *J. Appl. Cryst.* 22:64

Fewster, P.F., 1991a, Combining high resolution X-ray diffractometry and topography, *J. Appl. Cryst.* 24:178

Fewster, P.F., 1991b, Multicrystal X-ray diffraction of heteroepitaxial structures, *Appl. Surf. Science* 50:9

Fewster, P.F., 1992, The simulation and interpretation of diffraction profiles from partially relaxed layer structures, *J. Appl. Cryst.* 25:714

Fewster, P.F., 1993, Review article: X-ray diffraction from low dimensional solids, *Semicond. Sci. Technol.* 8:1915

Fewster, P.F., and Andrew, N.L., 1993a, Diffraction from thin layers, in *Proceedings of the second European Powder Diffraction Conference*, R. Delhez and E.J. Mittemeijer, Ed., TransTech Publications, Switzerland

Fewster, P.F., and Andrew, N.L., 1993b, Determining the lattice relaxation in semiconductor layer systems by X-ray diffraction, *J. Appl. Phys.* 74:3121

Fewster, P.F., and Andrew, N.L., 1993c, Interpretation of the diffuse scattering close to Bragg peaks by X-ray topography, *J. Appl. Cryst.* 26:812

Fewster, P.F., and Andrew, N.L., 1995a, Applications in multiple-crystal diffractometry, *J. Phys. D* 28:A97

Fewster, P.F., and Andrew, N.L., 1995b, Absolute lattice-parameter measurements, *J. Appl. Cryst.*, 28:451

Fewster, P.F., and Andrew, N.L., 1996 reciprocal space mapping and ultra high resolution diffraction of polycrystalline materials, *Microstructure Analysis from Diffraction*, R.L. Snyder, H. Bunge and J. Fiala, Ed., Oxfors University Press, in preparation

IIda, A. and Kohra, K., 1979, Seaparate measurements of dynamical and kinematical X-ray diffractions from silicon crystals with a triple crystal diffractometer, *Phys. Stat. Solidi A*, 51:533

Kato, N., 1980, Statistical dynamical theory of crystal diffraction. I. General formulation, *Acta Cryst. A* 36:763

Takagi, S., 1962, Dynamical theory of diffraction applicable to crystals with any kind of small distortions, *Acta Cryst.* 15:1311

Takagi, S., 1969, A dynamical theory of diffraction for a distorted crystal, *J. Phys. Soc. Japan* 26:1239

Taupin, D., 1964, Théorie dynamique de la diffraction des rayons X par les cristaux déformés, *Bull. Soc. Fr. Minéral. Cristallogr.* 57:467

SUPERLATTICES

Paul F Fewster

Philips Research Laboratories
Cross Oak Lane
Redhill RH1 5HA
U.K.

1. INTRODUCTION

The physical properties of materials can be modified by creating an additional structural periodicity by the deposition of alternate thin layers. The wavelength should be such that the physical property of interest is influenced by this periodicity and this differentiates a superlattice from a multilayer. Unfortunately this definition is not so simple, since the physical property (electron confinement for example) may see a periodic structure as a series of individual layers (individual quantum wells for example), whereas some other physical property may see this as a superlattice. A superlattice will therefore be defined in terms of an "X-ray diffraction superlattice" (i.e. it depends on the nature of the diffraction pattern from a structure). This can therefore be defined as a structure that averages the scattering in some way for the periodic unit and does not consist of individual peaks, associated with the individual components, but appears as an average peak and a perturbation in terms of satellite reflections. The limits of the term "superlattice" also requires clarification by considering how few repeats will constitute a superlattice and how thick the layers can be before these simple averaging effects are no longer valid.

The analysis of superlattice structures using both kinematical and dynamical diffraction theories will be discussed including the bounds of their validity. The important parameters to be considered are the superlattice period, the individual layer thicknesses and the interfacial roughness. The latter consists of two aspects; correlation lengths in the plane of the interface and grading that can be a consequence of roughening and intermixing. The use of high resolution techniques reveals greater complexity in the diffraction pattern and this will be discussed. X-ray diffraction can be so sensitive that it can really detect the deviation from perfection therefore a compromise for a practical analysis is necessary that depends on the information required.

2. INVESTIGATING THE X-RAY DEFINITION

Consider Figure 1, which consists of a series of simulated diffraction profiles using dynamical theory, of alternating layers of AlAs and GaAs both 50Å thick deposited on a GaAs substrate.

Figure 1(a) is that from a single bilayer where there is significant fringing as though it behaves as some form of average. The broad average peak of the bilayer is offset to a smaller ω value from the sharp substrate peak because of the larger lattice parameter of the AlAs increases the average "superlattice" lattice parameter. By comparison with Figure 1(b) for four bilayers the first order satellites appear as shoulders of the average peak. As the number of bilayers increases the satellites become sharper, although the pattern has become more complex in detail with high frequency fringing occurring throughout the pattern.

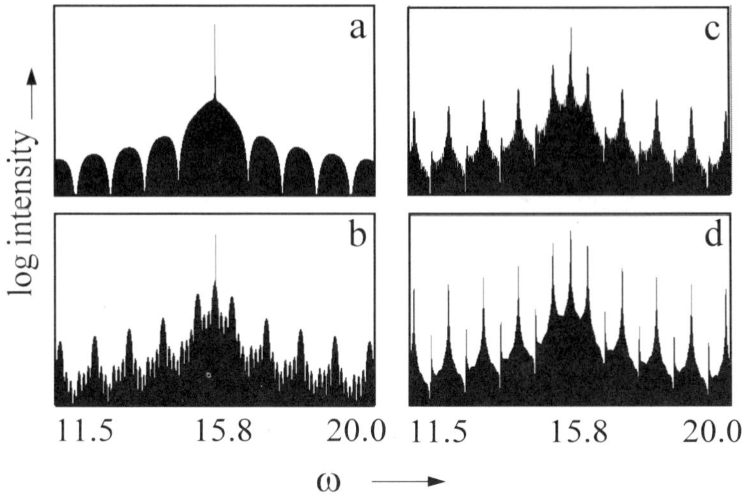

Figure 1. The changes in the diffraction pattern as the number of bilayers is increased with (a) n=1, (b) 4, (c) 10 and (d) 50, where (AlAs [50Å] GaAs [50Å]) x n for the 002 reflection CuKα radiation.

Consider the situation where the layers constituting the superlattice become thicker, Figure 2, then the satellites become more closely spaced. The substrate peak can be used as a marker in the following discussion and is marked in Figure 2. The peak immediately to the low ω side of the very sharp substrate peak in Figure 2(a) is the strongest and in fact can be considered as the average superlattice peak and the remaining prominent peaks are the satellites that are approximately equidistant from this position. The same is true when we increase the bilayer thickness to 400Å. As we move to 800Å and especially 1600Å periods the pattern is no longer a clear large average peak and weaker satellites; in fact the largest peak is now in the position of the -1 order satellite and the +1 satellite has become very weak! Also from the diffraction patterns for the structures of Figure 2(a) and Figure 2(b) it is clear that the even order satellites are weak yet this is not the case as the thicknesses increase. In Figure 2(d) the even order satellites on the low ω side of the average peak become the strongest, whereas the satellites on the high angle side start off being dominated by the odd order reflections and eventually the even order satellites dominate. The concept of the average peak has now become rather dubious although the diffraction cannot be considered in simple terms of independent scatters but clearly the interaction is such that this is still an x-ray diffraction "superlattice," (Fewster, 1993).

The kinematical diffraction that simply sees this modulation as a perturbation, (Guinier, 1963) cannot be applied to these extreme cases and therefore dynamical modelling of the intensity profile is necessary. Hence there are three distinct regions at this stage;

1) very few periods, one or two that can only be interpreted in terms of the diffraction profile,

2) an intermediate region when the profile appears like a true average and the satellite intensities can be used to extract information rather easily and

3) a region when the individual layers are so thick that significant complexity occurs in the diffraction pattern that only profile modelling with dynamical theory would suffice.

Largely the regions 1 and 3 create no real choice although perhaps in the former the kinematical theory approximation would be sufficient, but in general dynamical diffraction theory should be applied using an iterative process of changing the parameters. Therefore this paper will concentrate on the intermediate region and consider the advantages and disadvantages of the various methods.

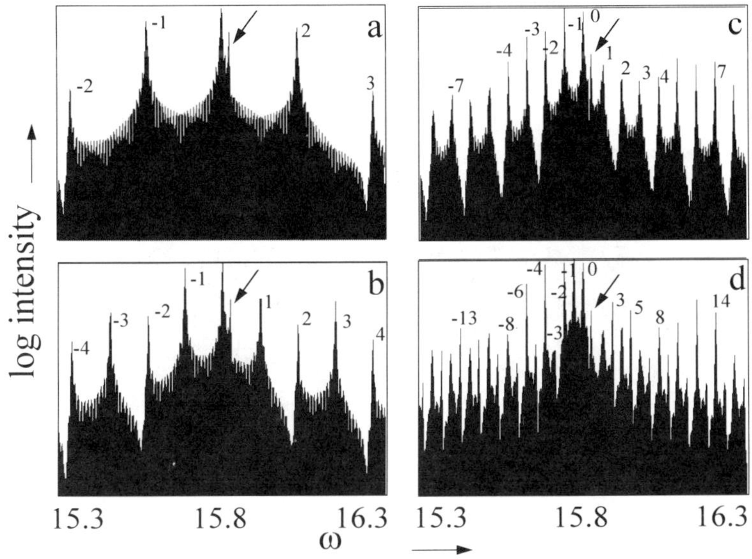

Figure 2. The changing complexity of the dynamical diffraction profile of an ($AlAs_m$ + $GaAs_m$) x 50 superlattice as the period is increased from (a) 200, (b) 400, (c) 800 to (d) 1600Å with the 002 reflection and CuKα radiation.

3. ORIGIN OF SATELLITES

There are several ways in which to interpret how the satellites originate and consequently how the structural parameters can be deduced. The bilayer can be considered as a large unit cell and this is valid within the kinematical theory approximation and provided that the structure is commensurate. Alternatively the bilayer periodicity can considered as a modulation of the lattice parameter as well as the structure factor and this is valid provided that the assumption that it as a perturbation on the average structure is correct. The other approach is to just put in the structural parameters into a dynamical diffraction simulation program and see what comes out; unfortunately this latter approach is not so transparent in understanding their origin.

In the first instance consider the large unit cell that is composed of several unit cells of each component. From this the structure factor for any reflection, **H** can be calculated:

$$F_H = \sum_j{}^N f_j \exp(-2\pi \boldsymbol{H}.\boldsymbol{r}_j)$$

where N is the number of atoms in this large unit cell and $\boldsymbol{r}_j$ is the position of the j th atom in that cell and f_j is its scattering factor. In the kinematical theory approximation this can be equated to the scattered intensity for the reflection by the square of this quantity:

$$I_H = |F_H^* F_H|$$

and the scattering angle for this reflection is given by:

$$2\theta_H = 2\sin^{-1}\left(\frac{\lambda}{2d_H}\right)$$

where d_H is the interplanar spacing and λ the wavelength. In the above equation for the structure factor we have two modulations, one associated with the difference in scattering and the second the difference in lattice parameter by virtue of the different location of the atoms within the two regions of the unit cell. By calculating the intensity and positions of the satellites it is possible to obtain very good agreement with experiment (Keravec, Baudet, Caulet, Auvray, Emery and Regreny, 1984).

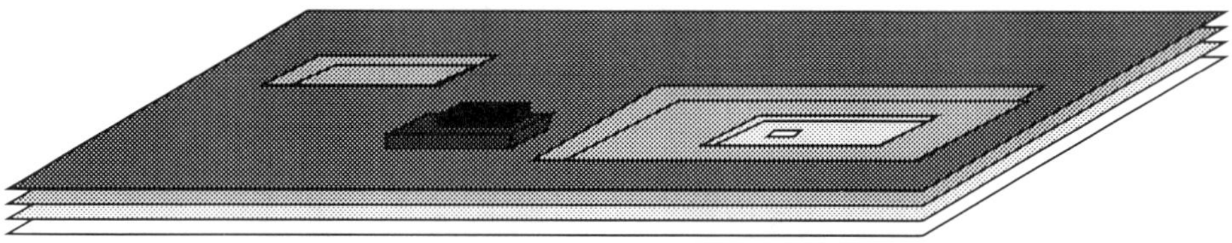

Figure 3. The creation of an incommensurate superlattice and consequential interfacial roughness.

The above analysis is readily transparent but the control over growth of layers is often insufficient to create a bilayer and individual layers within a bilayer to have an integer number of atomic layers. We therefore have an incommensurate structure, (Fewster, 1986). Suppose that the layer growth begins from a perfectly flat surface and then a few atomic layers are deposited, Figure 3. There will always be a finite roughness unless an exact number of atoms have been deposited and the atoms are able to diffuse across the surface. The effective layer thickness will therefore in general be a non-integer number of atomic layers and therefore must be rough at the atomic level. The x-ray diffraction probe will average within the lateral coherence length and will therefore see the situation in Figure 3 as an interfacial grading. The diffraction profile of an incommensurate superlattice is given in Figure 4, where the satellite peaks give rise to a periodicity that is close to an integer number of atomic layers and intensities that do not match those of $(GaAs)_3(AlAs)_3$ as intended and therefore must also included combinations of $(GaAs)_4(AlAs)_2$ and $(GaAs)_2(AlAs)_4$, etc. A commensurate superlattice on the other hand has a complete and continuous set of satellites as shown in Figure 5. However the scattering close to the interface will now have regions of finite correlation lengths, i.e. between regions of roughness and if these are less than the coherence length then this will cause diffraction profile broadening, described later. This incommensurability must be accounted for in the analysis, especially in short period superlattices, by including a range of large unit cells that straddle the "average" incommensurate structure, (Fewster, 1986).

The second approach is to consider the scattering and lattice parameter modulation as a perturbation on the average structure, (Guinier, 1963). In this case the scattering factor is assumed to vary from $f(1-\eta)$ to $f(1+\eta)$ and the interplanar spacing by $d(1-\epsilon)$ to $d(1+\epsilon)$. From this the intensity ratio is derived:

$$\frac{I_{H\pm m}}{I_H} = \left(\frac{a_m}{2}\left[\left(\frac{H}{d} \pm \frac{m}{\Lambda}\right)\Lambda\epsilon \mp \eta\right]\right)^2$$

where **H** and **m** are the diffraction orders for the average lattice periodicity and the superlattice periodicity of wavelength Λ. a_m is the Fourier coefficient for the satellite reflection of mth order. The satellites appear at distances $\mathbf{m}/\Lambda$ from the main "average" diffraction peak which is at **H**/d in reciprocal space. This equation is very useful for visualising the satellite intensities; i.e. suppose that the strain modulation is negligible, $\epsilon\sim0$ or alternatively the scattering factor modulation is negligible, $\eta\sim0$ then the satellite intensities either side of the average peak should be equal. We can also consider the amplitude of the satellite reflections as Fourier coefficients and then this equation creates a very simple procedure to move from reciprocal space to real space provided that we have an estimate of the phase component of the Fourier coefficients.

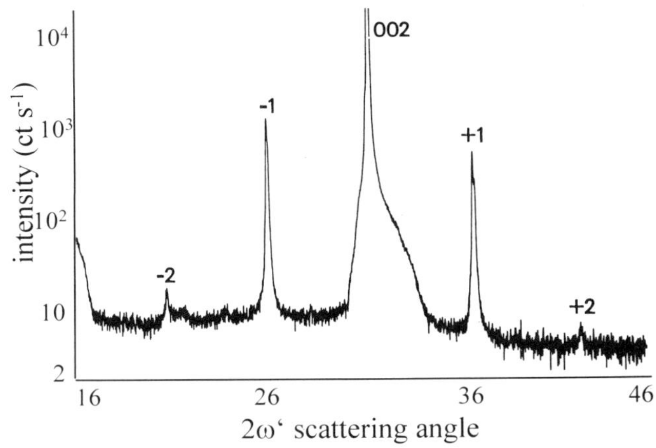

Figure 4. Satellites from an incommensurate superlattice with intensities and positions not matching that of a structure with an integer number of monolayers in a period (6.03). Collected on a powder diffractometer with CuKα radiation.

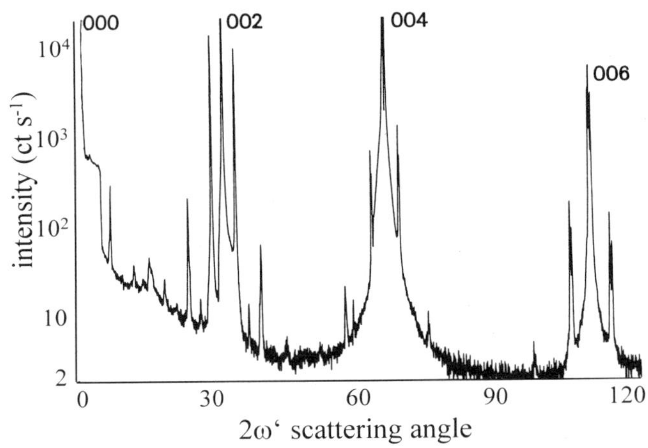

Figure 5. A complete set of satellites out to the 006 peak from a complex laser structure whose thicknesses are an integer number of atomic layers.

Both the methods above can prove very convenient provided the warnings of the earlier section are heeded, because they only require determination of the integrated intensities for comparison or Fourier transforming. The profile fitting methods, e.g. dynamical theory rely on very good quality data for comparison, unless it is used as a method of determining the integrated intensity for comparison, however this method can be rather slow. Also when using the dynamical theory model it is important to remember it is only strictly "exact" close to the diffraction for which the parameters are defined unless all parameters are maintained as angular variables. The deviation parameter $\alpha_H(\omega)$ is also used in very approximate forms in many models and a more exacting form as that given below should be used, (Fewster, 1996):

$$\alpha_H(\omega)=(1+4\sin\frac{\omega-\omega_0}{2}[\sin\frac{\omega-\omega_0}{2}+\sin(2\theta-\frac{\omega-\omega_0}{2})])^{\frac{1}{2}}-1$$

where θ is the Bragg angle for the reflection being simulated. ω is the angle of incidence and ω_0 is the angle of incidence in the Bragg condition. This is the most exact derivation of $\alpha_H(\omega)$ and could be further simplified by assumptions to the more familiar forms:

$$\alpha_H(\omega)\simeq-2\sin\frac{\omega-\omega_0}{2}\sin2\theta\sim-(\omega-\omega_0)\sin2\theta$$

for small deviations, i.e. ω-ω_0 small.

4. MEASURING THE PERIOD

This may seem a rather trivial section since one would assume that the satellites that arise from this modulation can be directly transformed using Bragg's law and creating an expression as given here:

$$\Lambda=\frac{(i-j)\lambda}{2(\sin\theta_i-\sin\theta_j)}$$

where i and j are the satellite orders with Bragg angles θ_i and θ_j. This is correct in the kinematical theory approximation and will give a very good estimate of the period within about 1%. The complication arises with the dynamical diffraction effects that treat the diffraction process as a whole and this can lead to unexpected interactions. The above expression has been used very successfully for rapid routine analysis of superlattice structures, (Fewster, 1987), by including all the measurable satellites and determining all the various combinations. In this case the dynamical diffraction effects are almost averaged out. Further hazards in determining the period will be discussed later that result from further complicating material uncertainties.

Firstly consider the dynamical diffraction effects with reference to Figure 1. The above equation is based on kinematical theory and it is possible to compare the validity of this expression by applying it between different pairs of satellites as the number of periods changes. The period of the superlattice of Figure 1 is 100Å in all cases yet consistently satellite peak separation on the low ω angle side of the "average" peak is larger than expected. For example the -10 to -8 satellite separation produces a modulation wavelength of ~98Å. On the other hand the +8 to +10 satellite peak separation gives a modulation wavelength of ~100.6Å. Whereas the -1 to +1 satellite peak separation gives rise to periods of 101.9Å, 100.3Å and 99.9Å for the structures with 4, 10 and 50 repeats respectively. This may appear rather strange but on closer inspection it is clear that diffraction profile is far from symmetric about the "average" peak, especially in the region close to the weaker odd order satellite reflections and therefore the "apparent" modulation wavelength asymmetry is not so surprising. The first order satellite peak separation is perhaps more obvious, in that it has been known for sometime that diffraction peak "pulling" effects can arise in thin layer structures, because the wavefield that has insufficient periodicity to lock into and diffract independently, (Fewster, 1987; Fewster and Curling, 1987).

Therefore in this particular case the period can be determined within a few Ångstroms with the above equation using a single measurement. Averaging several satellites above and below the "average" peak will improve the accuracy.

5. PERIOD VARIATIONS

Period variations can be interpreted as interfacial roughness, (Fullerton, Sculler, Vanderstaeten and Bruynseraede, 1992), provided that these are known to be uncorrelated, i.e. the fluctuations are randomly distributed throughout the structure. The simplest way to visualise the influence of these variations on the diffraction profile is to consider a structure having satellites with periods of Λ_1, Λ_2, Λ_3, ...Λ_n then the diffraction will be composed of satellites at different positions, (Fewster, 1986). Therefore the width of an individual satellite $\Delta\omega$ can be related to the period variation by differentiating the above equation, (Fewster, 1988). If the profiles are approximated to Gaussian shapes and integral breadths are used, then a quantity termed the Gaussian distribution of periods can be determined:

$$\Delta\omega_\Lambda = \frac{(i-j)\lambda\overline{\beta_S}}{\omega_{ij}^2\cos\theta}$$

where β_S is the integral breadth as a statistical average for all the measured satellites and ω_{ij} is the angle between two satellite peaks. The advantage of approximating the profile to a Gaussian is the rather trivial method for deconvoluting the intrinsic profile width for the superlattice.

A structure with 4% random variations in the period has been simulated and displayed in Figure 6(a) and from this the satellites can be seen to broaden progressively with order although the complexity of the peak shapes are complex. In practice therefore the randomness should be "averaged" to arrive at the practical situation. However it is important to be aware of the difference between random fluctuations and progress period variations that may occur if the growth rates are changing in a systematic way during deposition. The profile of such a variation appears as in Figure 6(b). Holý, Kuběna and Ploog (1990) have calculated the profile using a semi-kinematical approach to illustrate how the different forms of thickness fluctuations influence the profile shape.

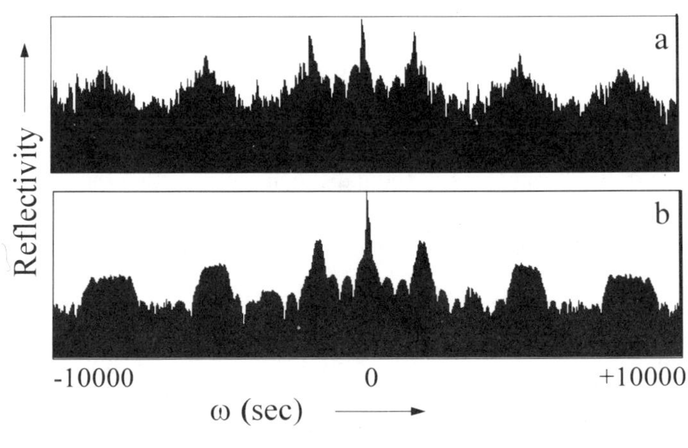

Figure 6. The influence of variations in the periodicity (a) random variations of ±4% and (b) systematic increase of 8% for a (AlAs [50Å] GaAs [50Å]) x 50 superlattice 002 reflection and CuKα radiation.

6. MEASURING INDIVIDUAL THICKNESSES

The individual thicknesses of a superlattice can be determined in a number of different ways. The approaches described above use the concept of a large unit cell, the idea of a perturbation and the full profile simulation method using dynamical theory. The most rapid methods make use of the two former approaches since the thicknesses can be obtained by collecting the integrated intensities, whereas the latter require a good profile shape. The former greatly simplifies the instrumentation required.

In the equation above an expression was derived for the strain, ϵ, and the scattering modulation, η, in terms of the intensity ratio, (Fleming, McWhan, Gossard, Wiegmann and Logan, 1980). Of course a good estimate of the thicknesses would be required to start with to include a value for a_m since this quantity involves a phase component, i.e. the relative positions of the sine waves in the reconstruction of the modulation. It is most important also to be aware of the dangers of Fourier transforms as here, in that it is strongly dependent on the number of observed satellites. If there are only the first order satellites measured then the modulation will be a simple sine wave and the two thicknesses determined will be equivalent. Only when a reasonable number of satellites are included in this determination can there be sufficient confidence in the result. Usually the waveform can be considered as a rectangular wave with perfect symmetry of the interfaces then a cosine function with simple phases of 0 or π need be included.

The large unit cell approach makes fewer assumptions with regard to the number of satellites since the problem of termination effects are not present and the fit is only as good as the data, which is not unreasonable. In this case we should calculate the structure factor for large unit cells that straddle the determined value of the period and add the structure factors before converting them to intensities for comparison with the measured integrated intensities. The low order intensities are less sensitive to interface imperfections, (low order Fourier coefficients), and therefore the fit to these should be good to be confident of the result. To obtain the individual thicknesses depends on whether the composition of the layers within the superlattice are known, if they are then only one parameter, the ratio of the thicknesses for a bilayer, needs to be known, although more complex structures can be modelled, (Fewster, 1987). If for simplicity it is assumed that the structure is composed of binary (AB) and ternary ($A_{1-x}B_xC$) semiconductor superlattice then some simple relationships can be stated that again bring the solution to a simple one parameter problem:

$$\Lambda = t_1 + t_2$$

and

$$x_2 = \frac{x_\Lambda \Lambda}{t_2}$$

where t_1 and t_2 are the thicknesses of the individual layers within the structure. x_Λ is the average composition through the superlattice. This requires the determination of this average composition, which was discussed earlier and will relate in some way to the location of the "average" peak.

For both these methods we do require this "average" scattering (perturbation method) and the "average" composition. To obtain this the central region of the diffraction profile should be modelled with dynamical theory, (Fewster and Curling, 1987) and not rely on the peak separation. It is also very important to model the whole structure not just the average equivalent structure. If it is assumed that the diffraction averages the composition to give an equivalent peak splitting as in the superlattice then there will an error, (even if the reduction in number atoms in the layers that have a larger lattice parameter is taken into account),

Figure 7.

It may appear obvious that dynamical theory should be applied and these approximate theories forgotten, but this does rely on obtaining very good data by mapping the intensity in reciprocal space and projecting onto a radial direction.

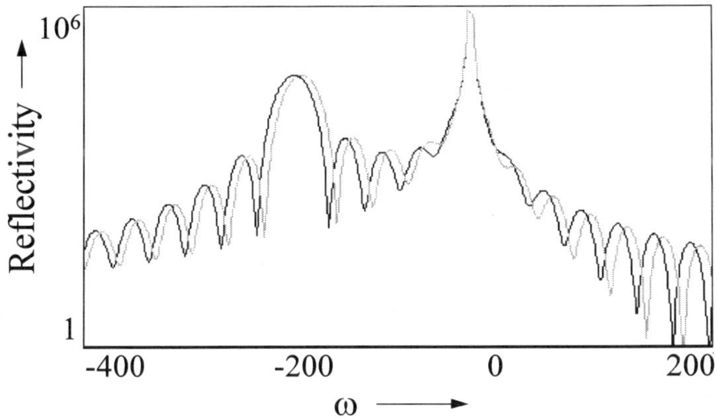

Figure 7. The effect of assuming that the superlattice behaves as an average structure will give the wrong "average" composition (dot) compared with the simulation of the actual structure (line), (004 reflection and CuKα radiation).

7. INTERFACE ROUGHNESS

Interfacial grading can be tackled in exactly the same way as described above by either of the methods. However a good estimate of the individual thicknesses and compositions should be determined first before introducing more parameters. A report of the predictive Fourier method has been given by Fleming et al (1980) and of the large unit cell method by Fewster (1988). In the latter paper a comparison was made between the two methods and it was shown that provided sufficient numbers of satellites are included in the Fourier method then an equivalent average interface could be determined. However in most structures the interfaces are not equivalent and the predictive Fourier method is no longer viable because of the lack of phase information to determine the asymmetry. Fewster, Andrew and Curling (1991) have carried out an extensive study of the AlGaAs system to ascertain the interfacial perfection and differences between the GaAs to AlGaAs and AlGaAs to GaAs interfaces. The large unit cell method will show a difference between interfaces but cannot ascertain which is which with these experiments. The mixing of elements at the interface to create new compounds can also have dramatic effects on the satellite intensities, for example when the gas valve sequencing is not instantaneous as in growth from the gas or vapour phase, (Vandenberg, Bean, Hamm, and Hull, 1988; Vandenberg, Panish, Temkin and Hamm, 1988; Lyons, Scott, and Halliwell, 1989)

Lateral interface roughness exhibits itself in the diffraction pattern as a broadening of the superlattice peaks parallel to the interface plane. Of course since the interface is only a small proportion of the scattering of the superlattice will appear as shoulders on the satellite profiles, (Fewster, 1991), from which we can estimate a lateral correlation length. A more exacting method is to simulate the diffraction effect to extract this information either in high angle diffraction, (Holý, Kuběna, Abranof, Lischka, Pesek and Koppensteiner, 1993) or with grazing incidence reflectometry, (Phang, Savage, Kariotis and Lagally, 1993) .

Reflectometry is sensitive to the x-ray density as a function of depth and is another very useful tool and should be considered in the same way as diffraction except that the strain component is negligible. The advantage here is that we can use a very simple optical theory,

(Parrat, 1954), however the sensitivity to interfacial roughness can be severe. This of course make it very good for modelling roughness yet the diffuse scattering from lateral correlation lengths and correlated thickness variations can have a significant impact on the specular scattering, (de Boer, Leenaers and Wolf, 1995). Combining the merits of diffraction and reflectometry can of course isolate and increase the number of measurable parameters, for example measuring the individual lattice parameters or elastic parameters in a superlattice, (Birch, Sundgren and Fewster, 1995).

8. ADDITIONAL COMPLICATIONS IN ANALYSIS

Apart from the complexities arising from the assumptions concerning the relevant theories, the sensitivity of x-ray diffraction to minor structural imperfections can also create problems. It is possible to simply state that the assumptions of the kinematical theory are adequate and certainly x-ray diffraction in general can boast greater precision than other analytical methods. However for those who wish to strive for perfection then the presence of diffraction "wiggles" necessitates full diffraction space mapping and dynamical simulation to be certain of the parameters of interest, (Fewster, 1993). The reason for this is that the satellites do not lie on a radial direction from the origin of reciprocal space, thus giving varying satellite spacings with open detector techniques or missing satellites with radial scans. To reconstruct the radial scan to fit with the parameters that influence $\alpha_H(\omega)$ the diffraction space map should be projected onto a radial direction and then modelled. These "wiggles" contain extra structural information that will be described in a later publication.

If the modulation is not perpendicular to a low index atomic plane then the implied interfacial roughness should be considered, or region of homogeneous composition and of course the inclination of the satellites. The above methods are all applicable although the repeat unit may now become rather large laterally. The terrace width can be determined if the coherence length is larger than this width, Neuman, Zabel and Morkoç (1983).

9. REFERENCES

Birch, J, Sundgren, J -E and Fewster, P F, 1995, Measurement of the lattice parameters in the individual layers of single-crystal superlattices, *J Appl. Phys.* 78:6562

de Boer, D K G, Leenaers, A J G and Wolf, R M (1995, X-ray reflectometry from samples with rough interfaces: an oxidic-multilayer study, *J Phys. D: Appl. Phys.* 28:A227

Fewster, P F, 1993, Review article: X-ray diffraction from low dimensional solids, *Semicond. Sci. Technol.* 8:1915

Fewster, P F, 1991b, Multicrystal X-ray diffraction of heteroepitaxial structures, *Appl. Surf. Science* 50:9

Fewster, P F, 1986, X-ray diffraction from multiple quantum well structures, *Philips J Res.* 41:268

Fewster, P F, 1987, Probing semiconductor structures by X-ray diffraction, in: *Thin Film Growth Techniques for low dimensional Structures,* R F C Farrow, S S P Parkin, P J Dobson, J H Neave and A S Arrott, Ed., Plenum Press, New York

Fewster, P F, 1988, Interface roughness and period variations in MQW structures determined by X-ray diffraction, *J Appl. Cryst.* 21:524

Fewster, P F, 1993, Characterisation of quantum wells by X-ray diffraction, *J Phys.D: Appl. Phys.*26:A142

Fewster, P F, 1996, unpublished work.

Fewster, P F, Andrew, N L and Curling, C J, 1991, Interface roughness and period variaions in the AlGaAs system grown by molecular beam epitaxy, *Semicond.Sci.Technol.* 6:5

Fewster, P F and Curling, C J, 1987, Composition and lattice-mismatch measurement of thin semiconductor layers by X-ray diffraction, *J Appl. Phys.* 62:4154

Fleming, R M, McWhan, D B, Gossard, A C, Wiegmann, W and Logan, R A, 1980, X-ray diffraction study of interdiffusion and growth in $(GaAs)_n(AlAs)_m$ multilayers, *J Appl. Phys.* 51:357

Fullerton, E E, Sculler, I K, Vanderstaeten, H and Bruynseraede, Y, 1992, Structural refinement of superlattices from X-ray diffraction, *Phys. Rev.* B45:9292

Guinier, A, 1963, *X-Ray Diffraction, in Crystals, Imperfect Crystals, and Amorphous Bodies,* W H Freeman, San Francisco

Holý, V, Kuběna, J, Abranof, E, Lischka, K, Pesek, A and Koppensteiner, E, 1993, X-ray double and triple

crystal diffractometry of mosaic structure in heteroepitaxial layers, *J Appl. Phys.* 74:1736
Holý, V, Kuběna, J and Ploog, K, 1990, X-ray analysis of structural defects in a semiconductor superlattice, *Phys. Stat. Sol.* (b) 162:347
Keravec, J, Baudet, M, Caulet, J, Auvray, P, Emery, J Y and Regreny, A, 1984, Some aspects of the X-ray structural characterisation of $(Ga_{1-x}Al_xAs)n_1(GaAs)n_2$/GaA(001) superalttices, *J Appl. Cryst.* 17:196
Lyons, M H, Scott, E G and Halliwell, M A G, 1989, Investigation of interfaces in GaInP/InP superlattices by X-ray multiple crystal diffractometry, in: *Microscopy of Semiconducting Materials 1989*, A G Cullis and J Hutchinson , Ed., Institute of Physics, Bristol
Neuman, D A, Zabel, H and Morkoç, H, 1983, X-ray evidence for a terraced GaAs/Alas superlattice, *Appl. Phys. Lett.* 43:59
Parrat, L G, 1954, Surface studies of solids by total reflection of X-rays, *Phys. Rev.* 95:359
Phang, Y H, Savage, D E, Kariotis, R and Lagally, M G, 1993, X-ray diffraction measurement of partially correlated interfacial roughness in multilayers, *J Appl. Phys.* 74:3181
Vandenberg, J M, Bean, J C, Hamm, R A and Hull, R, 1988, Kinematic simulation of high-resolution x-ray diffraction curves of Ge_xSi_{1-x} strain layer superlattices: a structural assessment, *Appl. Phys. Lett.* 52:1152
Vandenberg, J M, Panish, M B, Temkin, H and Hamm, R A, 1988, Intrinsic strain at lattice-matched $Ga_{0.47}In_{0.53}As$/InP interfaces as studied with high-resolution x-ray diffraction, *Appl. Phys. Lett.* 53:1920

CHARACTERIZATION OF LATTICE DEFECTS IN ION–IMPLANTED SILICON

Marco Servidori, Franco Cembali and Silvia Milita

CNR–Istituto LAMEL
Via Gobetti 101
I–40129 Bologna, Italy

1. INTRODUCTION

The wide use of ion implantation as a doping step in electronic device manufacturing stimulated a large number of scientific papers aimed at understanding the physical effects of the process on the target substrate. Among the different structural investigation methods, such as transmission electron microscopy techniques and Rutherford back-scattering/channelling of high energy light ions, X–ray diffraction in multicrystal optical geometries is preferable because of its high sensitivity to weak lattice strains.

In Sect. 2 an X–ray diffraction formalism will be briefly outlined for multiple layers. This because simulation procedures of calculated kinematical (Speriosu, 1981), semi–kinematical (Kyutt et al., 1980) and dynamical (Wie et al., 1986) intensity profiles (rocking curves, RCs) to experimental data demonstrated that the depth–dependent structural damage distribution caused by ion bombardment is well described by step–like functions.

In Sect.3 the most suitable diffraction optics, either dispersive or non–dispersive, to make measurements in the highest resolution mode will be described and the calculation procedures to best fit the experimental RCs will be the subject of Sect. 4. Finally, examples of application will be reported in Sect. 5 to show how information on the features of the implant process can be extracted from RC analysis. In particular, attention will be drawn to (i) the formation of lattice defects and their evolution after thermal treatments, (ii) the effects of the nuclear and electronic energy losses of the ions on penetrating the substrate and (iii) the transition from damaged to amorphized sample surface.

2. DYNAMICAL DIFFRACTION FROM MULTIPLE LAYERS

The theory presented here follows the formalism developed by Zachariasen (1945) and Taupin (1964), and assumes that

- the incidence angles of the beam on the crystal surface are far from the critical angle for total external reflection,
- the beam is a plane wave,
- only one beam, in addition of the incident beam, is produced by diffraction (two-beam condition),
- the lattice deformation is purely elastic,
- the strain is much less than 1,
- the deformation gradient, within the distance of one lattice spacing, is much less than the deformation itself,
- the lattice parameter varies in the crystal only with depth (mono-dimensional case).

From this theory several approximated models were derived for thin crystals with zero absorption (kinematical models), in which the interaction of the wave fields associated with the incident and diffracted beams inside the crystal is neglected. However, the exact (dynamical) diffraction model is reported here, valid for a cubic, generally absorbing and non-centrosymmetric crystal of any thickness.

For such a crystal, the Fourier coefficients of H–order of 4π times the polarizability per unit volume are complex quantities and are defined as

$$\psi_H = -\Gamma \sum_j f_j exp(i2\pi \mathbf{H}.\mathbf{r_j}) \ ,$$

where $\Gamma = (e^2/mc^2)\lambda^2/\pi V$, e^2/mc^2 is the classical electron radius, λ the X–ray wavelength, V the volume of the unit cell, $\mathbf{H}$ and $\mathbf{r_j}$ are the reciprocal lattice vector and the atomic position vector, respectively,

$$f_j = (f_j^0 + \Delta f_j' + i\Delta f_j'')\ exp(-M)$$

is the atomic scattering factor, f_j^0 is the atomic scattering factor at zero absolute temperature, $\Delta f_j'$ and $\Delta f_j''$ are the anomalous dispersion corrections, and $exp(-M)$ is the temperature factor. The ψ_H coefficients fulfil the conditions that $\psi_H \neq \psi_{\overline{H}}$ and $\psi_{\overline{H}} \neq \psi_H^*$ (star denotes the complex conjugate). Since $exp(i2\pi\mathbf{H}.\mathbf{r_j})$ is in general complex, the following relations (Cole and Stemple, 1962) hold

$$\psi_H = \psi_H' + i\psi_H'' \ , \quad \psi_H' = (\psi_H')_r + i(\psi_H')_i \ , \quad \psi_H'' = (\psi_H'')_r + i(\psi_H'')_i \ , \qquad (1)$$

where the subscripts r and i indicate real and imaginary parts, respectively. From (1) we can define the modulus of the real part of ψ_H as

$$|\psi'_H| = \sqrt{(\psi'_H)^2_r + (\psi'_H)^2_i} \tag{2}$$

and the absorption coefficients of the diffracted wave as

$$k = |\psi''_H|/|\psi'_H| \ , \quad p = [(\psi'_H)_r(\psi''_H)_r + (\psi'_H)_i(\psi''_H)_i]/|\psi'_H|^2 \ . \tag{3}$$

The differential equation describing the variation with the crystal thickness of the amplitude of the external diffracted wave (X), normalized to that of the external incident wave, is

$$-i(dX/dA) = BX^2 + 2CX + B \tag{4}$$

with

$$B = \sqrt{1 - k^2 + i2p}, \quad C = y + ig, \quad A = \pi K|\psi'_H|e^{-L_H}z/(\lambda\sqrt{|\gamma_0\gamma_H|}). \tag{5}$$

Here A is the dimensionless depth parameter; $exp(-L_H)$ is the static Debye–Waller factor (the meaning of this factor and the expression for L_H will be given in Sect. 4); K is the polarization factor ($= 1$ or $|cos(2\theta_B)|$ for polarization normal (σ) or parallel (π) to the diffraction plane, respectively); θ_B is the kinematical Bragg angle; $|\psi'_H|$, k and p are defined in (2) and (3); z is the depth coordinate; $\gamma_0 = sin(\theta_B - \alpha)$ and $\gamma_H = -sin(\theta_B + \alpha)$ (negative in the Bragg case) are the direction cosines of the incident and diffracted wave vectors with respect to the inward normal to the surface (α is the angle between surface and reflecting planes, positive or negative for grazing incidence or emergence), and

$$y = [(1-b)\psi'_0 + b\alpha_H](2K\sqrt{|b|}|\psi'_H|e^{-L_H})^{-1} \ , \tag{6}$$
$$g = [(1-b)\psi''_0](2K\sqrt{|b|}|\psi'_H|e^{-L_H})^{-1}$$

are the incidence parameter and the absorption coefficient for the incoming beam, respectively. ψ'_0 and ψ''_0 are the real and imaginary parts of Fourier coefficient of 0–order of 4π times the polarizability, $b = \gamma_0/\gamma_H$ is the asymmetry factor of the reflection and

$$\alpha_H = 4sin(\theta_B)[sin(\theta_B) - sin(\theta)] \ , \tag{7}$$

where θ is the incidence angle on the diffraction planes. The solution of (4) is

$$X(A_{top}) = \frac{DX(A_{bottom}) + i[B + CX(A_{bottom})] \ tan[D(A_{top} - A_{bottom})]}{D - i[C + BX(A_{bottom})] \ tan[D(A_{top} - A_{bottom})]} \ , \tag{8}$$

where $D = (C^2 - B^2)^{1/2}$. This solution holds for X depending only on A and for the incidence parameter y (6) independent of A (flat crystal). Solution of the differential equation allows the calculation of the amplitude X of the diffracted wave at the top of a parallel plate perfect crystal, if the amplitude X at its bottom is known. For a sequence of laminae, (8) can be used as a recursion formula starting from the bottom lamina. If the crystal is made by laminae on a substrate of infinite thickness, the initial condition for the recursive calculation is the knowledge of the X value at the surface of the substrate. This value is the infinite crystal solution of (8) for X independent of A and is given by $X_{sub} = -(C+D)/B$. Starting from this initial condition, the reflectivity at the surface ($A = 0$) of the composite crystal is calculated by (8) as $|X(0)|^2$, in the useful range of values of the incidence parameter y, by properly combining the σ and π polarization components. How to combine them depends on the crystal optics in which the sample crystal is inserted and will be explained later on. We remind the reader that $p = k$ for an absorbing crystal with an inversion centre.

If a sequence of mismatched overlayers is present on a substrate (we recall that mismatch does not necessarily imply deformation, but deformation always implies mismatch), the term α_H (7) in the incidence parameter (6) must be written in a more general form. For a relaxed (fully incoherent) surface layer

$$\alpha_H = 4sin(\theta'_B)[sin(\theta'_B) - sin(\theta)] \ ,$$

where

$$\sin(\theta'_B) = sin(\theta_B)(1+m_r)^{-1} \ ,$$

θ'_B and θ_B are the Bragg angles of overlayer and substrate, respectively, and m_r is the relaxed lattice mismatch. If the surface layer is deformed by partial coherency with the substrate, then

$$\alpha_H = 4sin(\theta'_B)[sin(\theta'_B) - sin(\theta + \alpha' - \alpha)] \ , \qquad (9)$$

where

$$sin(\theta'_B) = \frac{sin(\theta_B)}{1+m_\perp}\sqrt{\frac{1+[(1+m_\perp)/(1+m_{||})]^2 \ tan^2(\alpha)}{1+tan^2(\alpha)}} \ , \qquad (10)$$

$$tan(\alpha') = (1+m_\perp)/(1+m_{||}) \ tan(\alpha) \ , \qquad (11)$$

$m_\perp$ and $m_{||}$ (both different from m_r) being the normal and parallel mismatches to the surface, respectively. (10) describes the variation in the Bragg angle for planes inclined to the surface due to the mismatches, while the difference $\alpha' - \alpha$ in (9) is the variation in the incidence angle θ due to the rotation of the these planes with respect to those of the same family in the perfect substrate. The primed quantities in (10) and (11) will operate not only in (9), but also in the direction cosines γ_0 and γ_H and in the polarization factor $|cos(2\theta_B|$. The relaxed mismatch is related to the perpendicular and parallel mismatches by

$$m_r = (1-\nu)/(1+\nu) \ m_\perp + 2\nu/(1+\nu) \ m_{||} \ ,$$

where ν is the Poisson's ratio. For ion implanted silicon, the mismatch m is commonly replaced by the lattice strain ε.

3. OPTICS FOR HIGH RESOLUTION MEASUREMENTS

The theory outlined above enables the calculation of the intrinsic σ and π RCs of absorbing and non–centrosymmetric sample structures when the radiation is generally unpolarized. Since the experimental set up never provides the sample with a perfectly parallel and monochromatic beam, instrumental broadening and beam chromaticity have to be taken into account. The procedure to do this allows the proper combination of the two polarization profiles and hence the comparison of the calculation with the experiment. The ways by which this procedure is done differ for the different sequences and geometries of single crystals making up the optics of the diffractometer. We can distinguish between two broad categories, *i.e.* dispersive and non–dispersive arrangements, consisting, in general, of a group of crystals acting as the monochromator or the collimator, the sample and a group of crystals acting as the analyzer. The use of analyzer crystals distinguishes between triple and double crystal optics. The advantages of a triple crystal set up resides in, e.g., the separation of the diffuse scattering from the diffracted intensities, the limitations of the effects due to curved samples and the mapping of the intensity distribution in the reciprocal space.

3.1. Dispersive Geometries

In these geometries the lattice spacings of the crystals participating in the diffraction are different and hence the corresponding Bragg curves are not overlapped in the λ–θ_B plane. As can be deduced from the excellent old paper by DuMond (1937), when sample and analyzer are rotated together in γ:2γ coupling to produce a RC, the intensity profile $R(\gamma)$ collected by the detector is affected by the horizontal and vertical divergences of the beam diffracted from the monochromator or collimator and the wavelength distribution in the source. To reproduce by calculation such a profile, a three–dimensional (3D) integration is necessary, which can be written in the following compact form

$$R(\gamma) = I_0^{-1} \int_\phi F(\phi)\, d\phi \int_\lambda L(\lambda)\, d\lambda \int_{\Delta\theta} G(\Delta\theta)\, d\Delta\theta\; R^{\sigma,\pi}_{(out,m/c)}(\phi,\lambda,\Delta\theta) \times R^{\sigma,\pi}_{(in,s)}(\phi,\lambda,\gamma+\Delta\theta) \times R^{\sigma,\pi}_{(in,a)}(\phi,\lambda,2\gamma+\Delta\theta)\;, \qquad (12)$$

where

$$I_0 = \int_\phi F(\phi)\, d\phi \int_\lambda L(\lambda)\, d\lambda \int_{\Delta\theta} G(\Delta\theta)\, d\Delta\theta\; R^{\sigma}_{(out,m/c)}(\phi,\lambda,\Delta\theta) + \int_\phi F(\phi)\, d\phi \int_\lambda L(\lambda)\, d\lambda \int_{\Delta\theta} G(\Delta\theta)\, d\Delta\theta\; R^{\pi}_{(out,m/c)}(\phi,\lambda,\Delta\theta)\;. \qquad (13)$$

In (12) and (13) $F(\phi)$, $G(\Delta\theta)$ and $L(\lambda)$ represent the distribution functions of the beam divergence ϕ and $\Delta\theta$ normal and parallel to the diffraction plane, respectively, and of the wavelength in the source. Moreover, the three R functions under the integrals are the intrinsic profiles of the monochromator or collimator (m/c), sample (s) and analyzer (a), the superscripts σ and π indicate the polarization states, and the subscripts $(out, m/c)$, (in, s) and (in, a) mean exit from the monochromator or collimator, entrance on the sample and entrance on the analyzer, respectively. For a crystal (or a group of crystals) exit and entrance profiles are calculated from α_H in (7), by considering that the corresponding exit and entrance angles, referred to θ_B, must fulfil the relation $\Delta\theta_{out} = -b\ \Delta\theta_{in}$. In (12), the superscript σ, π indicates that the σ 3D integration must be summed to that of the π component. It is clear from (12) that the 3D convolution is not practicable for sample analysis by best fit with minimization routines, because it is highly time consuming for conventional computers. Therefore, some precautions must be taken from the experimental point of view. For instance, the effect of ϕ can be minimized by narrow beam limiting slits. Also, monochromators can be used in antiparallel (n, m) configuration (with $n \neq m$ or $n = m$), to reduce strongly peak broadening due to λ dispersion in the source (DuMond, 1937). The effect of the λ distribution is then minimized, thus reducing the 3D convolution to a monodimensional (1D) integration, if the Bragg angle of the sample is small (low order reflections) or very near to those of the monochromator crystals. However, for large Bragg angles of the sample the peak broadening can become dramatic, because the effects of λ, as of ϕ too, depend linearly on $tan(\theta_B)$. Irrespective of these precautions (negligible ϕ and small Bragg angle of the sample), the reduction from the 3D integration to a 1D case has recently been proposed (Servidori et al., 1994) for a Bartels (1983) germanium monochromator and 8 keV X–rays, by means of parametric Lorentzian functions. This approach proved itself to work quite well up to $\theta_B \lesssim 90$ deg. Bartels and Bartels–like crystal configurations, making use of symmetric or asymmetric reflections in antiparallel geometry from channel cut silicon or germanium single crystals, are nowadays well known and available commercially. These monochromators have the big advantage that they can be used with good optical resolution for any reflection from samples of any material.

3.2. Non–Dispersive Geometries

These geometries are non–dispersive because the Bragg angles of all the crystals involved in the optical path are exactly the same. This implies that the Bragg curves in the λ–θ_B plane (DuMond diagram) are perfectly overlapped, so that the effects of ϕ and λ on the broadening of the intrinsic RC of the sample vanish (Godwod et al., 1974). 3D convolution then reduces to a 1D integration over the beam divergence in the diffraction plane. However, a further simplification is possible if the reflections on collimator and analyzer are asymmetric. In fact, a value of $|b| \ll 1$ greatly decreases the angular width of the range of total reflection (Darwin width) for emergence from a crystal (see expression (6) for y). If collimator and analyzer are then asymmetrically cut in such a way that the emergence Darwin width from the first and the Darwin width for incidence on the latter are much narrower than the intrinsic one for incidence on the sample, (12) and (13) can be replaced by the very simple relation

$$R(\Delta\theta) = \frac{I^{\sigma}_{(in,a)} R^{\sigma}_{(in,s)}(\Delta\theta) + I^{\pi}_{(in,a)} R^{\pi}_{(in,s)}(\Delta\theta)}{I^{\sigma}_{(out,c)} + I^{\pi}_{(out,c)}} = w^{\sigma} R^{\sigma}_{(in,s)}(\Delta\theta) + w^{\pi} R^{\pi}_{(in,s)}(\Delta\theta) \; , \quad (14)$$

where w^{σ} and w^{π} are the weights of the σ and π integrated intensities of the collimator/analyzer, respectively, and

$$I^{\sigma(\pi)}_{(in,a)} = \int_{\Delta\theta} R^{\sigma(\pi)}_{(in,a)} d\Delta\theta \; , \qquad I^{\sigma(\pi)}_{(out,c)} = \int_{\Delta\theta} R^{\sigma(\pi)}_{(out,c)} d\Delta\theta \; . \quad (15)$$

In (15), $I^{\sigma(\pi)}_{(in,a)}$ are the σ or π integrated intensities calculated at the entrance on the analyzer when the collimator and analyzer crystals are in parallel arrangement in the absence of the sample, while $I^{\sigma(\pi)}_{(out,c)}$ are the same quantities calculated at the exit from the collimator in the absence of sample and analyzer. It can be seen that (14) is not as complex as (12), thus implying a much faster computing time. In fact, the RC of the sample is the average of its σ and π polarization profiles over w^{σ} and w^{π}. In a double crystal set up without analyzer, $I^{\sigma(\pi)}_{(in,a)}$ is replaced by $I^{\sigma(\pi)}_{(out,c)}$.

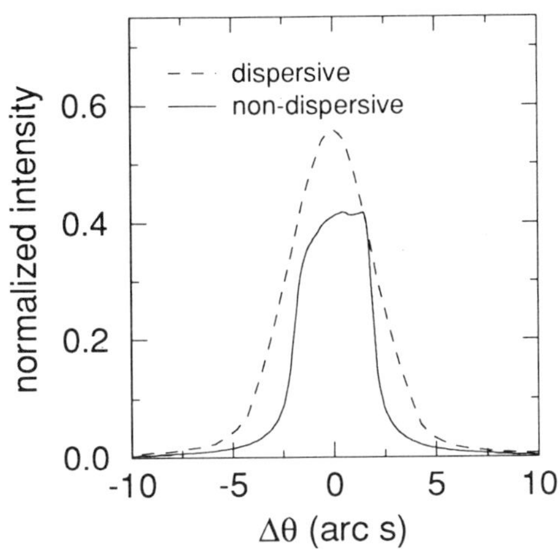

Figure 1. Longitudinal scan of $\mathbf{H}_{004}$ of Si in dispersive (broken line) and non–dispersive (full line) geometry.

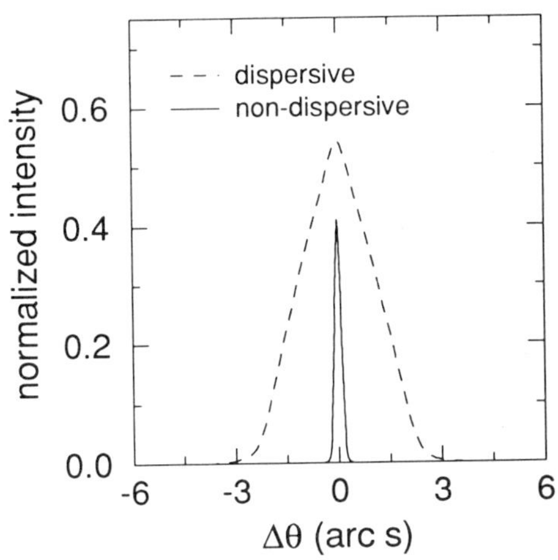

Figure 2. Transverse scan of $\mathbf{H}_{004}$ of Si in dispersive (broken line) and non–dipersive (full line) geometry.

Figures 1 and 2 shows the longitudinal (along the reciprocal lattice vector **H**) and transverse (perpendicular to **H** through the 004 reciprocal lattice point) scans, respectively, for the symmetric 004 reflection from a silicon sample with 8 keV X–rays. The broken lines in Figures 1 and 2 refer to a dispersive geometry (Figure 3), in which the sample is preceded by a Bartels (1983) monochromator made by four 440 symmetric reflections from germanium in (n,-n,-n,n) arrangement and followed by two (n,-n) 440 symmetric reflections from a germanium analyzer. The full lines refer to a non–dispersive geometry (Figure 4), in which the same sample is preceded by a pair of 004 asymmetric Si reflections ($|b_1| = |b_2|^{-1} = 33.84$) and followed by a pair of identically asymmetric Si reflections ($|b_3| = |b_4|^{-1} = |b_1|$). The narrower peak widths for the perfectly parallel optics are evident. The higher resolution of the parallel configuration is confirmed by the intensity distributions (Figures 5 and 6) around the 004 reciprocal

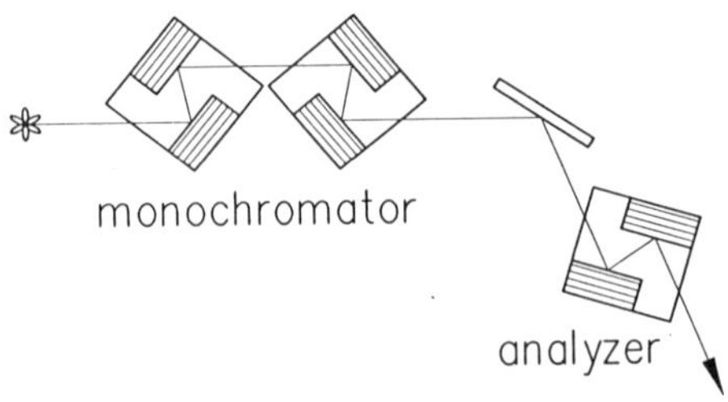

Figure 3. Dispersive geometry for a four reflection Ge monochromator, Si sample and a two reflection Ge analyzer.

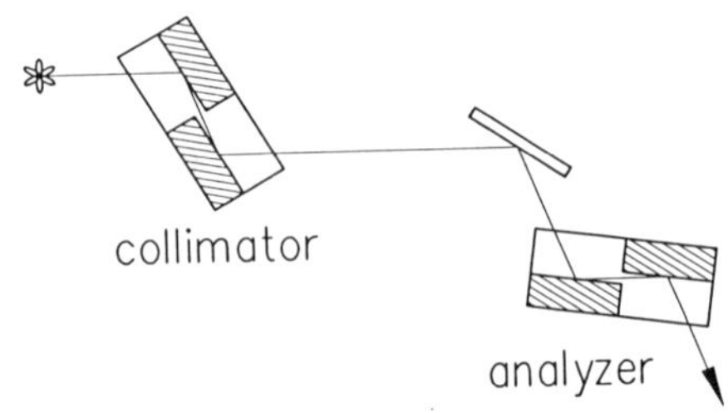

Figure 4. Non–dispersive geometry for a two reflection Si collimator, Si sample and a two reflection Si analyzer.

lattice point of the silicon sample, whose contour levels are plotted in the $q_{||}$–$q_{\perp}$ plane. Here, for symmetrical reflections $(q_{||}, q_{\perp}) = [\gamma\ cos(\theta_B), (2\Delta\theta - \gamma)\ sin(\theta_B)]/\lambda$ are the longitudinal and transverse components in the diffraction plane of the vector $\mathbf{q}$, defined

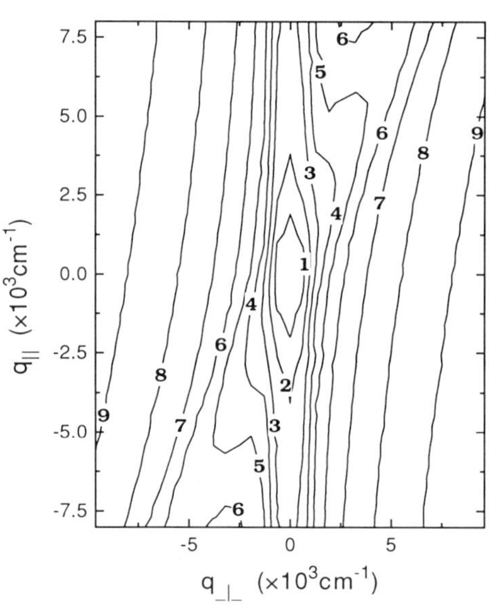

Figure 5. Normalized intensity distribution around the 004 Si reciprocal lattice point for the dispersive geometry. Level curves are shown for $n = -log_{10}I/I_0$.

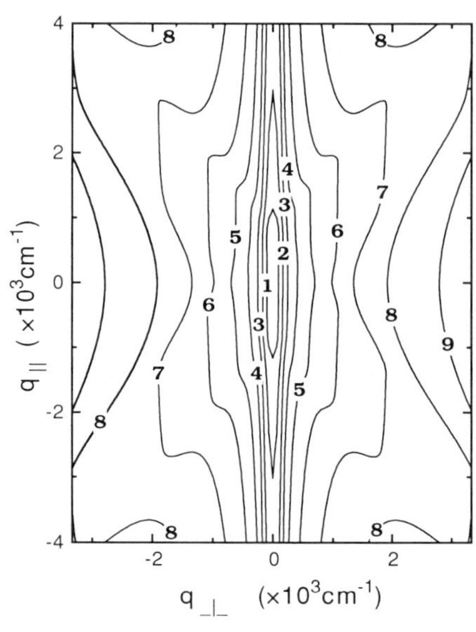

Figure 6. Normalized intensity distribution around the 004 Si reciprocal lattice point for the non–dispersive geometry. Level curves are shown for $n = -log_{10}I/I_0$.

as the deviation of the diffraction vector from $\mathbf{H}$. $\Delta\theta$ and γ are the rotation angles of sample and analyzer, respectively. The streaks of intensity in the figures arise from the terminations of the bulk crystals by their surfaces (crystal truncation rods). That along $q_{||}$ is the sample streak, while the monochromator (collimator) and the analyzer streaks are inclined anticlockwise and clockwise, respectively, to that along $q_{||}$ by the Bragg angle θ_B of the sample. More complex intensity maps are obtained for asymmetric sample reflections and for crystal geometries which do not sufficiently limit the wavelength spread (Gartstein and Cowley, 1990).

In both cases of Figures 5 and 6, the low tail intensities of the monochromator (collimator) and the analyzer, due to multiple reflections, permit a better analysis of the distribution of the diffuse scattering in the reciprocal space.

The non–dispersive geometries offer advantages over the dispersive configurations. The calculation of the RC of the sample is faster not only because the convolution can be avoided, but also because avoiding convolution implies that the experimental step width of this curve is not subjected to the constraint of being smaller than the angular width of the convolving function. Therefore, fewer experimental points speed up the best fit procedure. Also, they offer higher optical resolution than that produced by dispersive geometries. However, drawbacks are inherent to the non–dispersive geometries. In fact, its use is limited to the case where the sample material is of such a good crystal perfection (as in the case of silicon and silicon based heterostructures), that high performance collimators and analyzers can be manufactured. Moreover, each reflection from the sample needs the same reflection on collimator and analyzer.

4. ANALYSIS OF DIFFRACTION DATA

Analysis of the diffraction data involves the reproduction of experimental RCs by a diffraction model like that described in Sect. 2. Since the depth profile of the structural damage caused by ion implantation can be well described by step–like functions, in this model the varying parameters, strain ε and exponent L_H of the static Debye–Waller factor, are constant inside each lamina of variable thickness t. As to ε, only its component perpendicular to the surface, $\varepsilon_\perp$, is different from zero for silicon implantation, unless subsequent high temperature thermal treatments are made to remove radiation damage produced by ions whose size is quite different from that of the matrix atoms. The Debye–Waller factor describes the displacements of the atoms from the sites of the deformed lattice and lowers consequently the polarizability of the crystal. For a spherically symmetric Gaussian distribution of displacements not confined to small values or for any type of distribution of small displacements, L_H can be written as

$$L_H = 8\ [\pi\ sin(\theta_B)/\lambda]^2 < u^2 > , \tag{16}$$

where $< u^2 >$ is the mean square atomic displacement parallel to the diffraction vector.

Two ways are generally used to reproduce an experimental RC. The first is based on a trial and error method, which foresees that the parameters t, $\varepsilon_\perp$ and L_H are changed manually for each lamina until the goodness of simulation satisfies the user of the calculation program. This approach is time consuming and practicable only when the varying parameters are few. The second is a best fit procedure assisted by routines which minimize automatically the differences between experimental (I_i^{exp}) and calculated (I_i^{calc}) intensities. In this case, the efficiency of the best fit depends on the performances of the minimization routine. The criterion

$$\chi^2 = (1/l) \sum_{i=1}^{l} [log(I_i^{calc}/I_i^{exp})]^2 , \tag{17}$$

where l is the number of data points in the experimental RC, has always been used by the authors of the present paper.

The initial $\varepsilon_\perp$ distribution is tailored on the basis of the implant parameters (ion dose and energy), by using a structural model of layer sequence for which the strain values in the laminae are constrained to follow a physically plausible depth profile. This precaution strongly reduces the probability that unreasonable saw–toothed profiles are obtained when the lamina strains are free to vary independently of one another. As to the L_H profile, a two–step minimization was adopted. During the first step L_H is linked to $\varepsilon_\perp$ by a simple relation of proportionality. A sufficiently low value of χ^2 is then reached. After removal of this link, the program goes into the final minimization step by optimizing the L_H profile independently of that of $\varepsilon_\perp$.

The values of the parameters determined in this way are affected by errors coming essentially from the experimental errors in angles and intensities of the RCs. The method we use to calculate the errors in the parameters is a Monte Carlo analysis very similar to that applied by Ellis and Freeman (1995) to extended X–ray fine–structure data. Briefly, it consists in assuming the best fitted RC as the true one. Each point of this RC is replaced by a data point randomly chosen inside the window determined by the errors in angle and intensity of the corresponding experimantal point. Every set of replaced points is a new RC which will give, after best fit, different parameter values. The procedures of data replacement (*i.e.* the construction of new RCs) and best fit will be repeated for a sufficient number of times, enabling the average values of the parameters with their standard deviations to be obtained. This method proved itself more reliable, though more time consuming, than the analytical approach of least squares based on the calculation of the curvature matrix including the $\partial^2(\chi^2)/(\partial p_i \partial p_j)$ (Press et al., 1992), where $p_{i,j}$ are the parameter values. The calculation of the mixed second derivatives is necessary because the parameters are not uncorrelated from one another.

The automatic process of parameter optimization requires the values of w^σ and w^π in (14), to combine the σ and π intrinsic profiles of the sample. These weights are determined before the beginning of the best fit by a companion computer program, able to calculate the optical features of a beam produced by an indefinite crystal sequence of interest for laboratory and synchrotron facility experiments. This software has been developed for energy distributions of X–rays coming from conventional and synchrotron light sources, for symmetric and asymmetric Bragg and Laue diffraction from Si, Ge, diamond, Be and α–quartz, and for reflection from mirrors of any single– or multi–layer material. This program permits the calculation of w^σ and w^π as well as of the shape of the convolving functions in the case where dispersive geometries are used.

5. DEFECT ANALYSES IN IMPLANTED SILICON

Some significative examples of the variety of analyses which can be done to characterize the lattice damage in implanted semiconductors will be reported here for the silicon case. We emphasize that with defect analysis we mean the study not only of well defined crystal defects, such as dislocations, but also the analysis of the depth distributions of distortions in crystalline samples and of alternate crystalline and amorphous layers on a crystalline substrate. The last two aspects of damage investigations are important, because they help the understanding the physics of the implant process on one side and the interference phenomena resulting from the interaction of X–rays with implanted structures on the other.

5.1. Determination of Dislocation Loop Size and Density

It is well known that ion implantation is accompanied by the production of vacancy and interstitial point–like defects, whose depth profiles are slightly displaced with respect to one another (Servidori, 1987; Servidori, Zaumseil et al., 1987). Due to the mechanism of interaction between ions and target, the ions undergo energy loss when penetrating the substrate, mainly by nuclear collisions if the energy of the incident ions does not exceed some hundred keV. The atomic density of the implanted material then varies with depth, because atomic depletion and accumulation occur in the near surface layer and at a deeper region, respectively. When the implant dose is sufficiently high, a continuous amorphous layer can form at the surface. If Si^+ ions are used to amorphize silicon, post–annealing thermal treatments at $T \geq 550$ °C induce recrystallization of the amorphous layer as almost perfect crystal by epitaxial regrowth from the bulk. In addition to a heavy recombination between vacancies and interstitials, vacancy aggregates form at the surface, where the atomic depletion was maximum after the implant, and interstitial point defects collapse into dislocation loops in the damage tail where their local concentration dominates that of the vacancies. These defects induce deformation in the silicon matrix, so that lattice contraction ($\varepsilon_\perp < 0$) and expansion ($\varepsilon_\perp > 0$) are expected in the vacancy– and interstitial–rich regions, respectively. Profiles of negative and positive strains are shown in Figure 7 for a Si sample implanted with 1×10^{16} Si^+ cm^{-2} at 100 keV and furnace heated at 750 °C for 30 min. These profiles are the result of the best fit of the (n,-n,n) triple crystal 004 symmetric RC of Figure 8. In Figure 7, the L_H distribution is absent at the surface, indicating that the displacements of silicon atoms in the distortion field around vacancy clusters are much weaker, and hence hardly detectable, than the ones associated with the strain field of interstitial dislocation loops. This result agrees well with theoretical calculations in diamond cubic structures.

According to the theory developed by Dederichs (1973), for interstitial prismatic dislocation loops the displacement field in the surrounding crystal is such that the lattice strain and the static atomic disorder is given by

$$\varepsilon_{ij} = \frac{1}{2} C_L (|\mathbf{b_i}||\mathbf{F_j}| + |\mathbf{b_j}||\mathbf{F_i}|) , \quad L_H = \frac{1}{2}(|\mathbf{H}||\mathbf{b}|)^{1/2} C_L R_L^3 / V_a , \tag{18}$$

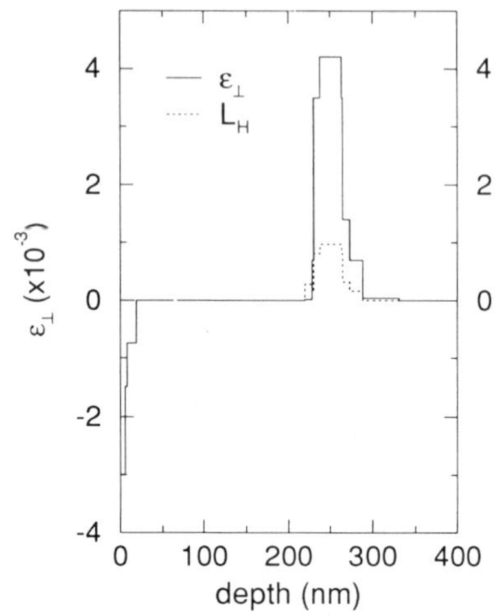

Figure 7. $\varepsilon_\perp$ and L_H profiles in (001) silicon after implant with 1×10^{16} Si^+ cm^{-2} at 100 keV and heating at 750 °C for 30 min.

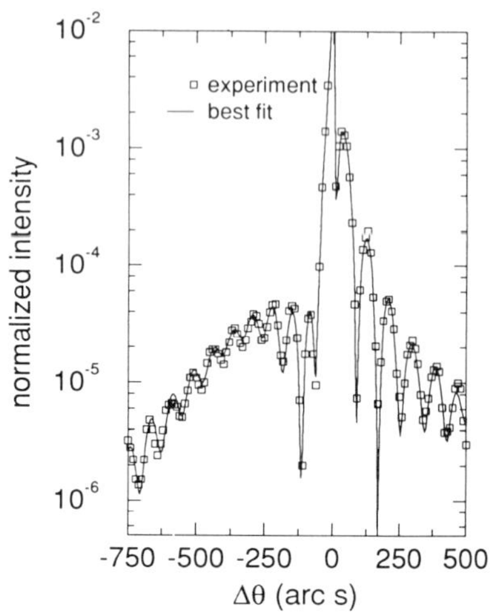

Figure 8. Experimental (symbol) and best fitted (line) (n,-n,n) triple crystal 004 symmetric RCs for Figure 7. $\lambda = 0.15406$ nm.

respectively, where C_L is the mean concentration of the loops (loops/cm^3 in ε_{ij} and loops/atom in L_H), $|\mathbf{b}|$ is the the modulus of the Burgers vector of the loops, $|\mathbf{F}|$ is the modulus of the vector normal to the loop plane (its value is the loop area), $|\mathbf{H}|$ is the modulus of the reciprocal lattice vector, R_L is the mean radius of the loops, and V_a is the volume per atom. A symmetric reflection is sensitive only to the component of the strain tensor normal to the surface ($\varepsilon_{11} = \varepsilon_\perp$). Its value is the average over all the orientations that **b** and **F** can assume inside the crystal lattice. The calculation of $\varepsilon_\perp$ for all the possible loop configurations, assuming random distribution, leads to values very close to the ones resulting from a simpler expression than (18). In fact, for prismatic interstitial dislocation loops of concentration C_L, the fractional change in the lattice parameter they induce through the volume change $\Delta V = |\mathbf{b}|\pi R_L^2$ is

$$\varepsilon_\perp = (f/3) C_L |\mathbf{b}| \pi R_L^2 / V_a , \tag{19}$$

where $f = 1 + 2C_{12}/C_{11}$ for (001) oriented cubic crystals and C_{12} and C_{11} are elastic constants. This factor corrects $\varepsilon_\perp$ to take the effect of tetragonal distortion into account. If $f = 1$, (19) holds for a statistically uniform loop concentration in the whole crystal. Since Figure 7 shows that the loops are confined in a thin layer and since zero value of $\varepsilon_\parallel$ followed from the analysis of asymmetric reflections, (19) describes the fact that lattice expansion is active only in the direction normal to the surface. The system made by (19) and by the L_H expression in (18) can hence be solved for C_L and R_L in each lamina of the distortion profile. Figure 9 shows the trends with annealing temperature of loop density and size in an Si sample with 1×10^{16} Si^+ cm^{-2} at 100 keV and isochronally heated in the range 700-900 °C for 30 min. The dislocation

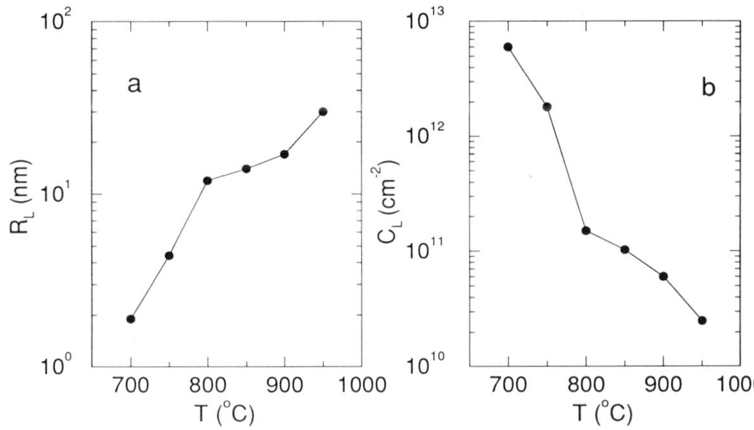

Figure 9. Temperature evolution of loop mean radius R_L (a) and concentration C_L (b) obtained by best fit of triple crystal rocking curves. The error bars coming from the errors in $\varepsilon_\perp$ and L_H are nearly as wide as the symbol size.

loops clearly increase in diameter at the expense of their concentration. This phenomenon, typical of implanted and annealed semiconductors, is dramatic in very large scale integration technology, because it results in an overall decrease in the interstitial concentration collapsed in the dislocation loops. Interstitials are then released into the lattice, favouring deepening (and widening for through–mask implants) of junction depths by enhanced dopant anomalous diffusion (Servidori, Sourek et al., 1987).

5.2. Atomic Displacements in High Energy Implants

In Sect. 5.1 it was claimed that the damage is essentially produced in a crystal by nuclear energy loss of ions when their energy is in the range up to some hundred keV. However, interaction also occurs between ions and the electrons of the target atoms, this phenomenon becoming more noticeable with increasing ion mass and energy. Therefore, with the recent entry of high energy implantation into device technology to make, for instance, buried doped layers or deep damaged regions for impurity gettering, the attention of scientists was drawn to the study of what occurs when MeV ions are implanted in semiconductors. Once again, and mainly because thicker layers are involved, X–rays play a relevant role in damage characterization.

Figure 10 shows the experimental and best fitted RCs of a (001) silicon sample implanted with 5×10^{14} B^+ cm^{-2} at 1.5 MeV. Figures 11 and 12 show the $\varepsilon_\perp$ and L_H profiles, respectively, coming from the best fit. The trends with depth of these parameters are strongly uncorrelated from one another and describe a situation very different from that usually encountered in implants made at medium-low energy. At these energies the L_H distributions are peaked at the maximum values of $\varepsilon_\perp$ whenever the displacement fields of the lattice defects are sufficiently strong and dense to produce static disorder (Servidori, 1987). Conversely, Figures 11 and 12 show that, while the lattice strain increases with depth and shows a peak centred at 2.25 μm, the static dis-

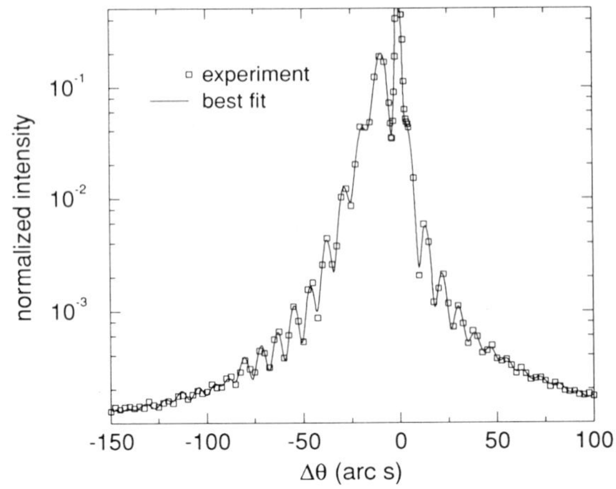

Figure 10. Experimental (symbol) and best fitted (line) (n,-n) double crystal 004 symmetric RCs of Si sample implanted with 5×10^{14} B^+ cm^{-2} at 1.5 MeV. λ=0.15406 nm.

order decreases monotonically from the surface and vanishes in correspondence of the strain peak. In other words, the lattice deformation is maximum where the atomic displacement is zero and, more important, the static disorder is maximum at the surface where the lattice parameter is nearly the same as that of the perfect crystal (Fabbri et al., 1992). This result, obtained thanks to the two–step procedure of best fit described in Sect. 4, is rather surprising, because, as underlined previously, common experience is

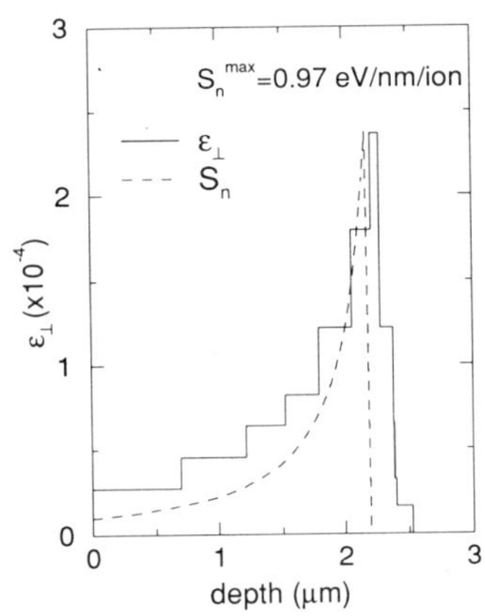

Figure 11. Profiles of lattice strain ($\varepsilon_\perp$) and nuclear energy loss (S_n).

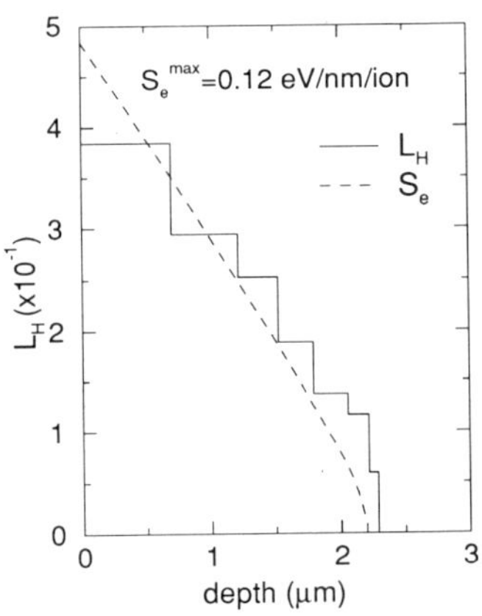

Figure 12. Profiles of static disorder (L_H) and electronic energy loss (S_e).

that the atoms are displaced from their sites in the regions immediately surrounding crystal defects. Subsequent calculations of the nuclear, S_n, and electronic, S_e, energy losses in Si implanted with 1.5 MeV B ions gave for these quantities the profiles reported in Figures 11 and 12, respectively. The agreement between the trends of the damage profiles and those of the energy losses is very good. Figure 11 can be interpreted according to the classical view that the energy loss by nuclear collisions creates interstitials in amounts directly related to the energy transferred to the nuclear subsystem. The matching between the distributions of L_H and S_e suggests that the energy contribution to the electronic subsystem produces atomic displacements. This picture confirmed a previous theoretical model (Mazzone, 1991) devised to describe relaxation processes of a hot electron gas produced by bombardment with high energy particles. The analysis of the lattice relaxation, resulting from the model, shows that the surplus energy supplied by the excited electrons leads the lattice to a stably disordered state and that this damage is of the same type as that given, in ordinary lattice dynamics, by an increased crystal temperature.

We emphasize that the observation of a wide spatial separation between the profiles of strain and static disorder was made possible by a RC best fit which assumes independent depth distributions of these parameters and by an implant of relatively low ion dose. In fact, the lattice defects which form at higher doses increase in size and density, thereby contributing with their own static disorder to the overall profile of L_H.

5.3. Analysis of the Amorphization Process

A common feature for all the implanted materials is that the crystalline substrate is amorphized when the ion dose reaches a critical value. If the dose is such that the amorphous layer appears on the surface, there is no chance for X–ray diffraction to give information on this layer. However, for immediately lower doses amorphization begins to form at the depth where the maximum of the $\varepsilon_\perp$ profile is found in crystalline substrates implanted at the same energy with still lower doses. In the situation of a buried amorphous layer, the Bragg (or Laue) case of diffraction is accompanied by interesting interference phenomena, whose study is useful to understand not only how interference occurs inside this structure, but also how the amorphization process evolves with increasing dose.

Figure 13 shows a series of experimental (symbols) and best fitted (lines) RCs from silicon implanted at 180 keV with Si^+ ions in the dose range from 2×10^{14} to 1.25×10^{15} cm^{-2} (Milita and Servidori, 1996). What is observed is that the shape of the RCs changes considerably with dose. The presence of well defined diffraction peaks (arrowed) for low doses suggests diffraction from highly strained layers. These peaks disappear in the range of intermediate doses, where only interference fringes are seen. They in turn vanish at the highest dose investigated. The profiles in Figure 14 confirm that the samples corresponding to the RCs (a) and (b) of Figure 13 are crystalline and that the distributions of either $\varepsilon_\perp$ or L_H, peaked at a depth slightly larger than 200 nm, increase with dose.

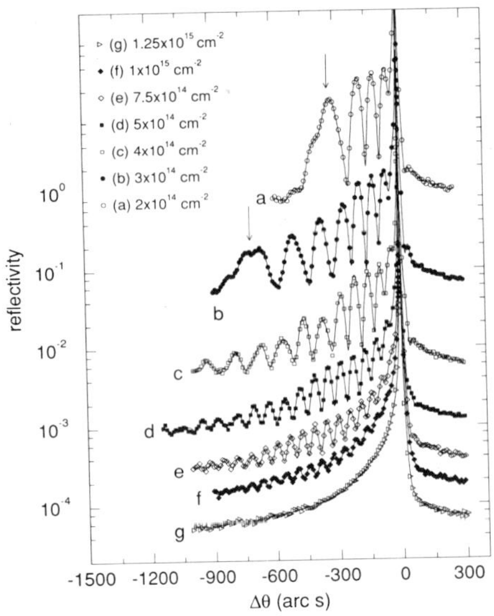

Figure 13. Experimental (symbols) and best fitted (n,-n) 004 symmetric RCs of Si self–ion implanted at 180 keV. λ=0.15406 nm. Curves (a) to (f) are shifted upwards.

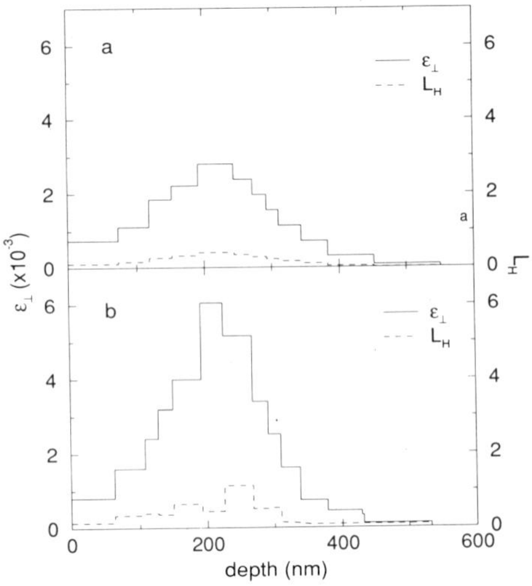

Figure 14. Profiles of lattice strain ($\varepsilon_\perp$) and static atomic disorder (L_H) in the samples implanted with 2×10^{14} cm^{-2} (a) and 3×10^{14} cm^{-2} (b) at 180 keV.

The profiles resulting from the best fits of the RCs (c), (d), (f) and (g) are shown in Figure 15. Qualitative remarks lead us to say that a region of very small $\varepsilon_\perp$, confined

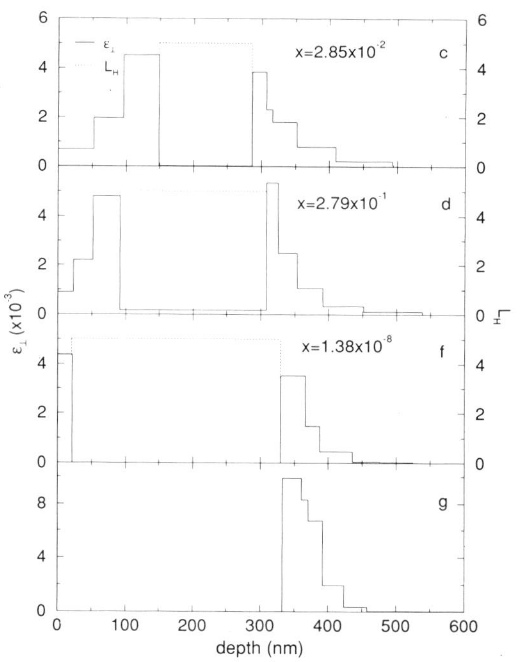

Figure 15. Profiles of lattice strain ($\varepsilon_\perp$) and static disorder (L_H) resulting from best fits of the RCs (c), (d), (f) and (g) in Figure 13.

within two distributions of deformed layers and characterized by a large value of L_H, begins to form at the same depth where the deformation peaks are located in Figure 14. This region widens with dose until it reaches the surface. Recalling that the static Debye–Waller factor $exp(-L_H)$ multiplies the polarizability (structure factor) $|\psi'_H|$ in (5) and (6), the vanishing of the diffraction power of the embedded layer means the presence of amorphous (α) material. The diffraction properties of a structure made by two perfect crystal layers separated by a buried amorphous lamina are more easily quantified if resort is made to a simpler diffraction approach than the recursive dynamical formalism outlined in Sect. 2.

5.3.1. Interference in a Perfect Crystal with Buried α–layer. Starting from the semi-kinematical model reported previously by Kyutt et al. (1980) for crystalline structures, a mathematical expression was recently developed by Milita and Servidori (1995) to describe diffraction from a perfect, infinitely thick substrate covered by an α–layer and a perfect crystalline cap (lamina 1). This expression was derived by treating formally the α–layer as a crystalline layer in the limit case of $L_H \to \infty$, so that all the quantities defined for a crystal in the dynamical model hold also for a non–diffracting material. Moreover, it was assumed that (i) the two layers on the substrate are so thin that absorption can therein be neglected, (ii) the relative linear expansion in the α–layer (the equivalent of $\varepsilon_\perp$ in a crystal) is $\varepsilon_{\perp\alpha} \ll 1$, (iii) the incidence angle $\theta = \theta_B + \Delta\theta$ is not too far from θ_B of the substrate, and (iv) the reflection is symmetric ($b = -1$). Under these simplifying assumptions, the intrinsic RC of the composite results

$$R(\Delta\theta) = R_p(\Delta\theta)|1 + exp(i2y_\alpha \Delta A_\alpha)[exp(i2y_1 \Delta A_1) - 1]|^2 \ , \tag{20}$$

where $R_p(\Delta\theta)$ is the reflectivity profile of the perfect substrate. The main quantity for the surface layers operating in the RC of the structure is the product of y, the incidence parameter (6), and ΔA, the dimensionless lamina thickness from (5). y includes the term α_H (9) which, for $m_\parallel = \varepsilon_\parallel = 0$ and $m_\perp = \varepsilon_\perp$ in the case of implanted silicon, reduces simply to $\alpha_H = -2sin(2\theta_B)[\Delta\theta + \varepsilon_{\perp\alpha} tan(\theta_B)]$ for the α–layer. From (20) we obtain

$$R(\Delta\theta) = R_p(\Delta\theta)[3 + 2cos(F)[cos(D) - cos(E)] + 2sin(F)[sin(D) - sin(E)] - 2cos(D - E)] \ , \tag{21}$$

where

$$F = 2\pi\varepsilon_{\perp\alpha} t_{0\alpha}/d \ , \tag{22}$$

$$D = [2\pi(\psi'_0 + \Delta\theta sin(2\theta_B))/\lambda sin(\theta_B)](t_1 + t_{0\alpha}) \ ,$$

$$E = [2\pi(\psi'_0 + \Delta\theta sin(2\theta_B))/\lambda sin(\theta_B)]t_{0\alpha} \ ,$$

t_1 and $t_{0\alpha}$ being the thicknesses of the surface perfect crystalline lamina and the buried α–layer prior to deformation, respectively, and d the spacing of the diffracting planes in the perfect crystal (one fourth of the lattice spacing for the 004 reflection). For straightforward reasons, $R(\Delta\theta) = R_p(\Delta\theta)$ for $t_1 = 0$ or $t_{0\alpha} = 0$. (21) describes an interference pattern which depends on both t_1 and $t_{0\alpha}$. The appearance in (22) of the product $\varepsilon_{\perp\alpha}t_{0\alpha}$ is the consequence of this approximated treatment. It assumes the meaning, on the analogy of the crystalline case, of the increment in the initial thickness $t_{0\alpha}$ pre–existing the deformation $\varepsilon_{\perp\alpha}$. The outward expansion $\varepsilon_{\perp\alpha}t_{0\alpha}$ of the sample surface can be measured mechanically by the difference in height of implanted and unimplanted surfaces after a through–mask process. This step can be larger than d and then is generally written as $\varepsilon_{\perp\alpha}t_{0\alpha} = (n + x)d$, where n is an integer (0,1,2,..) and $0 \leq x < 1$. Due to (22), the RC as a whole will exactly repeat itself for different n values. If we consider that n can in principle vary in an infinite range of values, this implies that the thickness $t_\alpha = t_{0\alpha}(1 + \varepsilon_{\perp\alpha})$ is not measurable by X–ray diffraction. On the other hand, the experimental RC is by itself produced by a definite thickness of the α–layer. Therefore, we have to impose that $t_{0\alpha}$ includes the integer fraction nd (if any) of the rigid outward translation and that $\varepsilon_{\perp\alpha} = xd/t_{0\alpha} = x/n$. Hence, xd is the quantity responsible for the actual lattice dephasing of the perfect lamina 1 from the perfect substrate. In this way, $\varepsilon_{\perp\alpha}$ will assume the minimum value possible and $t_{0\alpha}$ will be the thickness of the α–layer not necessarily coincident with the thickness pre–existing the amorphization process.

The oscillatory terms in (21) show that interference occurs also when x=0 (F=0), *i.e.* when $t_\alpha = t_{0\alpha} = nd$. This condition is commonly referred to as the *in–phase* condition, for which the interference is maximum. Unlike the case for $t_{0\alpha}$, whose determination is affected by errors much larger than d, the sensitivity to minute variations in x is very high, as they produce considerable changes in the *out–of–phase* condition. The maximum of the out–of–phase condition is for x=0.5.

The appearance of the key term $2\pi\varepsilon_{\perp\alpha}t_{0\alpha}$ in (22) follows from the restrictions introduced in the semi–kinematical model of $\varepsilon_{\perp\alpha} \ll 1$ and $\theta \simeq \theta_B$. These restrictions are not a mere tool to make the model easier from the mathematical point of view. In fact, the $\varepsilon_\perp$ values in as–implanted silicon hardly exceed 1%. Moreover, an accurate analysis of the RC at angles very far from θ_B is hindered by the presence of diffuse scattering in double crystal RCs or by the reduction in intensity usually encountered when a crystal analyzer is placed in front of the detector to remove this background. The analytical approach given here excludes the possibility that the whole RC could cycle with n. This cycling was excluded also by Parisini et al., (1996), who demonstrated on the basis of a kinematical treatment that this does not occur for a typical value of t_α, even when t_α equals an integer multiple of d.

Finally, we underline that, from the above analysis, the true value of $\varepsilon_{\perp\alpha} = (n + x)d/t_{0\alpha}$ cannot be measured by X–ray diffraction, because the true value of n is unknown. For the α–layer the value of n=0 is then assumed, because a different choice of n is arbitrary.

5.3.2. Interference in a Deformed Crystal with Buried α–layer. The X–ray interference for a sequence of deformed laminae separated by an α–layer is more complex than that reported in the previous paragraph. In fact, we can speak of in–phase or out–of–phase conditions only for layers i and j having spacings $d_i \equiv d_j$ and separated by an amorphous region of $t_\alpha = nd_i$ or $t_\alpha = (n + x)d_i$. In general, in the case of ion implantation this never occurs. Nevertheless, interference phenomena are expected whenever $d_i \simeq d_j$. Moreover, the phenomenon is enhanced on increasing the thickness of the interfering layers and when the further condition $t_i \simeq t_j$ is met. When $d_i \neq d_j$, the mere definition of rigid outward translation is more appropriate than de-phasing for $\varepsilon_{\perp\alpha} \neq 0$ (since different values of d_i are involved in a layer sequence after ion implantation, in the following $\varepsilon_{\perp\alpha}$ will be referred to the d of the perfect substrate).

However, these remarks do not mean that interference cannot be appreciable between pairs of deformed layers separated by the α–layer. In the case of Figure 15, the weights by which these layers contribute to the overall interference pattern can be evaluated by comparing the experimental RC with the diffraction [rofile given by selected deformed layers at the sides of the α–lamina. This type of calculation confirms that not only is interference observed for layers of comparable values of strain and thickness (not necessarily those enclosing the α–layer) but also that variations in $t_{0\alpha}$ and/or $\varepsilon_{\perp\alpha}$ strongly influence the intensity profiles.

Figure 15 can now be interpreted in the light of the arguments of the previous paragraph. The strain profiles in Figures 15 (c) to (f) show the thicknesses $t_{0\alpha} = nd$ and the values of $\varepsilon_{\perp\alpha} = x/n$. The fractions x of d are also reported for the α–layers. The vanishingly small value of x in Figure 15 (f) indicates that the rigid outward translation of the surface crystalline layers is zero or an integer multiple of d.

As to the profiles of the static Debye–Waller factor, values of $L_H > 0$ were obtained in all of the implanted region. However, only those much larger and more significant for the α–layers are shown for clarity in Figure 15. The value of 5 was imposed inside these layers, though L_H's as large as 100 were obtained after the two–step best fits of the RCs. In fact, L_H=5 is sufficiently high to reduce to zero the structure factor in the amorphized material.

Thickening of the α–layer and thinning of the surface crystalline region are evident in Figures 15 (c) to (f). This explains the lowering with increasing dose of the amplitude and angular spacing of the interference fringes in Figure 13, because they depend on both these thicknesses (see (21)). The thickening of the α–layer is faster towards the surface than away from it. This results from the process of the ion dechannelling in disordered crystals. In fact, during the implant the incoming ions are progressively stopped, and hence produce progressively larger amounts of damage, at depths immediately above heavily distorted or amorphous layers.

Figure 15 (g) needs a comment. Fringe vanishing at the dose of 1.25×10^{15} cm^{-2} (Figure 13) means that the Bragg–case interferometric structure is destroyed, *i.e.* that the α–layer appeared on the surface. In this case, the RC best–fit is able to describe only the damage tail beneath the α–layer. However, according to what said in paragraph 5.1, epitaxial regrowth of the α–layer from the substrate occurs after heating, leaving a sharp interface in the $\varepsilon_\perp$ profile between regrown layer and substrate. The depth of this interface is reported in Figure 15 (g).

6. CONCLUSIONS

Multicrystal X–ray diffraction is a powerful technique to investigate implanted semiconductors. Information on the lattice defects are obtained in terms of lattice strain and static atomic disorder. The latter quantity (too often neglected) plays a relevant role in, for example, (i) the characterization of dislocation loops which form as the consequence of post–annealing treatments, (ii) the description of the effects of the electron energy loss of the ions in high energy implants, and (iii) the study of the transition from surface damaged crystals to fully amorphized overlayers.

REFERENCES

Bartels, W.J., 1983, Characterization of thin layers on perfect crystals with a multipurpose high resolution diffractometer, *J. Vac. Sci. Technol. B* 1:338.

Cole, H. and Stemple, N.R., 1962, Effect of crystal perfection and polarity on absorption edges seen in Bragg diffraction, *J. Appl. Phys.* 33:2227.

Dederichs, P.H., 1973, The theory of diffuse X–ray scattering and its application to the study of point defects and their clusters, *J. Phys. F: Metal Phys.* 3:471.

DuMond, J.W.M., 1937, Theory of the use of more than two successive X–ray crystal reflections to obtain increased resolving power, *Phys. Rev.* 52:872.

Ellis, P.J. and Freeman H.C., 1995, XFIT – an interactive EXAFS analysis program, *J. Synchrotron Rad.* 2:190.

Fabbri, R., Servidori, M., Cembali, F., Nipoti, R. and Bianconi, M., 1992, X–ray determination of lattice damage depth profiles due to electronic and nuclear energy losses in silicon implanted with MeV boron ions, *Nucl. Instr. and Methods B* 66:511.

Gartstein, E.L. and Cowley, R.A., 1990, The intensity distribution observed with a multi–crystal X–ray diffractometer, *Acta Cryst. A* 46:576–584.

Godwod, K., Kowalczyk, R. and Szmid, Z., 1974, Application of a precise double X–ray spectrometer for accurate lattice parameter determination, *Phys. Stat. Sol. A* 21:227.

Kyutt, R.N., Petrashen, P.V. and Sorokin, L.M., 1980, Strain profiles in ion–doped silicon obtained from X–ray rocking curves, *Phys. Stat. Sol. A* 60:381.

Mazzone, A.M., 1991, Processes of hot electrons and lattice relaxation in ion implantation, *Phys. Stat. Sol. B* 166:79.

Milita, S. and Servidori, M., 1995, X–ray rocking curve analysis of crystals with buried amorphous layers. Case of ion implanted silicon, *J. Appl. Cryst.* 28:666.

Milita, S. and Servidori, M., 1996, Damage in ion implanted silicon measured by X–ray diffraction, *J. Appl. Phys.*, in the press.

Parisini, A., Milita, S. and Servidori, M., 1996), Bragg–case X–ray interference in multilayered structures, Comparison between kinematical approximation and dynamical treatment, *Acta Cryst. A*, in the press.

Press, W.H., Teukolsky, S.A., Vetterling, W.T. and Flannery, B.P., 1992, *Numerical Recipies in Fortran*, Cambridge University Press, New York.

Servidori, M., 1987, Characteriztion of lattice damage in ion implanted silicon by multiple crystal X–ray diffraction, *Nucl. Instr. and Methods B* 19/20:443.

Servidori, M., Cembali, F., Zazzetti, L. and Balboni, R., 1994, The problem of convolution in the simulation of multicrystal X–ray rocking curves of semiconducting materials, *Materials Sci. and Engineering B* 28:523.

Servidori, M., Sourek, Z. and Solmi, S., 1987, Some aspects of danage annealing in ion implanted silicon: discussion in terms of dopant anomalous diffusion, *J. Appl. Phys.* 62:1987.

Servidori, M., Zaumseil, P., Winter, U., Cembali, F. and Mazzone, A.M., 1987, Defect distribution in ion implanted silicon: comparison between Monte Carlo simulation and triple crystal X–ray measurements, *Nucl. Instr. and Methods B* 22:497.

Speriosu, V.S., 1981, Kinematical X–ray diffraction in nonuniform crystalline films: strain and damage distributions in ion–implanted garnets, *J. Appl. Phys.* 52: 6094.

Taupin, D., 1964, Théorie dynamique de la diffraction des rayons X par les cristaux deformés, *Bull. Soc. Fr. Minéral. Cristallogr.* 87:469.

Wie, C.R., Tombrello, T.A. and Vreeland Jr., T, 1986, Dynamical X–ray diffraction from nonuniform crystalline films: application to X–ray rocking curve analysis, *J. Appl. Phys.* 59:3743.

Zachariasen, W.H., 1945, *X–ray Diffraction in Crystals*, John Wiley and Sons, New York.

MULTIPLE BRAGG SCATTERING AND THE PHASE PROBLEM IN X-RAY DIFFRACTION: PERFECT CRYSTALS

R. Colella

Department of Physics
Purdue University
West Lafayette, IN 47907 U.S.A.

1. INTRODUCTION

A crystal set in such a way as to excite a Bragg diffraction from the (hkl) planes, may under certain circumstances be oriented so that another set of planes (HKL) forms the correct Bragg angle with the incident beam. In reciprocal space, this corresponds to the situation in which three points, the origin, hkl, HKL, all lie simultaneously on the Ewald sphere. In this case we can say that an incoming x-ray photon has a choice of being diffracted either by the (hkl) or by the (HKL) planes. Clearly, both diffracted beams exist in the crystal, and some intensity exchange must take place among the various diffraction channels. More specifically, the intensity of a given beam, say, hkl, must be affected (that is to say: increased or decreased), by the presence of the HKL beam.

This phenomenon was recognized by M. Renninger in 1938, and is commonly referred to as the "Renninger effect". The technique used by Renninger, subsequently duplicated in virtually all multiple diffraction experiments, consisted in keeping one reflection always excited, for instance, hkl. We will call this one the "main" reflection, called $\vec{P}$ in this paper. The crystal is then spun around the scattering vector $\vec{P}$, so that the incident beam always forms the correct Bragg angle with the (hkl) planes. For certain values of the azimuthal angle ψ, which measures the rotation around $\vec{P}$, another reflection HKL may be excited. This will be called the "simultaneous reflection" $\vec{H}$. Renninger pointed out that in this case a third set of crystallographic planes is participating in the diffraction process: the planes with Miller indices: $(h-H, k-K, l-L)$ or: $\vec{P}-\vec{H}$. This will

X-ray and Neutron Dynamical Diffraction: Theory and Applications
Edited by Authier *et al.*, Plenum Press, New York, 1996

be called the "coupling reflection". The reason for considering this third set of planes is that the beam produced by the HKL planes acts as a new incident beam inside the crystal. The Ewald construction then tells us that the HKL node may be considered a new origin, so that all nodes can now be referred to this new origin. The hkl node can then be renamed $(h-H, k-K, l-L)$. Since the Miller indices of the coupling reflection also correspond to a reciprocal lattice vector, a third reflection is present, redirecting the twice scattered x-ray photon along the direction it would have taken if scattered only once by the (hkl) planes (Figure 1). We can then say that a photon has a choice of being: i) either directly scattered by the (hkl) planes in a single step, or, ii) scattered in a two step process by the (HKL) and by the $(h-H, k-K, l-L)$ planes. In either case, the x-ray photon emerges from the crystal *exactly* along the same path.

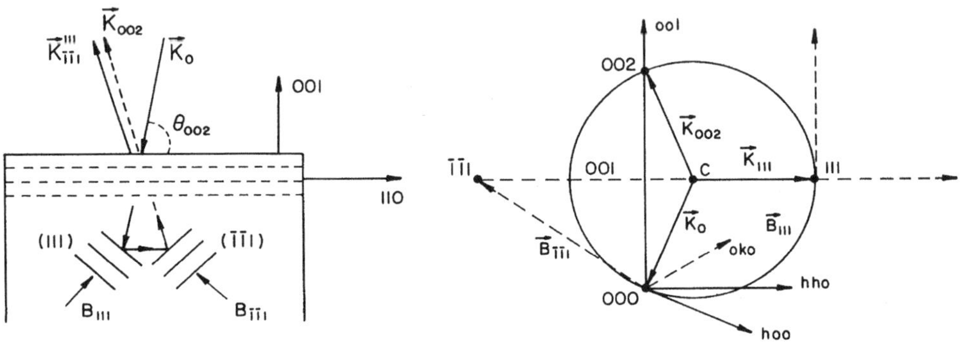

Figure 1. Simultaneous Bragg diffraction. In part (a) the real space situation is represented. The incident beam $\vec{K}_o$ satisfies Bragg's law for two sets of planes simultaneously, the (002) and the (111). For convenience, the "coplanar" case is considered, in which the normal to the (111) planes lies in the plane defined by $\vec{K}_o$ and the normal to the (002) planes. In part (b) the same situation is represented in reciprocal space.

Since both processes are possible, the outgoing beam is then the coherent superposition of two beams, and interference effects between the two beams are expected. Since one of the two beams involves double scattering, the phase shift of this beam is the sum of the two individual phase shifts associated with diffraction by the $\vec{H}$ and $\vec{P}-\vec{H}$ reflections. The doubly scattered beam can be turned on and off by varying the azimuth ψ, therefore the phases of the individual reflections play a role in the intensity of the outgoing beam.

This diffraction technique gives information about the *phases* of the reflections involved. A plot of the hkl intensity vs. ψ is called "Renninger plot", or "azimuthal plot", and contains phase information.

In ordinary two-beam diffraction the phase shift of the scattered photons is not experimentally accessible, because we can only measure *intensities*, not phases.

This is commonly referred to as the "*phase problem*" in diffraction. It is a very old and crucial problem which had been recognized since the early days of x-ray crystallography.

2. EARLY WORK

Initially attention was concentrated on the geometrical aspects of multiple diffraction. A complete study of the Renninger plots of germanium was performed and published in the early sixties (Cole, Chambers and Dunn, 1962). This study was restricted to the case in which the main reflection $\vec{P}$ was the "quasiforbidden" 222.* Cole, et al., (1962) present in their paper a detailed study of 222 in germanium. They verified, for example, the symmetry properties of the Renninger plots, which is generally greater than the symmetry properties of the $\vec{P}$ axis. In the case of germanium, in fact, the [222] direction has 3-fold symmetry, with three mirror planes parallel to the [222] axis. The Renninger plot, however, has a 60° periodicity, with mirror symmetry within each 60° sector, on the ψ scale, which is the consequence of particular geometric features typical of the Renninger experiment.

An interesting result mentioned in this paper is shown in Figure 2. It is apparent in this figure that the "background" on the right side decreases to almost zero and then goes back to the asymptotic value. It will be recalled here that the "background" in Figure 2 is the "true value" of the 222 reflection, without contributions from multiple reflections.

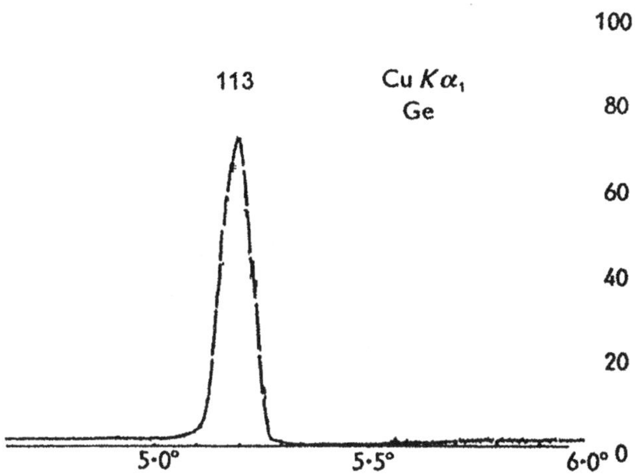

Figure 2. The strongest line on the Renninger plot of Ge 222 using Cu$K\alpha_1$ radiation. Note the suppression of the background on one side of the peak. From: (Cole, Chambers and Dunn, 1962).

The authors make an explicit comment on this curious feature of Figure 2, without giving any interpretation. It turns out that this decrease of background is an interference effect. It is what we now call the "asymmetry effect", which has been understood several years later, and it is the source of phase information. (Chapman, Yoder and Colella, 1981). It will be discussed in more detail in Sect. 4.

*A list of pertinent references of early work on the 222 in diamond structures and multiple diffraction in general is given in (Colella and Merlini, 1966).

3. N-BEAM DYNAMICAL THEORY

The theory of n-beam diffraction has been developed by several authors.†

The work described by Colella (1974) is the first complete and most general treatment of n-beam Bragg diffraction, free from restrictions, capable of handling any number of beams, and special situations such as grazing incidence, or Bragg diffracted beams almost parallel to the surface of the crystal. We will briefly summarize here the essential steps of the theory.

Following Ewald's original approach (1916a, b; 1917) the problem is first solved *within* the crystal, supposed to be infinite and unbounded. The starting point is the vector form of Bragg's law:

$$\vec{\beta}_j = \vec{\beta}_o + \vec{B}_j \tag{3.1}$$

where $\vec{\beta}_o$ and $\vec{\beta}_j$ are crystal wavevectors, ($\vec{\beta}_o = \vec{\beta}_1$; $j = 2, 3, \ldots n$) and $\vec{B}_j$ are reciprocal space vectors. The physical significance of Eq. (3.1) is the following: if a plane wave $\vec{\beta}_o$ for some reason exists in a crystal, then, by necessity, an infinite set of plane waves $\vec{\beta}_j$ will also exist, as a result of Bragg diffraction from lattice planes normal to $\vec{B}_j$. Using Maxwell's equations, it is possible to express the condition of self-consistency for all plane waves existing in the crystal:‡

$$(k_o^2 - \beta_i^2)\vec{D}_i - \sum_1^n {}_j\psi_{i-j}[(\vec{\beta}_i \cdot \vec{D}_j)\vec{\beta}_j - \beta_i^2 \vec{D}_j] = 0 \tag{3.2}$$

where $k_o = 1/\lambda$, λ being the wavelength of X-rays in vacuum, $\vec{D}_j$ is the displacement vector amplitude of the plane wave $\vec{\beta}_j$, and ψ_j is the Fourier component of the polarizability per unit volume (times 4 π) associated with the node $\vec{B}_j$. i can be equal to $1, 2 \ldots n$, so that Eq. (3.2) represents n linear homogeneous equations for the amplitudes $\vec{D}_j$. In principle the system (3.2) should consist of an infinite number of linear equations containing an infinite number of terms, corresponding to the nodes in reciprocal lattice. In practice, only the nodes on or close to the Ewald sphere are considered, so that n will typically be a small number (3 or 4).

A system of $2n$ homogeneous linear equations for the unknown amplitudes $\vec{D}_j$ is obtained from Eqs. (3.2). Meaningful solutions are obtained when the determinant of the system is zero, which happens for special values of the magnitude β_o, which is the only disposable parameter in Eqs. (3-2). Since these equations involve the square of β_o, there will be $4n$ solutions.

When the $4n$ eigenvalues are determined, the total displacement vector in the crystal can be written as:

$$\vec{D}(\vec{r}) = \sum_1^{4n} {}_\ell q_\ell \sum_1^n {}_j \vec{D}_j^\ell \exp(-2\pi i \vec{\beta}_j^\ell \cdot \vec{r}) \tag{3.3}$$

where: $\vec{\beta}_j^\ell = \vec{\beta}_o^\ell + \vec{B}_j$, $\vec{D}_j^\ell$ are the eigenvectors, and q_ℓ are the strengths of the various plane waves (they would be called Bloch waves in the case of electrons), to be determined by the boundary conditions.

---

†A number of pertinent references is given in: (Colella, 1974).

‡See Eq. 3-105b of (Zachariasen, 1945).

The crystal is supposed to be in form of a slab, infinitely extended, of given thickness t_o. The boundary conditions are written in accord with the criteria given by Lamla (1939). In each half-space, on both sides of the crystal slab, two propagating plane waves are introduced, with the same tangential component wavevector and opposite normal components, for each reciprocal lattice node $\vec{B}_j$ The amplitudes of the $2n$ plane waves ($\vec{E}$ and $\vec{H}$) leaving the crystal are unknown, and are to be determined by the boundary conditions. The amplitudes of the plane waves impinging on the crystal are all set $= 0$, except for $j = 1$ ($\vec{k}_o$), set arbitrarily equal to 1.

Outside of the crystal, the vector form of Bragg's law (3.1) is not valid on account of refraction. A similar condition exists, however, for the tangential components:

$$\begin{aligned} \vec{k}_j^t &= \vec{k}_o^t + \vec{B}_j^t \\ |\vec{k}_j| &= |\vec{k}_o| = 1/\lambda \end{aligned} \tag{3.4}$$

The boundary conditions are obtained by imposing continuity for the tangential components of the electric and magnetic intensities ($\vec{E}$ and $\vec{H}$) and for the normal components of the electric displacement and magnetic induction ($\vec{D}$ and $\vec{B}$). The latter conditions, however, are not independent from the former, so a total of four scalar equations are to be satisfied for each plane wave j. Ultimately, only the components of the displacement vectors will be involved in the final equations, through the relations:[§]

$$\begin{aligned} \vec{H}_j &= (k_o/\beta_j)\vec{U}_j \times \vec{D}_j \\ \vec{U}_j &= \vec{\beta}_j/\beta_j \\ \vec{E}_j &= (k_o^2/\beta_j^2)\vec{D}_j \end{aligned} \tag{3.5}$$

Within the crystal, for each beam j, there are in general $4n$ distinct directions given by the eigenvalues of the dispersion equations (3.2). Since there are four equations for each beam j on both surfaces of the crystal slab, we have a total of $8n$ equations with $8n$ unknowns: the $4n$ coefficients q_ℓ in Eq. (3-3) and the $4n$ components of the electric displacement vectors projected on the entrance and exit surfaces of the crystal. A linear system of simultaneous equations $8n \times 8n$ can therefore be written, whose solutions give at the same time the strengths q_ℓ of the Ewald waves in the crystal, and the amplitudes of the vacuum waves. Within each group of the four equations associated with the same beam j on each surface it is possible, however, to eliminate the two components of the electric displacement, so that ultimately we are left with a $4n \times 4n$ system whose unknowns are the q_ℓ's. The equations are give explicitly in (Colella, 1974).

4. VIRTUAL BRAGG SCATTERING

In 1981 a new aspect of multibeam diffraction was discovered that greatly enhanced the possibilities of applications to structural problems.

[§]See, for instance, (Jeans, 1933)

The theory summarized in Sect. 3 is only applicable to perfect crystals. However, most crystals have some mosaic spread. So, it seemed that the range of applications would have to be limited to crystals like silicon, germanium, some III-V or II-VI semiconductors, for which there is no phase problem to talk about because those structures are very well known.

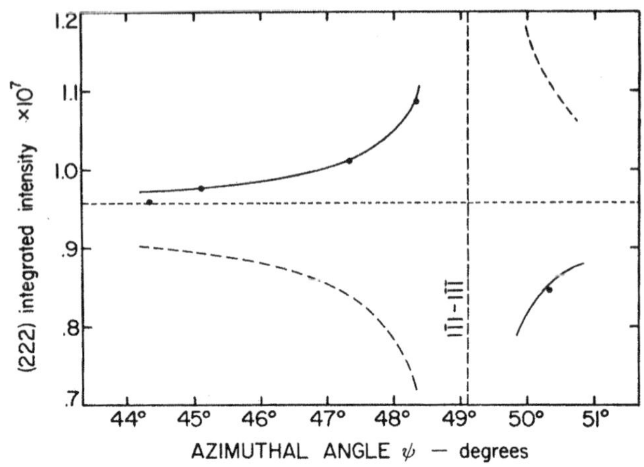

Figure 3. Intensity of the 222 reflection as a function of azimuthal angle. Each point represents an integrated intensity with respect to θ, angle of incidence. Two strong Renninger peaks, with angles 3' apart, are located at a position marked by a vertical dotted line at $\psi \cong 49.08°$. The two peak positions are not resolvable in this figure. Experimentally, one single peak is observed, 0.7° wide, with a peak intensity about a factor of 6.5 greater than the two-beam 222 value. The solid line is given by theory. The dotted line is also given by theory, in which the sign of one of the structure factors has been changed. From: (Chapman, Yoder and Colella, 1981).

A careful examination of the 222 azimuthal plot of silicon revealed that the asymmetry effect, referred to in Sect. 2, was visible over a range of a few degrees. Since the asymmetry effect owes its origin to the perturbation introduced in 2-beam diffraction by a third node approaching the Ewald sphere, this result was found somewhat surprising, because the rocking widths of a perfect crystal are generally of the order of seconds of arc. The effect was analyzed in detail. (Chapman, Yoder and Colella, 1981). It was found that the asymmetric shape of the azimuthal plot near a strong multibeam diffraction peak was an interference effect, and therefore it was inferred that phase information could be extracted from it. (Overhauser, 1981). We emphasize that the mechanism responsible for appreciable deviations from the two beam value at angles ψ that are 2° or 3° off the multibeam peaks is basically different from that involved in the tails of an ordinary two-beam diffraction peak. In the latter, momentum is not conserved, and the intensity falls off very rapidly with θ, the angle of incidence on the lattice planes. In a typical two-beam diffraction peak the intensity falls to 5% of the maximum value at an angle $\theta \simeq 5" - 10"$ off the peak value, which corresponds to a distance between the Ewald sphere and the hkl node of the order of 0.001% of the radius. It turns out that Δp_x and Δx as calculated from dynamical theory (x is a direction normal to the surface and Δx is the penetration depth, *i.e.*, the extinction length of the x-ray beam) satisfy the uncertainty principle $\Delta x \cdot \Delta p_x \simeq \hbar$. In the experiment described in by Chapman, Yoder and Colella (1981), which is a 4 beam experiment, the main reflection, the 222, is always fully excited, so momentum is always conserved. The participating nodes perturb the 222 intensity at large distances from the Ewald sphere. For example, when ψ is 4° off the value for full 4-beam excitation, the 222 intensity is about 2% greater than the two beam value, (see Figure 3) and the participating nodes have distances from the Ewald sphere amounting to a few percent of the radius, which is enormous compared to the 0.001% we found in the two-beam case. We have, therefore, a situation in which Bragg reflections that cannot be excited because energy would not be conserved in the process are able to appreciably affect the 222 intensity. In this sense we introduce the notion of *Virtual Bragg Scattering*, (Overhauser, 1981), in analogy with virtual transitions in atomic and nuclear physics that do not conserve energy. The deviations from the two beam value observed in Figure 3, positive on the left side of the multibeam excitation point, and negative on the right side, *do contain phase information*, as proved by the calculations. In fact, while the solid line was obtained by the theory given by Colella (1974), the dotted line was obtained by the same theory, after changing sign to one of the participating structure factors.

It was immediately realized that the great advantage of working with the tails of a multibeam azimuthal plot, rather than with the peak itself, would make it possible to apply this method to mosaic crystals. In fact, the overall scattering cross section in the tails region is weak, because the primary reflection is weak and the strong simultaneous reflections are weakly excited. Since it is a well known result, in the two beam case, that dynamical theory can be applied to mosaic crystals for weak reflections, it appears that the theory described in Sect. 3 can be applied to mosaic crystals under the conditions of "Virtual Bragg Scattering". This point will be further elucidated in (Colella, 1996).

5. AN EXPERIMENTAL PROOF OF PHASE INVERSION

In order to convince ourselves that the ideas expressed in Sect. 3 and 4 are indeed correct, we felt that a full experimental verification was needed, in which the structure factor of a reflection could be varied in a known and reproducible way. Since the phase of a reflection depends on the atomic sites in the crystal, it is not easy to change phases. However, such an unusual opportunity is offered by the 442 reflection in silicon. The 442, like the 222, is a forbidden reflection, in that the two f.c.c. sublattices of the diamond structure are exactly out of phase. However, like for the 222, some weak intensity may be expected for the 442, as a result of tetrahedral distortion of the valence electrons. The 442 was indeed detected and measured, and found exceedingly weak, (Trucano and Batterman, 1972) but definitely different from zero. Forbidden reflections like the 222 or the 442 may also be contributed by anharmonic thermal vibrations, with terms of opposite sign. Trucano and Batterman (1972) showed that the 442 decreased from the value found at room temperature, as the temperature was increased, became zero, and then increased again. This curious behavior was correctly interpreted as a sign inversion of the structure factor, resulting from a competition between bonding charge effects and anharmonicity.

We found that this was an ideal situation for testing our ideas and methods on determining phases via multibeam diffraction.

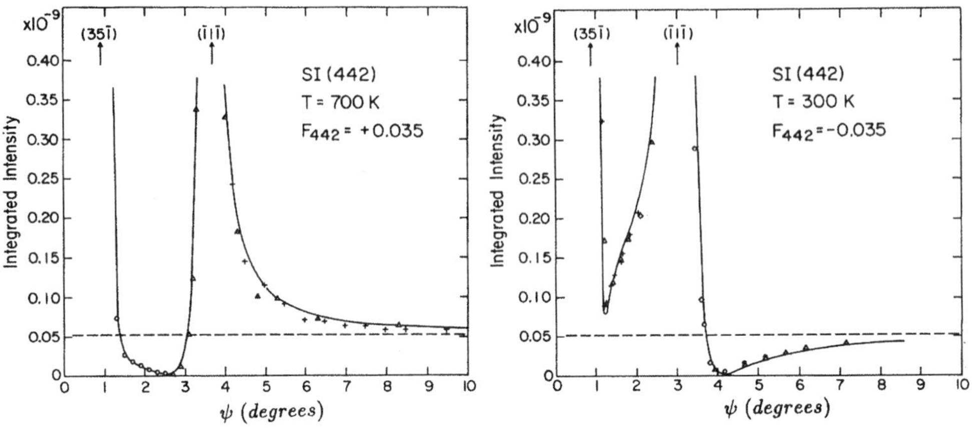

Figure 4. Integrated intensities (with respect to the angle of incidence θ) of the 442 reflection *vs* ψ, azimuthal angle, at 300K and 700 K. The horizontal dotted line corresponds to the two-beam intensity, which has been used for standardizing all points on an absolute basis. The ordinate values become very large near two ψ values, around 1° and 3°, corresponding to two strong simultaneous reflections, the $35\bar{1}$ and $\bar{1}1\bar{1}$ respectively. The continuous solid line is calculated from theory, with $F_{442} = -0.035$. Different symbols correspond to separate runs. Note how the integrated intensity almost vanishes at $\psi \simeq 4.2°$. From: (Tischler, Shen and Colella, 1985).

The experiment was repeated at the Cornell High Energy Synchrotron Source, at room temperature and at 700 K (Tischler, Shen and Colella, 1985). The latter temperature was chosen so that the same intensity was measured as at room temperature. In this way the magnitudes of the structure factors in the two cases were the same.

The results are reproduced in Figure 4. It is clear that the asymmetry effect is different in the two cases, in fact it is reversed, and that the theory (solid line) perfectly reproduces the experimental results.

We conclude, therefore, that the asymmetry effect on the wings of a multibeam peak in an azimuthal plot does contain phase information, and that the theory used to analyze the experimental results of Figure 4 is indeed correct.

REFERENCES

Chapman, L.D., Yoder, D.R., and Colella, R., 1981, Virtual Bragg scattering: a practical solution to the phase problem, *Phys. Rev. Lett.* 46:1578.

Cole, H., Chambers, F.W., and Dunn, H.M., 1962, Simultaneous diffraction: indexing Umweganregung peaks in simple cases, *Acta Cryst.* 15:138.

Colella, R. and Merlini, A., 1966, A study of the 222 "forbidden" reflection in germanium and silicon, *Phys. Stat. Sol.* 18:157.

Colella, R., 1974, Multiple diffraction of X-rays and the phase problem. Computational procedures and comparison with experiment, *Acta Cryst. A* 30:413.

Colella, R., 1996, Multiple Bragg scattering and the phase problem in x-ray diffraction. Mosiac crystals, in: *X-ray and Neutron Dynamical Diffraction: Theory and Applications*; A. Authier, S. Lagomarsino and B. K. Tanner, eds., Plenum Press, New York, U.S.A.

Ewald, P.P., 1916a, Zur Begründung der Kristalloptik (Part 1), *Ann. d. Physik*, 49:1.

Ewald, P.P., 1916b, Zur Begründung der Kristalloptik (Part 2), *Ann. d. Physik*, 49:117.

Ewald, P.P., 1917, Zur Begründung der Kristalloptik (Part 3), *Ann. d. Physik*, 54:519.

Jeans, J., 1933, *The Mathematical Theory of Electricity and Magnetism*, Chapt. XVIII, Cambridge Univ. Press.

Lamla, E., 1939, Zur frage der Umweganregung bei Röntgenstrahlinterferenzen, *Ann. Phys.* (5) 36:194.

Overhauser, A.W., 1981, Private communication.

Renninger, M., 1937, "Umweganregung", eine bisher unbeachtete Wechselwirkungserscheinung bei Raumgitterinterferenzen, *Z. Phys.* 106:141.

Tischler, J.Z., Shen, Q., and Colella, R., 1985, Phase determination of the forbidden reflection 442 in silicon and germanium using multiple Bragg scattering, *Acta Cryst. A* 41:451.

Trucano, P. and Batterman, B.W., 1972, Bonding, electron distributions, anharmonicity, and the temperature dependence of the forbidden Si-442 reflection, *Phys. Rev. B* 6:3659.

Zachariasen, W.H., 1945, *Theory of X-ray Diffraction in Crystals*, John Wiley and Sons, Inc., New York.

MULTIPLE BRAGG SCATTERING AND THE PHASE PROBLEM IN X-RAY DIFFRACTION: MOSAIC CRYSTALS

R. Colella

Department of Physics
Purdue University
West Lafayette, IN 47907 U.S.A.

1. A NEW APPROACH BASED ON PERTURBATION THEORY

The scattering of electromagnetic radiation from a finite object can be treated with complete generality by making use of Maxwell equations (Jackson, 1975). In this section we will describe a perturbation theory of n-beam diffraction, due to Q. Shen (1986), which has been extremely useful for the application of multiple scattering to the solution of the phase problem. Approximate analytical solutions using the Bethe approximation have been obtained for the three-beam case, by treating the third Bragg reflection as a perturbation (Hümmer and Billy, 1986). Applications of this method of attack have been demonstrated for protein crystals (Hümmer, Schwegle and Weckert, 1991). When the incident beam is described as a plane wave:

$$\vec{D}_o(\vec{x}) = \vec{D}_o \exp(-i\vec{k}_o \cdot \vec{x}) \tag{1.1}$$

we get the following equation

$$(\nabla^2 + k_o^2)\vec{D} = -\nabla \times \nabla \times (\delta\epsilon\vec{D}) \tag{1.2}$$

$\vec{D}_o$ and $\vec{D}$ are the displacement vectors in vacuum and in the crystal respectively, $\vec{k}_o$ is the incident wavevector, whose magnitude is $2\pi/\lambda$, and $\delta\epsilon = \delta\epsilon(\vec{x})$ is the deviation of the dielectric constant function in the crystal from its value in vacuum.

A formal solution of Eq. (1.2) can be obtained by using Green's function methods. The solution to (1.2) is in the form:

$$\vec{D} = \vec{D}^{(0)} + \vec{D}^{(1)} + \vec{D}^{(2)} + \dots \tag{1.3}$$

X-ray and Neutron Dynamical Diffraction: Theory and Applications
Edited by Authier *et al.*, Plenum Press, New York, 1996

where: $I_\perp$ is the scattered intensity relative to the 2-beam case (O and H beams), $\gamma_{HL} = \Gamma|F_{H-L}F_L/F_H|$, $\Gamma = r_e\lambda^2/\pi V_c$, r_e = classical radius of electron, λ = X-ray wavelength in vacuum, V_c = volume of the unit cell, F_H = structure factor of the main reflection $\vec{H}$, F_L = structure factor of the simultaneous reflection $\vec{L}$, F_{H-L} = structure factor of the coupling reflection $\vec{H} - \vec{L}$, δ_{HL} is the *triplet invariant*, $\delta_{HL} = \alpha_{H-L} + \alpha_L - \alpha_H$ where the α_H's are the phase angles of the F_H's, $k_L = |\vec{k}_o + \vec{L}|$, and L_σ is the component of $\vec{L}$ normal to the scattering plane. $A(\psi)$ is a slowly varying function of ψ, the azimuthal angle.

Eq. (1.4) can be easily evaluated numerically, even for a large number of simultaneous reflections, and directly compared to the azimuthal plots experimentally obtained. Eq. (1.4) has been used, for example, to interpret the 442 results in Si (Sect. 5, Fig. 4 of (Colella, 1996)) and found extremely accurate even in close proximity of the exact ψ-value corresponding to the multibeam excitation point. Eq. (1.4) of course diverges at that point, because the denominators become all zeros.

One important point emerges from a cursory inspection of Eq. 1.4: there are no boundary conditions. The crystal shape does not play any role in Eq. 1.4. The important consequence is that the azimuthal plots obtained under the conditions in which Eq. 1.4 is valid do not depend on crystal shape, *except for the absorption correction*, which is however constant over the small range of a typical azimuthal plot (5-6 degrees or less). The fact that the boundary conditions play no role in Virtual Bragg Scattering situations opens up the possibility of using crystals in the form of small fragments of irregular shape, which greatly extends the range of applications of n-beam diffraction. This important feature of Virtual Bragg Scattering has been verified experimentally (Shen and Colella, 1987).

The perturbation theory obtained by Q. Shen is of great value for the computational speed with which an azimuthal plot can be calculated using Eq. (1.4). Moreover, being the final result written in closed analytic form, it allows one to understand the physics of n-beam diffraction at a deeper level than a purely numerical approach.

2. MULTIPLE DIFFRACTION IN MOSAIC CRYSTALS

The theory of n-beam diffraction described in Sect. 3 of Ref. 3 is called "Dynamical Theory", and, strictly speaking, is applicable only to perfect crystals without mosaic structure. The physical reason for this limitation is that the rocking width of a typical diffraction peak calculated from dynamical theory is of the order of a few arcseconds, much smaller than the typical value of the average misorientation of a mosaic structure, of the order of 120–180 arcseconds. It means that, while diffraction is dynamical within each mosaic block, long range coherence does not exist in a mosaic crystal, each block diffracts incoherently, and the diffracted intensity is simply the sum of the radiation diffracted by each individual block. The difference in the mechanism of diffraction between a perfect and a mosaic crystal is illustrated for the two-beam case in Figure 1. In (1) we see a perfect crystal. An X-ray photon is reflected back and forth several times by the diffracting planes before exiting the crystal. Dynamical theory takes into account this multiple scattering process. It is built in the very formalism of the theory.

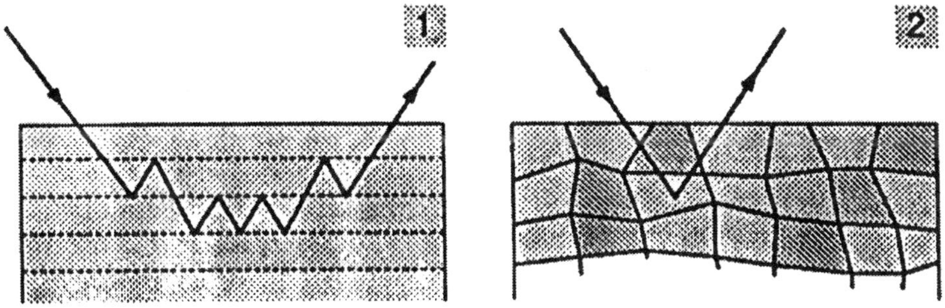

Figure 1. Bragg diffraction from: 1) perfect crystal (multiple scattering); 2) mosaic crystal (single scattering events).

In (2) a mosaic crystal is shown. A photon must penetrate to some depth before finding a crystallite correctly oriented for Bragg diffraction. After being diffracted, it is very unlikely that the same photon will find another crystallite perfectly oriented (within arcseconds) for a second Bragg diffraction. Most likely the photon will emerge from the crystal having suffered only one scattering event. This is the signature of "kinematic diffraction", as opposed to "dynamical diffraction". Kinematic theory is based on such an assumption, and the values predicted for the integrated intensity of a given Bragg reflection are vastly different in the two cases.

However, it is easy to realize that there is a situation in which the two theories converge to the same values. This happens when a Bragg reflection is very weak. In this case the reflectivity of the atomic planes is very small, and the probability of multiple scattering in a perfect crystal (part 1 of Figure 1) is very small. The scattering will be single in both cases.

The convergence of dynamical and kinematic theory to the same integrated intensities, for the standard two-beam case, has been formally derived for the Bragg case of diffraction (Hirsch and Ramachandran, 1950).

At this point we are tempted to extrapolate this criterion to the n-beam case. We can say that whenever the interaction between a photon and a crystal is weak, the scattering is single, and dynamical theory gives the right answer even in the case of a mosaic crystal. This is precisely the situation under which "Virtual Bragg Scattering" (VB Scattering) occurs. (see Sect. 4 of: (Colella, 1996)). The scattering is weak because the primary reflection is intrinsically weak, and the simultaneous and coupling reflections, though strong, are weakly excited because the crystal is deliberately misset from the exact position for multiple diffraction.

The first question in our mind is the following: will a mosaic crystal exhibit the asymmetry effect, which is a clear signature of VB Scattering?

Early experiments showed that this must be the case. That VB Scattering effects are visible in mosaic crystals is clearly evident, for example, in the data of Post and Gong for zinc tungstate (Post, 1983; Gong and Post, 1983).

To convince ourselves that phases can indeed be determined in mosaic crystals using the VB Scattering approach, a multibeam experiment was performed in V_3Si, using as primary reflection the 140 which is space-group-forbidden, and owes its existence to the non-cubic environment of a vanadium site (Schmidt and Colella, 1985). The experimental result is shown in Figure 2. It is clear that phase effects are visible in the data and that phases can be determined.

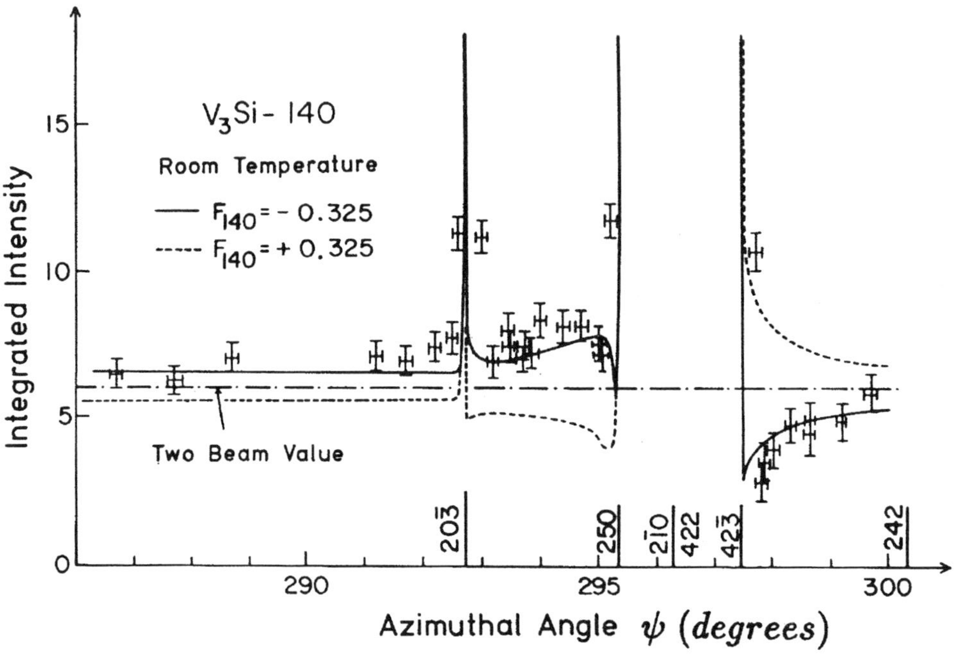

Figure 2. Intensity of the (140) reflection at room temperature *vs* ψ, the azimuthal angle. Each point here is an integrated intensity with respect to θ, the angle of incidence. The solid line is given by theory with $F_{140} < 0$, while the dotted line is calculated with $F_{140} > 0$. The horizontal dot-dashed line corresponds to the two-beam value. The vertical solid lines, normal to the ψ axis, mark the positions of simultaneous reflections. From: (Schmidt and Colella, 1985).

The last question to be resolved is about the *shape* of the crystal. Again, the theory of Sect. 3 in (Colella, 1996) is developed for the case of a nicely flat-shaped crystal in form of a parallelepiped, so that the boundary conditions can be defined in order to calculate the strengths of the various Ewald waves.

However, most crystals for which a phase problem exists are small fragments of spherical or irregular shape.

On the other hand, Shen's perturbation theory described in Sect. 1 does not make use of boundary conditions. It would seem that in a VB Scattering situation, in which perturbation theory applies, the boundary conditions are irrelevant. To prove this point the same multibeam experiment was repeated using benzil, a mosaic crystal, shaped in form of a parallel sided slab, and in form of a small sphere (0.3 mm diameter). Even though the intensities in the two cases were vastly different, the azimuthal profiles were identical, and the same phases could be obtained from the two experiments (Shen and Colella, 1987).

Having proved that multibeam diffraction in a VB Scattering situation provides phase information that does not depend on crystal perfection and shape paves the way to a large scale application of this method to all kinds of crystals.

3. QUASICRYSTALS

Quasicrystals are solid substances in which there is no long range periodicity, yet they are capable of producing sharp diffraction patterns. The most striking feature of quasicrystals is the presence of 5-fold symmetry axes, which are not compatible with long range periodicity. The reciprocal lattice of an icosahedral quasicrystal, like Al-Cu-Fe or Al-Pd-Mn, has the symmetry properties of an icosahedron, but it is not periodic.

In a periodic reciprocal space, every point $\vec{R}_{hkl}$ can be reached by forming a linear combination of basis vector using three integer numbers. This is not possible for a quasicrystal. However, a linear combination using six integer numbers with irrational coefficients can be used to reach every point in reciprocal space corresponding to an observable Bragg reflection. More specifically, every Bragg reflection $\vec{G}$ is given by:

$$\vec{G} = K \sum_{1}^{6} n_i \vec{e}^{\,i} \tag{3.1}$$

where: $K = 1/[2\pi a\sqrt{1+\tau^2}]$; $\tau = \frac{1+\sqrt{5}}{2}$; a is the quasilattice constant, and the xyz components of the $\vec{e}^{\,i}$ basis vectors are:*

$$\vec{e}^{\,1} = \begin{pmatrix} 0 \\ 1 \\ \tau \end{pmatrix}; \vec{e}^{\,2} = \begin{pmatrix} \tau \\ 0 \\ 1 \end{pmatrix}; \vec{e}^{\,3} = \begin{pmatrix} 1 \\ \tau \\ 0 \end{pmatrix}; \vec{e}^{\,4} = \begin{pmatrix} -1 \\ \tau \\ 0 \end{pmatrix}; \vec{e}^{\,5} = \begin{pmatrix} -\tau \\ 0 \\ 1 \end{pmatrix}; \vec{e}^{\,6} = \begin{pmatrix} 0 \\ -1 \\ \tau \end{pmatrix} \tag{3.2}$$

The expressions (3.2) are obtained under the assumption that the xyz axes are chosen to coincide with the 2-fold axes of the icosahedron. Eqs. (3.1) and (3.2) are commonly interpreted as "projections" from a 6-dimensional periodic space $n_1, n_2 \ldots n_6$ into a 3-dimensional non-periodic space xyz. It will be noted that one of the xyz coordinates of the basis vectors is $\pm\tau$, the golden mean.

The big question with quasicrystals is: where are the atoms? In other words, the real space structure of a quasicrystal is still a largely unresolved problem despite the large wealth of diffraction data available. The reason is that traditional crystallographic methods for solving structures are based on the assumption of periodicity, and therefore are not applicable to quasicrystals. A key step in the determination of a crystal struc-

*In this work we label Bragg spots with a modified version of a sixfold Miller indices due to Elser (1986).

ture is the solution of the *phase problem*, which is normally done using *direct methods* (Viterbo, 1992). With quasicrystals this is not done on a routine basis.†

Multiple Bragg diffraction has been used to obtain structural information in quasicrystals.

Since reciprocal space is not periodic, the first point to verify is that the difference between two reciprocal lattice vectors $\vec{P}$ and $\vec{H}$ (corresponding to the main and simultaneous reflections) is another reciprocal lattice vector. This is certainly the case, because every reciprocal lattice vector is a linear combination of basis vectors (see Eq. 3.1).

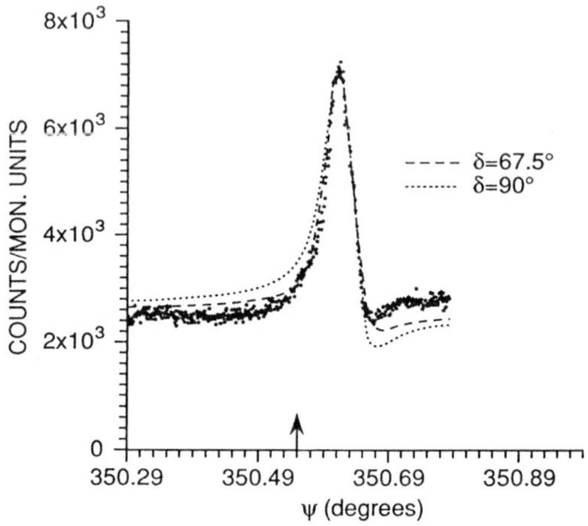

Figure 3. Intensity of the weak $(2\bar{4}044\bar{2} = \vec{P})$ reflection as a function of the azimuthal angle ψ, when the crystal is rotated around the scattering vector. The arrow indicates the position where the peak was expected. Two simultaneous reflections are excited at the same angle: the $(0\bar{4}042\bar{2} = \vec{H}_1)$ and the $(400242 = \vec{H}_2)$. This is a four beam case, and the asymmetry effect does not disappear at $\delta = 90°$, because two triplet invariants are involved. The dashed and dotted lines are theoretical fits with different values for the phase of the $\vec{H}_1$ reflection with respect to $\vec{P}$. From: (Lee and Colella, 1993).

Experimental results have been obtained with Al-Cu-Fe (Lee and Colella, 1993) and Al-Pd-Mn quasicrystals. Figure 3 is an example of a 4-beam experiment on Al-Cu-Fe. The asymmetry effect is clearly visible, therefore phase information is present in this plot. Figure 4 is another example, a 3-beam experiment, on Al-Pd-Mn.

The question arises, at this point, whether or not the theories developed for analyzing azimuthal plots generated by regular crystals (Colella, 1974; Shen, 1986) are applicable to quasicrystals.

Since Bragg diffraction exists, there must be enough space coherence between atomic planes so as to generate multiple interference effects, which is the only basis for the theories of Colella (1974) and Shen (1986), despite some "disorder", which allows the existence of 5-fold axes. The situation is reminiscent to that of thermal motion in a perfect crystal, which does not prevent, for instance, the onset of anomalous transmission. Indeed, the existence of anomalous transmission in a quasicrystal has been

†Some attempts have been made in this direction (Jarić and Qiu, 1993; Jarić and Qiu, 1994).

theoretically predicted (Berenson and Birman, 1986) and experimentally observed (Kycia, *et al.*, 1993). Since it is not possible to define a periodic unit cell in a quasicrystal, the definition of "structure factor" is not obvious. We will make use of the notion of *structure factor per unit volume*:

$$F_H^{QC} = \sum_{1}^{N} {}_i f_i e^{2\pi i \vec{B}_H \cdot \vec{r}_i} / V_{QC} \tag{3.3}$$

where the numerator on the right side is the standard expression for a structure factor, extended to *all* the atoms present in the particular quasicrystalline specimen used in the experiment, and V_{QC} is the volume of the same specimen.

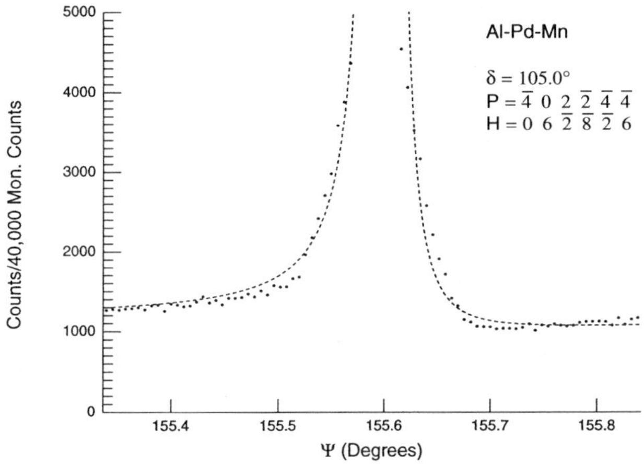

Figure 4. Three beam experiment with Al-Pd-Mn quasicrystal. The main reflection if the $\bar{4}02\bar{2}\bar{4}\bar{4}$ and the simultaneous reflection is the $06\bar{2}\bar{8}\bar{2}6$. The best fit is obtained for a triplet invariant $\delta = 105°$. The profile could not be fitted with $\delta = 0°$ or $180°$. The dotted line is calculated using n-beam dynamical theory, without smearing function (zero mosaic spread).

Having convinced ourselves of the legitimacy of using n-beam dynamical theory for quasicrystals, we have analyzed several azimuthal profiles similar to those given in Figures 3 and 4 with a view to obtaining phase information. In each case we were able to determine the *triplet invariant* $\delta = \phi_H + \phi_{P-H} - \phi_P$ (see Sect. 1). For the profiles of Figures 3 and 4 we found $\delta = 67.5°$ and $105°$, respectively.

All theoretical treatments of structure factors for quasicrystals are based on the assumption of centrosymmetry, for which only two values of δ are expected: 0° and 180°. The reason for this choice is that all electron and X-ray diffraction patterns exhibit inversion symmetry, both for Al-Cu-Fe and Al-Pd-Mn. However, the sensitivity of photographs to lack of inversion symmetry is limited. Friedel's law predicts the same intensity for $\vec{H}$ and $-\vec{H}$ reflections, apart from small effects due to the imaginary components of the scattering factors. In the case of Al-Pd-Mn, the contrast-variation method (Janot *et al.*, 1987) was used in neutron scattering experiments to provide proof of centrosymmetry. This result was arrived at by comparison of the intensities of selected Bragg reflections in powder samples in which manganese (a negative neutron scatterer) was replaced with Fe or FeCr (positive neutron scatterers). The same results (centrosymmetry) was found in several icosahedral structures, other than Al-Cu-Fe, investigated by the same method. Since the presence or absence of centrosymmetry is a minor perturbation on the intensities of most Bragg reflections, it is very difficult to draw firm conclusions from powder diffraction data.

A careful quantitative experiment has been done recently (de Boissieu *et al.*, 1994) in which Friedel pairs ($\vec{H}$ and $= -\vec{H}$ reflections) have been measured quantitatively using a spherical single crystal. Most pairs have almost identical intensities, which is in support of centrosymmetry. There are however a few exceptions, and some differences between $\vec{H}$ and $-\vec{H}$ reflections could be the consequence of weak deviations from centrosymmetry.

There is no question that the values we found for the triplet invariants in Figures 3 and 4 (67.5° and 105°) are very far from the centrosymmetric values (0° and 180°). We are not able at this stage to explain these seemingly contradictory results.

There may be a fundamental problem in defining the very meaning of centrosymmetry for a quasicrystal, a non-periodic structure. It seems that the notion of centrosymmetry (or lack of it) can only be established *on average.* It may be useful to make a comparison with a regular crystal subject to thermal motion. The "average" structure of a quasicrystal is the analog of an ensemble average, in the sense of statistical mechanics, of a crystal subject to thermal motion.

The problem has been analyzed in great detail by M. de Boissieu et al. (1994), who show a concrete example of a one-dimensional quasicrystal, projected from a two-dimensional periodic space. They consider the case of a chemical decoration, which breaks centrosymmetry if the two species A and B are different. The interesting conclusion of their discussion is that even when A and B are identical, the projected structure may not be centrosymmetric. M. de Boissieu *et al.* (1994) conclude that a centrosymmetric quasicrystal contains regions that are centrosymmetric within limited volume domains, quasiperiodically distributed over the whole quasicrystal.

4. CIRCULAR POLARIZATION AND INVERSION SYMMETRY

All of the experimental work described so far involves the use of linearly polarized X-rays, which is the normal case for synchrotron radiation. The X-rays generated by an X-ray tube or a rotating anode are *unpolarized*, that is to say, they consist of a mixture of linearly polarized photons randomly oriented around the direction of propagation.

A new twist to the interpretation of 3-beam diffraction experiments is obtained when circularly polarized X-rays are used. The theory shows that the asymmetry effect is sensitive to the helicity of the X-ray photons *when the specimen is a non-centrosymmetric crystal* (Shen and Finkelstein, 1990).

Shen and Finkelstein (1990) consider the case of the 442 reflection in GaAs as primary reflection, in combination with the (151) as simultaneous reflection. This combination yields a triplet invariant $\delta = 90°$, which means, according to Eq. (1.4) derived from Shen's perturbation theory for linear polarization, that the asymmetry effect is absent.

When the theory is modified for circular polarization, however, the asymmetry effect is restored, and *the sign of the perturbing term depends on the helicity of the X-ray photons*. This means that reversing the helicity of the incident beam has the effect of reversing the asymmetry effect. This prediction has indeed been verified by experiment. Figure 5 shows the effect for the 442–151 combination of GaAs. The difference between the two profiles is obtained by simply moving up and down a slit so that X-rays above and below the orbital plane of the synchrotron are used for the incident beam. It is known that, while the synchrotron X-rays in the orbital plane are linearly polarized in the plane, a circular component is added for X-rays outside of the orbital plane. For the parameters of the 6.5 GeV Cornell Synchrotron (CHESS), a vertical distance of ± 1 mm is enough to introduce a substantial circular component, so that the incident beam outside of the orbital plane is elliptically polarized.

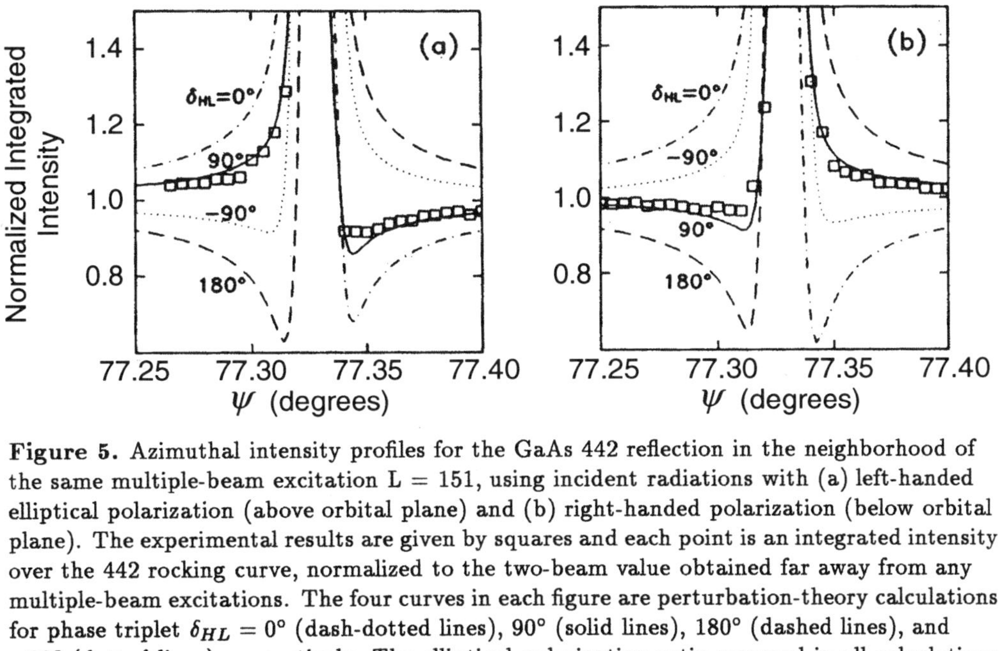

Figure 5. Azimuthal intensity profiles for the GaAs 442 reflection in the neighborhood of the same multiple-beam excitation L = 151, using incident radiations with (a) left-handed elliptical polarization (above orbital plane) and (b) right-handed polarization (below orbital plane). The experimental results are given by squares and each point is an integrated intensity over the 442 rocking curve, normalized to the two-beam value obtained far away from any multiple-beam excitations. The four curves in each figure are perturbation-theory calculations for phase triplet $\delta_{HL} = 0°$ (dash-dotted lines), 90° (solid lines), 180° (dashed lines), and −90° (dotted lines), respectively. The elliptical polarization ratio assumed in all calculations is (a) $b = +0.45$ and (b) $b = -0.45$, and is consistent with the theoretical predictions for the experiment setup. From: (Shen and Finkelstein, 1990).

Since the structure of GaAs is well known, all triplet invariants can be calculated. Shen and Finkelstein (1990) propose to use this method *to measure* the elliptical polarization ratio of an X-ray beam. It turns out that, in view of the growing interest in using circularly and elliptically polarized X-rays for studying magnetic properties of crystalline solids, the method proposed by Shen and Finkelstein is one of the best for characterizing the parameters of an elliptically polarized X-ray beam. Their method has been extended and generalized (Shen and Finkelstein, 1992; Shen, 1993) by making use of a density matrix formalism, in such a way that by performing three-beam experiments all the Stokes-Poincaré parameters can be determined.

The important point that emerges from the theory is that a reversal of the asymmetry effect is expected when the helicity is changed from left to right, but *no effect is expected when the specimen is centrosymmetric.* This is what perturbation theory predicts, which only applies to the wings of the azimuthal plots. A rigorous numerical treatment (Shen, 1993) of the problem by making use of the NBEAM program (Colella, 1974) shows that even for a centrosymmetric crystal changing the helicity affects the peak region of the azimuthal plot, in close proximity of the exact 3-beam interaction.

It appears that the use of circular, or elliptically, polarized X-rays adds a new dimension to multiple Bragg scattering as a technique for phase determination. Since the shape of the wings of an azimuthal profile is different for left and right X-ray photons, in the case of a non-centrosymmetric crystal, and does not depend on helicity in the case of centrosymmetry, it appears to be an useful test applicable to quasicrystals. It would a matter of replacing the GaAs crystal used in the experiment of Shen and Finkelstein (1990) with a quasicrystal, and obtaining azimuthal plots with left and right handed photons.

Some very preliminary data have been obtained, for Al-Pd-Mn quasicrystal, indicating in at least one case a small but perceptible difference between the two helicities, favoring the hypothesis of non-centrosymmetry (Lee, Colella and Shen 1996). However, it is too soon to draw firm conclusions from these results. More experiments and more analyses are needed for a coherent picture to be drawn. The work is in progress, and hopefully more definitive results will be available in the near future.

ACKNOWLEDGMENTS

The author is indebted for many invaluable discussions to M. de Boissieu (Grenoble), who has provided several specimens of Al-Pd-Mn quasicrystals, and has made available the crystallographic data files (calculated and measured structure factors) of Al-Pd-Mn.

This work has been supported by the National Science Foundation, grant no. 9301004–DMR.

REFERENCES

Berenson, R., and Birman, J.L., 1986, Anomalous transmission of X-rays through a quasicrystal, *Phys. Rev. B* 34:8926.

Colella, R., 1974, Multiple diffraction of X-rays and the phase problem. Computational procedures and comparison with experiment, *Acta Cryst. A* 30:413.

Colella, R., 1996, Multiple Bragg scattering and the phase problem in X-ray diffraction. Perfect crystals, in: *X-ray and Neutron Dynamical Diffraction: Theory and Applications*; A. Authier, S. Lagomarsino and B.K. Tanner, eds., Plenum Press, New York, U.S.A.

de Boissieu, M., Stephens, P., Boudard, M., and Janot, C., 1994, Is the Al-Pd-Mn icosahedral phase centrosymmetrical?, *J. Phys. Cond. Matter* 6:363.

Elser, V., 1986, The diffraction pattern of projected structures, *Acta Cryst. A* 42:36.

Gong, P.P., and Post, B., 1983, The experimental determination of phases of reflections from mosaic crystals, 1. $ZnWO_4$,

Hirsch, P.B., and Ramachandran, G.N., 1950, Intensity of X-ray reflexion from perfect and mosaic absorbing crystals, *Acta Cryst.* 3:187.

Hümmer, K., Schwegle, W., and Weckert, E., 1991, A feasibility study of experimental triplet-phase determination in small proteins, *Acta Cryst. A* 47:60.

Hümmer, K., and Billy, H., 1986, Experimental determination of triplet phases and enantiomorphs of non-centrosymmetric structures. I. Theoretical considerations, *Acta Cryst. A* 42:127.

Jackson, J.D., 1975, *Classical Electrodynamics*, 2nd Ed., Sect. 9.7, pp. 418–421. John Wiley, New York.

Janot, C., Pannetier, J., de Boissieu, M., and Dubois, J.M., 1987, Contrast variation effects on neutron diffraction patterns with quasi-periodic structures, *Europhys. Lett.* 3:995.

Jarić, M.V., and Qiu, S.Y., 1993, On the solution of the phase problem in quasiperiodic crystals, *Acta. Cryst. A* 49:576.

Jarić, M.V., and Qiu, S.Y., 1994, Inner-space reconstruction of quasicrystal structure factors, *Phys. Rev. B* 49:6614.

Kycia, S.W., Goldman, A.I., Lograsso, T.A., Delaney, D.W., Black, D., Sutton, M., Dufresne, E., Brüning, R., and Rodricks, R., 1993, Dynamical X-ray diffraction from an icosahedral quasicrystal, *Phys. Rev. B* 48:3544.

Lee, H., and Colella, R., 1993, Phase determination of X-ray reflections in a quasicrystal, *Acta Cryst. A* 49:600.

Lee, H., Colella, R., and Shen, Q., 1996, Multiple Bragg diffraction in quasicrystals: The issue of centrosymmetry in Al-Pd-Mn, *in press, Phys. Rev. B.*

Post, B., 1983, The experimental determination of the phases of X-ray reflections, *Acta Cryst. A* 39:711.

Schmidt, M.C., and Colella, R., 1985, Phase determination of forbidden X-ray reflections in V_3Si by virtual Bragg scattering, *Phys. Rev. Letters*, 55:715.

Shen, Q., 1986, A new approach to multibeam X-ray diffraction using perturbation theory of scattering, *Acta Cryst. A* 42:525.

Shen, Q., and Colella, R., 1987, Solution of phase problem for crystallography at a wavelength of 3.5 Å, *Nature* 329:232.

Shen, Q., and Finkelstein, K.D., 1990, Solving the phase problem with multiple-beam diffraction and elliptically polarized X-rays, *Phys. Rev. Lett.* 65:3337.

Shen, Q., and Finkelstein, K.D., 1992, Complete determination of X-ray polarization using multiple-beam Bragg diffraction, *Phys. Rev. B* 45:5075.

Shen, Q., 1993, Effects of a general X-ray polarization in multiple-beam Bragg diffraction, *Acta Cryst. A* 49:605.

Viterbo, D., 1992, Solution and refinement of crystal strutures, in: *Fundamentals of Crystallography*, C. Giacovazzo, ed., International Union of Crystallography, Oxford University Press, Oxford, U.K.

DETERMINATION OF REFLECTION PHASES BY THREE-BEAM DIFFRACTION

Kurt Hümmer and Edgar Weckert

Institut für Kristallographie
Universität Karlsruhe (TH)
76128 Karlsruhe, Germany

1. INTRODUCTION

The basic idea how three-beam diffraction can be used for physical determination of phase relations originates from Lipscomb (1949). He proposed to exploit the diffracted intensity when two Bragg reflections are simultaneously excited. This situation is called three-beam diffraction since besides of the forwardly transmitted two additional diffracted rays, in total three strong rays, are simultaneously be propagated (*cf.* Figure 1). More generally, N-beam diffraction occurs when besides of the origin N-1 nodes of the reciprocal lattice lie very close to or on the Ewald sphere.

Experimentally, N-beam diffraction can be adjusted systematically by the azimuthal Ψ-scan technique. In this experiment the crystal is first aligned for a selected reflection, called primary reflection, to generate an ordinary two-beam case. Then the crystal is rotated about that reciprocal lattice vector (r.l.v.), that means rotated about the normal of the corresponding lattice planes, keeping the Bragg angle of the primary reflection constant and monitoring its intensity. As the crystal is rotated, in general, several nodes of the reciprocal lattice simultaneously pass through the Ewald sphere. In favorable cases exactly one additional secondary reflection is excited, which leads to a distinct three-beam case.

To specify nomenclature, in the following the primary reflection is denoted by its r.l.v. $\mathbf{h}$, around which the Ψ-azimuthal scan is carried out. The additional secondary reflection is denoted by $\mathbf{g}$, the related three-beam case by $\mathbf{0}/\mathbf{h}/\mathbf{g}$.

A rough estimation of the interaction of the excited waves in multiple diffraction shows that the intensities of the diffracted rays are affected by the relative phases of the structure factors involved. We concentrate on a three-beam case (Lipscomb, 1949) shown in Figure 1 where for simplicity a coplanar geometry will be discussed. The incident beam simultaneously excites two diffracted waves $\mathbf{h}$ and $\mathbf{g}$ denoted by their corresponding r.l.v.'s. With respect to an arbitrarily chosen origin of the unit cell their phases are given by $\varphi(\mathbf{h})$ and $\varphi(\mathbf{g})$. Since the difference vector $\pm(\mathbf{h}-\mathbf{g})$ has to be also a vector of the reciprocal lattice the $\mathbf{g}$-wave is diffracted into the direction of the $\mathbf{h}$-wave by $\mathbf{h}-\mathbf{g}$, and vice versa. Focusing on the $\mathbf{K}(\mathbf{h})$-direction two waves are superimposed, namely the primary diffracted wave $\mathbf{h}$ and a wave which is diffracted at the lattice planes of $\mathbf{g}$ and $\mathbf{h}-\mathbf{g}$ called detour excited (Umweg)

X-ray and Neutron Dynamical Diffraction: Theory and Applications
Edited by Authier *et al.*, Plenum Press, New York, 1996

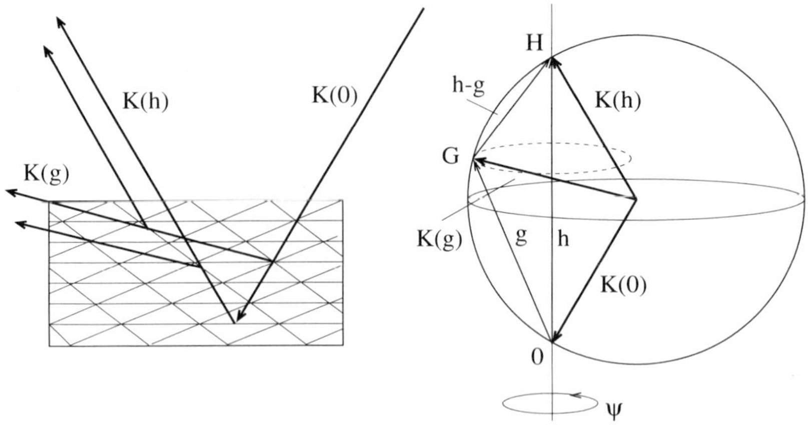

Figure 1. Three-beam case: schematic representation in crystal and reciprocal space with primary reflection **h** and secondary reflection **g**

wave, whose phase is consequently given by $\varphi(\mathbf{g})+\varphi(\mathbf{h}-\mathbf{g})$. The existence of this Umweg wave has already been proved by Renninger (1937). He observed that simultaneously excited reflections give rise to Umweganregung peaks by monitoring the very weak intensity of the 'forbidden' 222 reflection of diamond as primary reflection. Interference of these coherent waves leads to a resultant intensity which depends on their phase difference:

$$\Phi_{3\pm} = \pm(\varphi(\mathbf{g}) + \varphi(\mathbf{h}-\mathbf{g}) - \varphi(\mathbf{h})), \quad \Phi_{3+} = -\Phi_{3-}. \tag{1}$$

as it is well known that the resulting intensity due to the interference of two waves with amplitudes A_1 and A_2 and phases α_1 and α_2 is governed by the cosine of $\pm(\alpha_2 - \alpha_1)$

$$I = \mid A_1 \exp i\alpha_1 + A_2 \exp i\alpha_2 \mid^2 = A_1^2 + A_1^2 + 2A_1A_2\cos(\alpha_2 - \alpha_1) \tag{2}$$

It turns out that the phase difference Φ_3, a so-called triplet phase relationship, is independent of an arbitrarily chosen origin. Therefore, it is also called structure invariant. In fact, only structure invariants are measurable quantities. Hence, it should be possible to exploit the three-beam intensities for experimental determination of triplet phases.

Now, the distinct advantage of the Ψ-scan technique is obvious. As the primary reflection remains in its reflection position the two-beam intensity serves as a reference level which is modulated when the Umweg wave is continuously turned on and off scanning through a three-beam position. Thus, the intensity change due to the interference contrast of the primary reference beam contains information on the triplet phase. This experimental technique is very similar to the technique used to produce a hologram, where the intensity of a reference beam is modulated by the scattered waves. The interference pattern contains the phase information.

These simple considerations do not explain the experimentally observed three-beam diffraction profiles. In order to understand the actually observed profiles and the underlying diffraction physics, a modified two-beam approximation is applied, the so-called first-order Bethe approximation, commonly used in electron multi-beam diffraction. It should be pointed out, however, that it is very difficult to use this approximation for a quantitative analysis of multiple-diffraction effects, in particular, for the quantitative analysis of the various properties, like phase independent Aufhellung and Umweganregung effects, Pendellösung effects and the effects of different polarization states. There are several authors who spent considerable amount of work for the elaboration of analytical solutions of multiple-beam diffraction using Bethe approximation (Juretschke, 1982a; Juretschke, 1982b; Hoier and Marthinsen, 1983; Chang, 1984). The present authors prefer computational simulations. However, simple versions of the two-beam approximation are used to get insight into the underlying diffraction physics. Quantitative computer analysis discussed later is based on the plane-wave dynamical theory using boundary conditions for parallel sided crystal slabs.

Thorkildsen (1987) published the solution of dynamical three-beam diffraction by means of Takagi-Taupin equations for a parallelepiped shaped crystal. The extension of this work could be a way to solve the problem for arbitrarily shaped crystals.

2. TWO-BEAM APPROXIMATION

In order to have a better understanding of three-beam diffraction and the exploitation of the three-phase invariant involved, it is worth discussing their main properties by means of first-order Bethe solution of the fundamental equations of dynamical theory.

$$\mathbf{D}(\mathbf{h}_m) = \frac{\mathbf{K}(\mathbf{h}_m)^2}{\mathbf{K}(\mathbf{h}_m)^2 - \mathbf{k}_0^2} \sum_n \chi(\mathbf{h}_m - \mathbf{h}_n)\mathbf{D}(\mathbf{h}_n)_{[m]}. \tag{3}$$

They can be rewritten in the form of eigenvalue equations

$$\frac{\mathbf{k}_0^2 - \mathbf{K}(\mathbf{h}_m)^2}{\mathbf{K}(\mathbf{h}_m)^2}\mathbf{D}(\mathbf{h}_m) + \sum_n \chi(\mathbf{h}_m - \mathbf{h}_n)\mathbf{D}(\mathbf{h}_n)_{[m]} = 0 \tag{4}$$

where the usual structure factors are related to the Fourier components of the dielectric susceptibility by

$$\chi(\mathbf{h}_n) = -\Gamma F(\mathbf{h}_n). \tag{5}$$

All the symbols have their well known usual meaning. For a three-beam case $\mathbf{h_m}, \mathbf{h_n}$ run over $\mathbf{0}, \mathbf{h}, \mathbf{g}$.

To make the discussion as simple as possible without losing general physical arguments the coupling between the two mutually perpendicular π and σ polarization components of the wavefield amplitudes is neglected, i.e. $\boldsymbol{\pi} \cdot \boldsymbol{\sigma}$ scalar products appearing in (4) are set equal to zero. This holds for a coplanar three-beam case. Then (4) are reduced to equivalent non-coupled equations for only one polarization component either π or σ (Hümmer and Billy, 1986). With the approximation

$$R(\mathbf{h}_m)^{-1} = \left(\frac{k_0^2}{\mathbf{K}(\mathbf{h}_m)^2} - \left(1 - \chi(\mathbf{0})\right)\right) \approx \frac{\mathbf{K}_0^2 - \mathbf{K}(\mathbf{h}_m)^2}{\mathbf{K}(\mathbf{h}_m)^2} \tag{6}$$

equations (4) are rewritten in matrix notation:

$$\begin{pmatrix} R(\mathbf{0})^{-1} & \alpha_{0h}\chi(-\mathbf{h}) & \alpha_{0g}\chi(-\mathbf{g}) \\ \alpha_{0h}\chi(\mathbf{h}) & R(\mathbf{h})^{-1} & \alpha_{hg}\chi(\mathbf{h}-\mathbf{g}) \\ \alpha_{0g}\chi(\mathbf{g}) & \alpha_{hg}\chi(\mathbf{g}-\mathbf{h}) & R(\mathbf{g})^{-1} \end{pmatrix} \cdot \begin{pmatrix} \mathbf{D}(\mathbf{0}) \\ \mathbf{D}(\mathbf{h}) \\ \mathbf{D}(\mathbf{g}) \end{pmatrix} = 0. \tag{7}$$

The α_{nm} represent geometrical coupling factors which result from the scalar products $\boldsymbol{\pi}_n \cdot \boldsymbol{\pi}_m$ or $\boldsymbol{\sigma}_n \cdot \boldsymbol{\sigma}_m$.

To solve these equations for the ratio $\mathbf{D}(\mathrm{h})/\mathbf{D}(\mathbf{0})$ a perturbational approach, called Bethe potential method (Bethe, 1928) is adopted. The amplitude of $\mathbf{D}(\mathrm{g})$ may be expressed in terms of $\mathbf{D}(\mathbf{0})$ and $\mathbf{D}(\mathrm{h})$ using the third equation of (7). Upon insertion of this expression in the second equation of (7), for instance, and solving for $\mathbf{D}(\mathrm{h})/\mathbf{D}(\mathbf{0})$ using (5) we get:

$$\frac{\mathbf{D}(\mathrm{h})}{\mathbf{D}(\mathbf{0})} = N^{-1}R(\mathrm{h})\left(\alpha_{0h}\Gamma F(\mathrm{h}) + R(\mathrm{g})\alpha_{0g}\alpha_{hg}\Gamma^2 F(\mathrm{g})F(\mathrm{h}-\mathrm{g})\right) = N^{-1}R(\mathrm{h})F_{eff} \quad (8)$$

where $N = 1 - \alpha_{hg}^2(\Gamma F(\mathrm{h}-\mathrm{g}))^2 R(\mathrm{g})R(\mathrm{h})$. This result has to be interpreted as follows. The amplitude in the two-beam case, given by $\mathbf{D}_2(\mathrm{h})/\mathbf{D}(\mathbf{0}) = N^{-1}R(\mathrm{h})\alpha_{0h}\Gamma F(\mathrm{h})$ the first term of (8) (i.e. no secondary reflections are excited) is modified by higher order terms due to excitation of other reflections. Obviously, if $R(\mathrm{h})$ is negligibly small, i.e. Bragg's law for the h-reflection is not fulfilled, then no intensity can be observed in the direction of $\mathbf{K}(\mathrm{h})$. This is also true even though other wave fields are excited. In this case Bragg's law for the scattering of the secondary g-reflection into the h-reflection is not fulfilled, since the point of the coupling vector $\mathrm{h} - \mathrm{g}$ does not lie on the Ewald sphere. Thus, the basic requirement in order to observe the modification of the intensity of the **h**-reflection by additional excitation of other reflections is to keep **h** precisely on the Ewald sphere during the Ψ-scan.

(8) confirms the basic considerations by Lipscomb (1949). The resulting amplitude of the wave $\mathbf{D}(\mathrm{h})$ is given by a superposition of two waves: the directly diffracted wave governed by the structure factor $\mathbf{F}(\mathrm{h})$ and the Umweg wave governed by the product of structure factors $\mathbf{F}(\mathrm{g})\mathbf{F}(\mathrm{h}-\mathrm{g})$. However, the resonance term $R(\mathrm{g})$ determines not only the amplitude but also the phase of the Umweg wave. It causes a phase shift of 180° by scanning through a three-beam position. This behaviour can be seen by means of the Ewald construction, where the radius of the Ewald sphere is given by $\mid \mathbf{K}_0 \mid$. We assume that the azimuthal scan is carried out so that the point of g passes the Ewald sphere from inside to outside, in short 'in-out' scan. At the beginning when g terminates inside then $\mid \mathbf{K}(\mathrm{g}) \mid < \mid \mathbf{K}_0 \mid$ since $\mathbf{K}(\mathrm{g}) = \mathbf{K}_0 + \mathrm{g}$ (*cf*. Figure 2) and R(**g**) is positive. When the point of **g** approaches the Ewald sphere R(**g**) gets larger as the the denominator of $R(\mathrm{g})$ gets smaller, i.e. the amplitude of the Umweg wave increases. It has its maximum value when the point of g exactly lies on the Ewald sphere. When it leaves the Ewald sphere again the magnitude of $R(\mathrm{g})$ and therefore the amplitude of the Umweg wave decreases. However, $R(\mathrm{g})$ has changed its sign, since $\mid \mathbf{K}(\mathrm{g}) \mid > \mid \mathbf{K}_0 \mid$ when g terminates outside (*cf*. Figure 2). Changing its sign $R(\mathrm{g})$ causes an additional phase shift $\Delta(\Psi)$ by 180°, which is called resonance phase shift. $R(\mathrm{g})$ is a lorentzian where it is taken into account that the wave vectors are complex quantities due to absorption. The behaviour of the amplitude and the phase shift of the Umweg wave during the Ψ-scan is shown in Figure 3.

To specify nomenclature, henceforth, all the Ψ-scan profiles refer to an 'in-out' scan: $\Psi = 0$ marks the exact geometrical three-beam position, for $\Psi < 0$ the point of **g** is inside, for $\Psi > 0$ it is outside the Ewald sphere. $\Delta(\Psi)$ varies from 0 to 180° for an 'in-out' Ψ-rotation sense. If the rotation sense is reversed (out-in) the resonance phase shifts from 180 to 0°. Therefore, it is important to know the rotation sense for the exploitation of the triplet phase from the three-beam diffraction profiles (Chang, 1982).

Now, the fundamental features of the integrated intensity of a Ψ-scan profile can easily be calculated by means of (8). In a first-order approximation, except for a range very close to the three-beam position, N may be taken as a constant, $N \approx 1$. The range of validity for the first-order solution has been discussed in detail by Juretschke (1984), where ξ_L in Juretschke's paper is equivalent to $R(\mathrm{g})^{-1}$. It depends on the magnitudes of the structure factors involved. Then, the integrated three-beam intensity is approximately given by:

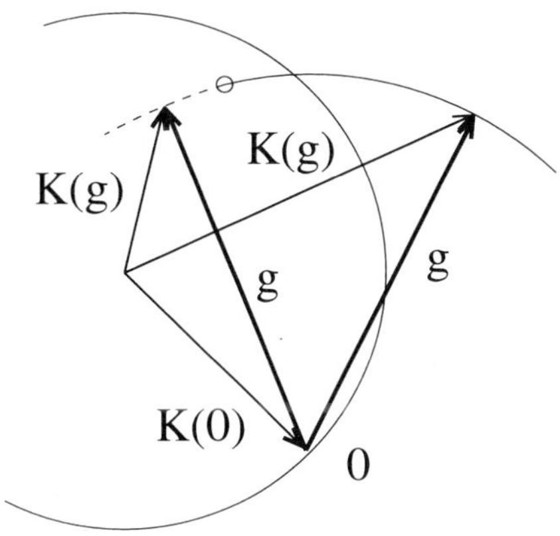

Figure 2. Schematical drawing illustrating the change of sign of the resonance term R(**g**).

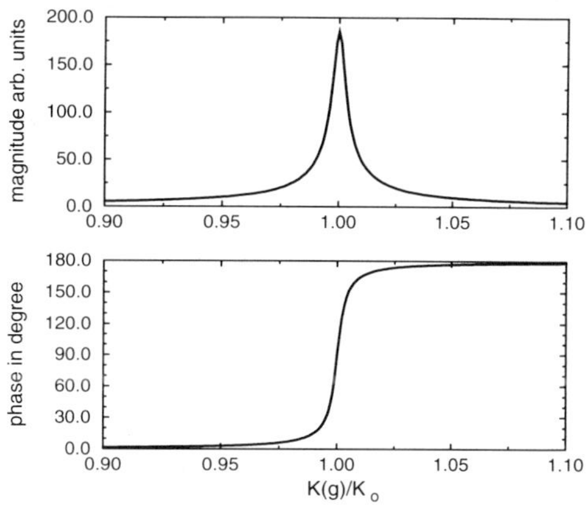

Figure 3. Magnitude and phase of the Umweg wave scanning through a three-beam position.

$$I_h(\Psi) \sim | F_{eff} |^2 \tag{9}$$

Introducing the magnitude $| R_g(\Psi) |$ and the phase $\Delta(\Psi)$ of the resonance term

$$R(\mathbf{g}) = | R_g(\Psi) | \, exp[i\Delta(\Psi)] \tag{10}$$

and the magnitude and the phase of the structure factors involved

$$F(\mathbf{h}_n) = | F(\mathbf{h}_n) | \, exp[i\varphi(\mathbf{h}_n)], \ \mathbf{h}_n = \mathbf{h}, \mathbf{g}, \mathbf{h} - \mathbf{g} \tag{11}$$

$I_h(\Psi)$ using (2) and (8) is given by:

$$I_h(\Psi) \sim A_1^2 + A_2^2 + 2A_1A_2cos[(\varphi(\mathbf{g}) + \varphi(\mathbf{h} - \mathbf{g}) + \Delta(\Psi)) - \varphi(\mathbf{h})] \tag{12}$$

where

$$\begin{aligned} A_1 &= \alpha_{0h}\Gamma \mid F(\mathbf{h}) \mid \\ A_2 &= \alpha_{0g}\alpha_{hg}\Gamma^2 \mid R_g(\Psi) \mid\mid F(\mathbf{g}) \mid\mid F(\mathbf{h} - \mathbf{g}) \mid . \end{aligned} \tag{13}$$

Hence, the interference of the primary wave and the Umweg wave is governed by the total phase $\Phi_{tot}(\Psi)$ which according to (1) and (12) is given by:

$$\Phi_{tot}(\psi) = \pm(\Phi_{3+} + \Delta(\Psi)) = \mp(\Phi_{3-} - \Delta(\Psi)). \tag{14}$$

It should be noticed that the resonance phase shift was not considered by Lipscomb (1949). It is a result of the self-consistent dynamical interaction. In fact, such a phase shift occurs also in two-beam diffraction rocking the crystal through a two-beam diffraction position. Already Ewald (1917, 1965) in his work on crystal optics of X-rays has pointed out that Bragg diffraction is a spatial resonance phenomenon. The resonance term may be regarded as the efficiency of the crystal for converting a given amplitude of polarization **P** into the field amplitude **D**. If Bragg's diffraction condition is fulfilled, i.e. $\mathbf{K}(\mathbf{h})^2 = \mathbf{K}_0^2$, this efficiency will be optimum. Thus, the diffraction will be optimum when the spatial periodicity of the incoming wave matches the spatial periodicity of the lattice.

At this stage of discussion it should be stressed once again, that the three-beam intensity does not depend on the cosine of the triplet phase but on the cosine of the total phase $\Phi_{tot}(\Psi)$ which contains the resonance phase shift $\Delta(\Psi)$. Due to this fact the three-beam intensity depends also on the sign of $|\Phi_3|$ with $0 \leq |\Phi_3| \leq 180^o$.

Three-beam Interference Profiles

In the previous section we have already considered the three-beam interference between the directly diffracted wave and the Umweg wave and their phase difference. Let us now find the principal features of three-beam profiles for different triplet phases using the azimuthal Ψ-scan technique. Suppose the triplet phase of a three-beam case $\mathbf{0}/\mathbf{h}/\mathbf{g}$ is zero: $\Phi_{3+} = 0^o$. Then, at the beginning of the Ψ-scan $\Delta(\Psi) = 0$ and $\Phi_{tot}(\Psi) = 0$. The amplitude of the Umweg wave is very small and the two-beam intensity $I_h^{(2)}$ is observed. Scanning towards the

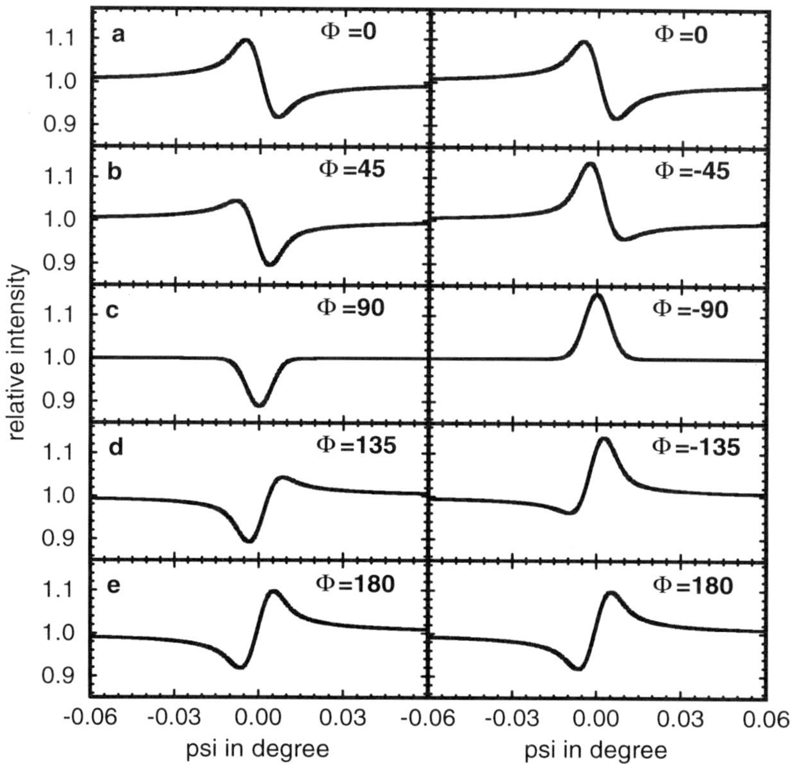

Figure 4. Calculated three-beam interference profiles for different triplet phases

three-beam position the amplitude of the Umweg wave increases. The primary wave and the Umweg wave interfere in a constructive way which leads to an increase in the resultant amplitude of $\mathbf{D}(\mathbf{h})$. Thus, the intensity is increased. Very near to the three-beam position $\Delta(\Psi)$ shifts very rapidly from 0 to 180°, then $\Phi_{tot}(\psi) = 180^o$. That means that the interference becomes destructive and the two-beam intensity is decreased. At the end of the Ψ-scan when the amplitude of the Umweg wave gets smaller the two-beam intensity is observed again. A calculated profile of this type is shown in Figure 4a. It reflects the fact that $\cos[\Phi_{tot}(\Psi)]$ changes its sign as $\Phi_{tot}(\Psi)$ varies from 0 to 180°. In the case that $\Phi_{3+} = 180^o$ the Ψ-scan profile will be reversed with respect to $\Psi = 0$, as $\Phi_{tot}(\Psi)$ varies from 180 to 360° and thus $\cos[\Phi_{tot}(\Psi)]$ changes from negative to positive values (*cf.* Figure 4b).

For $\Phi_{3+} = +90$ or -90° it is important to notice that different Ψ-scan profiles result. Therefore, it is possible to distinguish both cases experimentally. Scanning through a three-beam position $\cos[\Phi_{tot}(\Psi)]$ is always negative for $\Phi_{3+} = +90^o$, for $\Phi_{3+} = -90^o$ it is always positive. That means that the interference term, the third term of (12), is symmetric around the three-beam position $\Psi = 0$. Therefore, a symmetric decrease or increase of the two-beam intensity is expected due to destructive or constructive interference, respectively (*cf.* Figure 4c).

If we had taken the negative sign in (14), we would have got the same type of profiles. This is also true if we had taken the negative sign for the definition of the triplet phase in (1), i.e. $-\Phi_{3-}$ instead of Φ_{3+}. Both definitions lead to the same type of profiles. Hence, there is no ambiguity with respect to the experimental phase determination from the Ψ-scan profiles, since one gets either Φ_{3+} or $-\Phi_{3-}$. Therefore, in the following we use Φ_{3+} as a definition for the triplet phase (*cf.* (1)), in that case we use the short hand notation Φ_3.

It remains to discuss the diffraction profiles if Φ_{3+} equals ± 45 or $\pm 135^o$. Their principal features must be something intermediate between 0 and ± 90 or ± 90 and 180°, as can be seen from Figures 4b and d. For instance, if $\Phi_{3+} = +45^o$, then at first a smaller region of constructive interference is observed, as long as $\cos(45^o + \Delta(\Psi))$ is positive. Since the amplitude of the Umweg waves reaches its maximum for $\Phi_{tot}(\Psi) = 45 + 90 = 135^o$, in that case, a larger region of destructive interference is to be expected and the increase of the two-beam intensity is weaker than the decrease.

A summary outline of integrated three-beam profiles is given in Figure 4. Because of its experimental relevance profiles of triplet phases having opposite sign are compared.

3. COMPUTATIONAL ANALYSIS OF THREE-BEAM DIFFRACTION

The solutions of (4) give the displacement vectors only on a relative basis (Colella, 1974; Weckert and Hümmer, 1990). In order to get their absolute values the boundary conditions for electromagnetic waves have to be applied. They require continuity of the tangential components of the electric and magnetic fields across the boundary vacuum-crystal. Then the transmitted and reflected intensities in Laue case and Bragg case geometry are calculated by means of the known amplitudes outside the crystal at the exit and entrance surface, respectively. In case of asymmetric diffraction geometries, where the lattice planes are neither parallel nor normal to the crystal surface, the cross-section of the reflected and transmitted beams will be changed. In order to satisfy the condition of conservation of energy, as the intensity is defined as the energy flow density, asymmetry factors $\gamma(\mathbf{h})_n$ have to be introduced which take this change into account (von Laue, 1960). Accordingly, the reflected or transmitted intensity is calculated as

$$I_{\mathbf{h}_n} = \frac{|\gamma(\mathbf{h}_n)|}{\gamma_0} \frac{\mathbf{E}(\mathbf{h_n})\mathbf{E}^*(\mathbf{h}_n)}{\mathbf{E}_0\mathbf{E}_0^*} \tag{15}$$

$\gamma(\mathbf{h}_n)$ are the direction cosines between the vacuum wave vectors $\mathbf{k}(\mathbf{h}_n)$ and the surface normal, γ_0 that of the incident beam. $\mathbf{E}_0$ is the amplitude of the incident beam.

3.1. Dispersion Surface

In a three-beam case the dispersion surface consists of 6 sheets due to the π and σ components for each of the **0**-, **h**- and **g**-waves whose wave vectors terminate at the corresponding reciprocal lattice points 0, H and G. Since the three dimensional dispersion surface for a non-coplanar three-beam case is very difficult to survey it is appropriate to discuss intersections adapted to the azimuthal ψ-scan technique where the diffraction condition for the two-beam case **0/h**, for instance, is always satisfied. In the two-beam case the dispersion surfaces are surfaces of revolution around the diffraction vector **h**. Rocking the crystal about **h** by small angles ψ the actual tie points move approximately along straight lines perpendicular to **h**. An intersection of the dispersion surfaces perpendicular to **h** through the diameter points of the two-beam hyperbola for different triplet phases of the same three-beam case is shown in Figure 5. They are calculated as the eigenvalues of (4). Near the three-beam setting the two-beam hyperbola, i.e. the vertical straight lines, is splitted due to the interaction with the secondary reflection **g**. The nearly horizontal parts are the g-like branches. It can be seen that the dispersion surface is centrosymmetric with respect to the three-beam Lorentzpoint Lo_3 for $\Phi_3 = 90°$ (Ewald and Heno, 1968). It is asymmetric for Φ_3 equal to 0 and $180°$. Contrarily to the three-beam Ψ-scan profiles calculated by the eigenvectors of (4) it only depends on $\cos \Phi_3$.

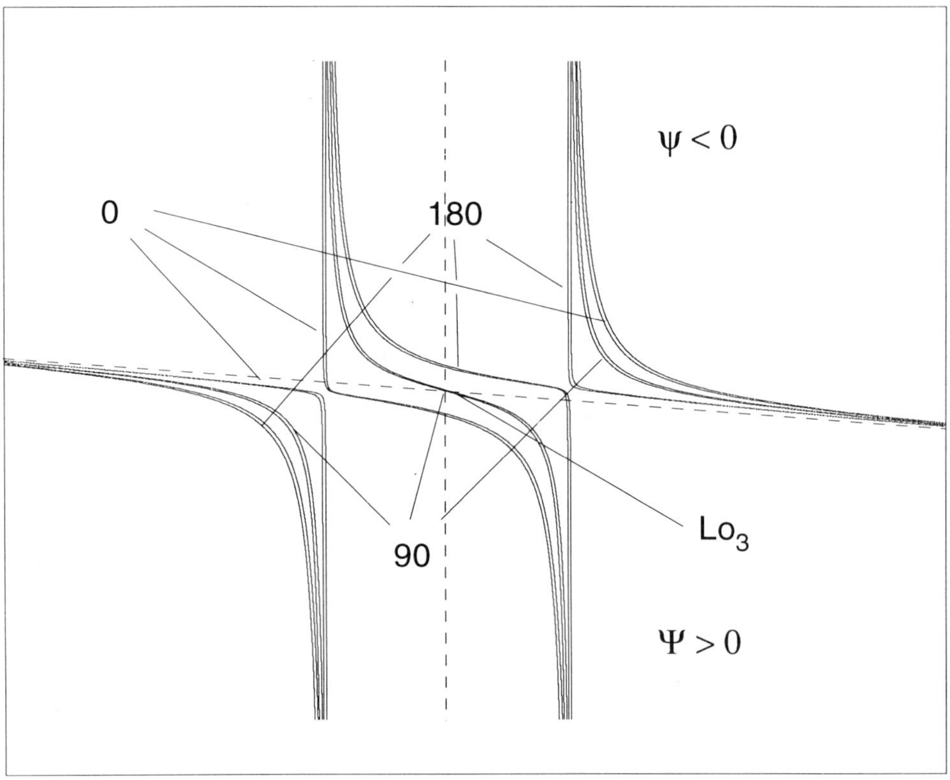

Figure 5. Intersection through the three-beam dispersion surface for different triplet phases

3.2. Integrated Three-Beam Ψ-Scan Profiles

In the framework of plane wave dynamical theory, i.e. taking ideally monochromatic and ideally non-divergent collimated beams, the N-beam Ψ-scan profiles are calculated from the reflectivity or transmissivity of the primary diffracted ray using (15) as a function of two angular variables ψ and ω. The ω-rotation axis is perpendicular to the Ψ-rotation axis (*cf.* Figure 1). Hence, away from the N-beam setting ($|\ \psi\ | > 0$) an ω-scan generates an unperturbed two-beam rocking curve of the primary reflection (*cf.* for instance Figure 6).

In order to take into account experimental parameters, like divergence, spectral bandwidth and mosaicity, which lead to a broadening of the plane-wave dynamical profiles, the latter have to be convoluted with a suitable gaussian broadening function G.

$$I_{\mathbf{h}}^{int}(\psi,\omega) = constant \int_{-\infty}^{+\infty} dv \int_{-\infty}^{+\infty} du I_{\mathbf{h}}(\omega - v, \psi - u) G(v,u) \tag{16}$$

In the following figures the integrated intensity $I_{\mathbf{h}}^{int}(\psi)$ is normalized with respect to the unperturbed integrated two-beam intensity $I_{\mathbf{h}}^{(2)}$ of the primary reflection.

$$I_{\mathbf{h}}^{rel}(\psi) = I_{\mathbf{h}}^{int}(\psi)/I_{\mathbf{h}}^{(2)} \tag{17}$$

Therefore, these diagrams show the relative change of the two-beam intensity by simultaneous excitation of other beams.

3.3. Reflection and transmission geometries

We are now going to consider different three-beam diffraction geometries. If the crystal is supposed to be a parallel sided slab four distinct situations may occur. First of all the diffraction geometry of the primary reflection has to be considered. It can be a Bragg diffraction case or a Laue transmission case. Since the two-beam intensities in both cases show different behaviour the question arises whether there are basic differences with respect to the evaluation of triplet phases from three-beam diffraction profiles. In particular in the Laue case the influence of Pendellösung effects has to be considered. In this case the intensity of the primary reference beam depends critically on the thickness of the crystal and it is expected that additional phase effects occur.

Moreover, the secondary reflection can be either a Bragg or Laue case. However, it turns out that the diffraction geometry of the secondary reflection is not a crucial point. Therefore, the diffraction geometry is denoted by that of the primary reflection, e.g. Bragg or Laue case; if necessary the name of the diffraction geometry of the secondary reflection is appended.

An additional parameter in our analysis for different diffraction geometries will be the thickness of the crystal. To have a short hand notation, we speak of a thin or thick crystal if the thickness is smaller or larger than the Pendellösung distance, respectively.

Bragg Case. A calculated three-beam diffraction profile $I_{\mathbf{h}}(\omega,\psi)$ with the primary reflection in Bragg geometry in case of a thick crystal is shown in Figure 6.

Near the three-beam setting $\psi \approx 0$ the two-beam is modified due to the excitation of a secondary reflection. The excitation of **K(h)** waves along the **g**-like part of the dispersion surface is clearly been seen. The two-beam profile $I_{\mathbf{h}}(\omega, |\ \psi\ | = constant > 0)$ far from the three-beam setting shows the typical features of a Bragg case rocking curve. The reflectivity is about 90% in the total-reflection domain since absorption is taken into account. The relative integrated three-beam reflectivity is shown in Figure 7. It shows the typical asymmetry of a

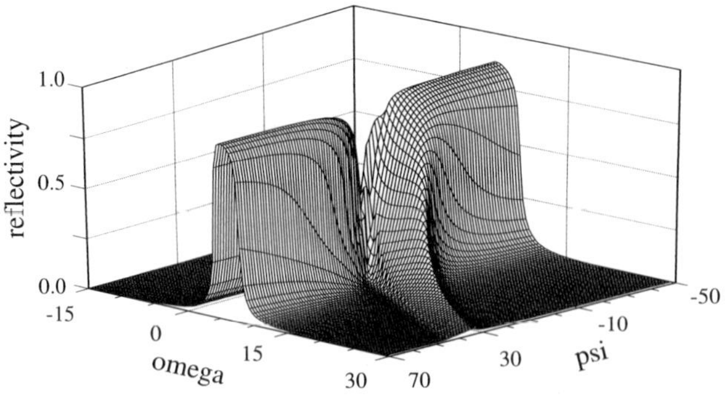

Figure 6. Three-beam reflectivity for the thick-crystal Bragg-case $\Phi_3 = 0°$; GaAs: 311/220; (angle units in arcsec)

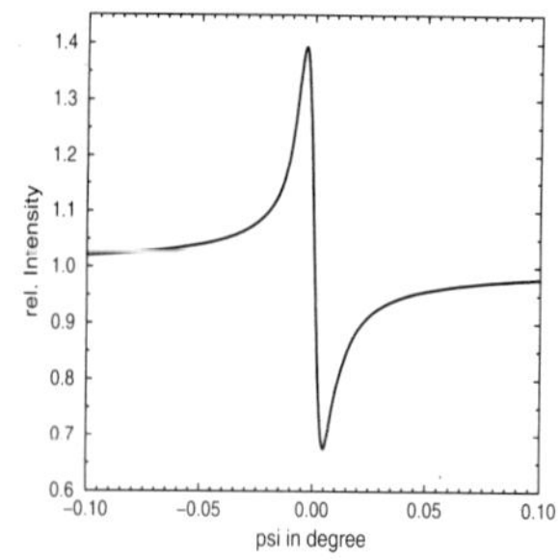

Figure 7. Integrated three-beam ψ-scan profile of Figure 6

Ψ-scan profile for $\Phi_3 = 0°$. This is not immediately obvious in Figure 6, since the increase of the integrated reflectivity is due to a broadening of the two-beam profile near the three-beam setting for $\psi \leq 0$. In the Bragg case for thick crystals the maximum height of the two-beam rocking curve cannot be increased since it is essentially governed by total reflection. Contrarily, in the destructive interference range for $\Psi < 0$ the reflection width and so the integrated intensity are decreased.

For a thin crystal in Bragg reflection geometry, there exists no total reflection even in absence of absorption. Figure 8 shows the three-beam reflectivity of $\pm 90°$ triplet phases where the thickness of the plate was chosen so that the maximum reflectivity of the two-beam case is about 30%. In this case the maximum two-beam reflectivity is symmetrically increased for $-90°$ due constructive interference and symmetrically decreased for $+90°$ due to destructive interference. The integrated profiles are also symmetrical in both cases. Figure 8 demonstrate that three-beam profiles for Φ_3 equal to $+90$ and $-90°$ also depend on the sign of the triplet phase. These numerical results clearly confirm the results of the two-beam approximation.

To summarize the results of this section it should be noticed that if the primary reflection is in Bragg geometry then there is no distinct difference between the thick- and thin-crystal case with respect to the integrated three-beam intensity which is usually observed in a real experiment. The important differences for Φ_3 equal to $+90$ or $-90°$ can be used for the experimental determination of the absolute structure.

Laue Case. The essential difference compared to the Bragg case is that Pendellösung effects occur in the Laue case for thick crystals which seriously affect phase determination by N-beam diffraction. It is well known from the two-beam case that these effects are owing to the interference of the waves which belong to tie points at different branches of the dispersion hyperbola. The diffracted power at the exit surface oscillates between the two transmitted waves in dependence on the crystal thickness. In the three-beam Laue-Laue case, i.e. the primary **h**-reflection as well as the secondary **g**-reflection are in transmission geometry, the transmitted power oscillates between three transmitted rays. In a Laue-Bragg case the situation is comparable to that of a two-beam Laue case. In any case, there are additional phase effects which, drastically influence the phase indication of the three-beam profiles. Therefore, we focus on the thin-crystal case where the thickness is well below the first Pendellösung period of the primary **h**-reflection where Λ^{-1} is given by the distance of the diameter points of the two-beam hyperbola.

Figure 9 demonstrates that also in the Laue case the three-beam intensity depends on the sign of the triplet phase. A comparison of Figures 9 and 8 reveals that in the thin-crystal

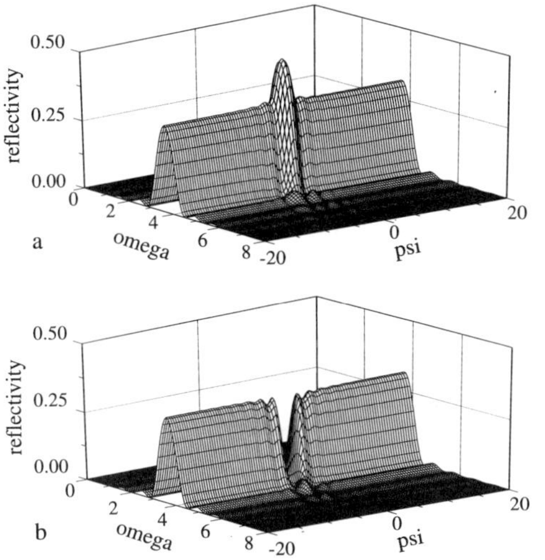

Figure 8. Three-beam reflectivity for the thin-crystal Bragg case: a: $\Phi_3 = -90°$, b: $\Phi_3 = +90°$ (angle units in arcsec)

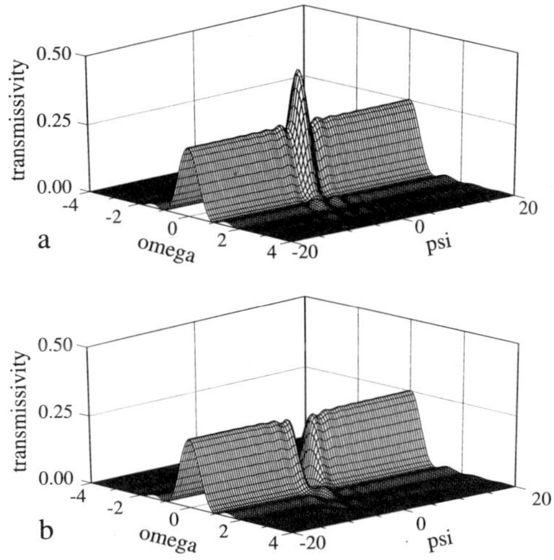

Figure 9. Three-beam transmissivity for the thin-crystal Laue case a: $\Phi_3 = -90°$, b: $\Phi_3 = +90°$ (angle units in arcsec)

case the three-beam profiles show no essential difference between the Laue and Bragg case geometry. However, for thick crystals in the Laue case Pendellösung effects come into play. Then, the three-beam profiles give no unique phase indication.

A summary outline of integrated three-beam Ψ-scan profiles for different triplet phases which are relevant in a real experiment is given in Figure 4. The respective profiles of triplet phases with opposite sign are compared. This comparison is also of experimental relevance since, as will be understood later, we always measure the couple of three-beam cases **0/h/g** and **0/-h/-g**. They have opposite sign of the triplet phase, however, the magnitudes of the involved structure factors are the same if anomalous scattering can be neglected.

3.4. Phase-Independent Umweganregung and Aufhellung Effects

In calculating the integrated three-beam Ψ-scan profiles of Figure 4 the magnitudes of the involved structure factors are chosen so that the constructive and destructive interference effects are approximately of equal size comparing the profiles of triplet phases with opposite sign. Ideally, as can be understood by means of the two-beam approximation and as can be seen in Figure 4 the relative intensity change $I_{\mathbf{h}}^{rel}(\psi)_+$ for a triplet phase with positive sign taken at a certain azimuth angle ψ is inverted with respect to that taken at the corresponding negative angle $-\psi$ of the profile for the negative triplet phase $I_{\mathbf{h}}^{rel}(-\psi)_-$. That means, if we sum up (Weckert and Hümmer, 1990)

$$\Delta I(\psi) = \frac{1}{2}[I_{\mathbf{h}}^{rel}(\psi)_+ + I_{\mathbf{h}}^{rel}(-\psi)_-] \tag{18}$$

and if we assume that the intensity modulation is completely governed by three-beam interference according to (8) then $\Delta I(\psi)$ should remain constant equal to one. This condition is approximately satisfied by the profiles of Figure 4.

However, it turns out that in case of n-beam diffraction there are also phase-independent contributions which lead also to a modification of the two-beam intensity. Such effects have been well known for a long time. Enhancement of the two-beam Bragg intensity was called *Umweganregung* (Renninger, 1937). Attenuation was called *Aufhellung*. These effects are due to the dynamical self-consistent balance of the energy flow. For example, if in a three-beam case the structure factor of the primary **h**-reflection is much weaker than those of the secondary **g**- and of the coupling **h-g**-reflection then part of the intensity of the **g**-reflection is coupled into the **h**-reflection via **h-g**. As a result, enhancement of the two-beam intensity is observed, as that was the case in the Renninger experiment, independently of the phase relationships. In contrast to this situation, if the intensity of the primary **h**-reflection is much stronger than that of the secondary **g**-reflection then a considerable amount of the intensity of the first is diffracted into the second via -(**h-g**). This loss of intensity is not compensated by the diffraction power of the **g**-reflection into the **h**-reflection. Consequently, the intensity of the primary reflection is attenuated.

The mean energy flow due to this mutual coupling is described by the so-called energy transfer equations (Zachariasen, 1967; Moon and Shull, 1964) which is a system of coupled equations that must satisfy the condition of conservation of energy. The interaction terms are taken to be proportional to the respective intensities of the individual reflections. Thus, the interaction is assumed to be phase independent. If in a three-beam case this mutual coupling of the intensities is well balanced, then no changes of the two-beam intensity would result, except for those due to interference effects which modulate the mean energy flow. This situation would lead to so-called *ideal* three-beamΨ-scan profiles where $\Delta I(\psi) = 1$ (*cf.* Figure 11).

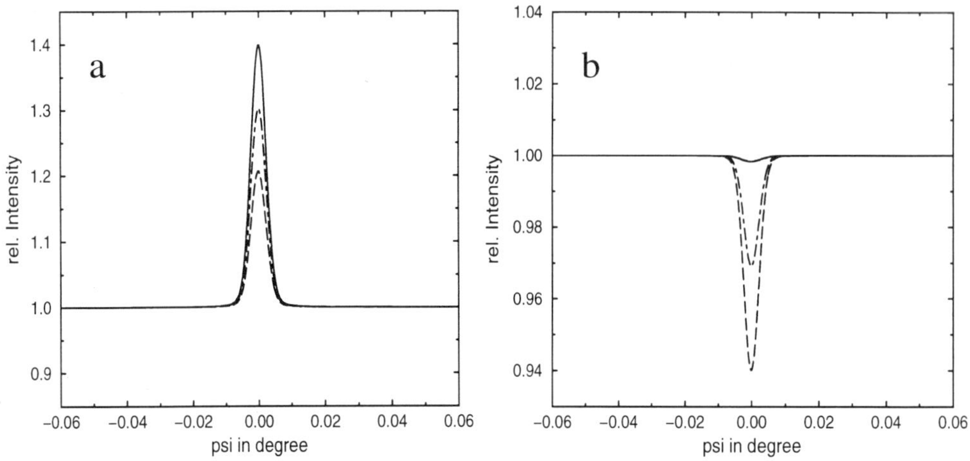

Figure 10. Integrated three-beam Ψ-scan profiles with (a) Umweganregung and (b) Aufhellung; solid: $\Phi_3 = 90°$, dashed: $\Phi_3 = -90°$, dot-dashed: $\Delta I(\psi)$

In Figure 10 typical *Umweganregung* and *Aufhellung* profiles are shown. In each case the $\Delta I(\psi)$ curve is plotted. According to the definition (18) the difference between a Ψ-scan profile with *Umweganregung* or *Aufhellung* and the corresponding $\Delta I(\psi)$ curve gives the ideal profile. Thus, in general, each Ψ-scan profile can be separated into two parts:(1) the symmetrical $\Delta I(\psi)$ curve which represents the phase independent part due to *Umweganregung* and *Aufhellung* and (2) the ideal profile which bears the information on the triplet phase.

This is the reason why we always measure the couple of centro-symmetrically related three-beam cases **0/h/g** and **0/-h/-g**. They differ in the sign of their triplet phase, however, the magnitudes of the involved structure factors being the same. Therefore, *Umweganregung* or *Aufhellung* have the same magnitude in both cases. By calculating the $\Delta I(\psi)$ curve the phase independent effects can be separated. If the phase independent effects are strong compared

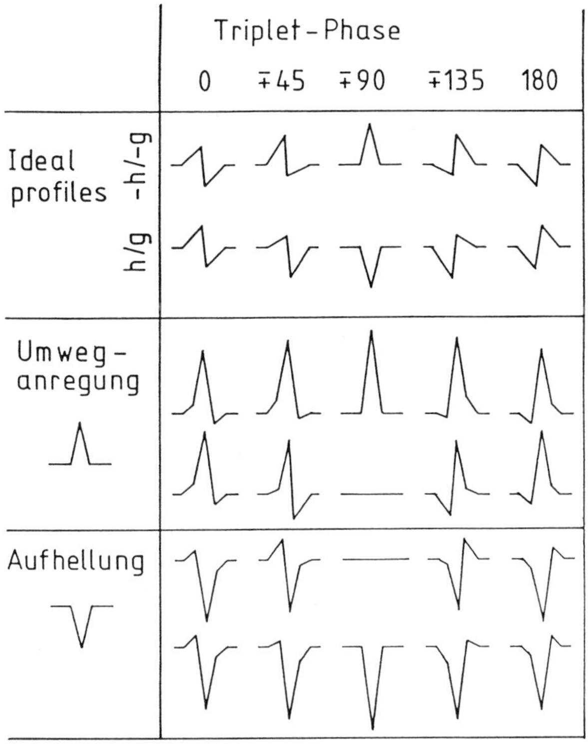

Figure 11. Schematic drawings of observed three-beam Ψ-scan profiles

to the interference effects then the evaluation of the phase relationships gets more and more difficult. Particularly, this happens if the primary reflection is weak. Then the phase information, if at all there is any, is contained in the asymmetry of the wings of an *Umweganregung* peak (Shen and Colella, 1988; Tang and Chang, 1988).

In Figure 11 a survey of observed couple of three-beam Ψ-scan profiles are given. To each of the ideal profiles a symmetrical *Umweganregung* or *Aufhellung* profile is added, respectively. This clearly demonstrates that the centers of the phase octants can be determined if the phase independent effects are comparably strong with the interference effects (Hümmer et al., 1990).

It should be pointed out that these phase independent effects are inherently contained in the solutions of the dynamical N-beam theory. Obviously, they depend on the ratio of the magnitudes of the structure factors $\mid F(\mathbf{h}) \mid$ and $\mid F(\mathbf{g})F(\mathbf{h}-\mathbf{g}) \mid$ which determine the amplitudes of the primary reflection and the $Umweg$ wave, respectively. As to theoretical results and our experimental experience (Weckert and Hümmer, 1990) the ratio

$$Q = \frac{\mid F'(\mathbf{g})F'(\mathbf{h}-\mathbf{g}) \mid}{\mid F'(\mathbf{h}) \mid^2} \tag{19}$$

should cover the range between $2 < Q < 6$ in order to keep *Umweganregung* and *Aufhellung* small provided that $|F'(\mathbf{g})|$ and $|F'(\mathbf{h}-\mathbf{g})|$ are also in the same order of magnitude. The F' are the structure factors corrected for the geometrical polarization factors. It has turned out that this rule-of-thumb is not independent of the absolute magnitudes of the structure factors compared to $F(\mathbf{0})$ (Weckert et al., 1993).

4. EXPERIMENTAL DETERMINATION OF TRIPLET PHASES

It has been shown, that the intrinsic dynamical width of the three-beam interference profiles is of the order of some arc seconds. It depends essentially on the relative magnitude of the structure factors, on the modulus of the triplet phase and on the diffraction geometry. The experimentally observed width is given by the convolution of the dynamical profiles with a broadening function which depends on the angular divergence and the spectral bandwidth of the incident beam as well as the mosaic spread of the investigated sample. These instrument and sample smearing effects reduce the interference contrast. Consequently, the use of a well collimated incident beam and crystals with high perfection is advantageous.

An other point to be regarded concerns the density of three-beam positions as the crystal is rotated about a scattering vector of the primary reflection. A rough estimation for a small molecule structure where the unit cell volume is about 600 Å^3 shows that for a reasonable length of $|\mathbf{h}| = 0.3\ \text{Å}^{-1}$ using $CuK\alpha$ radiation the mean angular distance between two three-beam positions is approximately $0.05°$ on the ψ-scale. As these are not equi-distant it is therefore possible for structures of that size to find for a selected three-beam case gaps in the sequence of three-beam positions by tuning the wavelength so that the angular distance to its neighbours is larger than $0.1°$. Then overlap of adjacent profiles can be avoided. This is shown in Figure 12 for a small molecule compound. Thus, tunability of wavelength is another requirement for triplet-phase data collection with regard to solution of unknown structures that needs a large number of triplet phases. The problem of inevitable overlap of three-beam reflections for macromolecular structures will be discussed later.

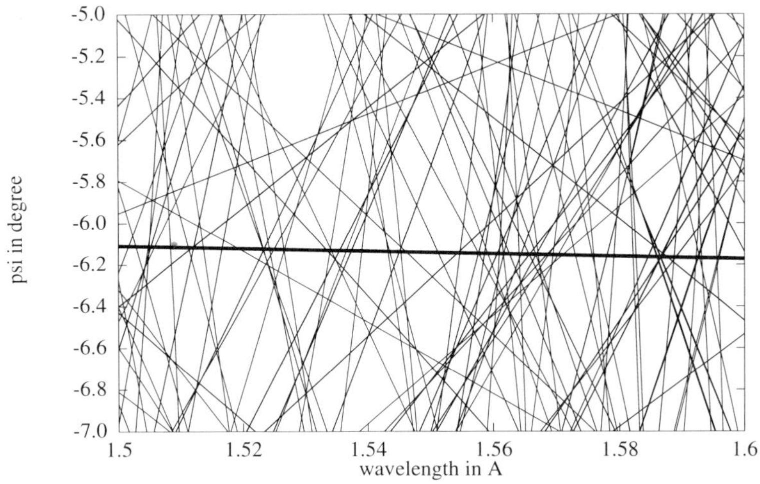

Figure 12. Three-beam positions in dependence on ψ and λ for L-asparagine monohydrate with $V_c = 646\text{Å}^3$. Each line represents the position of a three-beam case for the primary reflection 2 0 0. The thick line shows the position of the three-beam case 2 0 0/1 0 2/1 0 $\bar{2}$. There is a gap for $\lambda = 1.54\text{Å}$ of about $\pm 0.2°$, for instance.

4.1. X-ray Sources

The preceding considerations show that high brilliant sources should be used. Therefore, for home laboratory experiments a high brilliant rotating-anode equipment is advantageous. A rough estimate for crystals with dimensions of 0.3 mm using the bathed crystal technique, an effective source size of $0.3 \times 0.3\ mm^2$ and a source to crystal distance of 1 meter an appropriate divergence of 0.42 *mrad* ($\sim 0.024°$) is achieved.

The major advantages using synchrotron radiation (SR) are the possibility to tune the wavelength and several orders of magnitude higher brilliance comparing the characteristic line of a 6 kW/mm^2 rotating anode with SR from a bending magnet at the ESRF. Experiments were carried out with synchrotron radiation either from a bending magnet (BM) of DORIS (HASYLAB, Hamburg, Germany) or from a bending magnet of the ESRF (Swiss-Norwegian Beamline, Grenoble, France). Very commonly a fixed-exit double crystal Si (111) monochromator operating in vertical mode giving a bandwidth of approximately 0.01% to 0.03% is used. The selected wavelengths for our experiments cover the range from 0.5 to 2.5Å.

An indispensable demand on the quality of SR is beam position stability. For good quality crystals the typical FWHM of reflections is $0.005°$. That means that short term beam instabilities in the order of seconds lead to serious noise of the reference two-beam intensity. Long term instabilities can be averaged out, since in our measuring routines each ψ-scan profile is the sum of repeated fast scans with a typical measuring time of about 0.1 seconds per step. That is why monitoring of the long term decrease of the SR intensity due to the decay of the ring current is not necessary. First multiple-beam diffraction experiments at the ESRF have shown that beam stability is not a serious problem with this source and this should be the case for storage rings of the new generation.

4.2. The Ψ-Circle Diffractometer

Measurement of interference profiles does not only make high demand on the X-ray sources but also on the diffractometer. High precision of the angular resolution and of the ψ-scan accuracy is required. This means that the scattering vector of the basic reflection must not perform any staggering motion during the ψ scan, i.e. it has to be aligned very accurately with the ψ-axis so that it always lies exactly on the Ewald sphere. In our experimental experience exact ψ scans are difficult with a conventional four-circle diffractometer (but see also Mo et.al., 1988;). Therefore, a special ψ-circle diffractometer has been constructed (Figure 13). This

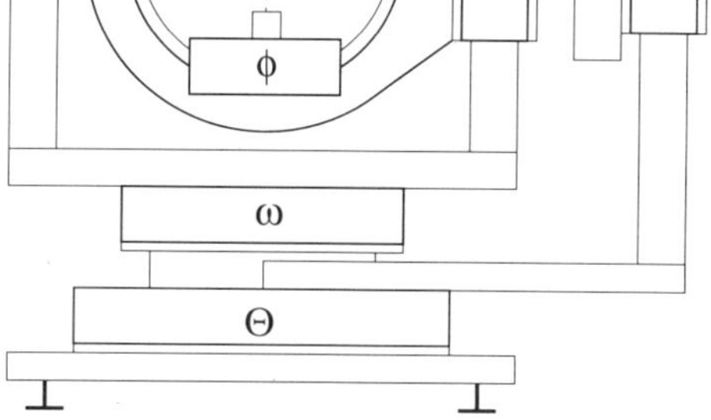

Figure 13. ψ-circle diffractometer

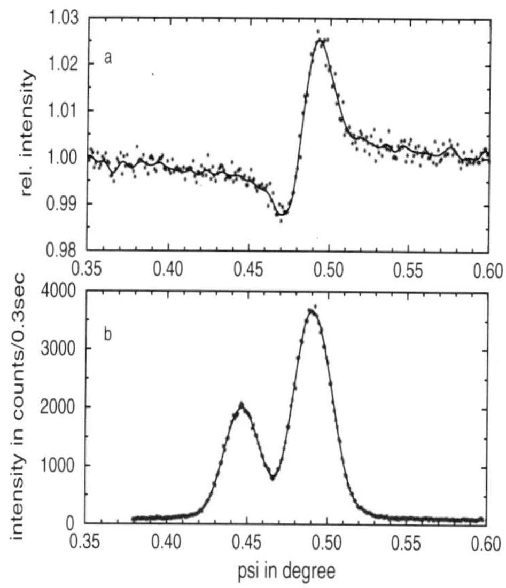

Figure 14. (a) Three-beam profile of the primary reflection (b) Two-beam profile of the secondary reflection

instrument contains two circles θ, ν for the detector with axes perpendicular to each other, and four circles for the crystal motion. The ψ axis, perpendicular to the ω axis, bears an Eulerian cradle with motions χ and φ. Thus an arbitrary scattering vector **h** can be aligned with the ψ axis and a ψ-scan is possible by moving only one circle. With the ν circle the detector can be moved to any position on a half sphere above the horizontal diffraction plane of the primary reflection. In this way the ψ angle for the three-beam position can be controlled measuring the second Bragg reflection by means of a ψ-scan about the primary reflection.

All circles are driven by stepper motors which are computer controlled. The angular resolution of the detector axes is $0.001°$, that of the crystal axes at least $0.0002°$.

4.3. Crystals

The crystals usually used for experimental phase determination have dimensions in the range from 0.1 to 0.5 mm. No special preparation technique is applied. They will be used as they are grown. Sometimes their habit was quite irregular. The investigated protein crystals have been mounted in closed capillaries together with some mother liquor so that they do not lose their solvent.

It should be mentioned that absorption in case of irregularly shaped crystals is not critical since the total $\psi-$rotation range for each three-beam setting is about $0.1°$. As already mentioned above the perfection of the crystals should be as high as possible. However, ideal perfection is not required. Most of the investigated crystals showed some mosaicity. It can be estimated from the width of two-beam profiles measured with a highly collimated beam. As a rule-of-thumb, if the FWHM exceeds $0.05°$ using SR of typically $0.01°$ divergence then it is difficult to measure any interference contrast.

4.4. Basics

In order to select suitable three-beam cases of a given crystal for the determination of triplet phases the following data are needed: *(i)* the metric parameters of the unit cell, *(ii)* a F_{obs} data set as complete as possible including also the low resolution reflections in case of protein crystals, *(iii)* the orientation parameters for each individual crystal.

Selected three-beam cases have to satisfy the following conditions: *(i)* $2 < Q < 6$ (*cf.* (19)) in order to avoid *Umweganregung* and *Aufhellung* effects, *(ii)* selection of a wavelength where the angular distance to nearest neighbours is large enough in order to avoid overlap of adjacent three-beam reflections.

It should be noticed, that by the use of a $\psi-$circle diffractometer any error with respect to the ψ rotation sense, i.e. whether it is an in-out or out-in ψ scan, is excluded, because it can be read directly from the rotation sense of the ψ axis. No further geometrical analysis is needed. All experimental profiles will be drawn with in-out rotation sense.

Figure 14 shows that the two-beam intensity of the primary reflection is modified due to the excitation of a secondary reflection. As an example, the three-beam case $0\,0\,0/\overline{2}\,0\,0/\overline{1}\,0\,\overline{2}$ of l-asparagine monohydrate is shown. In Figure 14a the three-beam interference profile of the primary $\overline{2}\,0\,0$-reflection is plotted. Figure 14b shows the two-beam intensity of the secondary $\overline{1}\,0\,\overline{2}$-reflection where the $Mo - K\alpha_1 - K\alpha_2$ splitting is clearly resolved. The intensities of both reflections are measured scanning ψ through the three-beam position and keeping the primary reflection in its diffraction position. They are plotted on the same angular scale for ψ. It is seen that interference takes place within the bandwidth accepted by the three-beam case, which in that case is the $MoK\alpha_1$ emission line of a rotating anode for which the three-beam case is aligned.

4.5. Quantitative Phase Determination

It has been already discussed that *Umweganregung* and *Aufhellung* effects may drastically influence the $\psi-$scan three-beam profiles and thus the indication with respect to the triplet phase involved. Therefore, these phase independent effects have to be subtracted from the measured profiles in order to get the pure phase dependent ideal interference profiles which then give an unambiguous phase indication. One possible way to evaluate these effects is to measure the two centrosymmetrically related three-beam cases **0/h/g** and **0/-h/-g** which ideally should show identical phase independent profiles since the they depend on the magnitudes of the structure factors. The triplet phases in both cases have different signs. Thus by calculating the $\Delta I(\psi)$ curve as defined in (18) it is possible to obtain the undisturbed interference profiles (ideal profiles). However, since in both measurements the crystal is differently orientated with respect to the incident beam and usually non-cut crystals with grown faces are used, the magnitude of the phase independent intensities as well as the interference contrast may be different. This considerably complicates the numerical evaluations. For the same reasons theoretical fitting of the profiles is only possible in exceptional cases where the geometry is accurately known. Therefore, as to our experience only the center of the phase octants of the triplet phases can be determined, i.e. generally, phase determination will be restricted to the values $0\ mod \pm 45°$, by means of comparison of the couple of measured profiles with the types given in the catalogue of Figure 11. However, this accuracy will be sufficient for structure determination.

5. APPLICATIONS

5.1. Determination of the Absolute Structure

For each non-centrosymmetric space group there are two enantiomorphic forms that can be mapped onto each other by a center of symmetry, i.e. the two forms differ in their handedness,

Figure 15. Measured three-beam Ψ-scan profiles of $C_{28}H_{20}N_2$ with an estimated triplet phase $\mp 110°$ (a) $0\,3\,2/\bar{2}\,2\,\bar{2}$, $\Phi_3^{calc} = -107.8°$ (b) $0\,\bar{3}\,\bar{2}/2\,\bar{2}\,2$;

which cannot be distinguished from ordinary X-ray diffraction patterns if Friedel's law is valid. This ambiguity has different meanings for different point groups. For enantiomorphic merohedral point groups its resolution means the determination of the absolute configuration for chiral species or the determination of the absolute conformation for achiral species. For polar point groups it means fixing the structure with respect to the polar direction. For non-centrosymmetric roto-inversional point groups, with roto-inversions $\overline{4}$ or $\overline{6}$, it means assignment of absolute axes, for example for ZnS-type structures (Burzlaff and Hümmer, 1988). Jones (1986) summarized the resolution of these ambiguities by the term: determination of the absolute structure. It ultimately reduces to a determination of the signs of structure-factor phases. One possible way for the determination of the absolute structure is to exploit the violation of Friedel's law due to anomalous scattering comparing the intensities of suitable Bijvoet pairs (Bijvoet et al., 1951). However, difficulties arise for light-atom structures. In contrast with this method, three-beam diffraction provides a means of resolving the enantiomorphism problem without the need of anomalous scattering (Hümmer and Weckert, 1995).

Applying the operation of inversion which maps the two enantiomorphic forms, let say *A* and *B*, onto each other, i.e. all the coordinates (x_j, y_j, z_j) are changed to $(-x_j, -y_j, -z_j)$, each structure factor phase changes its sign. Since a triplet phase is structure invariant, i.e. independent of the choice of the origin, the triplet phase of both enantiomorphs differ by their sign

$$\Phi_{3A} = -\Phi_{3B} \tag{20}$$

provided that the set of atomic coordinates refers to a coordinate system of the same hand. Therefore, best selectors for distinguishing *A* and *B* are triplet phases with $\Phi_3 = \pm 90^\circ$ or close to this value (Rogers, 1980). It should be pointed out again that relation (20) holds without anomalous scattering effects. Therefore, if the sign of a triplet phase with $\mid \Phi_3 \mid \simeq 90^\circ$ can be determined by three-beam diffraction, the absolute structure is unambiguously fixed. That is the distinct advantage of the three-beam method over anomalous dispersion methods in determinating of the absolute structure of light-atom compounds. An example where the determination of the absolute structure by means of anomalous dispersion effects would be very difficult if not impossible is given in Figure 15 for $C_{28}H_{20}N_2$ with space group $P2_12_12_1$.

5.2. Three-beam Diffraction of Non-Periodic Structures

There are classes of crystals, which are no longer translational periodic in three dimensions:incommensurately modulated structures and quasicrystals.

Incommensurately Modulated Structures. These structures show additional incommensurate modulation of some structural details, i.e. atomic position or occupancy. It is not possible to index their reflections with integers using a three-dimensional basis. According to de Wolf et al.(1981) in case of one-dimensional modulation all the diffraction vectors **h** can be indexed by four integers.

$$\mathbf{h} = h\mathbf{a}^* + k\mathbf{b}^* + l\mathbf{c}^* + m\mathbf{q} \tag{21}$$

with h, k, l, m integer and $\mathbf{q} = \alpha_1\mathbf{a}^* + \alpha_2\mathbf{b}^* + \alpha_3\mathbf{c}^*$ where at least one of the coefficients α_1, α_2 and α_3 has to be irrational. All the reflections with $m \neq 0$ are considered to be satellite reflections with respect to their main reflections hkl0. The satellite reflections correspond to the incommensurate modulation.

Three-beam interference experiments were carried out with incommensurately modulated calaverite ($AuTe_2$), one example is shown in Figure 16 with satellite reflections of the third order (m=3). Since the structure is supposed to be centrosymmetrical with superspace-group $P_{\overline{1}s}^{C2/m}$ (Schutte and de Boer, 1988) only triplet phases of 0 and 180° occur. All the measured triplet phases were in agreement with the phases calculated of the known structure model (de Boer, 1993). The mosaic spread of both crystals was about 0.01°. The experiments clearly show that in case of incommensurate structures triplet phases which include phases of satellite reflections can be determined provided that the crystal quality is sufficiently good.

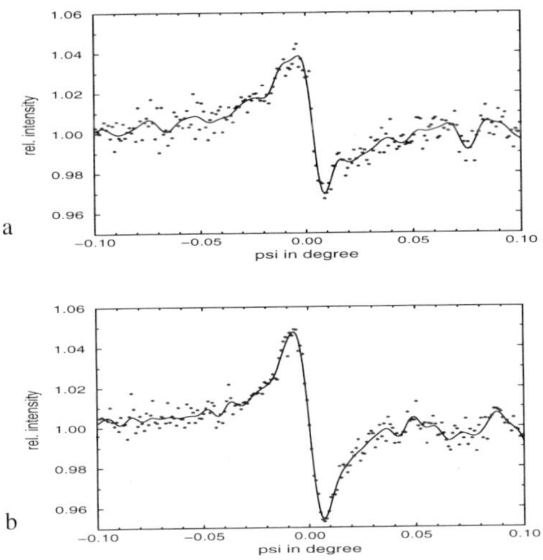

Figure 16. Measured three-beam Ψ-scan profiles of Calaverite ($AuTe_2$); $\Phi_3^{est} = 0°$; (a) **h/g/h-g**: $2\,6\,\overline{4}\,3/5\,5\,\overline{1}\,3/\overline{3}\,1\,\overline{3}\,0$, $\Phi_3^{calc} = 0°$; (b) **-h/-g/-(h-g)**

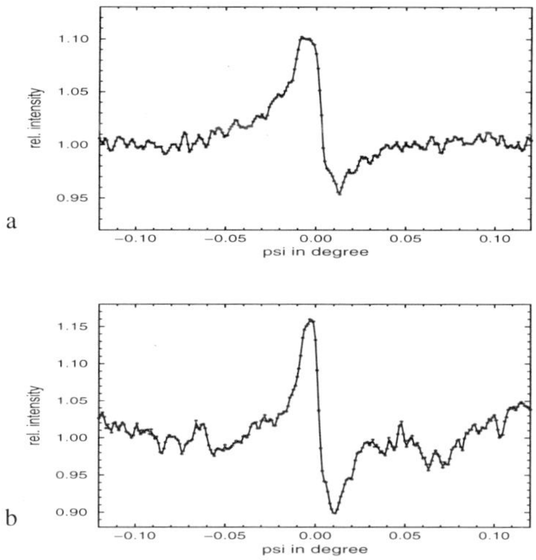

Figure 17. Measured three-beam Ψ-scan profiles of $Al_{70}Pd_{20}Mn_{10}$; $\Phi_3^{est} = 0°$; (a) **h/g/h-g**: $\overline{1}\,3\,1\,\overline{3}\,\overline{3}\,1/\overline{1}\,\overline{3}\,\overline{1}\,1\,1\,\overline{1}/0\,6\,2\,\overline{4}\,\overline{4}\,2$ (b) **-h/-g/-(h-g)**

Quasicrystals. Icosahedral quasicrystals show no translation periodicity at all in three dimensions. They can be described periodically in six dimensions, that means six indices are needed to index the reflections with integers. In Figure 17 examples of three-beam interference profiles of the icosahedral phase $Al_{70}Pd_{20}Mn_{10}$ are shown. The presence of interference effects is obvious. The indices for the three-beam case given in the caption refer to a six-dimensional coordinate system with the axis in the directions of the fivefold axis of the icosahedron (Cahn et al., 1986). All the measured three-beam interference profiles indicate either triplet phases close to 0 or 180°. This is a strong indication for the centrosymmetric point group: $m\,\overline{3}\,\overline{5}$. However, a final decision on the basis of the available triplet-phase data is not possible.

5.3. Direct Phasing of Macromolecular Structures

Experimental phase determination of protein crystals by means of three-beam diffraction reveals some specific differences compared to small molecule structures. This concerns the weak scattering power, the very dense reciprocal lattice and therefore unavoidable overlapping of multiple-beam interference profiles and for nearly all structures radiation damage. These problems will be discussed in the following.

Weak Scattering Power. The magnitude of structure factors for macromolecular structures even of strong reflections are small compared to the total number of electrons in the unit cell. In case of tetragonal lysozyme for example the strongest structure factors of low resolution reflections are about 3 to 5% of the maximum scattering power. This number has to be compared with about 30 to 50% for small organic compounds and sometimes even more for inorganic compounds. The extinction length in case of protein crystals is therefore in the or-

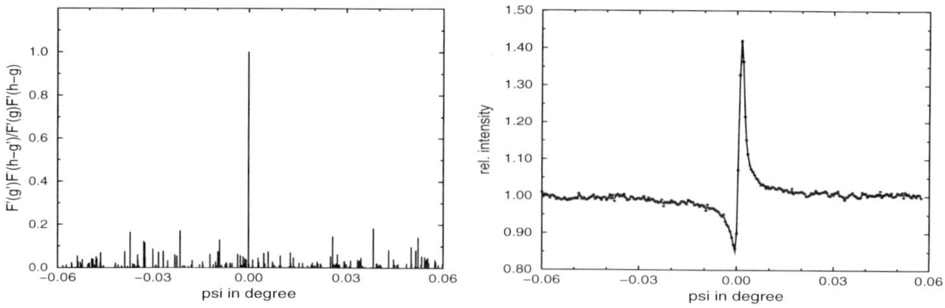

Figure 18. Three-beam case $\overline{7}\,4\,0/\overline{5}\,2\,1$ for tetragonal lysozyme at λ = 1.3302 Å; left: Distribution of the q-values of neighbours. Only three-beam cases with reflections up to 2Å resolution are involved; right: measured interference profile.

der of millimeters. In this case diffraction is considered to behave kinematically for crystal or mosaic blocks of usual sizes of some tenth of a millimeter. Kinematic scattering in this context means that $I(\mathrm{h}) \propto |F(\mathrm{h})|^2$. The reflectivity or transmissivity in this range is far below the maximal possible values for an arbitrary thick perfect crystal (Weckert et al., 1993). Nevertheless three-beam interference effects occur as it is shown theoretically and experimentally. (Hümmer et al., 1991; Weckert et al., 1993). The advantage in this regime is that the interference effects are independent of the diffraction geometry, i.e. independent on whether it is a primary Laue or Bragg case (*cf.* Figures 8 and 9). As shown below for protein crystals only three-beam cases where reflections with strong structure factors are involved can be used for phase determination. These are usually reflections at small diffraction angles where Bragg case reflection is hardly possible.

The modulation of the intensity of the primary reflection due to three-beam interference is normally in the range of 2 to 20%. In order to obtain a reliable estimate for the triplet phase the interference profiles have to be integrated up to a sufficient statistics. Since, additionally, a high number of triplet-phase data are necessary for structure solution a high brilliant synchrotron radiation source is needed to accomplish this in reasonable time.

Pseudo Three-Beam Diffraction. A serious problem for structures with large unit cells as proteins is the very crowded reciprocal lattice. Therefore, with finite experimental resolution it is not possible to excite a single three-beam case. It has been found, however, that in spite of the high number of overlapping multi-beam cases the exploitation of triplet phases is possible (Hümmer et al., 1991; Weckert et al., 1993). Configurations can be found at a specific wavelength for which there are only two strong reflections h and g simultaneously excited besides other weaker reflections with their reciprocal-lattice vectors g′ inevitably at or near that diffraction position The contributions of these weaker reflections to the interference profile of the 'strong' three-beam case if also the coupling reflection $\mathrm{h}-\mathrm{g}$ is comparably strong, do not disturb the estimation of the triplet phase. The weaker reflections produce a sort of noise on the recorded profile. Their influence is negligible if the amplitude of their *Umweg* wave governed approximately by $|F'(\mathrm{g}')\,F'(\mathrm{h}-\mathrm{g}')|$ is weaker than that of the strong **g**-reflection. As a rule found experimentally, this is the case if

$$q = \frac{|F'(\mathrm{g}')\,F'(\mathrm{h}-\mathrm{g}')|}{|F'(\mathrm{g})\,F'(\mathrm{h}-\mathrm{g})|} \lesssim 0.25. \tag{22}$$

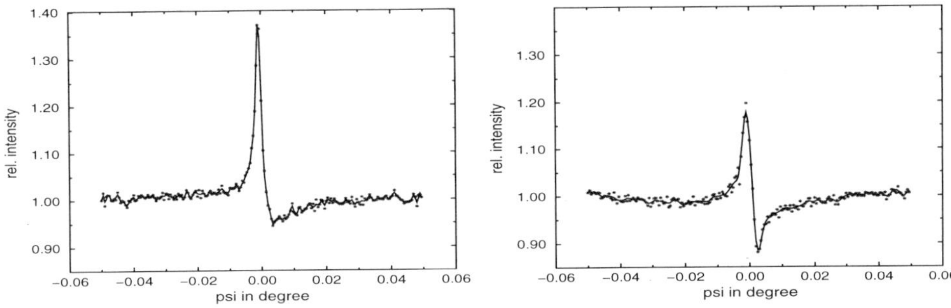

Figure 19. Measured three-beam Ψ-scan profiles of tetragonal lysozyme; estimated triplet phase $\mp 45^{\circ}$ (a) $4\ 15\ \bar{4}/2\ \bar{1}\ 0$, $\Phi_3^{calc} = -35^{\circ}$ (b) $\bar{4}\ \overline{15}\ 4/\bar{2}\ 1\ 0$

The $F'(\mathrm{h})$ mean structure factors corrected for polarization. By tuning the wavelength it is possible to find gaps so that the profiles of the strong three-beam cases $(q > 0.25)$ do not overlap. Such a situation is shown in Figure 18 where the distribution of the q-values of neigbouring three-beam cases for a selected strong one is plotted. Its experimental interference profile is shown on the right. As can be seen, the weak neighbouring three-beam cases do not seriously affect the interference profile of the strong one.

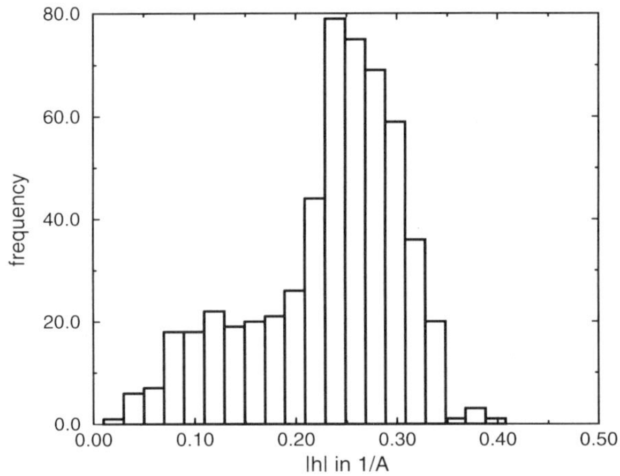

Figure 20. Distribution of resolution of triplet reflections measured of tetragonal lysozyme.

An advantage of the over-all weak scattering power of proteins is that *Aufhellung* effects are smaller as extinction is small (Weckert et al., 1993). The disadvantage is that in order to obey condition (22) only reflections with large structure factors are suitable for phase determination.

Triplet-Phase Data Collection. Due to the stable beam conditions at the ESRF it is possible to measure 4 to 6 three-beam profiles within 1 hour. A further example for lysozyme is shown in Figure 19. This is a typical profile with *Umweganregung* for a triplet phase of $\mp 45^{\circ}$ according to the catalogue of Figure 11. In total about 600 triplet phases have been measured of tetragonal hen-egg lysozyme having a mean phase error of about 20° compared to the phases calculated from the known structure (entry 1lse of PDB, Kurinov and Harrison, 1995). The distribution of resolution of the 550 different single reflections involved are plotted in Figure 20. It can be seen that most reflections belong to the range of 4Å resolution. Because of the reasons discussed above the number of reflections drastically decays with increasing resolution.

A possible way to solve macromolecular structures where direct methods do not work without further experimental information, could be to combine measured triplet phases with statistical methods like direct and maximum entropy methods. This work is in progress.

6. REFERENCES

Bethe , H. A., 1928, Theorie der Beugung von Elektronen an Kristallen, *Ann. Phys.(Leipzig)* 87:55.

Bijvoet , J. M., Peerdeman , A. F., and Bommel van, A. F., 1951, Determination of the absolute configuration of optically active compounds by means of X-ray, *Nature (London)* 168:271.

Boer de, J. L., 1993, private communication.

Burzlaff , H. and Hümmer , K., 1988, On the ambiguities in merohedral crystal structures, *Acta Cryst.* A44:506.

Cahn , J. W., Shechtman , D., and Gratias , D., 1986, Indexing of icosahedral quasiperiodic crystals, *J.Mater.Res.* 1:13.

Chang , S. L., 1982, Direct determination of X-ray reflection phases, *Phys.Rev.Lett.* 48:163.

Chang , S. L., 1984, *Multiple Diffraction of X-rays in Crystals*, Springer Verlag, Berlin, Heidelberg, New York.

Colella , R., 1974, Multiple diffraction of x-rays and the phase problem. computational procedures and comparison with experiment, *Acta Cryst.* A30:413.

Ewald , P., 1917, Zur Begründung der Kristalloptik. III. Röntgenstrahlen, *Ann. Phys. (Leipzig)* 54:519.

Ewald , P. P., 1965, Crystal optics for visible light and X-rays, *Acta Cryst.* 37:46.

Ewald , P. P. and Heno , Y., 1968, X-ray diffraction in the case of three strong rays. I. Crystal composed of non-absorbing point atoms, *Acta Cryst.* A24:5.

Hoier , R. and Marthinsen , K., 1983, Effective structure factors in many-beam X-ray diffraction - use of the second Bethe approximation, *Acta Cryst.* A39:854.

Hümmer , K. and Billy , H., 1986, Experimental determination of triplet phase and enantiomorphs of non-centrosymmetric structures. I. Theoretical considerations, *Acta Cryst* A42:127.

Hümmer , K., Schwegle , W., and Weckert , E., 1991, A feasibility study of experimental triplet-phase determination in small proteins, *Acta Cryst.* A47:60.

Hümmer , K. and Weckert , E., 1995, Enantiomorphism and three-beam X-ray diffraction: determination of the absolute structure, *Acta Cryst.* A51:431.

Hümmer , K., Weckert , E., and Bondza , H., 1990, Direct measurements of triplet phase relationships of organic non-centrosymmetric structures using synchrotron radiation, *Acta Cryst.* A46:393.

Jones , P. G., 1986, The determination of absolute structure. III. An ambiguity table for the non-centrosymmetric crystal classes, *Acta Cryst.* A42:57.

Juretschke , H. J., 1982a, Invariant-phase information of X-ray structure factors in the two-beam Bragg intensity near a three-beam point, *Phys.Rev.Lett.* 48:1487.

Juretschke , H. J., 1982b, Non-centrosymmetric effects in the integrated two-beam Bragg intensity near a three-beam point, *Phys.Lett.* A92:183.

Juretschke , H. J., 1984, Modified two-beam description of X-ray fields and intensities near a three-beam diffraction point. general formulation and first-order solution, *Acta Cryst.* A40:379.

Kurinov , I. V. and Harrison , R. W., 1995, The influence of temperature on lysozyme crystals, structure and dynamics of protein and water, *Acta Cryst.* D51:98.

Laue von, M., 1960, *Röntgenstrahl-Interferenzen*, Akademische Verlagsgesellschaft m. b. H., Frankfurt/Main.

Lipscomb , W. N., 1949, Relative phases of diffraction maxima by multiple reflection, *Acta Cryst.* 2:193.

Mo , F., Hauback , B. C., and Thorkildsen , G., 1988, Physical estimation of X-ray triplet phases in a centrosymmetric, mosaic crysal with unit cell volume $\sim$ 3000 $Å^3$, *Acta Chem. Scand.* A42:130.

Moon , R. M. and Shull , C. G., 1964, The effects of simultaneous reflection on single-crystal neutron diffraction intensities, *Acta Cryst.* A17:805.

Renninger , M., 1937, Röntgenometrische Beiträge zur Kenntnis der Ladungsverteilung im Diamantgitter, *Z. Kristallogr.* 97:107.

Rogers , D., 1980, Definition of origin and enantiomorph and calculation of $|E|$ values, in:*Theory and Practice of Direct Methods in Crystallography*, M. F. C. Ladd and R. A. Palmer, ed., Plenum Press, New York.

Schutte , W. J. and Boer de, J. L., 1988, Valence fluctuations in the incommensurately modulated structure of calaverite $AuTe_2$, *Acta Cryst.* B44:486.

Shen , Q. and Colella , R., 1988, Phase observation in an organic crystal (benzil: $C_{14}H_{10}O_2$) using long-wavelength X-rays, *Acta Cryst.* A44:17.

Tang , M. T. and Chang , S. L., 1988, Quantitative determination of phases of X-ray reflections from three-beam diffractions. II. Experiments for perfect crystals, *Acta Cryst.* A44:1073.

Thorkildsen , G., 1987, Three-beam diffraction in a finite perfect crystal, *Acta Cryst.* A43:361.

Weckert , E. and Hümmer , K., 1990, On the quantitative determination of triplet phases by X-ray three beam diffraction, *Acta Cryst.* A46:387.

Weckert , E., Schwegle , W., and Hümmer , K., 1993, Direct phasing of macromolecular structures by three-beam diffraction, *Proc. R. Soc. Lond. A* 442:33.

Wolf de, P. M., Janssen , T., and Janner , A., 1981, The superspace groups for incommensurate crystal structures with a one-dimensional modulation, *Acta Cryst.* A37:625.

Zachariasen , W. H., 1967, A general theory of X-ray diffraction in crystals, *Acta Cryst.* 23:558.

X-RAY AND NEUTRON INTERFEROMETRY: BASIC PRINCIPLES AND APPLICATIONS

R. Colella

Department of Physics
Purdue University
West Lafayette, IN 47907 U.S.A.

1. GENERAL PRINCIPLES

Interferometry is based on the idea of creating spatially separated coherent beams, which can be brought together and produce interference effects. Optical interferometry is a well established field. Coherent beams can be realized by using half-reflecting glass slides, and easily manipulated and directed along assigned directions by mirrors. The same optical elements cannot be used in the X-ray region, and this is the reason why we had to wait until 1965 before we could see an X-ray interferometer in operation (Bonse and Hart, 1965a; 1965b).

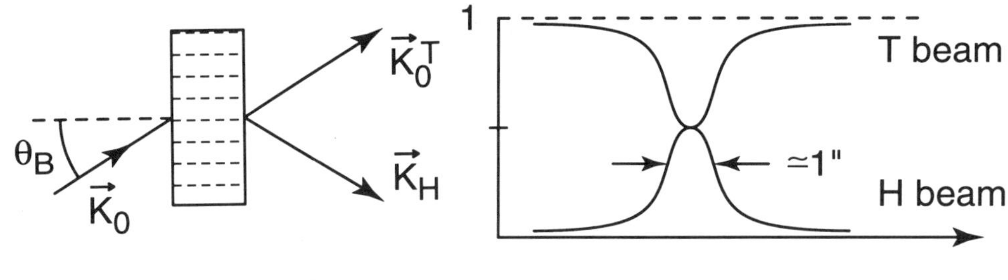

(a) (b)

Figure 1. a) Laue case diffraction. The lattice planes are perpendicular to the slab. b) Intensities of the T and H beams *vs.* θ, the angle of incidence.

A key ingredient in the first X-ray interferometer developed by Bonse and Hart (1965a) was *diffraction in the Laue case* (Figure 1). On the right part of the figure we see the rocking curves, as given by dynamical theory, for the transmitted and diffracted beam (the T and H beam, respectively), under the assumption of zero absorption, and of a perfectly parallel incident beam $\mathbf{K_0}$. While the H beam has a peak, the T beam

X-ray and Neutron Dynamical Diffraction: Theory and Applications
Edited by Authier *et al.*, Plenum Press, New York, 1996

has a dip. Their intensities always add up to 1, because of zero absorption (a realistic assumption in the case of neutrons in silicon). Note that the width of both peaks is extremely narrow, typically of the order of 1 arcsecond.

It is intuitively clear that both T and H beams are *coherent*. Their exact phase relationship can be derived from dynamical theory.

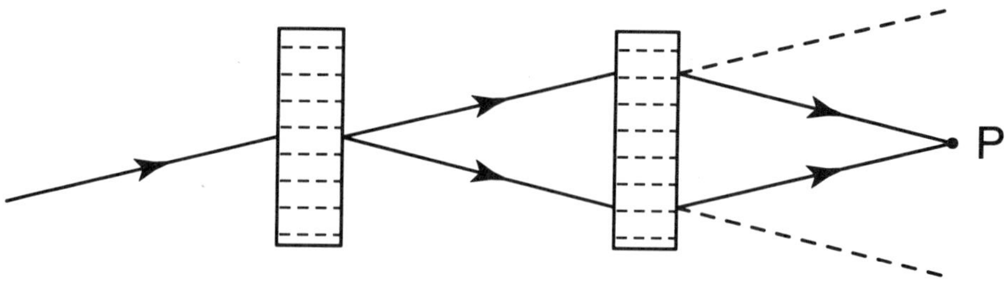

Figure 2. Two identical crystal slabs set up for Laue diffraction.

Now, let us combine two identical crystals set for Laue case, as in Figure 2. We would expect to observe interference effects at point P, where the two beams cross each other on the right side. Actually, no interference effects can be observed in this way. There are two reasons. The first one is that the two crystals should be aligned with angstrom accuracy. This is because the X-rays are scattered by *atoms*, which must be in perfect registry when the same photon jumps from the first to the second crystal. The second problem is that, even if we were able to align the two crystals with angstrom accuracy, there would be no way to observe the interference fringes generated in the volume region of the two intersecting beams, because they would be too close, about 1 Å apart, to be resolved on a photographic film.

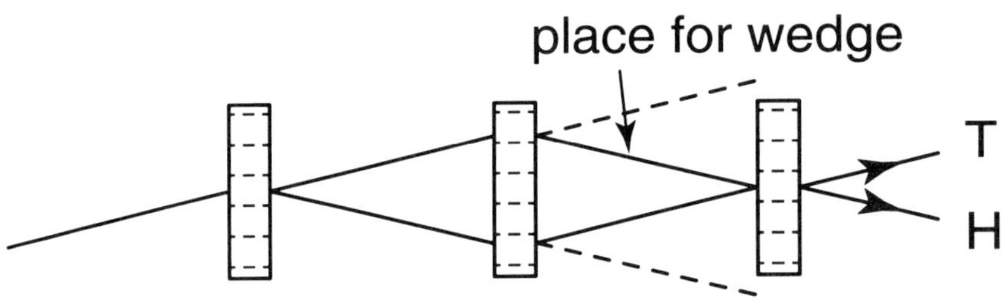

Figure 3. Schematics of the Bonse-Hart interferometer (top view). The three slabs are part of the same monolithic block.

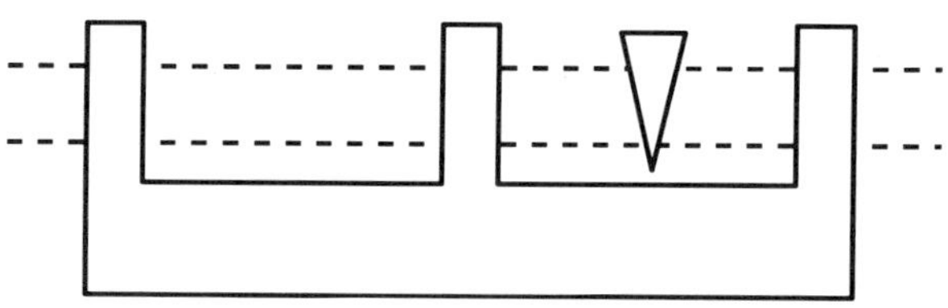

Figure 4. Side view of the Bonse-Hart interferometer. A plastic wedge is inserted on one leg of the interferometer in order to introduce a variable phase shift along a direction normal to the base.

Both difficulties were overcome in a single stroke of genius by Bonse and Hart (1965a) who carved the first X-ray interferometer out of a monolithic silicon block, and added a third slab, identical and parallel to the first two. (Figures 3 and 4). The third crystal acts as a "receiver". It mixes two coherent inputs, the two incident beams to be diffracted by the same lattice planes, and produces two output beams (T and H in Figure 3), which are expected to be sensitive to the phase relationships between the two incident beams. A convenient way to observe interference effects is to insert a lucite wedge along one of the beams (Figures 3 and 4) which introduces a variable optical path difference across the X-ray beam, and produces a series of black and white fringes, visible on a photographic film. (Figure 5).

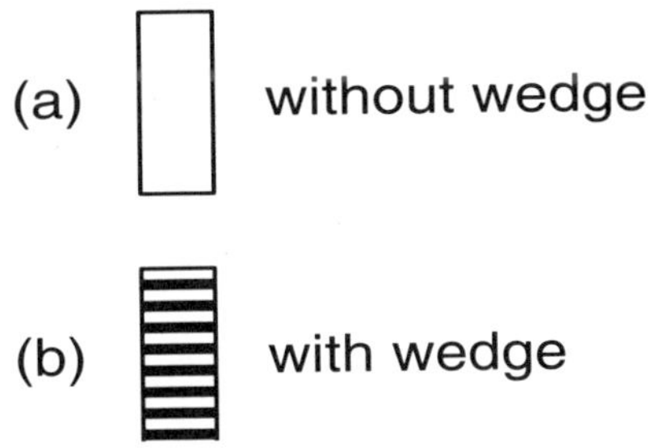

Figure 5. Images produced by the X-ray beam on a film placed behind the third slab of the interferometer. The region corresponding to part(a) is uniform because the relative phases of the two interfering beams are the same over the whole cross sections of the beams.

Other subtleties are to noted. The effect of surface roughness, for example. In optical interferometry the rule to follow is that every optical element, mirrors and beam splitters, must be polished to an accuracy of a fraction of the wavelength, say, 5000 angstroms. There is no way to accomplish this result in the X-ray region. What Bonse and Hart realized is that this condition is not needed in their interferometer. Here the X-rays are scattered by *atoms*, not by surfaces. The effect of a surface step (Figure 6) whose thickness is Δt is to introduce a phase shift given by $(2\pi/\lambda)\Delta t(n-1)$ which is quite small in view of the fact that $(n-1)$ is of the order of 10^{-6}. It is easy to verify that surface steps of a few microns do not appreciably spoil the coherence of a beam.

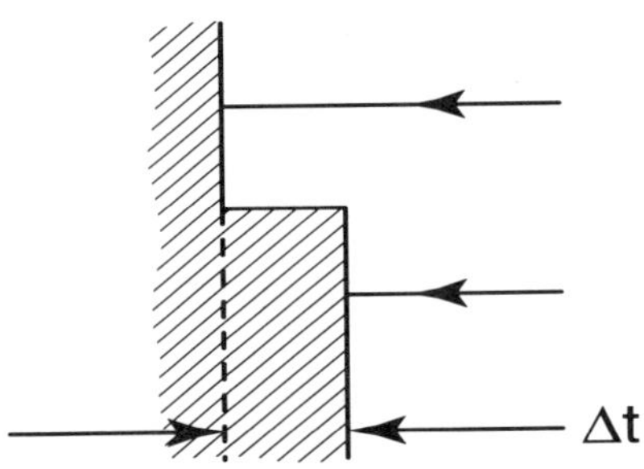

Figure 6. Effect of a surface step on coherence.

Another source of concern is the monochromaticity of a beam. To produce interference effects in the third slab of the interferometer (Figure 3), the X-ray photons must meet exactly at the entrance surface. This requires that the two distances between the central and the lateral slabs be exactly equal. How exactly? The answer to this question involves a discussion of the longitudinal and transverse coherence length of the X-ray photons reaching the third slab. The longitudinal coherence length of the neutron wave packet (Figure 7) is given by $\lambda^2/\Delta\lambda$. It turns out that it is not important to use a highly monochromatic X-ray source. If a white source were used, a multiplicity of Bragg angles would be observed, and the single rays drawn in Figure 3 would be replaced by fans. Figure 8 shows two trajectories, for two distinct monochromatic beams. However, *each photon interferes only with itself.* Therefore, what counts is *the* $\Delta\lambda$ *allowed along each trajectory.* This is controlled by the width of the rocking curves in Figure 1 (b), which is very small. It turns out that for typical reflections in silicon, the longitudinal coherence length of the X-ray photons amounts to 10–20 μm, which represents the tolerance with which the two distances between slabs must be equal. The transverse coherence length is controlled by the divergence of the beams between slabs. It is given by $(0.2\lambda/\theta)$ (for 10% dephasing) where θ is the divergence of the beams.

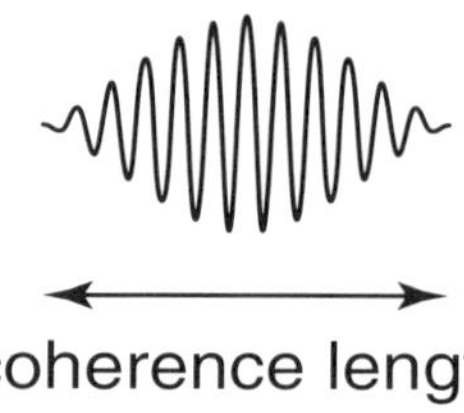

Figure 7. Wave packet produced by lack of monochromaticity in the incident X-ray beam.

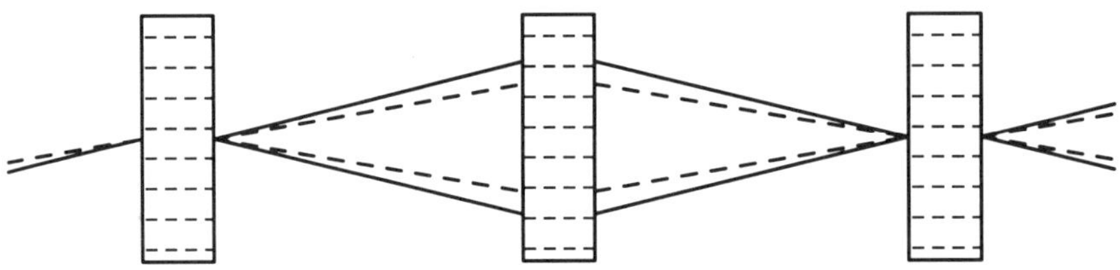

Figure 8. Situation of a non-monochromatic incident beam. Two monochromatic components are present.

For a 1 Å X-ray beam, with a 2 arcsecs divergence, the transverse coherence length is about 4 μm, which is again within the reach of machining tolerances.

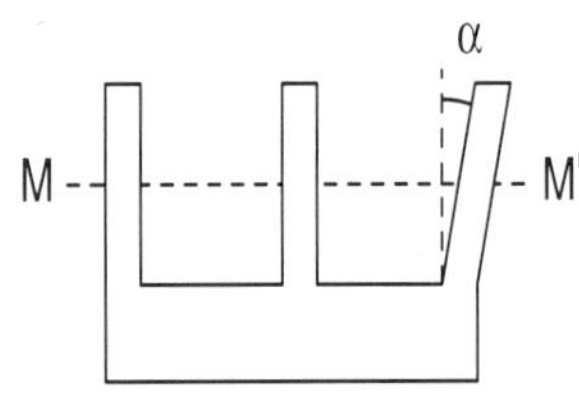

Figure 9. Faulty interferometer. The third slab has been cut on purpose at an angle α with respect to the first two slabs. It shows the effect of different distances between the slabs.

All these effects have been carefully analyzed by Bonse and te Kaat (1971), who analyzed the effect of introducing a difference in the two distances, between first and second slab, and between second and third slab. To this effect, they used an interferometer in which the third slab was oblique, forming an angle α with the first two (Figure 9). They found that, in this situation, no uniform image is obtained, like in Figure 5 (part (a)). Instead, a complicated set of fringes in form of hyperbolae is obtained. When a wedge is interposed, like in Figure 4, a new source of fringes is introduced. The result is a beating effect between fringes of different origin and periodicity, whose spacing becomes infinitesimal at some distance from the center (line MM' in Figure 9), where the two slabs are equally spaced. Interference fringes are only visible in the central region. This gradual transition from widely spaced fringes in the central region, to infinitesimally spaced fringes away from the central region, offers an excellent clarification of the physical meaning of "*loss of coherence*" when geometrical errors are present.

2. APPLICATIONS TO PHYSICAL PROBLEMS

Neutron interferometry was first demonstrated by Rauch, Treimer and Bonse in 1974. They used the same scheme described in Figure 3. They showed that interference fringes could be observed in form of intensity oscillations when the thickness of low absorption materials, such as aluminum, was varied along one of the interfering beams. A curious phenomenon was observed. While in the X-ray case the fringe pattern (Figure 5) is the same for both, the T and H beam of Figure 3, the patterns are *complementary* in the case of neutrons. When the T beam has a maximum the H beam has a minimum, and *vice versa*. The neutron result is obvious. Since absorption is virtually zero, neutrons must be conserved. What about the X-ray case? It turns out that, with $Mo - K\alpha$ radiation ($\lambda = 0.711$ Å) and 1 mm slabs, absorption is important, and therefore X-ray photons do not need to be conserved. Rigorous calculations based on dynamical theory have confirmed this interpretation (Colella, 1975).

Neutrons are much slower particles than photons. Moreover, they carry a finite mass. It is expected that neutron interferometry may lead to unusual and interesting new experiments. Such has indeed been the case.

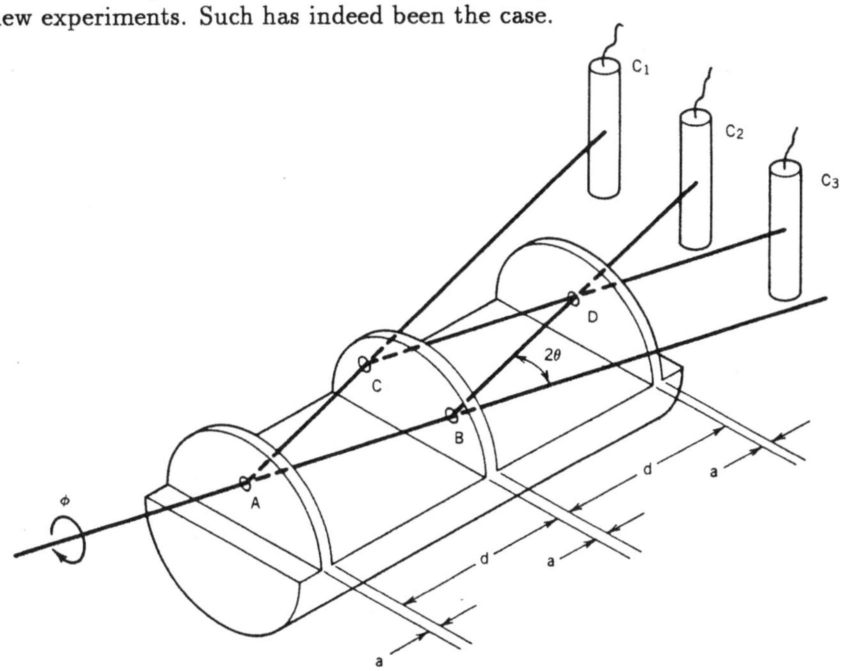

Figure 10. Schematic diagram of the neutron interferometer and ^{3}He detectors used in the gravity experiment. From: (Colella, Overhauser and Werner, 1975).

The first experiment we will discuss is the observation of gravitationally induced quantum interference (Colella, Overhauser and Werner, 1975). In this experiment no wedges are used. The interferometer is simply rotated around the incident beam, so as to preserve the Bragg condition always satisfied (Figure 10). Rotating the interferometer is like rotating the incident beam around itself. Nothing unusual is expected. But neutrons have mass, and gravity will affect differently the two legs of the interferometer (paths ACD and ABD in Figure 10) when the plane ACDB is tilted with respect to gravity. The difference in average elevations, measured from the center of the Earth, between the two interfering paths, will introduce different phase shifts, which may cause periodic intensity oscillations as a function of the angle ϕ in Figure 10. In other words, the existence of a gravitational acceleration is equivalent, for thermal neutrons, to a transparent medium with a gradient in the index of refraction, for light photons.. The index of refraction for neutrons in a gravitational field is in fact given by $n = (1 - mgh/E)^{1/2}$ where m is the neutron mass, g is the gravitational acceleration, and h is the distance of the neutron from the center of the Earth (or from some other reference point), and E is the neutron energy.

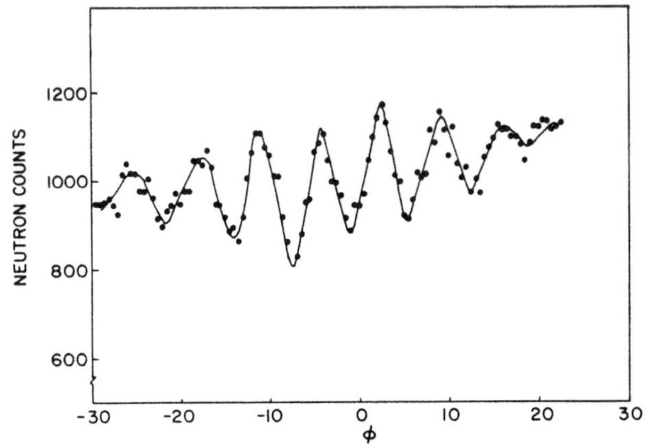

Figure 11. Laue case Intensity oscillations in the neutron counting rate as a function of the interferometer rotation angle ϕ. From: (Colella, Overhauser and Werner, 1975).

The effect has indeed been observed. (see Figure 11). The phase shift predicted is given by:

$$\beta_g + 2\pi\ m_i\ m_g\ \sin\phi\ \lambda\ A\ g/h^2 \tag{1}$$

where m_i and m_g are the inertial and gravitational neutron masses, respectively, ϕ is the rotation angle, λ is the neutron wavelength, A is the area defined by the four legs of the interferometer, g is the gravitational acceleration, and h is Planck's constant. It is noteworthy to observe, from Eq. (1), that this is the only experiment in physics, so far, whose outcome depends on Planck's constant *and* the gravitational acceleration.

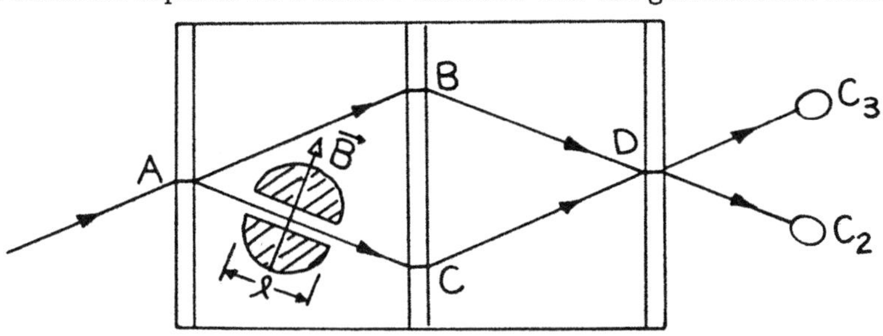

Figure 12. A schematic diagram of the neutron interferometer. On the path AC the neutrons are in a magnetic field B (0 to 500 G) for a distance l (2 cm). From: (Werner *et al.*, 1975).

At about the same time another experiment was done to verify an important prediction of quantum mechanics. Neutrons, like all fermions (particles with half-integer spin) are expected to exhibit a sign change in their wavefunction under a 2π rotation

(Bohm, 1951). In other words, they do not look the same to an observer whose frame of reference has been rotated by 2π. A sign change, however, is a subtle effect. It is a *phase* change. It is normally unnoticed in all experiments which are not sensitive to phases, that is to say, all experiments in which *intensities* are measured, with loss of information about phases. A convenient scheme to produce a 2π rotation is to insert a small magnet along one of the two interfering beams (Figure 12). The magnet can be adjusted so as to produce a magnetic field of controlled intensity in the gap shown in Figure 12. Neutrons travelling in a magnetic field are forced to rotate, as a result of Larmor precession due to their magnetic moment. The length of the path in the magnetic region and the strength of the field can be related in such a way as to produce one, two, three or more rotations. When the neutrons recombine on the third slab, a change in sign on one of the two neutron beams will result in a minimum of intensity. It turns out that over a path length of 20 mm a field of about 20 gauss is sufficient to produce a 2π rotation. This effect has been quantitatively verified (Figure 13). (Werner *et al.*, 1975; Rauch *et al.*, 1975).

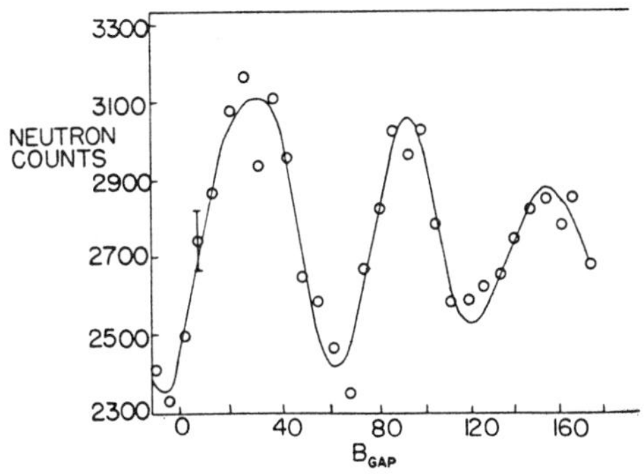

Figure 13. The neutron counting rate as a function of the magnetic field in the magnet air gap in gauss. Approximate counting time was 40 min per point. From: (Werner *et al.*, 1975).

The last experiment we want to discuss in this lecture is the neutron Sagnac effect due to the Earth's rotation.

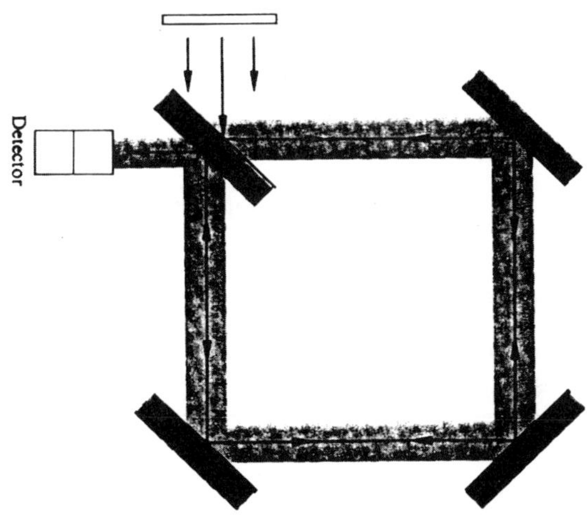

Figure 14. Schematic diagram of the Sagnac interferometer. The four mirrors are set into rotatory motion, while the source and the observation point are at rest. From: (Hecht, 1987).

The "Sagnac effect" is the fringe shift observed in a rotating ring interferometer (Hecht, 1987). In this device, depicted in Figure 14, the light photons travel along two counterpropagating paths. Actually, the two beam traveling in opposite directions along each leg of the interferometer originate from the same beam at the point where the incident beam goes through the beam splitter. When the whole apparatus is set into rotatory motion, a difference in optical path results between the two counterpropagating beams, producing a set of fringes visible by an observer. The fractional displacement of fringes is given by:

$$\Delta N = \frac{4A\omega}{c\lambda} \tag{2}$$

where A is the are of the ring, ω is the angular velocity of rotation, c is the speed of light, and λ is the wavelength of the light. The effect has been observed by Sagnac (1913) and it has found practical applications today in inertial navigation devices, as a replacement or in addition to gyroscopes.

The same kind of experiment was repeated on a grandiose scale by A. Michelson (1925), who kept the mirrors stationary, at distances of several hundred meters apart, with the rotatory motion being provided by the Earth's rotation. It is intuitively clear that the Sagnac effect is much more conspicuous with thermal neutrons than photons, since their velocity, close to the speed of sound, is orders of magnitudes less than that of photons. It turns out that in the gravity experiment (Colella, Overhauser and Werner, 1975) the Sagnac effect was a 2% correction needed for reconciling the observed fringe pattern with Eq. (1). If neutrons are viewed as massive particles rather than waves, the physical origin of the Sagnac effect for neutrons, if it exists, can be explained by considering the Coriolis acceleration arising as a result of the Earth's rotation. It is the same effect which would induce a slight eastern deviation on a neutron beam, a well known effect in ballistics.

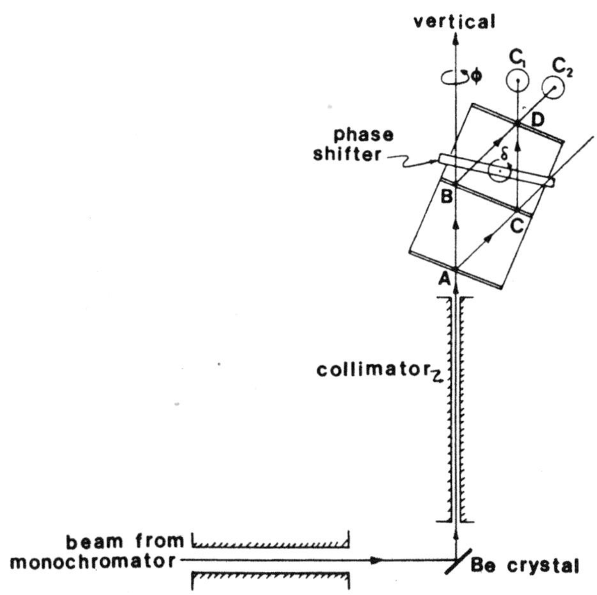

Figure 15. Schematic diagram of the apparatus. The drawing is not to scale. The collimator is approximately 1 m in length and the interferometer is approximately 8 cm long from point A to point D. The angle δ of the phase-shifting slab is zero when it is parallel to the three interferometer slabs. From: (Werner, Staudenmann and Colella, 1979) .

Neutrons have of course all the properties of classical massive particles. The gravitational fall due to parabolic trajectory has been experimentally observed (McReynolds, 1951; Dobbs *et al.*, 1965). The vertical drop over a path of 100 m amounts to 12.3 mm for 2 Å neutrons, and the eastern deviation for the same neutron would be 0.028 mm, too small to be observed. While the vertical drop has been experimentally observed, the eastern deviation has not been reported in either (McReynolds, 1951) or (Dobbs *et al.*, 1965), probably because of experimental difficulties.

It turns out that the phase effect due to the Coriolis force is more experimentally accessible than the eastern deviation. In our case, however, the effect was essentially masked by the overwhelming gravity effect. A 2% effect in Eq. 1 was within the error bars in the gravity experiment.

The solution to this difficulty was to suppress altogether the gravity effect by using a vertical neutron beam (Figure 15). In this way the gravity effect is the same for any angle ϕ of the interferometer around the incident neutron beam (Werner, Staudenmann and Colella, 1979).

The Hamiltonian of the neutron has a term which gives rise to the Coriolis force of the form: $\omega \cdot \mathbf{L}$ where ω is the angular rotation velocity of Earth and $\mathbf{L}$ is the angular momentum of the neutron's motion about the center of Earth. Since the angular momentum $\mathbf{L}$ is horizontal, it is the horizontal component of ω that is utilized in this experiment. Figure 15 shows the presence of a phase shifting slab, whose effective thickness is different for the two interfering beams BD and CD, in such a way that the relative phase shifts introduced in the two beams can be varied by turning the slab around an axis perpendicular to the diffraction plane of the interferometer by an angle δ.

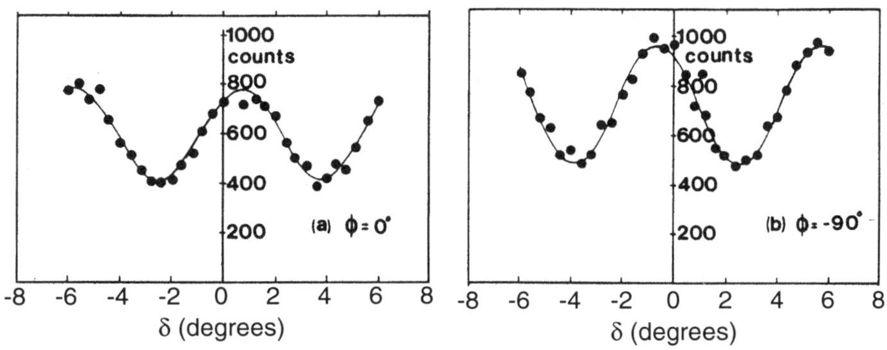

Figure 16. Typical oscillating counting rates observed in detector C_1 at two orientation settings ϕ of the interferometer. The counting time for each datum point was approximately 600 sec. From: (Werner, Staudenmann and Colella, 1979).

If the Earth were not rotating, the oscillations observed in counters C_1 and C_2 as a function of δ would be identical, irrespective of the rotation of the interferometer around the vertical neutron beam. In effect this is not the case. We see in Figure 16 that a phase shift is introduced when the angle ϕ is changed from $-90°$ to $0°$. A plot of the observed phase shift *vs.* ϕ is shown in Figure 17.

It shows, as expected, that the effect vanishes when the vector $\hat{n}$ normal to the parallelogram ABCD in Figure 15 is directed to west, and it reaches extreme values when $\hat{n}$ is parallel or anti-parallel to the horizontal component of **omega**, the angular rotation velocity of Earth. To our knowledge, this is the only verification of the existence of a Coriolis force on neutrons, a force arising from relative motion, not from masses. It is another case in which neutron interferometry has allowed us to get results that could not be obtained by any other means.

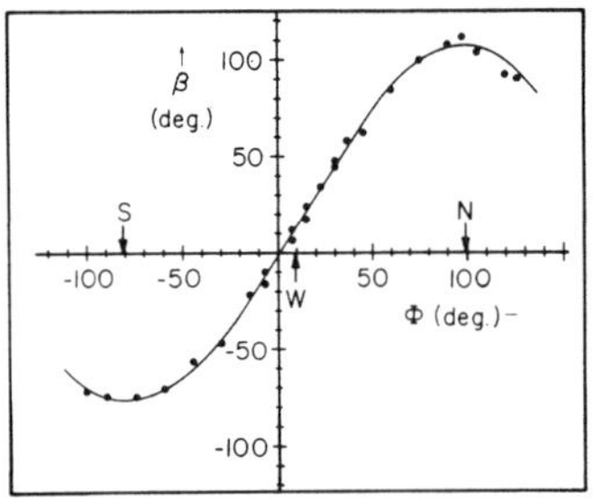

Figure 17. A plot of the phase shift β due to Earth's rotation as a function of the orientation ϕ of the normal area **A** of the interferometer about a vertical axis. The symbols N, W, and S indicate north, west, and south. From: (Werner, Staudenmann and Colella, 1979).

3. PRECISION MEASUREMENT OF NEUTRON SCATTERING LENGTHS

The index of refraction of a neutron in a medium whose scattering length is b, is given by:

$$n = 1 - \frac{\lambda^2 N b}{2\pi} \tag{3}$$

where N is the number of nuclei per unit volume and b is the coherent scattering length. If a material of varying thickness, for example, a wedge, is inserted in one of the interfering beams, a fringe pattern is produced, whose spacing is directly correlated with the index of refraction of the material. While angular deviations of neutrons due to refraction are difficult to measure because they are very small – the index of refraction n given by Eq. (3) differs from 1 by a number of the order of 10^{-6} – an interferometer can provide a precise determination of n because fringe spacings can be measured accurately.

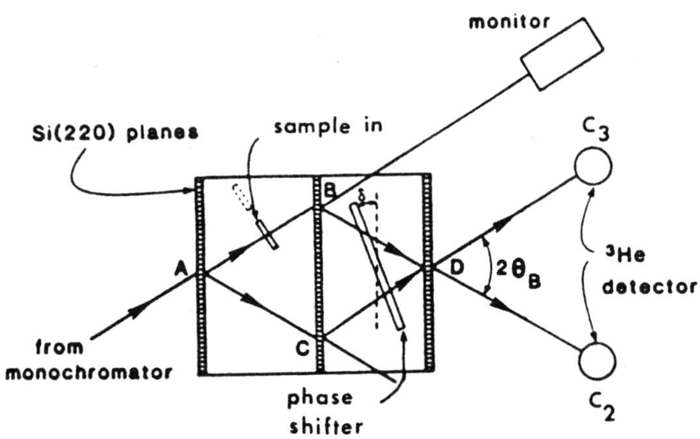

Figure 18. Experimental geometry to measure the phase shift. From: (Arif *et al.*, 1987).

This technique has been used to measure the coherent scattering length of ^{235}U, the isotope used for producing a chain reaction by nuclear fission, in an energy region in which theoretical and experimental data are not very reliable (Arif *et al.*, 1987). The experimental set-up is shown in Figure 18. In the case of neutron interferometry it is more practical and precise to measure intensity oscillations in the counting rate of detectors rather than take photographs. The absorption of most materials is virtually zero, therefore the undesirable loss of contrast in fringe patterns observed with X-rays, when slabs are used, is absent here. In practice, one single aluminum slab, accurately machined to a thickness of 10 mm, was used to vary the optical paths along both beams, through a rotation δ around an axis perpendicular to the base of the interferometer. A thin ^{235}U foil, about 0.5 mm thick, was inserted along one of the beams before recombination (Figure 18) and the oscillation plots (intensity *vs.* δ) were compared with the sample in and out. A typical experimental result is shown in Figure 19.

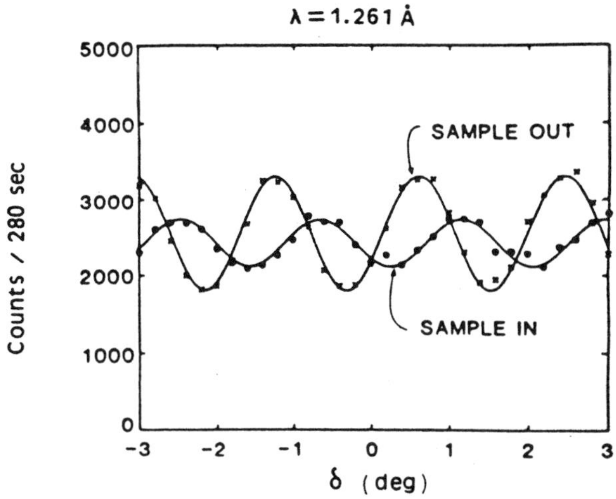

Figure 19. Typical data for experiments carried out for sample in (out) geometry as illustrated in Figure 18. From: (Arif *et al.*, 1987).

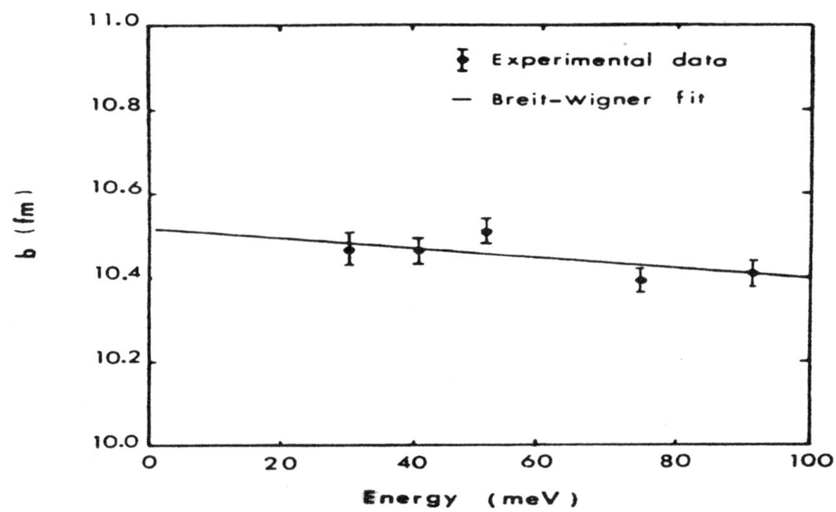

Figure 20. Comparison of the measured scattering length values (in fermis) with the calculated values from the Breit-Wigner formalism. The solid line represents the calculated values. From: (Arif *et al.*, 1987).

The phase shift between the two profiles is directly related to the scattering length in the following way:

$$b = -\frac{\Delta\Phi}{\lambda NT} \tag{4}$$

where $\Delta\Phi$ is the observed phase shift, λ is the neutron wavelength, N is the number of nuclei per unit volume, and T is the thickness of the uranium foil. The measurements were repeated for several neutron energies, in a region not easily accessible by the other techniques. The data are shown in Figure 20. It is interesting to see how well the experimental data fit the theoretical prediction. In this energy region the only available experimental value was at around 25 meV, with an error bar about ten times greater than those shown in Figure 20. It is clear that with that kind of error any attempt to correlate theory with experiment was meaningless.

This example shows that neutron interferometry provides values for coherent scattering lengths about ten times more accurate than those presently available in the low energy region.

ACKNOWLEDGEMENTS. This work has been supported by NSF grant 9301004-DMR.

REFERENCES

Arif, M., Kaiser, H., Werner, S.A., and Willis, J.O., 1987, Precision measurement of the bound-coherent-neutron scattering length of ^{235}U, *Phys. Rev. A* 35:2810.

Bohm, D., 1951, *Quantum Theory*, Prentice Hall Inc., Englewood Cliffs, New Jersey, Sect. 17–7.

Bonse, U., and Hart, M., 1965a, An X-ray interferometer, *Appl. Phys. Lett.* 6:155.

Bonse, U., and Hart, M., 1965b, An X-ray interferometer with long separated interfering beam paths, *Appl. Phys. Lett.* 7:99.

Bonse, U., and te Kaat, E. 1971, The defocussed X-ray interferometer, *Z. Physik* 243:14.

Colella, R., 1975, unpublished.

Colella, R., Overhauser, A.W., and Werner, S.A., 1975, Observation of gravitationally induced quantum interference, *Phys. Rev. Lett.* 34:1472.

Dobbs, J.W.T., Harvey, J.A., Paya, D., and Horstmann, H., 1965, Gravitational acceleration of free neutrons, *Phys. Rev. B* 139:756.

Hecht, E., 1987, *Optics*, Addison Wesley, Reading, Massachusetts, Sect. 9–4.

McReynolds, A.W., 1951, Gravitational acceleration of neutrons, *Phys. Rev.* 83:172.

Michelson, A.A., Gale, H.G., and Pearson, F., 1925, The effect of the Earth's rotation on the velocicty of light, *Astrophysics Journal* 61:140.

Rauch, H., Treimer, W., and Bonse, U., 1974, Test of a single crystal neutron interferometer, *Phys. Lett. A* 47:369.

Rauch, H., Zeilinger, A., Badurek, G., Wilfing, A., Bauspiess, W., and Bonse, U., 1975, Verification of coherent spinor rotation of fermions, *Phys. Lett. A* 54:425.

Sagnac, M.G., 1913, L'éther lumineux démontré par l'effet du vent relatif d'éther dans un interféromètre en rotation uniforme, *C. R. Acad. Sci. Paris,* 157:708.

Werner, S.A., Colella, R., Overhauser A.W., and Eagen, C.F., 1975, Observation of the phase shift of a neutron due to precession in a magnetic field, *Phys. Rev. Lett.* 35:1053.

Werner, S.A., Staudenmann, J.L., and Colella, R., 1979, Effect of Earth's rotation on the quantum mechanical phase of the neutron, *Phys. Rev. Lett.* 42:1103.

APPLICATIONS OF X-RAY INTERFEROMETRY

D.K. Bowen

Centre for Nanotechnology and Microengineering
University of Warwick
Coventry CV4 7AL, UK

1. INTRODUCTION

Optical interferometry has been a mainstay of precision instrumentation since the time of Newton. Its achievements are remarkable. Various instruments have been built with resolutions less than a nanometre, ranges of tens or hundreds of metres, speeds suitable for machine tools, with miniaturisation, portability, simplicity of operation and so on. At first sight there seems little to add. However, there are two fundamental limitations of optical interferometry. One is the wavelength; if sub-nanometre lengths are to be measured, the fringe division is of order $\lambda/1000$. Whilst this is certainly possible, it is much harder to be sure of the linearity and calibration of such extreme division. The second is the optical properties of air, whose refractive index varies with temperature, pressure and humidity and, thanks to turbulence, is never constant in any ordinary environment. Many ingenious methods have been designed to work round this problem, such as common-path interferometers, dual-wavelength interferometers and weather stations for measuring all the parameters influencing the optical properties. Nevertheless, it is evidently attractive to consider an X-ray method, in which the fundamental wavelength is sub-nanometre and for which air properties and turbulence have negligible effect. Moreover, novel aspects of X-ray and materials physics become accessible through the medium of X-ray interferometry.

Several problems had to be overcome before X-ray interferometry became possible. The two critical physical ideas were

X-ray and Neutron Dynamical Diffraction: Theory and Applications
Edited by Authier *et al.*, Plenum Press, New York, 1996

[1] That beam splitters and reflectors could be realised by means of dynamical X-ray diffraction on thin perfect crystals, and

[2] That the X-ray interference pattern could be interrogated by interference with another thin, perfect crystal.

The critical engineering idea was that controlled motions at the 0.01 nm level could be achieved by use of elastic flexure design applied to single crystals of silicon.

2. DEVELOPMENT OF X-RAY INTERFEROMETRY

Bonse and Hart (1969a) were the first to realise a practical X-ray interferometer. The X-ray optics are illustrated in Figure 1. It is the equivalent of a Mach-Zender optical interferometer. The first blade acts as a beam splitter, and we note that it is essential that both the diffracted and the forward-diffracted beams are in a phase relationship. Hence there must be sufficient thickness of crystal for the dynamical amplitudes to become established - kinematic interferometers will not work. The second crystal acts as a mirror. Two of the four beams arising from the second crystal will recombine to form a standing wave field which will have the periodicity of the crystal lattice. If the third crystal be placed at this point and translated through this standing wave field, the standing electron wave field otherwise known as the crystal lattice will interfere with the X-ray wave field. If the third crystal be translated, the resulting diffracted and forward-diffracted waves will oscillate in intensity with the period of the crystal lattice.

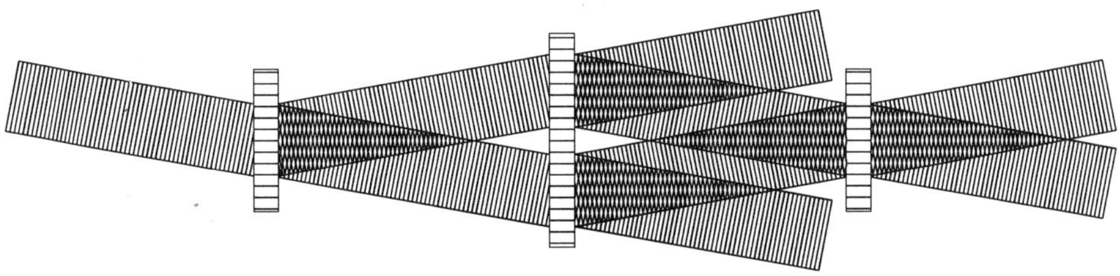

Figure 1. The X-ray optics of the Bonse-Hart interferometer, drawn to scale so that the standing waves (horizontal interference patterns) are visible.

The practical realisation of this simple optical principle is not trivial. The blades have to be held in alignment to nanoradians, relative vibration of more than picometres must be suppressed at all frequencies, and a drive capable of resolution and noise in the picometre regime must be provided. Bonse and Hart solved these problems with a classically simple concept, that of cutting the whole interferometer and its mechanism out of a monolithic single crystal of silicon - a method that is still preferred today for most applications. This preserves the intrinsic alignment of the crystal, provides sufficient stiffness between components that relative vibration is not difficult to suppress, and through the use of a spring flexure design, provides the 'slideway' for the translation. The drive method itself was a simple coil driving a magnet glued to the moving component. This is still used in many modern designs, though piezoelectric drives - which have higher noise but also higher stiffness - are also now used. The design is shown schematically in Figure 2.

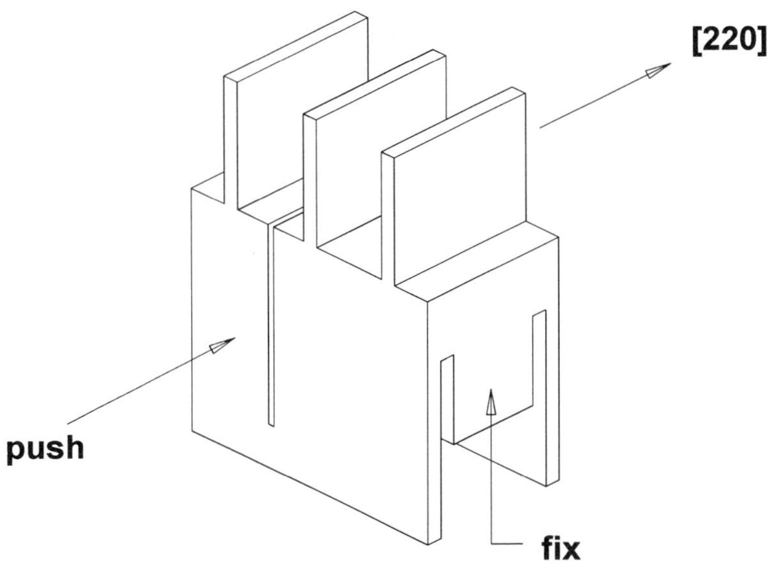

Figure 2. The monolithic design of the Bonse-Hart interferometer.

3. THEORY OF X-RAY INTERFEROMETRY

We shall briefly summarise the plane-wave description, originally given by Bonse and Hart (1969b) and Bonse and Graeff (1977), using concepts and formalisms developed earlier in this book. Consider the beam paths given in Figure 3.

The dynamical theory is applied straightforwardly to each crystal in turn. A great simplification is obtained by assuming a strong Borrmann effect, that is, that only Bloch waves arising from the low-absorption, σ polarisation branch of the dispersion surface will survive. This is reasonable for the common practical case of silicon crystals of about 1 mm thickness, with say MoK_α radiation, but we note that it is never the case for practical neutron interferometers. The amplitudes and phases of the exit beams of the first crystal are obtained by applying the dynamical theory to the entrance and exit surfaces together with the

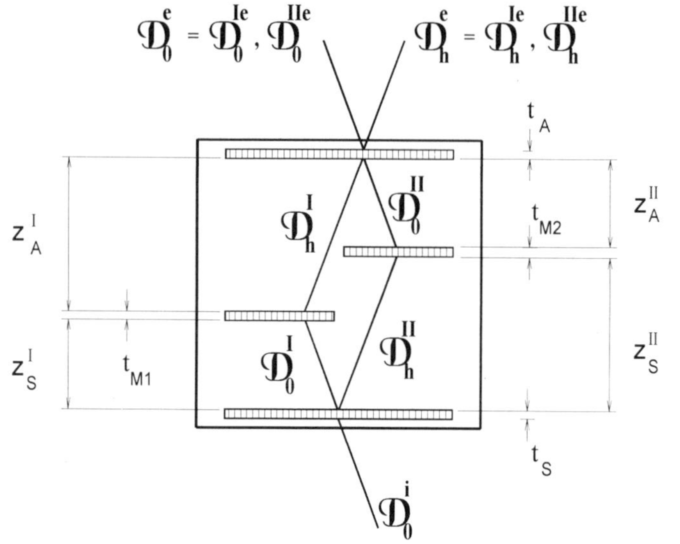

Figure 3. Beam paths in the X-ray interferometer. The general skew-symmetric case is shown; for the more usual symmetric case all beam path branches are made equal.

phase shift introduced by the crystal thickness. The outputs of the first crystal are the inputs for the second crystal, at which each path is treated separately. Phase shifts are introduced by the air gap and by the crystal thickness, which on the X-ray wavelength scale are each necessarily different for each branch. The outputs of the second crystal are the inputs for the third crystal, but now the two sides are recombined. The final amplitude ratio of the components of the two output beams is:

$$\mathrm{D}_0^{Ie} / \mathrm{D}_0^{IIe} = \mathrm{D}_h^{Ie} / \mathrm{D}_h^{IIe} = \exp\left\{-2\pi\,\mathrm{i}\,k\left[\delta_2'\left(z_A^I - z_S^{II}\right) + \delta_2\left(z_S^I - z_A^{II}\right)\right]\right\} \tag{1}$$

Clearly, any phase shifting object introduced into any of the paths will affect the measured intensity of either of the output beams, as will scanning of the third crystal in the direction normal to its diffracting planes. The contrast with respect to phase shifting of the two output beams is subtly different, however. If we use the symbol *t* to denote a forward-diffracted (transmitted) beam and *h* to denote a diffracted beam then the D_0^e beam is composed of *thh* + *hht* paths whereas the D_h^e beam is composed of *tht* + *hhh* paths. The balanced composition of the former results in higher contrast. Theoretically it is possible to obtain complete extinction with the correct phase shift but this is not observed in practice.

The remaining theory required is the examination of the practical realization of the X-ray interferometry with imperfect machining of the crystals and spherical wave rather than plane wave sources.

The effect of roughness of the crystal surfaces will be to introduce phase shifts owing to the different crystal thicknesses traversed by different parts of the beam. If we allow a maximum phase variation of 2π/10 to occur, then from [1],

$$1/10 = k\left[\delta_2'\left(z_A^I - z_S^{II}\right) + \delta_2\left(z_S^I - z_A^{II}\right)\right] \qquad [2]$$

The δ terms are of order 10^{-5} to 10^{-6}, so a roughness of 10^4 - 10^5 wavelengths or 1 - 10 μm may be tolerated. This is quite easy to achieve with optical and chemical polishing methods.

The effect of manufacturing tolerances on blade thickness and position will be to reduce the coherent overlap of the beams at the third crystal. Fortunately the conditions are not too severe. Figure 4 shows the geometry. Since X-ray rocking curves are very narrow and the Borrmann effect is very strong, the divergence of the incident radiation that can arrive at the exit surface of the interferometer is only about 1 arc second. Therefore, any point of the entrance surface sees only a small region of the focus, of order 2.5 μm with typical parameters. The Fresnel zone from a source this small, using CuK_α radiation, is about 30 μm diameter (using a source distance of 500 mm), and any region of the entrance surface is therefore illuminated with a highly coherent beam. Furthermore, the slight divergence of the incident beam of 1 arc second is magnified to about 4° by the effect of the Borrmann fan. The coherent region is thus spread to about 130 μm at the exit surface. Thus with quite easy machining, to a few μm, it is straightforward to ensure sufficient overlap of the coherent regions. The coherency condition is identical with the focussing condition.

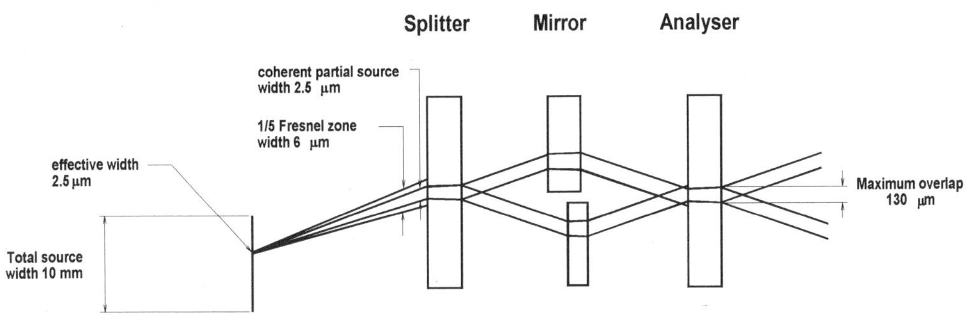

Figure 4. The coherency overlap in the LLL interferometer.

With sufficiently ideal geometry, the plane wave description is adequate, since each pair of interfering plane waves of the spectrum has the same amplitude ratio [1] and therefore the linear sum of plane waves that comprises a spherical wave will also have the same wavefield beam in the splitter, but since the phase varies only slowly across the Borrmann fan this is a minor effect. However, if the geometry is off ideal, as is deliberately so for the angular interferometer described in section 7.1, the (directional) δ parameters in [1] have a strong influence. A full spherical wave treatment must then be undertaken to calculate intensities and phase relationships in the exit beams.

4. DESIGN AND MANUFACTURE OF X-RAY INTERFEROMETERS

4.1. Basic designs and tolerances

The interferometer illustrated so far uses three transmission or Laue-case elements and is denoted a two-beam LLL interferometer. It has the excellent property of being non-dispersive. Any wavelength within the diffracting range will work, which makes the alignment of such interferometers very easy, typically within a few minutes of arc. Many other geometries are possible if one allows Bragg reflections and many-beam conditions, and one suitable for phase-contrast microscopy is discussed later. The small refractive index shift on Bragg reflection must be taken into account if Bragg and Laue-case reflections are mixed. Many-beam interferometers may be geometrically very simple but they are dispersive, with working wavelengths selected by the crystal and interferometer geometry. Only by chance will these wavelengths fall on characteristic X-ray lines, so these devices are much more suitable for use at synchrotron radiation sources. However, it should also be noted that many-beam interferometers are much more difficult to align, since the interfering beams must be very close in wavelength. This implies that at least one axis must be aligned within one second of arc. Intensity calculations are also much more complex and should use the many-beam theory discussed elsewhere in this book.

The static interferometer is one in which the diffracting elements are all fixed in relation to one another. In a scanning interferometer, the analyser (usually) crystal may move with respect to the other elements, introducing a phase shift of π for every shift of half an atomic plane. In the static interferometer, the phase shifting must be introduced by other means such as a wedge of material or, with neutrons, the gravitational potential. Scanning interferometers are much more versatile and provide greater possibilities for application, but, on the other hand, they are much more complicated to manufacture and they are more prone to instability against vibration.

The theory of the last section shows that surfaces must be smooth to microns and mechanical tolerances must be within 10 - 20 μm. In a LLL interferometer, the greatest problems lie in the uniformity of the central (mirror) blade and flaws such as grooves left by diamond saws will reduce the contrast. If the very best contrast is desired then the three blades must be as uniform in thickness and as smooth as possible. This can be achieved with polishing jigs and considerable care and patience (Krylova, 1993).

4.2. Materials and processing

Low defect density (nominally zero dislocation density) is essential to preserve the amplitudes of the dynamically-propagating wave fields in all the blades. The first and most subsequent interferometers have been made out of silicon. For X-rays this is the obvious choice. It has low X-ray absorption, is available in large crystals with very low defect density and is both brittle and strong at room temperature. If grown dislocation-free it therefore remains so. The contrast of the fringes depends amongst other factors upon the uniformity of the crystal lattice spacing. Float-zone silicon has a significantly lower oxygen content than Czochralski-pulled crystal and is much preferable for interferometers, since the higher oxygen content normally leads to fluctuations or lattice parameter. Germanium has often been used for neutron interferometers, because of its more suitable neutron cross-section, and it also can be be made very pure and homogeneous. Quartz interferometers have also been constructed.

The simplest method of shaping is to use a diamond saw. Working interferometers have been made using quite simple machines, but for the best accuracy and uniformity of the finished blades a number of refinements are possible. These are:

1. Use of a specialised machine with high precision and stiffness on the slideways
2. Use of the thickest possible blades to reduce vibration at the cutting surface
3. Balancing of the spindles and blades
4. Filtering of the coolant to, say, 0.3 μm to remove particles that can scratch the blade surfaces.

Before sawing, the crystal must be accurately oriented, and the Bond (1976) method, illustrated in Figure 5, is normally used. The crystal orientation is adjusted until the diffracted X-ray intensity is constant as the cylindrical jig carrying the crystal is rotated on an engineering V-block. The diffracting plane is then normal to the cylinder axis. The jig is then transferred to a V-block mounted perpendicular to a diamond sawing blade and the cut is made. Using standard workshop techniques, precisions of 0.1° are easily attained.

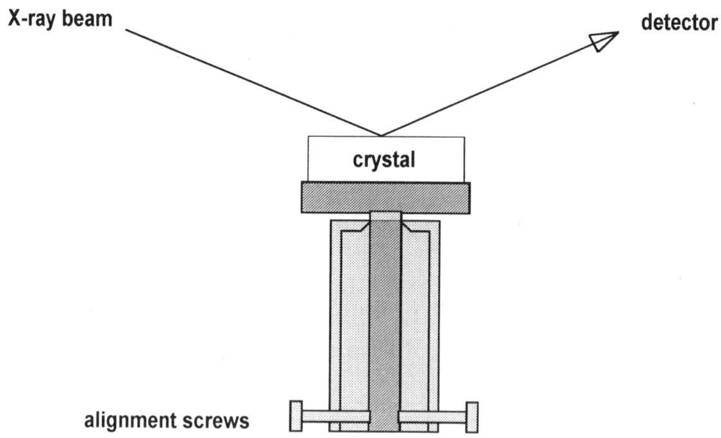

Figure 5. The Bond method for crystal orientation.

Spring flexure elements may also be cut with a saw as in the original design (Figure 1). However, there is more scope available in design if milling operations can be performed. A high speed axle and a diamond burr are required. Air turbine heads are available that fit onto standard machines, and provide up to 200,000 rpm. Diamond burrs with tungsten carbide shanks for added stiffness are used, and a CNC machine is highly desirable since cuts are of the order of 25 μm and much repetition is required to make a deep cut. The interferometer shown in Figure 6 can be cut by this method.

The final stage is etching, to remove strain and to polish the surface. The bulk of the strain in sawn silicon is removed after a few microns etching, but we normally remove about 100 μm since the interferometer is so sensitive to strain. The solution is typically concentrated HF:HNO_3 with ratios between 1:20 and 1:10 (the etching is faster with more HF but less controllable). To make the etching uniform, the interferometer should be totally immersed and not supported by fixtures or wires but gently tumbled in the etchant. The solution should be cooled to a few degrees C to slow down the reaction. The best way to terminate the etching is to flood the solution quickly with large amounts of water; this prevents excessive oxidation of the reactive freshly-etched silicon surface[1].

4.3. Scanning mechanisms

Spring flexures have been used in engineering at least since the time of the bow and arrow and the Roman ballista. In the last century, they were used for governor mechanisms

[1] It should be emphasized that the etching procedure is *hazardous*. It should only be undertaken by trained and licensed personnel in properly-equipped laboratories, with protective clothing and full safety systems and procedures to hand. In the European Community at least, these are legal requirements.

on steam engines. They provide a cheap and convenient means of making an accurate translation mechanism, in which the shape of the spring (or compound spring) gives high compliance in one direction and high stiffness in all others. Neither rotating nor sliding bearings are required, and the mechanism is free from backlash to quite remarkable levels as we shall see.

The use of spring flexures in instrumentation was pioneered in the 1940s by R.V. Jones and co-workers for military applications. A long period of development and analysis is summarised in Jones and Young, 1956. During this time many precision instruments were built using these principles, notably the Rank Taylor Hobson Talystep designed by R.E. Reason and later improved by Garratt and Bottomley (1990). These and other developments can be found in Whitehouse (1994).

Most instrument spring systems were made of metal such as phosphor-bronze or beryllium-copper. Nanometre precision can certainly be obtained with metal, but for the very highest precision, the plastic deformation and consequent hysteresis of metals is a limitation. Brittle materials have two advantages. One is that crystals such as silicon, or glasses well below their glass transition temperature are ideally brittle with no sources of internal friction. The second is that they either work perfectly or are broken; they never work badly, as might a metal stressed beyond its yield point. Silicon therefore is as ideal for its elastic properties as for its perfection and X-ray properties. It is, however, important to minimise strain in the X-ray optical elements, since strains as low as 10^{-8} can affect the performance of the interferometer. Strains are introduced at the roots of flexure elements and by adhesives. They are reduced by strain relief cuts, which allow the strain to decay over a long path before it affects the diffracting blades, and by adhesives with small volume change on setting.

Drive systems tried out include magnetic, piezoelectric, capacitive and gravitational. Almost all workers now use either magnetic or piezoelectric. The magnetic drive is a simple coil surrounding a small permanent magnet that is glued to the moving part of the interferometer and reacts against the spring force of the flexure. If properly designed (Smith and Chetwynd, 1990) this is a pure force drive that greatly assists in decoupling vibrations from the interferometer itself. It is free from hysteresis at the picometre level as will be seen in section 6.4. The heat input is low, since a few milliamps current will suffice. Its main disadvantage is its low bandwidth, since if stiff springs are made to increase the resonant frequency, the coil size and heat input become too large. Piezoelectric drives, which are displacement devices, are commercially available. They have similar precision, and a much higher stiffness and bandwidth but suffer from severe hysteresis. They must therefore be operated in closed-loop mode with (for example) capacitor gauges to sense the actual displacement.

There is no fundamental difference in the X-ray optics between monolithic and separated-blade interferometers, but the alignment problems are far more severe in the latter. The translation mechanism has almost always been a metal or glass-ceramic block with built-in elastic flexures. Recent developments in the use of independent long-range slideways are discussed in section 6.5. However, it is not obvious how to obtain a force or displacement drive that can move, say, 1 mm with resolution (smoothness) of 1 pm.

A diagram of a complex LLL interferometer with long traverse, a pre-monochromator, long strain-relief paths and large supporting platform for external systems is shown in Figure 6. This illustrates many of the features discussed above, and an interferometer to this design has been provided by the University of Warwick to the UK National Physical Laboratory as the national secondary length standard. The issues involved in the system design of X-ray interferometers and other ultraprecision mechanisms are reviewed in Smith and Chetwynd (1992).

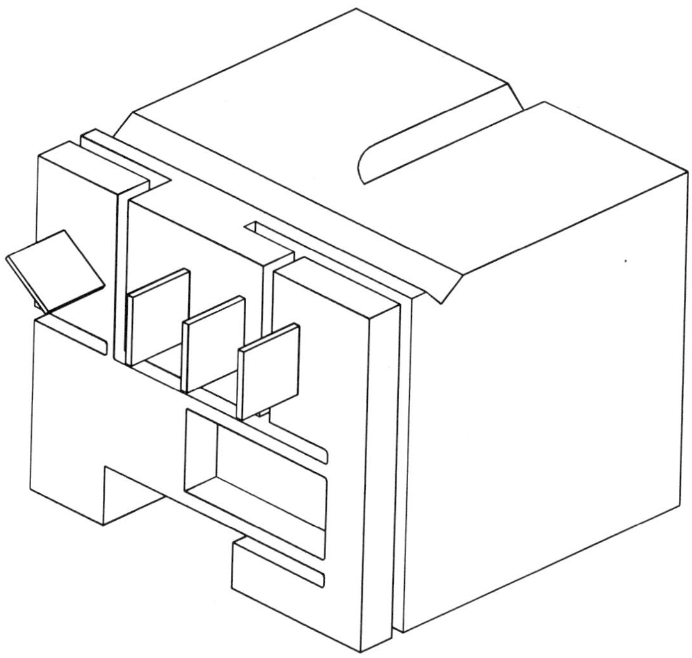

Figure 6. A complex LLL interferometer, designed as the UK national secondary length standard for the calibration of transducers at sub-nanometre levels.

5. APPLICATIONS TO X-RAY PHYSICS

5.1. X-ray scattering factors and dispersion corrections

X-ray scattering factors are fundamental to any calculation involving either kinematical or dynamical scattering theory Bonse and Hellkötter (1969) showed that scanning X-ray interferometry could be used to measure the refractive index and hence the dispersion corrections of a material. The principle is that a thin sheet of material introduced to either of the beams will introduce a phase shift in the output, which can be measured by scanning the third blade. At that time, uncertainties in material thickness and density and mechanical instabilities of the interferometer made these measurements no more accurate than others (e.g. wedge refraction) for this purpose. Cusatis and Hart (1977) refined the method in two ways: they reduced the mechanical drift of the interferometer to the order of 0.01 nm per day so that a resolution of two millifringes was achieved with bremsstrahlung radiation from a conventional tube, and they developed a two-wavelength method which removed uncertainties in the material parameters. The phase shift ratio P is defined as

$$P = {p_\lambda}/{p_{½\lambda}} \qquad [3]$$

where p_λ is the phase shift introduced at wavelength λ and $p_{½\lambda}$ is that introduced at the harmonic wavelength λ /2. Both wavelengths diffract simultaneously, and are sorted and counted separately by an energy-resolving detector. The desired dispersion correction f'_λ is given by

$$z + f'_\lambda = ½P(z + f'_{½\lambda}) \qquad [4]$$

where z is the atomic number. The term $f'_{½\lambda}$ at the wavelengths concerned is only about ½ % of the right hand side, and may be calculated or looked up to adequate accuracy. The results are unaffected by the thickness or density of the foil; the thickness is determined to about 0.1 μm in 100 μm by this method and the density to a similar accuracy.

Hart and Siddons (1981) further improved the accuracy of the technique, to the point where they were able both to show the accuracy of the Cromer-Libermann dispersion calculations and to detect computational errors in numerical solutions of the equations near absorption edges in zirconium, niobium and molybdenum. With the use of synchrotron radiation, Begum, Hart, Lea and Siddons (1986) measured both real and imaginary parts of the dispersion corrections in a number of elements.

5.2. Phase-contrast X-ray microscopy

The potential for X-ray microscopy using the interferometer was immediately recognised (Bonse and Hart, 1965b). Any object introduced into any of the beams inside the interferometer introduces a phase shift, as discussed in the last section. If the object is not homogeneous, then the shifts will differ across the object and if viewed with a phase-sensitive detector of sufficient resolution, phase-contrast X-ray microscopy results. The third blade of the interferometer is, of course, such a detector.

The phase shift is calculated from the refractive index n, and as seen earlier in Hart, this Volume:

$$n = 1 - \alpha - i\beta = 1 - \delta = 1 - \frac{e^2\lambda^2}{2\pi mc^2}\sum N_i\left(Z_i + f_i' + if_i''\right) \quad [5]$$

The real parts α of the refractive index decrement are listed in Table 1 for a few materials, and the materials thicknesses causing a phase advance of 2π are shown in Table 2 (Hart, 1975[2]) for a number of X-ray tube wavelengths.

Table 1. Examples of the real parts of the refractive index decrement, $10^6\alpha$

Material	W $K_{\alpha 1}$	Ag $K_{\alpha 1}$	Mo $K_{\alpha 1}$	Cu $K_{\alpha 1}$	Cr $K_{\alpha 1}$
C	0.1120	0.8022	1.290	6.101	13.52
H_2O	0.0655	0.4693	0.7549	3.575	7.931
Si	0.1369	0.985	1.586	7.562	16.81
Ge	0.2783	2.000	3.194	14.44	31.89
LiF	0.1444	1.035	1.666	7.89	17.51
NaCl	0.1224	0.881	1.418	6.728	14.94

Table 2. Examples of t_λ (thickness for phase advance of 2π), µm

Material	W $K_{\alpha 1}$	Ag $K_{\alpha 1}$	Mo $K_{\alpha 1}$	Cu $K_{\alpha 1}$	Cr $K_{\alpha 1}$
C	186.6	69.73	54.98	25.25	16.94
H_2O	319.1	119.2	93.96	43.09	28.87
Si	152.7	56.86	44.72	20.37	13.62
Ge	75.10	27.97	22.21	10.67	7.179
LiF	144.7	54.05	42.58	19.53	13.08
NaCl	170.8	63.53	50.02	22.9	15.33

Intensity changes of less than a percent can be detected on an image. Since a phase advance of π corresponds to 'complete' blackening, which with a reasonable interferometer is a contrast of three or four to one, it follows that the depth resolution is typically of the order of 1 µm. The case of carbon in water is of particular interest for cell biology and will be looked at more carefully. At the Cu $K_{\alpha 1}$ wavelength the contrast is 16 millifringes per mi-

[2] with the correction of the line for carbon at $\rho = 1.9$ g/cm^3 and the addition of values for water.

cron of carbon embedded in water. It is not difficult to obtain a scanning interferometer with peak:valley signals of 12000:3000 with conventional X-ray tube intensities. A 1 μm thickness of carbon embedded in water then gives a signal change near the fringe centre of about 450 cps or 6%. This is easily detectable on film, and by running at softer wavelengths using synchrotron radiation a depth resolution significantly better then 1 μm should be possible. Similar conclusions were arrived at by Snigirev et al (1995) in considering phase contrast imaging with highly coherent X-ray sources such as the ESRF.

Any image taken will contain both phase and absorption contrast. If a scanning interferometer is used, these may be distinguished by taking two images, with the analyser crystal displaced parallel to the diffraction vector by $\pm d/4$ relative to the positions of intensity maxima (Hart, 1975). The mean intensity is then pure absorption contrast and the difference intensity is the pure phase contrast map.

The lateral resolution is governed by (a) diffraction limitation, which is not serious at these dimensions and wavelengths, and (b) the width of the energy-flow triangle in the analyser. With a Laue-case analyser, the latter will cause a blurring of around 10 μm. It may be reduced by the use of an interferometer with a Bragg-case analyser, such as the three-beam design shown in Figure 7; the limitation now becomes the spreading of the beam within the extinction depth of the final crystal.

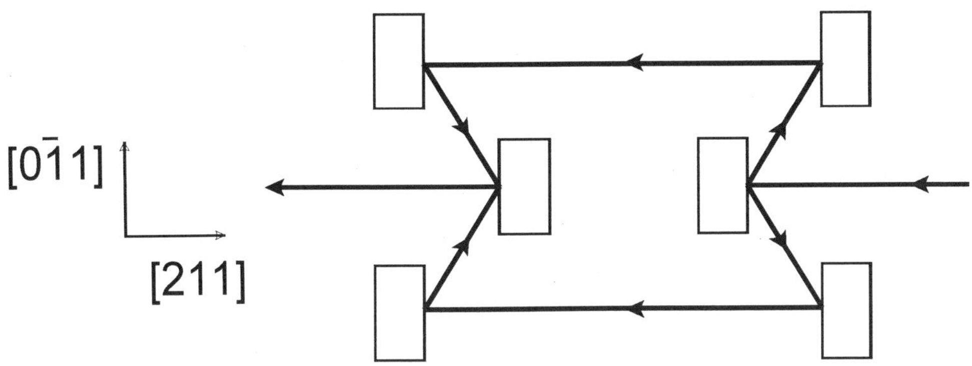

Figure 7. A three-beam, Bragg case interferometer using the 440 + $0\overline{44}$ reflections in Si working near Ni $K_{\alpha 2}$ (Graeff, 1976, Graeff and Bonse, 1977).

X-ray interferometry was first applied to phase-contrast microscopy by Bonse and Hart (1965b) who imaged the constant thickness contours of a phase plate. Ando and Hosoya studied various inorganic and biological materials and showed that phase-contrast could lead to substantial improvement in contrast over absorption contrast, and gave different contrast (and hence information) from that in optical microscopy. It is particularly advantageous for the lighter elements. The resolution was a little worse than optical microscopy in that case. Cloetens, Barrett, Baruchel, Guigay and Schlenker (1996) have used a type of phase contrast with highly coherent beams at the ESRF. This is discussed further by Baruchel in this Volume.

Up to the present, phase-contrast microscopy has not been greatly used. It does not seem likely that it will supplant optical microscopy as a routine method, or soft-X-ray microscopy in the 'water window' (the region between the oxygen and carbon absorption edges in which good contrast is obtained for biological specimens *in vitro*) since resolutions of 10 nm are possible in that case. However, soft X-ray microscopes even of 1 μm resolution are very complicated and expensive; an X-ray interferometer is a far simpler device, and for *in vitro* phase-contrast imaging of biological materials with, say, 0.5 μm resolution, a Bragg-case interferometer would seem to be worth revisiting. Multiple-beam interferometers with precision aligning devices are worth investigating for this purpose.

6. METROLOGY OF LENGTH: THE ANGSTROM RULER

Since its invention, X-ray interferometry has been used for a number of measurements with both very high precision and excellent absolute accuracy. The distinction is important and the emphasis on accuracy is essential. For example, methods of measuring displacement to sub-nanometre precision by optical interferometry or by differential transformer or capacitance transducers have been known for decades. However, all such methods suffer from problems of calibration and non-linearity. One never has exact knowledge of, for example, the dimensions, shapes and material constants of physical transducers, or the linearity of fringe division of interferometers. It was not until the application of X-ray interferometry that these methods could be calibrated with confidence.

The linearity and absolute accuracy of X-ray interferometry depend upon the regularity of perfect crystals, and the degree to which perfection can be approached. As discussed in the last chapter, silicon is the most practical material for X-ray interferometers, and with careful selection and impurity analysis or by comparison with a standard its lattice parameter in absolute terms may be relied upon to a few parts in 10^8. Whilst this is far less precise than the definition of the International Metre, which is defined to the order of 10^{-16} using the measurement of time, it is much more precise than any other metrological artifact in the nanometre regime.This accuracy has been exploited in a number of applications in X-ray physics and in engineering metrology, which are discussed in this section.

6.1. Absolute length measurement in the nanometre regime

Soon after Bonse and Hart (1965) invented the X-ray interferometer, Hart (1968) published a paper entitled 'The Angstrom Ruler'. In this he pointed out the possibility of using the interferometer as an encoder for measuring displacement, with a sub-nanometre fringe spacing that arises naturally from the crystal interplanar spacing. In order to exploit the accuracy available it was necessary to relate these spacings to the International Metre and to develop methods of using interferometers in practical metrology.

The first question to be settled is the value of the lattice parameter of silicon (or whatever crystal be used). As with all secondary standard determinations, there is no true answer, simply an answer that improves with technology and understanding.

At the time of invention of the X-ray interferometer, lattice parameters were determined by X-ray diffraction measurement of the Bragg angle of the reflection, using (normally) the method of Bond (1960). This requires knowledge of the X-ray wavelength and measurement of the angle. Optical angular interferometers and mechanical division tables, if calibrated, were capable of measuring angles up to about 30 degrees to an accuracy of 0.01 arc seconds, and 360 degrees to about 0.1 second, figures that are not very much better today. X-ray wavelengths were determined by diffraction from ruled gratings by both X-rays and light, and by density measurements combined with Avogadro's number. They were accurate only to about 1 part in 25,000 at this time, and were improved to 1 in 50,000 from 1967 (Bearden, 1967). Furthermore, one of the applications of lattice parameter measurement is to determine the Avogadro number more precisely, so an independent method was required.

The X-ray interferometer gave the opportunity directly to compare the lattice parameter with the optical wavelength secondary standard. The principle is to scan an interferometer by moving one of the blades, recording both the X-ray fringes and also optical fringes from an associated optical interferometer which independently measures the blade movement. The longer the range of the X-ray interferometer, the more precise can this determination be made. Naturally, careful attention to systematic and statistical errors is necessary. This fundamental work has been performed over the years by many national standards laboratories (Deslattes 1969, Deslattes and Henins 1973, Becker, Seyfried and Siegert 1982, Deslattes and Henins 1973, Becker, Dorenwendt, Ebeling, Lauer, Lucas, Probst, Rademacher, Reim, Seyfried and Siegert 1981, Basile, Bergamin, Cavagnero, Mana and Zosi 1989; Basile et al. 1993; Basile et al. 1995). It is also necessary to have 'standard silicon', and Windisch and Becker (1988) at the PTB have done valuable work on the effect of common impurities on the lattice parameter of silicon. These efforts have between them resulted in a value of $a_0 = 0.543\ 102\ 047\ \pm 0.000\ 000\ 014$ nm for the lattice parameter of pure silicon (Basile et al. 1995). The accuracy with which this is related to the International Metre is ±2.5 parts in 10^8. If float zone silicon is purchased and analysed for carbon content within a factor of 4, the lattice parameter may be predicted to an accuracy within a decade of this figure, ±1.7 parts in 10^7.

6.2. Determination of the Avogadro number

There are two independent methods of measuring N_A, the Avogadro number, one based upon electrical measurements and the other on physical dimensions of a pure crystal. In the electrical method (Taylor, 1994), depending upon measurement of Planck's constant h and the fine-structure α,

$$N_A = \frac{cm_e M_p \alpha^2}{2m_p h R_\infty} \qquad [6]$$

with the symbols having their usual meaning. In the dimensional method, first attempted by Deslattes *et al.* (1974), N_A is given by

$$N_A = \frac{M/\rho}{\sqrt{8}d_{220}^3} \qquad [7]$$

where d_{220}, ρ and M are the silicon (220) lattice spacing, density and molar mass, respectively. The numerator is also called V_{mol}, the molar volume. Density is measured by solid density standards in the form of optically-polished silicon spheres, molar mass by mass spectrometry and the silicon lattice spacing by means of simultaneous X-ray and optical interferometry as discussed in section 3.1. It is necessary to measure the small residual impurities in the best available silicon and to extrapolate the values of d_{220} and the molar volume to those for an ideally pure silicon crystal. Carbon and oxygen are the main impurities in float-zone material, occurring at the 10^{15} atoms/cm^3 level, around 1 ppb. Using the best available methodologies, historical values and impurity corrections and averaging over a variety of silicon crystals produced by both float-zone and Czochralski methods, Basile et al. (1995) arrived at the values:

$$N_A = (6.0221379 \pm 0.0000025) \times 10^{23} \quad \text{mol}^{-1} \qquad [8]$$

a claimed precision of $4.15 \times 10^{-7}\ N_A$. This is smaller by $(8.5 \pm 4.4) \times 10^{-7}\ N_A$ than the best measurements with the electrical method, whose best available result is:

$$N_A = (6.0221430 \pm 0.0000008) \times 10^{23} \quad \text{mol}^{-1} \qquad [9]$$

a claimed precision of 1.3 × 10^{-7} N_A. The error bounds do not overlap, therefore we must conclude that there are as yet unknown errors in one or both of the methods. Work is currently in progress on more rigorous crystal characterisation in the X-ray/density method.

6.3. Calibration of displacement transducers

Surprisingly many engineering transducers have sensitivity in the nanometre regime. These include optical interferometers, differential transformers, differential capacitors, and tunnelling probes. Many of these have been exploited in the design of engineering stylus profilometers and, more recently, in the design of scanning probe microscopes. However, all have some limitation which prevents us from obtaining an absolute measurement from the transducer without calibration. These are summarized in Table 3.

Table 3. Limitations of nanometre transducers

Transducer	Limitations in the nm regime
Optical interferometer	Air density and turbulence (refractive index) Accuracy of fringe division
Differential transformer	Coil geometry Magnetic properties of core
Differential capacitor	Shape, dimension and tilt of plates Permittivity of air
Tunnelling probe	Work function of surface Zero point
AFM cantilever	Stiffness of cantilever Calibration and linearity of optical or capacitance detector

These limitations are not trifling. We have detected calibration errors of 20% in a metrological instrument and of 40% in a commercial atomic force microscope, both from well-known manufacturers.

In order to apply the method of X-ray interferometry to transducer calibration, it was necessary to show that the devices could be operated simultaneously over long enough times for practical measurements, to determine the sense of movement of the interferometer blade from the X-ray readings alone and to define and to control the errors. This requires control of position, vibration and drift to the level of a few picometres. Chetwynd, Siddons and

Bowen (1984) showed that an X-ray interferometer could indeed be operated with a Rank Taylor Hobson Talystep transducer in contact with the silicon above the moving blade. Robust interferometers, capable of holding and calibrating sizeable engineering transducers were then constructed and applied to calibration of practical engineering systems such as laser interferometers and differential transformers (Bowen, Chetwynd, Schwarzenberger and Smith, 1990; Bowen, Chetwynd and Davies, 1985). A phase-stepping interferometer was used to sense the direction of motion of the blades. If a phase advance is introduced into one of the beams by the introduction of, for example, a film of plastic of thickness a few tens of μm, and two detectors or a switched phase advance are employed, the interferometer becomes a phase-quadrature device from which direction of motion may be sensed. A further improvement is to use three-phase stepping (Schwarzenberger, Chetwynd and Bowen, 1989) based upon the optical phase-stepping technique of Grievenkamp(1984), which allows for better fringe division and less sensitivity to X-ray source fluctuations. This permits the resolution and absolute accuracy obtainable using a conventional X-ray tube to be better than 5 pm.

The requirements for using X-ray interferometry with engineering transducers are essentially twofold: to obtain a long enough range and to couple the transducer without upsetting the operation of the interferometer. The range question is discussed in section 3.5. Bad coupling can introduce Abbe offset errors, vibration, thermal and mechanical drift and the introduction of blade misalignments.

6.4. Precision, accuracy and the error budget

The natural sub-nanometre pitch of the X-ray interferometer and its natural reproducibility are obvious reasons for using it in calibration applications. However, there are also practical advantages in the readout of the fringes in comparison to an optical interferometer. Table 4 gives a qualitative comparison of the two types of interferometry.

Schwarzenberger, Chetwynd and Bowen (1989) and Bowen, Chetwynd and Schwarzenberger (1990) have made a detailed treatment of errors in the use of X-ray interferometry for linear measurement, treating the cases of single beam and the double and triple phase-stepping beams. The principal sources are computational approximation in data analysis, statistical noise in intensity counting, X-ray generator and control circuit instability and thermal drift. The X-ray source and control stability problems are negligible with modern generators. The computational approximations are clearly coupled with the statistical noise. Thermal drift in the silicon itself is not a problem if the temperature is stabilised to about 0.1 K. However, experience has shown that engineering transducers under test are likely to be very much more susceptible to thermal drift. It is therefore important to make the measurements as quickly as possible, which conflicts with obtaining good statistics, therefore a data collection strategy is important.

Table 4. Types of error in encoders and optical and X-ray interferometry

Type of error	Optical interferometry	Moiré gratings, optical encoders	X-ray interferometry
Control of pitch	Set by optical wavelength, ~ μm	Depends on manufacturing capability, ~ 0.1 μm	Set by crystal lattice and temperature, ~ 0.3 nm
Manufacturing errors	Affect contrast, may introduce harmonics	Directly affect pitch and harmonic content	Affect contrast only; do not affect pitch nor introduce harmonics
Air turbulence	Seriously affect pitch, introduce noise	Little effect, though may introduce noise	Totally negligible effect.
Detector linearity	Requires careful design and verification; introduces harmonic errors	Requires careful design and verification	Photon counting, intrinsically linear.
Fringe division	Questionable accuracy at sub-nm level.	Questionable accuracy at sub-nm level.	Excellent to 1 pm
Noise	Introduced by amplifiers	Introduced by amplifiers	Only Poisson noise, may be integrated out.
Measurement speed	High, practical for metrological instruments and machine tools.	High, practical for metrological instruments and machine tools.	Low, only practical for calibration rather than measurement at present.

The authors conclude that whilst a computer analysis based upon the error equations is necessary to obtain the best strategy for a particular case, some guidelines may be deduced. These are, to use relatively few points per fringe and then maximise the counting time within the thermal drift constraints of the transducer under test. The three-point Stirling approximation is a good algorithm for finding peaks and zero crossings, and hence interpolation within the fringe, for sparse data. The phase-stepping methods are clearly superior, and the three phase method optimal. With a conventional non-focussed X-ray tube, and a fringe contrast of 0.2 (defined as amplitude/mean) the combined uncertainty of all the errors at

95% confidence level is slightly less than ±0.005 nm with a standard deviation about 0.0025 nm. These figures would improve approximately as the square root of the intensity and approximately linearly with the fringe contrast.

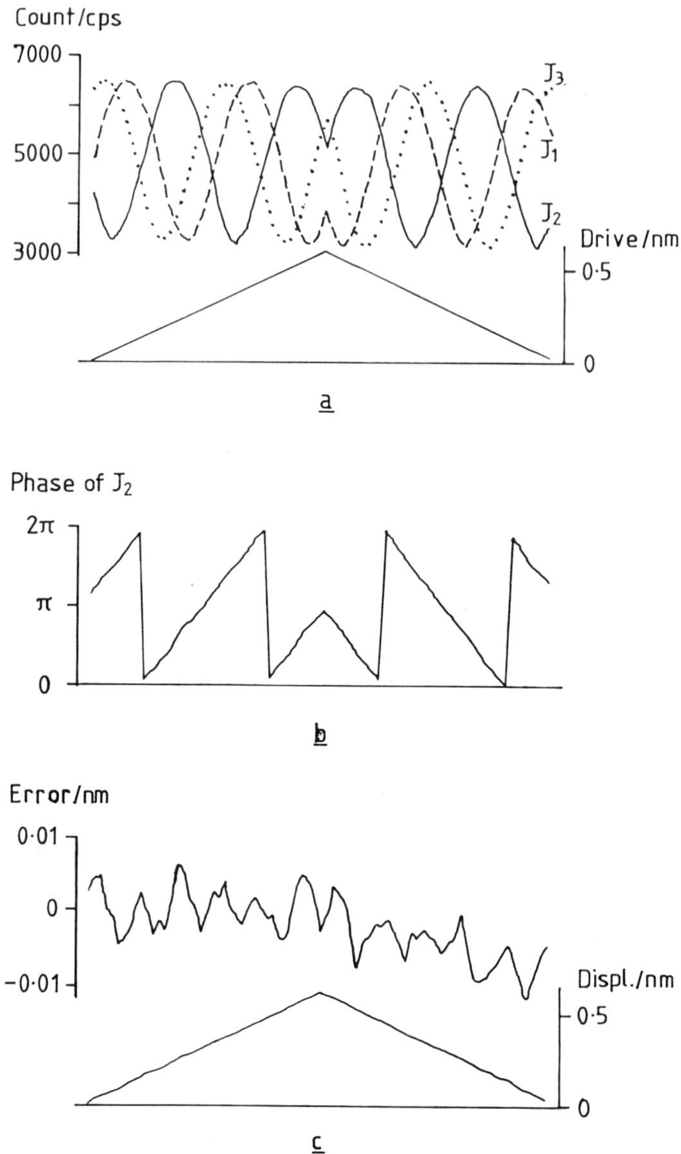

Figure 8. Experimental calibration of an open-loop drive system. (a) Direct plot of a three-phase scan over about 6 nm; the reversal of direction is in the middle of the plot. The demand ramp (coil voltage) is also shown. These data were collected in about 20 min. (b) Phase plot of the results in (a). (c) Displacement plot of results in (a) and the error of the reconstructed actual position from the demand position.

An example is shown in Figure 8. This shows the calibration of the open-loop magnetic drive itself with the three-phase interferometer. To test for thermal drift, it is usual to cycle the position up and down, and the clean reversal of fringe contrast is easily seen as the direction of motion is switched. The phase unwrapping and the final error signal are shown. In this case, the conclusion is that a simple open-loop design with cheap electronic components can provide reproducible displacement drive with errors below 0.01 nm.

The calibration system has been applied to other transducers: RTH Talystep heads, laser interferometers, differential capacitors and commercial AFM probes (Bowen, Chetwynd, Schwarzenberger and Smith, 1990). As emphasised in Table 4, the low speed of X-ray interferometry at present, as well as safety considerations, make it unrealistic to use the technique directly in measurement, for example on a coordinate measuring machine.

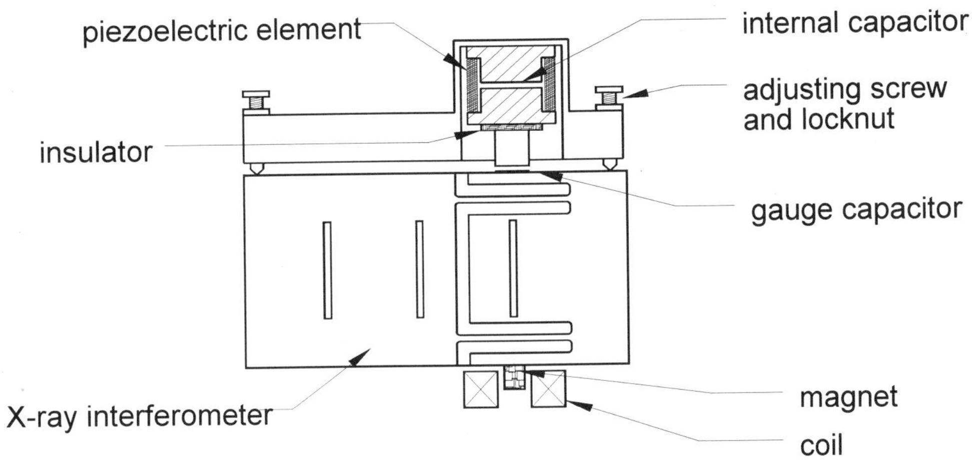

Figure 9. Schematic of a series capacitor calibrator and follower, after Smith, Chetwynd and Bowen (1991).

However, we have found much use for a transfer standard, which is fast and can readily be transferred between the X-ray interferometer and a practical instrument such as an AFM head. An example is shown in Figure 9. Here a Queensgate Instruments linear positioner, comprising a piezoelectric expansion element with an internal sensing capacitor to avoid the hysteresis inherent in open-loop piezo devices, is mounted directly on the X-ray interferometer with its moving end directly over the moving blade of the interferometer. A gauge capacitor is formed between these two components, by gold plating the top of the interferometer and also one of the springs in order to take out the electrical signal without the drag or vibration associated with wiring. The gauge capacitor is used as a nulling device. The interferometer is driven by an amount determined by the lattice plane spacing, then the actuator is driven until the gauge capacitor is again nulled. As shown in Figure 9, such capacitive-sensed actuators are easily capable of sub-nanometre performance, with the residual error after removal of thermal drift being in the region of 0.1 nm. The calibrated actuator can then be used in situations such as the control of a Fabry-Perot etalon, for which its high stiffness and bandwidth make it far more suitable than the X-ray interferometer itself.

In order to study the parasitic errors in elastic drives and to determine the requirements for twist correction, Chetwynd, Schwarzenberger and Bowen (1990) constructed a two-dimensional interferometer with three-phase stepping. Instead of using springs designed for compliance in one direction only, the flexure system for the third blade comprised four square-sectioned leaf springs placed at the corners of a rectangle, like the legs of a table. The blade could then be driven in any direction within the *xy* plane (with the springs in the *z* direction). The interferometer was cut so that {220} fringes were obtained in the *x* direction and {111} fringes in the *y*, using two X-ray sources. Detailed analysis of the fringe contrast in each direction using Moiré fringe contrast theory (Siegert, 1974), allowed extraction of the parasitic motions involved in each of the ultraprecision drives. An example of the analysis possible is shown in Figure 10, which gives the parasitic twist on each axis as a function of displacement. The sensitivities and absolute accuracies of this piece of nanometrology were about 0.01 nrad (0.002 arc seconds) and 0.01 nm simultaneously on two axes.

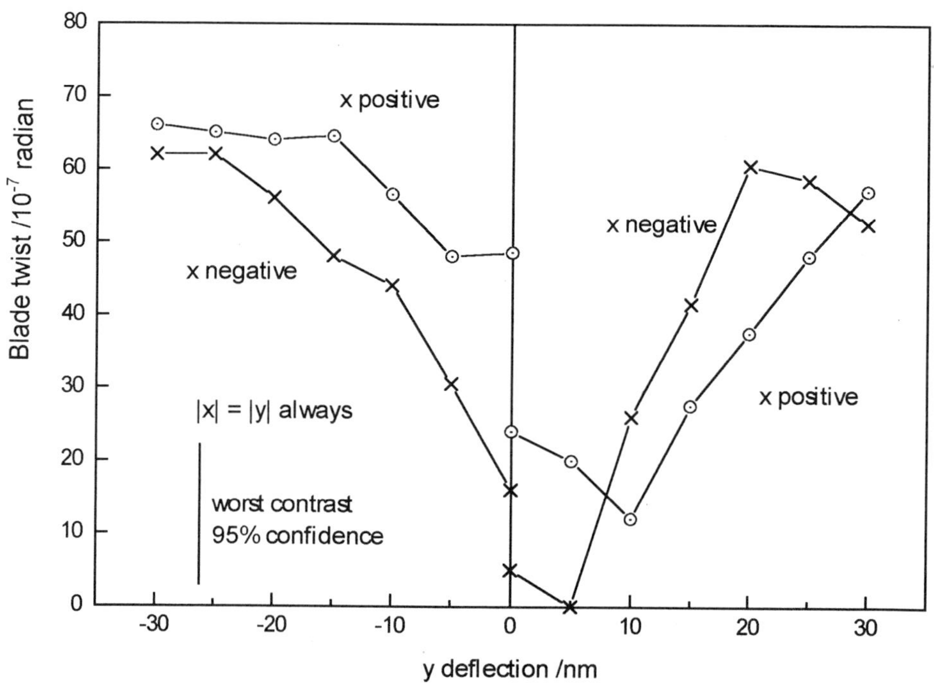

Figure 10. Parasitic twist in an x-y interferometer, derived from fringe contrast variation. Both signs of x are plotted for each sign of y.

A feasibility analysis for long-range X-ray interferometry using monolithic interferometers was published by Chetwynd, Harb, Krylova and Smith (1993). The parameters deduced were realised in a contract from the UK National Physical Laboratory to achieve the current state of the art with monolithic interferometers, placed with this group plus D.K. Bowen and D.C. Dyer. The UK National Secondary Length Standard Facility has now been constructed. This comprises a robust interferometer (Figure 6) able to take transducer loads of tens of grams, with a total range of ten micrometres, an ultimate resolution of one picometre, a 20-bit ultra-stable closed-loop drive and a system for twist correction to 0.001 arc seconds, all traceable to international standards. The twist correction utilises a dual magnet drive; with the magnets driven in parallel, translation is achieved, if in opposition, twist results. By monitoring the Moiré fringes with an X-ray TV detector, parasitic errors induced by relatively heavy transducer loads - up to some tens of grams - can be compensated.

6.5 Long-range X-ray interferometry

To go much further requires separated-blade interferometers with long-range slide movements. Although separated-blade interferometers have been made by a number of workers, these have always been cemented to an external monolith, usually made of Zerodur, for an elastic-flexure ‘slideway’. Alignment is difficult, but possible with systematic procedures (Mewes, Lehrke, Rademacher and Reim, 1974). However, Monteiro (1995) in collaboration with Krylova, Chetwynd and Bowen has recently succeeded in demonstrating the X-ray alignment and interferometric stability of a separated-blade interferometer mounted on a separate slide. The system is constructed entirely of the low-expansion glass-ceramic Zerodur, and the slideway is that developed by the NPL and the University of Warwick (Lindsey, Smith and Robbie 1988) for the NPL Nanosurf II instrument (now produced by Rank Taylor Hobson as the Nanostep, Garratt and Bottomley, 1990) which gives AFM sensitivity and stability over a 50 mm traverse. This is the proof of principle that long distance X-ray interferometry is possible; silicon crystals can be made over 1m long! It is unlikely that X-ray interferometry will become a direct measuring technique because of the low X-ray output, even with the new focussing sources. However, there will certainly develop a requirement for instruments such as nanometric coordinate measuring machines and production equipment with nanometric precision such as X-ray lithographic steppers. One may easily envisage such instruments equipped with a small X-ray interferometric calibration station, with range say 1 mm, which can be used repeatedly to calibrate the optical interferometers or precision scales that directly control the actual instrument.

7. METROLOGY OF ANGLE: THE ARCSECOND DIVIDER

There is no need to calibrate angle as such. Angular measurement has a calibration advantage over virtually all other calibration problems in that the fundamental unit of one revolution=2π radians is readily portable! It is rather the division of angle that is critical. The national standards laboratories all have dividers based upon the simultaneous calibration of two rotary tables driven in opposite directions, based on the principle that the sum of all interpolation errors when each has been driven through exactly one revolution must be zero.

However, these methods do not have the sensitivity to calibrate very small angular shifts, such as can be detected by the best autocollimators, say in the range of 0.001 to 2 arc seconds (5 to 2000 nanoradians). There are many examples where angles must be measured or asserted to very high precision and accuracy over limited ranges. Astronomical applications are star-pointing and segmented-mirror telescopes, satellite positioning and telecommunications. In materials research, sub-arcsecond measurements are routinely made in X-ray characterisation of materials and novel microscopies such as lateral-force AFM. In manufacturing engineering, the mechanisms for microcircuit lithography require straightness and freedom from parasitic rotations at sub-arc second levels, a requirement that will become more stringent still as wafer production sizes increase. Although very small angular shifts may be calibrated, slowly, by traditional sine-bar methods (in essence by knowing the lengths of three sides of a triangle), no continuously reading method was available.

Two methods based upon X-ray interferometry have now been developed. One is an analogue of the sine-bar method, and the other depends upon the Moiré fringes discussed in the last section.

7.1 The "sine bar" method.

Becker and Bonse (1974) developed a skew-symmetric interferometer, shown in Figure 11. They showed that both output beams gave oscillations as a function of angular displacement about the central pivot. The behaviour of this interferometer requires detailed analysis by dynamical scattering theory. Windisch and Becker developed further the theory of Becker and Bonse, using the spherical wave analysis, and showed that the period of oscillation, $P = 1/d(z_A^{II} + t_{M2}/2)$ (see the dimensions in Figure 11; it is interesting that the correction term involving the thickness of the final blade only appears in the spherical wave treatment). The amplitude of the fringes is a very complicated function of the dimensions of the interferometer and the absorption, since it is strongly influenced by spherical-wave Pendellösung fringes. The best geometry is given when $2t_s = t_{M1} = t_{M2} = 2t_A$, since then the focal point of the four interfering waves is shifted to the exit point of the interferometer (Bauspiess, Bonse and Graeff, 1976). In this case the theoretical contrast over an angular range of ±2 arc seconds is over 15%.

Experimentally, an interferometer was constructed to this model and fringes were detected, albeit with lower contrast than expected, in the region ±0.3 arc seconds. The period, calculated from geometric measurements to an accuracy of 2 μm, was slightly under 0.002", thus very good discrimination is seen at the fine end of the range. The interferometer was used to calibrate the PTB photoelectric autocollimator ACT. The authors state that if the phase stepping procedure of Schwarzenberger, Chetwynd and Bowen (1989) were applied, angular measurements with a standard deviation of 10^{-5} arc seconds should be routinely achieved.

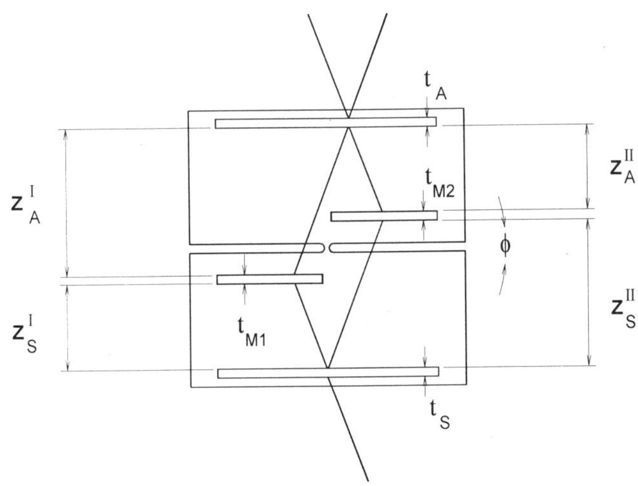

Figure 11. Skew-symmetric interferometer, for measurement of angle (Becker and Bonse, 1974).

7.2. The Moiré fringe method.

The Moiré fringe method was initially demonstrated in a pioneer experiment by Bradler and Lang (1968), using two slightly misoriented single crystals with a collimated beam direct from an X-ray tube. Clear rotational fringes were obtained over large areas in X-ray topographs, showing that high accuracy could be obtained in angular measurements at the sub arcsecond level. Two attractive features of this method are that there is a massive averaging in the production of the fringes, leading to reliable data, and that the accuracy of the angle measurement depends upon just two geometrical parameters, the *x* and *y* magnifications of the topograph.

Unfortunately, the contrast of the fringes rapidly degrades as the gap between the crystals increases and it is not obvious that a practical instrument can be achieved with such simple X-ray optics. However, a property of the X-ray interferometer, that of splitting and focussing of the X-ray beams, may be used in effect to project one lattice pattern onto another, removing the requirement that the blades be in close contact. Indeed, the use of fringe contrast in the two-dimensional interferometer was a demonstration of the practicality of small angular measurements by this method.

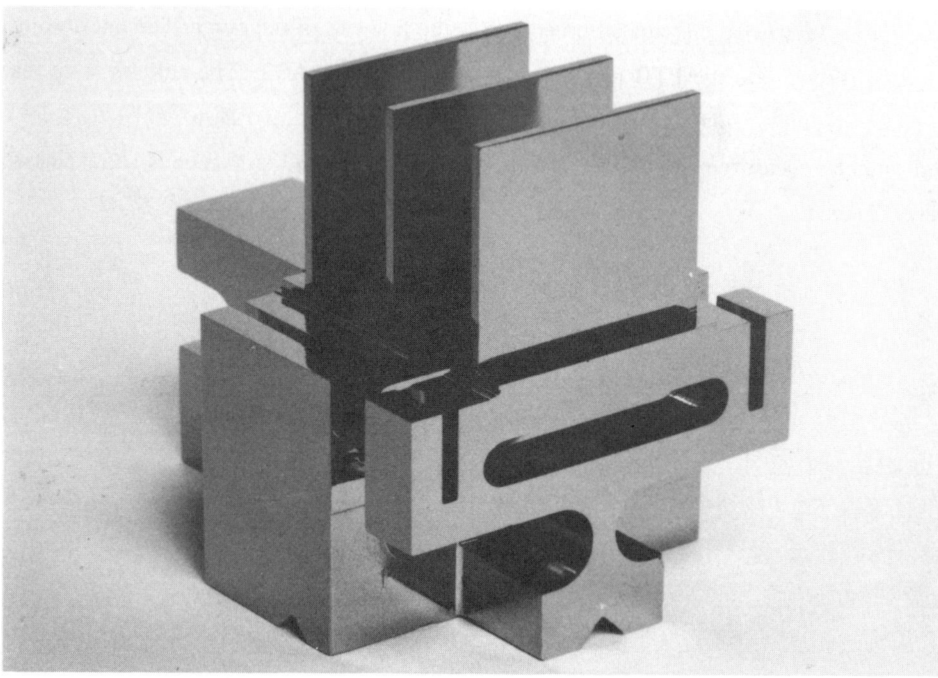

Figure 12. Moiré fringe interferometer, for measurement of angle. The rotating blade and its spring pivot are shown at the front. At the rear is another integral pivot, used for mounting a pre-monochromator to improve contrast. Strain relief cuts beside and below the moving blade minimise transmission of strains from the pivot and mirror mounting points beside the blade. Bowen, Chetwynd and Krylova, 1996.

Recently, Bowen, Chetwynd and Krylova (1996) have developed a robust angular interferometer based upon these principles, illustrated in Figure 12. Its large blades and single, stiff angular pivot provide optimum conditions for fringe contrast. Strong mounting points, separated from the X-ray optics by deep strain-relief cuts, are provided for mounting optical reflectors for autocollimators.

The interferometer performs well at any mounting angle. Its range is limited mainly by the ability to detect the fringes, i.e. the range/resolution of the X-ray imaging detector. Real-time output may be obtained from an 18 mm diameter X-ray TV with a 1000:1 range/resolution, for example over the range of about 0.002" to 2". With nuclear plates a range of 0.0001" to 10" is feasible. Autocorrelation and transform techniques being developed by Okuyama (1996) permit direct extraction of angular data. An example of the fringes is shown in Figure 13.

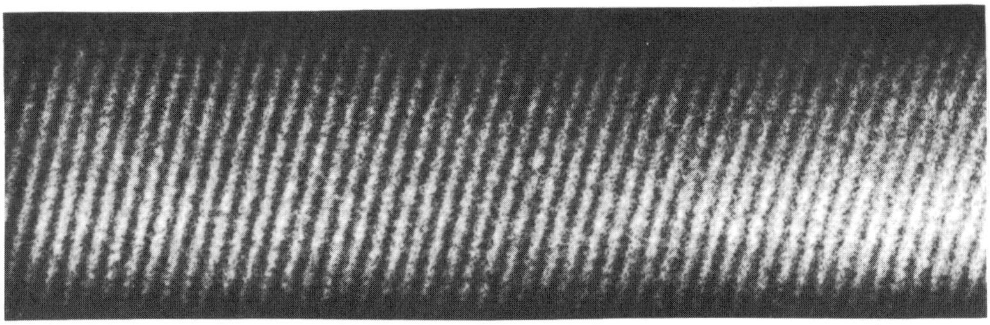

Figure 13. Moiré fringes produced at an angular setting of 0.7".

8. CONCLUSIONS

X-ray interferometry has achieved its high potential for measurement of X-ray dispersion corrections and for calibration of extremely small linear or angular displacements. Substantial improvements in scan length and in speed of measurement are likely, and it may become an on-line calibration method for the most accurate instrumentation. It has resulted in substantial improvement in measurement of the Avogadro number, though it is not yet clear whether it is superior to electrical methods. Its potential for phase-contrast microscopy remains largely unrealised.

9. ACKNOWLEDGEMENTS

It is a pleasure to acknowledge many years of collaboration with colleagues at the University of Warwick, in particular Dr. D.G. Chetwynd, Dr. N. Krylova, Dr. S.T. Smith, Mrs. D.R. Schwarzenberger, Dr. S.G. Harb and Dr. S.G. Cockerton. The invaluable advice of Professor M. Hart is gratefully acknowledged. The technical skills of Mr. R. Mortimore, Mr. S. Wallace and Mr. D. Robinson have been much appreciated. Much of the work at the University of Warwick was funded by the UK Department of Trade and Industry, the Engineering and Physical Sciences Research Council and the Royal Society.

REFERENCES

Ando, M. and Hosoya, S., 1972, in "Proc. 6th. Int. Congr. on X-ray Optics and Microanalysis" (eds. G. Shinoda, K. Kohra and T. Ichinokawa), Univ. Tokyo Press.

Basile, G., Becker, P. Bergamin, A., Bettin, H., Cavagnero, G., de Bievre, P., Kütgens, U., Mana, G., Mosca, M., Pajot, B., Panciera, R., Pasin, W., Pettorruso, S., Peuto, A., Sacconi, A., Stümpel, J., Valkiers, S. Vittone, E. and Zosi, G. 1995, A new determination of N_A, *IEEE Trans. Instrum. Meas.* 44:538

Basile, G., Becker, P., Bergamin, A., Cavagnero, G., Franks, A., Jackson, K., Kütgens, U., Mana, G., Palmer, E.W., Stedman, M., Stümpel J. and Zosi, G., 1993, COXI - a combined optical and X-ray interferometer for high precision dimensional metrology, in "International Progress in Precision Engineering" (eds. N. Ikawa, S. Shimada, T. Moriwaki, P.A. McKeown and R.C. Spragg), Butterworth Heinemann, Boston.

Basile, G., Bergamin, A., Cavagnero, G., Mana, G. and Zosi, G., 1989, Progress at IMGC in the absolute determination of the silicon d(220) lattice spacing, *IEEE Trans. Instrum, Meas.* 38:210

Bauspiess, W., Bonse, U. and Graeff, W. 1976, Spherical-wave theory of the zero-absorption LLL X-ray or neutron interferometer, *J. Appl. Cryst.* 9:68

Bearden, J.A., 1967, X-ray wavelengths and X-ray atomic energy levels, *Rev. Mod. Phys.* 39:78

Becker, P. and Bonse, U., 1974, The skew-symmetric two-crystal X-ray interferometer, *J. Appl. Cryst.* 7:593

Becker P., Dorenwendt, K., Ebeling, G., Lauer, R., Lucas, W., Probst, R., Rademacher, H-J., Reim, G., Seyfried, P. and Siegert, H., 1981, Absolute measurement of the (220) lattice plane spacing in a silicon crystal, *Phys. Rev. Lett.* 46, 1540-1543.

Becker, P., Seyfried, P. and Siegert, H., 1982, The lattice parameter of highly pure silicon crystals, *Z. Phys. B* 48:17

Begum, R., Hart, M., Lea, K.R. and Siddons, D.P., 1986, Direct measurement of the complex X-ray scattering factors for elements by X-ray interferometry at the Daresbury Synchrotron Radiation Source, *Acta Cryst. A* 42:456

Bond, W.L., 1960, Precision lattice constant determination, *Acta Cryst.* 13:814

Bond, W.L., 1976, "Crystal Technology", John Wiley, New York

Bonse, U. and Graeff, W., 1977, in "Topics in Applied Physics", vol 22: X-ray optics, p93

Bonse, U. and Hart, M., 1965a, An X-ray interferometer, *Appl. Phys. Lett.* 6:155

Bonse, U. and Hart, M. 1965b, Principles and design of Laue-case interferometers, *Z. Phys.* 188:154

Bonse, U. and Hart, M., 1965c, An X-ray interferometer with long separated working beam paths, *Appl. Phys. Lett.* 7:99

Bonse, U. and Hart, M. (1968), Combined Laue and Bragg case X-ray interferometers, *Acta Cryst. A*24:240

Bonse, U. and Hellkötter, H., 1969, Interferometrische Messung des Brechungsindex für Röntgenstrahlen, *Z. Phys.* 223:345

Bowen, D.K., Chetwynd, D.G. and Davies, S.T., 1985, Calibration of surface roughness transducers at Ångstrom levels using X-ray interferometry, *Proc. SPIE, 29th Symposium on Optical and Electro-Optical Engineering* 563:412

Bowen, D.K., Chetwynd, D.G. and Krylova, N. 1996, to be published.

Bowen, D.K., Chetwynd, D.G. and Schwarzenberger, D.R., 1990, Sub-Nanometre displacements calibration by X-ray interferometry, *Meas. Sci. Technol.* 1:107

Bowen, D.K., Chetwynd, D.G., Schwarzenberger, D.R. and Smith, S.T., 1990, Sub-Nanometre transducer calibration by X-ray interferometry, *Precision Engineering* 12:165

Bradler, J. and Lang, A.R., 1968, Use of the Ewald sphere in aligning crystal pairs to produce X-ray Moiré fringes, *Acta Cryst.* 24:246

Chetwynd, D.G., Cockerton, S.C., Smith, S.T. and Fung, W.W., 1991, The design and operation of monolithic X-ray interferometers for super-precision metrology, *Nanotechnology*, 2:1

Chetwynd, D.G., Harb, S.M., Krylova, N. and Smith, S.T., 1993, The feasibility of extended range monolithic X-ray interferometric calibrators, *Nanotechnology* 4:183

Chetwynd, D.G., Schwarzenberger, D.R. and Bowen, D.K., 1990, Two-dimensional X-ray Interferometry, *Nanotechnology* 1:19-26

Chetwynd, D.G., Siddons, D.P. and Bowen, D.K., 1983, X-ray interferometer calibration of microdisplacement transducers, *J. Phys. E* 16:871

Cloetens, P., Barrett, R., Baruchel. J., Guigay, J.P. and Schlenker, M., 1996, Phase objects in synchrotron radiation hard X-ray imaging, *J. Phys. D: Appl. Phys.* 29:133

Cusatis, C. and Hart, M., 1977, The anomalous dispersion corrections for Zirconium, *Proc. Roy. Soc. London A* 354:291

Deslattes, R.D., 1969, Optical and X-ray interferometry of a silicon lattice spacing, *Appl. Phys. Lett.*, 15:386

Deslattes, R.D., and Henins, A., 1973, X-ray to visible wavelength ratios, *Phys. Rev. Lett.* 31:972

Deslattes, R.D., Henins, A., Bowman, H.A., Schoonover, R.M., Carroll, C.L., Barnes, I.L., Machlan, L.A., Moore, L.J. and Shields, W.R., 1974, Determination of the Avogadro constant, *Phys. Rev. Lett.* 33:463

Garratt, J.D. and Bottomley, S.C., 1990, Technology transfer in the development of a nanotopographic instrument, *Nanotechnology* 1:38

Graeff, W., 1976, PhD Thesis, University of Dortmund.

Graeff, W. and Bonse, U., 1977, A three-beam case X-ray interferometer, Z. Phys. B27:19

Grievenkamp, J.E., 1984, Generalised data reduction for heterodyne interferometry, *Opt. Eng.*, 23:350

Hart, M. and Siddons, D.P., 1981, Measurements of anormalous dispersion made with X-ray interferometers, *Proc. Roy. Soc. London* A376:465

Jones, R.V. and Young, I.R.,1956, Some parasitic deflections in parallel-spring movements, *J. Sci. Instrum.* 33:11

Krylova, N, 1993, unpublished work (University of Warwick)

Lindsey K., Smith S.T. and Robbie C.J., 1988, Sub-nanometre surface texture and profile measurement with 'Nanosurf 2', *Annals of the CIRP*, 37:519

Mewes, E-R., Lehrke, K., Rademacher, H-J., and Reim, G., 1974, "Berichte über Arbeiten am Röntgen-Verschiebeinterferometer"; PTB, Braunschweig, Report APh-8

Monteiro, A., 1995, PhD thesis, University of Warwick, UK

Okuyama, E., 1996, to be published.

Schwarzenberger, D.R., Chetwynd, D.G. and Bowen, D.K., 1989, Phase Measurement X-Ray Interferometry, *X-Ray Science and Technology* 1:134

Siegert, H., 1974, Analyser gleichförmiger Kristallbewegungen in Moiré-Streifenbild in "Berichte über Arbeiten am Röntgen-Verschiebeinterferometer"; PTB, Braunschweig, Report APh-9

Snigirev, A., Snigreva, I., Kohn, V., Kuznetsov, S. and Schelokov, I., 1995, On the possibilities of phase contrast microimaging by coherent high energy synchrotron radiation, *Rev. Sci. Instrum.* 66:5486

Smith, S.T. and Chetwynd, D.G., 1990, An optimised magnet-coil force actuator and its application to precision elastic mechanisms, *Proc. Inst. Mech. Eng.* 204 C4:243

Smith, S.T. and Chetwynd, D.G., 1992, "Foundations of Ultraprecision Mechanism Design", Gordon and Breach, London.

Taylor, B.N., 1994, Determining the Avogadro constant from electrical measurements, *Metrologia* 31:181

Whitehouse, D.J., 1994, "Handbook of Surface Metrology", Institute of Physics Publishing, Bristol

Windisch, D. and Becker, P., 1988, Lattice distortions induced by carbon in silicon, *Phil. Mag. A* 58, 435

Windisch, D. and Becker, P., 1992, Angular Measurements with X-ray Interferometry, *J. Appl. Cryst.* 25:377

CONTRIBUTORS

Authier André, Laboratoire de Minéralogie-Cristallographie, Université P. et M. Curie, 4, Place Jussieu, 75252 Paris CEDEX 05, France

Baruchel José, Topography Group, Experiments Division, ESRF, BP 220, 38043 Grenoble CEDEX, France

Bowen Keith D., Department of Engineering, University of Warwick, Coventry CV4 7AL, United Kingdom

Colella Roberto, Department of Physics, Purdue University, West Lafayette, IN 47907, USA

Fewster Paul F., Philips Research Laboratories, Cross Oak Lane, Redhill, Surrey RH1 5HA, United Kingdom

Fontes E., CHESS, Cornell University, Ithaca, NY 14853, USA

Guigay Jean-Pierre, Laboratoire de Magnétisme Louis Néel, CNRS, BP 166 X, 38042 Grenoble CEDEX 9, France

Hart Michael, Building 725 B, National Synchrotron Light Source, Brookhaven National Laboratory, Upton, NY 11973-5000, USA

Holý Václav, Department of Solid State Physics, Faculty of Sciences, Masaryk University, Kotlárská 2, Cs-61137 Brno, Czech Republik

Hümmer Kurt, Institut für Kristallographie, Universität Karlsruhe, Kaiserstraße 12, D-76128 Karlsruhe, Germany

Kato Norio, Hoshigaoka Iris S512, Meito Honmachi, Nagoya 465, Japan

Klapper Helmut, Mineralogisches Institut, Universität Bonn, Poppelsdorfer Schloss, D-53115 Bonn, Germany

Lagomarsino Stefano, Istituto Elettronica Stato Solido, CNR, Via Cineto Romano 42, 00156 Roma, Italy

Malgrange Cécile, Laboratoire de Minéralogie-Cristallographie, Université P. et M. Curie, 4, Place Jussieu, Paris CEDEX 05, France

Mikulik Petr, Department of Solid State Physics, Faculty of Sciences, Masaryk University, Kotlarska 2, Cs-61137 Brno, Czech Republik

Patel Jamshed R., SSRL, Stanford University , Stanford, CA 94309-0210, USA

Schlenker Michel, Laboratoire de Magnétisme Louis Néel, CNRS, BP 166 X, 38042 Grenoble CEDEX 9, France

Servidori Marco, Istituto LAMEL, CNR, Via P. Gobetti 101, 40129 Bologna, Italy

Tanner Brian K., Department of Physics, University of Durham, South Road, Durham DH1 3LE, United Kingdom

Weckert Edgar, Institut für Kristallographie, Universität Karlsruhe, Kaiserstraße 12, D-76128 Karlsruhe, Germany

INDEX